ANTIPREDATOR DEFENSES IN BIRDS AND MAMMALS

INTERSPECIFIC INTERACTIONS *A Series Edited by John N. Thompson*

TIM CARO

ANTIPREDATOR DEFENSES IN BIRDS AND MAMMALS

WITH DRAWINGS BY SHEILA GIRLING

THE UNIVERSITY OF CHICAGO PRESS • CHICAGO AND LONDON

Tim Caro is professor in the
Department of Wildlife, Fish,
and Conservation Biology and
Center for Population Biology at the
University of California, Davis.
He has written more than one
hundred articles for the peer-
reviewed literature, has edited
*Behavioral Ecology and
Conservation Biology*, and is author
of *Cheetahs of the Serengeti Plains:
Group Living in an Asocial Species*,
the latter published by the University
of Chicago Press.

The University of Chicago Press, Chicago 60637

The University of Chicago Press, Ltd., London

© 2005 by The University of Chicago

All rights reserved. Published 2005

Printed in the United States of America

14 13 12 11 10 09 08 07 06 05 1 2 3 4 5

ISBN: 0-226-09435-9 (cloth)

ISBN: 0-226-09436-7 (paper)

Library of Congress Cataloging-in-Publication Data

Caro, T. M. (Timothy M.)

Antipredator defenses in birds and mammals / Tim Caro ; with
drawings by Sheila Girling.

    p. cm. — (Interspecific interactions)

Includes bibliographical references and indexes.

ISBN 0-226-09435-9 (hardcover : alk. paper) — ISBN 0-226-09436-7
(pbk. : alk. paper)

    1. Birds—Defenses.   2. Mammals—Defenses.   I. Title.   II. Series.

QL698.3.C36   2005

598.11—dc22

2004026464

CONTENTS

3   Behavioral mechanisms to avoid detection

4   Vigilance and group size

9   Morphological and physiological defenses

10   Nest defense

PREFACE, SCOPE, AND ACKNOWLEDGMENTS

I first considered writing a book about antipredator defenses in 1989, after my work on stotting had appeared and while my colleague in the Serengeti, Clare FitzGibbon, was writing up her doctoral dissertation on antipredator behavior of Thomson's gazelles, but it was not until 1994 that I put pencil to paper. My reason for writing a comprehensive book was that literature concerning antipredator defenses was very dispersed, and I wanted to pull it together; moreover, I had a fragmentary and partial knowledge of the field that I yearned to improve. My initial attempt at a first chapter incorporated the whole animal kingdom, but the taxonomic scale of the enterprise—by which I mean different forms of antipredator defenses seen in species ranging from zooplankton to butterflies to zebras and the amount of material that would have to be marshaled to avoid superficiality—was far too extensive; I thank Rick Grosberg for pointing this out to me in a refreshingly direct manner. I put my initial attempt aside for five years, and when I returned to read the manuscript, I realized that the best part was the mammal section; so at my second attempt, I restricted myself to this taxon. On approaching the topic of vigilance, however, I was quickly reminded that most of the critical studies were about birds, and so I expanded the taxonomic scope to homeotherms.

In retrospect, I think this decision makes a certain amount of sense. Birds and mammals are the taxonomic groups in which theory and empirical work have grown disproportionately. Antipredator defenses in reptiles and especially amphibians, with their two life history stages, are quite distinct from those of birds and mammals and feature phenomena, such as changing color, extrud-

---

Figure P.0 (*Facing page*) Harpy eagles build their nests in very tall trees. Birds of prey have formidable beaks and talons that can be employed in nest defense. In most species, females are larger than males and are present throughout incubation; large size may additionally help in deterring and fending off predators (reproduced by kind permssion of Sheila Girling).

ing toxic substances, and mimicry, that are rare in homeotherms. In contrast, central concepts in the homeotherm arsenal of antipredator defenses, such as group-related vigilance and nest defense, have never been examined so extensively in poikilotherms, although they do of course exist. Moreover, Harry Greene (1988) has written a wonderful review of antipredator mechanisms in reptiles. Yet, an impressive amount of analytical work has been carried out on antipredator behavior in fishes, and while I often felt remorse at leaving fish out of this treatise, I do refer to fish studies regularly. For logistical reasons, little work has been carried out on antipredator behavior of fishes in the wild, which limits the extent of comparisons that can be made with birds and mammals—but the truth is, I have no expertise with fish. Those interested in this class should read Magurran and Pitcher (1987), Pitcher and Parrish (1993), and R. J. F. Smith (1997). My knowledge of insects and invertebrates is similarly limited, but I have fewer qualms here because influential books have been written about antipredator defenses, using examples that focus principally on insects. These include monographs by Hugh Cott (1940) and Malcolm Edmunds (1974), which, in my opinion, still stand the test of time. Fortunately, a companion volume by Graeme Ruxton, Tom Sherratt, and Mike Speed (2004) dwells extensively on this taxon. That said, antipredator defenses in arthropods are far more spectacular, diverse, and intricate than those in birds or mammals, and I would advise any student of antipredator mechanisms to start with this group.

Aside from taxonomic considerations, the book focuses on methods of avoiding predators, not parasites. Regarding the former, I have been circumspect in limiting my discussion to methods of avoiding predation by members of other species, not conspecifics. Without this proviso, discussion would have sometimes led to infanticide and intrasexual contest competition, and although the means of avoiding these are often the same as those used to avoid predation, I would have been in danger of sacrificing depth for breadth. For similar reasons, I have not encompassed kleptoparasitism, or food piracy by other species. I have spent relatively little space on antipredator defenses that fall under the rubric of life history traits, such as clutch size and growth rates, which have been explored in birds by Thomas Martin (for example, Martin and Li 1992; Martin 1995; Martin et al. 2000), although addressed little in mammals. In addition, I have avoided the ecological consequences of antipredator defenses on habitat use and community structure, principally because they open up an enormous body of additional literature; fortunately, these indirect costs of predation have been covered by Steve Lima and Andy Sih in wide-ranging reviews (Sih et al. 1985; Sih 1987; Lima and Dill 1990; Lima 1998) that I could not hope to match. Last, I have touched but not dwelled upon ontogeny of antipredator behavior or morphology; the former is covered in Eberhard Curio's (1993) review. In short, I have tried to integrate all functional and evolutionary perspectives on antipredator defenses that have developed over the last century and exemplify them using two well-known groups of terrestrial vertebrates while attempting to keep pertinent aspects of their natural history in focus.

The book is chronologically structured along a hypothetical predatory sequence. Following John Endler (1991a), I begin with defenses that prey use to avoid being found by predators, and move to the ways that prey detect predators at a distance, warn each other of danger, and signal to predators. I then proceed to benefits of living in groups before turning to morphological and behavioral defenses that individuals and groups employ to thwart attack. Finally, I deal with flight and adaptations of last resort. There are other ways of structuring knowledge of antipredator defenses, such as distinguishing between primary defenses that operate regardless of whether a predator is in the vicinity and secondary defenses that are brought into play when a predator encounters the prey (Edmunds 1974), or separating individual and group-related benefits more explicitly (for example, Krause and Ruxton 2002); but the chronological approach has the advantage of running up a continuum of predation risk that I hope will yield insights into the evolution of antipredator defenses.

I would like to thank the University of California at Davis for continuing to support its tremendous collection of books and journals in the Peter J. Shields Library, without which I could not have written this book. My inability to use the computer program Endnote forced me to read every source I have listed in the references (except eight in German), and, incredibly, almost all are in the UC Davis library in hard copy. I doubt that I could have cast such a wide net if these sources were only online. I thank Tasila Banda-Sakala, Fiona Borthwick, Marisa Flores, Corine Graham, Christopher Holmes, and Ted Stankowich for obtaining sources from the library for me on occasions. I thank Steve Lima for discussion; Dick Coss, Mike Speed, Tom Martin, Peter Bednekoff, Will Cresswell, Klaus Zuberbuhler, Graeme Ruxton, Jens Krause, Peter Wainwright, Bob Montgomerie, John Quinn, Cilla Kullberg, and Andy Sih for commenting on chapters 1 to 13, in that order; and John Endler for remarks on the whole manuscript. Christie Henry and John Thompson provided encouragement during writing, and Sandy Hazel copyedited the manuscript. I am most grateful to my mother, Sheila Girling, for providing the lovely drawings at the start of each chapter, the appendix, the references, and the indexes; she is the person who sparked my interest in zoology when she gave me *The Observer's Book of Birds* when I was three. Finally, I thank Monique Borgerhoff Mulder and Barnabas Caro for letting me take over the living room table for five years.

1 Definitions and Predator Recognition

1.1 Introduction

Because most species are preyed upon by several others, they have been forced to evolve numerous protective responses. As a consequence, the number, subtlety, and complexity of antipredator defenses shown by even one prey individual can be as daunting as they are awe inspiring. Biologists in turn must face the considerable challenge of fathoming how natural selection has shaped mechanisms to avoid premature death at the hands of others. Attempts to understand these antipredator defenses in the animal kingdom have a long history that starts at the end of the nineteenth century, when naturalists working in the tropics described the form and behavior of the new species that they encountered while zoologists in universities and zoological gardens speculated on the function of these traits, particularly animal coloration (Poulton 1890; Beddard 1892; G. H. Thayer 1909). While interest in animal defenses lapsed during the first half of the twentieth century, aside from a scattering of North American work and Cott's (1940) synthesis, it picked up again in the 1950s with the advent of detailed behavioral studies of birds and insects that formed a springboard for modern behavioral ecology, which took off in the 1970s (Bertram 1978a; Harvey and Greenwood 1978). Since then, theoretical advances, especially cost-benefit analyses and models of trade-offs, along with empirical studies in both field and laboratory have grown enormously in number and sophistication to the extent that biologists are in a difficult position of having to synthesize a great deal of information spread over several subdisciplines. Viewed from the outside, it is difficult to be sympathetic to this embarrassment of riches; in truth, biologists simply need to take the time to

Figure 1.0 (*Facing page*) Female impalas observe a person at a close distance. Humans are routinely used in studies of anti-predator defense, raising several problems with data interpretation (reproduced by kind permission of Sheila Girling).

organize theory and evidence before moving forward. This book makes such an attempt.

This chapter begins by justifying the organizational structure of the text and clarifying terminology. The task of describing defenses quickly becomes burdensome during attempts to understand variation in these defenses across individuals or species. Consequently, antipredator responses to predation threat are often characterized either as a sequence of interactions between predator and prey or else as a list of antipredator defenses organized as stage-specific responses to predation threat. These classificatory schemes each have advantages and drawbacks, but I have adhered to the latter organizational structure in this book (section 1.2). The number of attempts to understand antipredator defenses exhibited by birds and mammals, the taxonomic groups that have been studied the most, has led to a proliferation of terms; it is necessary to elucidate the meanings of some important words, because many are applied in more than one way (section 1.3). After this etymological digression, I proceed to the mechanisms by which prey perceive that they are under threat from predation. The chapter, then, is awkwardly split into two disjunct sections.

Predator recognition is an important component of antipredator defense mechanisms because many, perhaps most, defenses necessitate prey first recognizing danger, that is, discriminating dangerous situations from harmless ones, and then taking evasive action (McLean and Rhodes 1991). Hence the topic must be dealt with early on. That said, a great many morphological characteristics and behavioral mechanisms used to avoid detection (chapters 2 and 3), as well as some aposematic traits (chapter 7), do not demand that prey recognize predators. Recognition refers to an animal taking note of a threat and can be measured using simple reflexes such as a change in cardiac response (Mueller and Parker 1980), although it is usually recorded as a behavioral change, notably enhanced vigilance, alarm calling, mobbing, or flight. In theory, a threatening stimulus may be classified at different levels of specificity. In decreasing order, these include recognizing a particular predator individual, a hungry individual, a particular species of predator, a class of predator such as a terrestrial or aerial, or simply a threatening situation that need not even involve a predator being present but just a situation in which the likelihood of predation is elevated (McLean and Rhodes 1991). Most analyses focus on prey recognizing a species or category of predator (see chapter 6).

Broadly, the study of predator recognition has centered on three areas of inquiry. The first simply demonstrates that a prey species is able to recognize danger either by observing prey in the wild or using experiments to match the salient features by which prey identify predators (section 1.4). The second area is developmental in nature, that is, understanding the extent to which predator recognition by naïve individuals is innate or learned and the mechanisms by which prey come to recognize dangerous situations accurately (section 1.5). The third and more recent arena of study concerns loss of predator recognition abilities in the absence of interactions with predators over historical or evolutionary time (section 1.6). Loss of ability to recognize predators bears on both ontogeny of recognition during an individual's lifetime and evolution of recognition systems over generations, and is of concern to conservation biologists attempting to understand the consequences of planned predator reintroductions or predator invasions into new or historical ranges (Wolf et al. 1996; Gittleman and Gompper 2001; Pyare and Berger 2003; Jones et al. 2004).

At the end of the chapter, the issue of observational bias in the study of antipredator defenses is broached (section 1.7).

1.2 The predatory sequence

Predation is sometimes portrayed as a series of interactions between two players, the predator and prey, in which the behavior of the predator leads to a defensive response by prey that may then be countered by the predator, and so on. This way of thinking about predation stems from observers collating all the possible products of interactions that they have witnessed, placing them in a logical sequence, and even assigning probabilities to them, as shown in figure 1.1. The advantage of a flow diagram such as this is that it is easy to grasp relatively complex interactions visually. One problem with these charts, however, is that they allow prey only one or a small number of possible responses to a given decision by the predator, thereby limiting the range of antipredator alternatives open to prey at each juncture. Moreover, flow charts can be misleading because they suggest a choreographed game composed of several interactions that are sequentially contingent on each other and thus imply that prolonged interactions are the rule. In nature, the number of interactions are often limited; for example, the predator appears suddenly, the prey sees it and flies off (Cresswell 1996). Furthermore, the prey is always intent on terminating an interaction as soon as possible. This is seen most readily using real data (fig. 1.2), in this case showing that moose (see the appendix for scientific names of vertebrates) are attempting to end the interaction with wolves through flight or intimidation at a number of junctures.

An alternative method of characterizing the details of predator-prey interactions consists of listing antipredator defenses in a chronological order derived from the predator's behavior and in a way that highlights their effectiveness in counteracting each particular phase of the predator-prey encounter (for example, Pitcher and Parrish 1993; Hileman and Brodie 1994). Phases vary in number and emphasis according to author. Vermeij (1982) separates the recognition or detection phase from the pursuit or escape phase, and again from the subjugation or resistance phase. Edmunds (1974) distinguishes

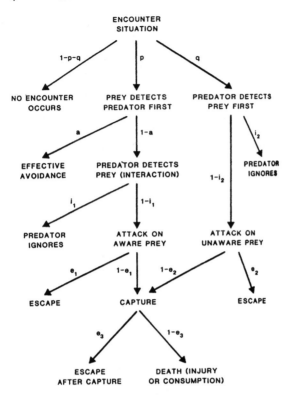

Figure 1.1 Flow chart showing the possible outcomes of an "encounter situation" between a prey and its predator. The symbols adjacent to arrows represent the conditional probabilities of following the labeled pathways. Only a proportion of all encounters lead to the death of the prey through consumption or fatal injury (Lima and Dill 1990).

between primary, or indirect, defenses, defined as defenses that operate regardless of whether a predator is in the vicinity, and secondary defenses that operate during an encounter with a predator (table 1.1.a). Endler (1991a) lists antipredator defenses as counterploys to predatory stages that progressively bring the predator closer to its quarry (table 1.1.b). A disadvantage of these sorts of classifications is that they mix evolutionary, ontogenetic, and proximate timescales, whereas flow diagrams usually adhere only to proximate timescales, that is, during the current interaction between predator and prey. Another difficulty is that although some defenses are tailored to a particular phase of encounter with a predator, the same antipredator defenses are sometimes used at more than one stage of predation, notably defenses such as grouping and immobility, although this problem is not insurmountable. The advantage of such a scheme is that it stresses antipredator defenses as alternatives rather than as a sequence of possibilities to which prey can resort only if a previous antipredator ploy has failed.

In this book, I have taken the second approach and constructed a modification of Endler's (1991a; see also Endler 1986) scheme, starting first with morphological (chapter 2) and then behavioral (chapter 3) defenses that prey use

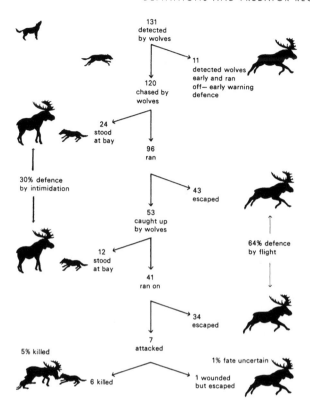

Figure 1.2 Behavior and fate of 131 moose detected by wolves on Isle Royale, USA (Edmunds 1974, based on Mech 1970).

to avoid being detected even when predators are not in the vicinity (classic primary defenses). I then examine behavior that prey use to determine whether predators are in the vicinity (chapters 4 and 5) and to warn conspecifics about predation risk (chapter 6); in the main, these behaviors occur whether or not a predator sees the prey. Next I proceed to situations in which the predator has noticed the prey and the prey is trying to avoid being the target of an attack, either when it is alone (chapter 7) or is a member of a group (chapter 8). Then I progress to defenses that the prey uses to fend off attack if it becomes the target, splitting these into morphological (chapter 9) and behavioral defenses, the latter of which is again split into defenses used by individuals (chapter 10) and those employed by individuals in groups (chapter 11). Finally, I examine how animals escape from imminent predatory attack (chapter 12). It is important to note that more than one of these defenses can be used by a prey individual to thwart a predator and that an interaction between prey and predator could start at the beginning, middle, or end of this predatory sequence. Having surveyed these antipredator defenses in this logical sequence, I try to provide a framework for understanding antipredator defenses in animals (chapter 13).

Table 1.1 Categories of defense used by Edmunds (1974) and Endler (1991a)

a. Edmunds

Primary (indirect) defenses
  Anachoresis (living in holes)
  Crypsis
  Aposematism
  Batesian mimicry
  Single and mixed species associations

Secondary (direct) defenses
  Withdrawal to prepared retreat
  Flight
  Diematic (frightening) behavior
  Thanatosis (death feigning)
  Deflection of attack
  Retaliation
  Single and mixed species associations

b. Endler

Encounter, or get within a distance in which the predator can detect prey
  Rarity
  Apparent rarity

Detection of prey as objects that are distinct from the background
  Immobility
  Crypsis
  Confusion
  Sensory limits and perception

Identification as profitable or edible prey and decision to attack
  Masquerade
  Confusion
  Aposematism
  Mullerian mimicry
  Batesian mimicry
  Honest signaling of unprofitability

Approach (attack)
  Mode of fleeing
  Unpredictable behavior
  Rush for cover
  Startle, bluffing, and threat behavior
  Redirection
  Encouraging premature attack
  Aggregation and predator saturation

Subjugation (prevent escape)
  Strength of escape
  Mechanical methods
  Noxiousness
  Lethality
  Group defense, mobbing, etc.
  Resistance to venom

Consumption
  Safe passage through the gut
  Emetic
  Poisonous
  Lethal

1.3 Definitions

This book is entitled *Antipredator Defenses in Birds and Mammals.* Defenses refer to suites of behavioral, morphological, and, less commonly, physiological and life history characters that these two taxa use to avoid predation. The word *defense* carries no connotation of intent on the part of animals, of consciousness or of military stratagem, but it is used to describe the major categories of resistance that prey mount against predators. In a colloquial sense, *defense* signifies warding off attack and therefore encapsulates some specificity, whereas the words *strategies* and *tactics* (that conveniently refer to both morphological and behavioral traits) lack specificity and, to some, imply intent. Moreover, strategies are used in the literature on alternative reproductive phenotypes within the sexes where they denote a genetically based decision rule that results in the allocation of resources to alternative ways of behaving or to morphological characters. These alternative phenotypes are termed *tactics* (Gross 1996). The special meanings of these terms in reproductive decision making makes them inappropriate to use in an antipredator context, so I have avoided them.

1.3.a Adaptation and evolution   The close resemblance between an animal's form and its environment, as between a stick insect and a twig, or between its form and that of another animal, such as the sphingid moth (*Leucorampha ornatus*) caterpillar and the head of a snake, is so surprising and wondrous that one cannot help but believe that natural selection has been instrumental in making these animals appear cryptic or mimetic to avoid attention of predators; but, from a scientific standpoint, it is troublesome to indulge in "uncritical guessing" about their function (Tinbergen 1963). The function of a character can be investigated in five ways: (i) by arguing that the trait is sufficiently well designed for a task that it must have been shaped by natural selection to solve a problem faced by an organism; (ii) by demonstrating that interspecific variation in a putative antipredator trait is associated with particular species-related ecological or social factors where it might be particularly useful; (iii) by showing that intraspecific variation in a trait that appears designed to avoid predation is correlated with a component of reproduction or mortality, or better still, with lifetime reproductive success; (iv) by using simple models to predict the optimum solution to a problem based on known constraints and a suitable currency, and then observing whether the animal's behavior matches the prediction quantitatively; and (v) by experimentally eliminating or reducing (or enhancing) the appearance of a morphological or behavioral trait and comparing survivorship or reproduction with a control sample (G. C. Williams 1966). Arguably, these five approaches can be ranked from least to most rigorous in the order given. In addition, the evolution of a trait can be documented by placing its appearance and loss on a phylogenetic tree, and by matching these events to the presence of social or ecological variables, evolutionary causation can be inferred (Sillen-Tullberg 1988). Historically, studies of antipredator defenses have principally used argument by design and experimental protocols to investigate function and have very rarely used optimality theory or tried to match defenses to lifetime reproductive success.

If a behavioral or morphological trait can be shown to serve a particular function using one or more of these approaches, it can be termed adaptive in the weak sense of bearers of the trait being more likely to leave more offspring in this and future generations than individuals that do not bear the trait (Clutton-Brock and Harvey 1979). Adaptation in the strong sense demands, in addition to conferring reproductive benefits now, the assumption that it did so in the past, knowing that the trait is heritable and that genetic differences between individuals are responsible for phenotypic differences in the appearance of the trait (Reeve and Sherman 1993). Unfortunately, there are very few examples in any vertebrate where we know that differences in antipredator defenses are determined even in part by genetic differences (Magurran et al. 1995), and no examples in homeotherms. Consequently, I have abstained from using the word *adaptation* (sensu stricto) in this book to avoid confusion between the two sorts of meaning.

One of the difficulties confronting researchers studying antipredator defenses is that a particular behavior may confer reproductive advantages on its bearer in several different ways, all related to avoiding predation. While the most important one of these mechanisms has, historically, been referred to as the primary function, in most cases it is extremely difficult to discern which are secondary or primary functions. Similarly, if a behavior helps prey escape predation at more than one phase in the predatory sequence, and this is evident in several instances (for example, grouping), then the efficiency with which predators capture prey at a particular phase in the sequence must be less than at other phases of the attack in order to be confident of the behavior's primary function (Vermeij 1982). Unfortunately, there are very few instances in which we have this sort of knowledge.

The book is principally about the evolution of antipredator defenses in the sense that I use a comparative perspective to highlight the diversity and distribution of antipredator defenses across homeothermic taxa. Note that *evolution* as understood here does not refer to the historical genesis of morphological traits within a clade, owing to the fact that remarkably little work has been done in this area. Modern comparative studies use statistical methods that reconstruct the phylogenetic history of a trait, but this method has been used only sparingly in antipredator studies and simply as a way to check whether shared ancestry could be responsible for the appearance of a trait in extant species (Caro et al. 2004) rather than as an investigative tool to determine when and where it arose over evolutionary time. Note also that *evolution* as used here does not refer to a change in gene frequencies, because we know so little about the genetics of any antipredator defense in warm-blooded animals despite classic evolutionary studies having been being conducted on invertebrates' defenses, including melanism in peppered moths (*Biston betularia*), crypsis in banded snails (*Cepaea nemoralis*), and polymorphisms in hawkmoth caterpillars (Ford 1964; Curio 1970).

1.3.b Antipredator terminology   There is a proliferation of terms used to describe the ways in which animals try to reduce risk of predation; as a result, many have been used inconsistently and in ways that pertain to more than one class of behavior. Sometimes a term is used descriptively, sometimes in a mechanistic (proximate) sense, other times in a functional or evolutionary (ultimate) fashion. Surprisingly, rather few authors of antipredator research papers have commented on these puns despite the danger of confusing description, cause, and function (but see Hennessy et al. 1981; Klump and Shalter 1984; Ficken and Popp 1996). The main difficulties are shown in table 1.2.

The act of avoiding a predator has a dual meaning in the study of antipredator defenses. On the one hand, it is used in a functional sense of prey changing its activity, often foraging behavior, so as to avoid being detected by predators. This carries no meaning of intent but implicitly suggests the change in behavior has been selected in response to predation. The second, more general, meaning simply refers to the prey having escaped the predator, as in "avoiding being eaten." *The Shorter Oxford English Dictionary* (*OED* 1973) defines *avoid* as "to move or go away, withdraw; to retire, retreat," as in the first usage, but also as "to leave alone; to have nothing to do with; to escape, evade; to keep out of the way of; to prevent, obviate," as in the second. I have restricted use of *avoid* and *avoidance* to the first sense of staying away from predators and thereby eluding detection, because *avoidance* in the second sense is so general as to lose special meaning. The shortcoming of this is that I assume that selection has acted on the prey's morphology and behavior to prevent detection, but this is often what researchers are trying to investigate.

Alarm signals or *alarm calls* is used by authors in several ways, one being the descriptive sense, as in behavior given in the presence of a predator. Some authors use them interchangeably with *mobbing calls*. In the psychological literature, *alarm calls* sometimes refers to the internal state of a stressed animal. Confusingly, *alarm signals* is additionally used in a functional sense of conveying knowledge about danger to conspecifics or rendering information to a predator; often these two are not distinguished. The *OED* defines *alarm* as "sound to warn of danger, or to arouse; a warning signal of any kind; a sudden attack,

Table 1.2 Common uses of terms in the study of antipredator defenses

Term	Definition	D[a]	P[a]	U[a]	Examples of this usage[b]
Avoidance	Prey limits its activities so as not be detected by predators			X	1
	An interaction between prey and predator results in the prey escaping	X			2
Alarm signals or calls	Any alarm signal given in presence of predator	X			3
	Mobbing call	X			4
	Calls when internally alarmed		X		5
	Informs conspecifics of danger			X	6
	Imparts information to predator			X	7
Warning signals or calls	Coloration or other distinctive mark	X			8
	Informs conspecifics of danger			X	9
	Imparts information to predator			X	10
Deterrence	Prey reduces probability of further advance by predator			X	11
	Prey stands its ground	X			12
	Intimidates predator or attacks predator	X			13
Distress calls	Fear screams	X			14
	Calls when internally distressed		X		15
	Calls when needs help			X	16
	Young call when need help			X	17
Defense calls	General term for calls given when in danger	X			18
	Prey calls loudly when predator nearby	X			19
Group defense	Prey prevents predation by bunching or mobbing	X			20
	Prey prevents predation by attacking predator	X			21
Alarm response	Becomes immobile	X			22
	Becomes agitated	X			23
	Elicits fear screams	X			24

[a] Definition is descriptive (D), proximate (P), or ultimate (U).
[b] 1. Ylonen et al. 1992; 2. Bertram 1978a; 3. Cresswell 1994b; 4. Maloney and McLean 1995; 5. Lishak 1984; 6. Sherman 1977; 7. Zuberbuhler, Jenny, and Bshary 1999; 8. Tullrot 1994; 9. Herzog and Hopf 1984; 10. Tilson and Norton 1981; 11. FitzGibbon and Fanshawe 1989; 12. Wenger 1981; 13. Edmunds 1974; 14. Hogstedt 1983; 15. Starkey and Starkey 1973; 16. Russ, Racey, and Jones 1998; 17. Gaioni and Evans 1986; 18. R. L. Knight and Temple 1986b; 19. Klump and Shalter 1984; 20. Shank 1977; 21. Hamilton, Buskirk, and Buskirk 1975; 22. Balph 1977; 23. Curio 1975; 24. E. D. Brown 1985.

a surprise; a state of excitement caused by danger apprehended." In short, all the behavioral uses of *alarm signals* and *alarm calls* are etymologically correct. Nonetheless, in science it is helpful to define terms precisely. As the function of most alarm signals or calls is unknown and, where they are known, several functions are sometimes imputed, it seems judicious to avoid functional definitions altogether. When they are known, it is helpful to distinguish conspecific warning signals (see below) from those addressed to predators. Similarly, it is an empirical issue as to whether alarm signalers are stressed or not, and difficult to measure routinely, so it is best to define *alarm calls* and *alarm signals* using a description of the frequency of calls or the animal's movement or posture.

Warning signal is enlisted to describe particular sorts of bright coloration, such as black and red stripes, as well as being applied as a less burdensome synonym for *alarm signal*, because it avoids innuendo about an individual's internal state. The first service poses little difficulty, as *warning colors* are usually defined by describing colors and patterns on an animal's body. Here it is often used interchangeably with *aposematism*, but this is incorrect, because the latter additionally means that the animal is distasteful, well defended, or unprofitable as well as being brightly colored. *Warning signals* or *warning calls* are also applied to conveying information to a conspecific or to a predator. The *OED* defines *warning* as "taking heed, precaution; previous intimation or threat of impending evil or danger, a portent of coming evil; advice to beware of a person or thing; cautionary advice against impudent or vicious action; a deterrent example; previous notice of an event whether good or bad." Thus, the aposematic use of *warning signal* leans on the "advice to beware" meaning, whereas the *warning call* refers to the "taking heed" usage. We often know so little about the consequences of warning signals for the signaler, conspecifics, or predator that a definition that simply describes the morphology, call, or movement pattern is again advisable.

Certain types of signals directed at predators inform the predator that pursuit would be unprofitable; these are called pursuit deterrent signals. In addition, *deterrence* is used in two other senses: when prey stands its ground and wields weapons or uses its body size to deter attack, and, loosely, when predators are intimidated. The *OED* defines *deter* as "to restrain from acting or proceeding by any consideration of danger or trouble; to terrify" and *deterrence* as "preventing by fear." The two biological usages follow the restraining meaning, where predators are persuaded not to pursue quarry because it would be unprofitable, whereas the third prescription relates to the fear connotation. Unfortunately, we normally have no independent evidence that the predator is frightened, so the word *deterrence* in the sense of intimidation should be dropped.

Distress calls are utilized in three ways: as a synonym for fear screams elicited during handling; when an individual is thought to be psychologically stressed; and when prey, especially offspring, are assumed to be calling for help. The *OED* defines *distress* as "anguish or affliction affecting the body, spirit or community"; this matches the first two practices, but the third, calling for help prescription, does not apply. As to whether distress calls do solicit help and to whom they might be directed is the subject of debate, and as psychological stress is difficult to measure in most species, it seems most sensible to restrict the definition of *distress calls* to calls given when animals are handled.

Defense calls are used generally to describe any one of several vocalizations made by prey under threat of predation. These include alarm calls, mobbing calls, and calls to quieten offspring, but they are also requisitioned more specifically to describe calls given when predators are closing in on prey and are thought to be attempts to startle predators. Regarding the first three applications, lack of specificity means that the term imparts little information and should not be used when these meanings are implied. It might be sensible to restrict using *defense calls* to situations in which the predator is close to prey, but this is a slippery, relative definition. The functional connotation is problematic, too, because there is no strong evidence that such calls do startle predators in any taxonomic group. Consequently, the term should be abandoned altogether.

Group defense is called forth to describe a stock of behaviors shown by grouped prey toward a predator. These include prolonged calling, bunching together, and rushing at and sometimes striking or biting the predator. Depending on circumstance, these behaviors are termed group mobbing, group defense, or group attack, but this is highly context specific, fastened on such issues as the identity of the predator and where the prey are in relation to it. *Defense* is variously defined in the *OED* as "the action of warding off; protecting from attack; resistance against attack, warding off injury, protection; capacity of defending, the art or science of defending oneself (with weapons or the fists), self defense; something that defends." That behavioral ecologists have applied the term *group defense* to include both broad and narrow meanings of the word is therefore correct. *Group defense* should continue to be defined by means of simple descriptions of the prey's behavior in the face of a predatory threat, but definitions should avoid incorporating the effect it may have on predators that may vary from situation to situation.

Alarm responses refer to animals becoming immobile in response to either the sight of a predator or the call of a parent. This usage stresses reactive mechanistic properties of behavior. Alarm responses also refer to animals becoming agitated, as observed by frequent or repetitive movement. In both cases, they are synonymous with *fright responses*. Finally, *alarm responses* are used as an alternative term for *fear screams*. The only one of several *OED* definitions that fits these is "a state of excitement caused by danger apprehended," but as *alarm* is such a well-exercised term already in the study of antipredator defenses, there are strong grounds for dropping the term *alarm response* altogether. *Fright* (defined in the *OED* as "fear in general; sudden fear, violent terror, alarm. An instance of this; anything that causes terror") *responses* should be applied instead, but defined strictly according to behavioral characteristics, be they increased or decreased movement, and not according to an imputed internal state. Alarm responses that are manifested as fear screams should simply be reported as fear screams or distress calls for reasons stated earlier.

In general, the function of many antipredator behavior patterns has proved difficult to decipher, so it is most cautious, although perhaps not so memorable, to define antipredator defenses in descriptive terms. Alas, a number of the studies quoted in this book do not make a clear distinction among the alternatives given in table 1.2.

1.4 Ability of prey to recognize predators

I now turn to the topic of predator recognition. I have placed it in chapter 1 because it is a necessary condition for many of the antipredator defenses that follow and because its mechanistic and developmental focus separates it from the evolutionary theme of the book. Predator recognition usually demands that prey can distinguish between predators and nonpredators, but most antipredator studies simply assume that prey are capable of recognizing predators and concentrate attention on prey responding to one or sometimes a handful of predator species. Thus, while it is unusual for a field study to document the relative strength of responses to dangerous and harmless animals, a few show that birds and mammals have sophisticated discriminatory abilities. For example, Belding's ground squirrels exhibit different behavioral responses to both aerial and terrestrial predators than to harmless species (table 1.3). In addition, they can discriminate between raptors (for example, northern harrier) and omnivores (for example, common raven) (table 1.3), and to a lesser extent between terrestrial predators such as large mammals, weasels, and snakes (not shown). They also react differently to harriers flying overhead and those flying south although not to harriers flying at different heights aboveground, despite being in danger only from low-flying birds (S. R. Robinson 1980). The degree to which the limits of discriminative abilities in this and other species are adaptive or are simply shackled by sensorimotor or time constraints is left unexplored in most studies (but see Evans, Evans, and Marler 1993).

Prey animals use several cues to recognize danger that include sight, sound, and odor of predators, or a combination of these sensory modalities. Studies that have investigated visual cues requisition models of predators with parts

Table 1.3 Responses of Belding's ground squirrels to aerial predators and terrestrial interactions with predators and harmless species

	Nonvocal and				Trills and			Chirps and		
	NR	P	E	A	P	E	A	P	E	A
Aerial encounters										
Predators (total)	63	36	19	na[a]	18	7	na	22	30	na
Raptors	6	8	17	na	5	6	na	10	29	na
Omnivores	57	28	2	na	13	1	na	12	1	na
Harmless birds	565	3	0	na	8	2	na	10	2	na
Terrestrial interactions										
Predators	11	4	5	2	10	30	3	1	3[b]	0
Harmless species	52	0	0	0	1	0	0	0	0	0

Source: S. R. Robinson 1980.
Note: Column headings are as follows: NR, no response; P, posting (standing upright and being vigilant); E, evasion; A, approach.
[a] Not applicable.
[b] These responses involved both trills and chirps.

of the body removed, rearranged, or colored differently, or have used various models that look increasingly unlike predators; they then compare prey's reactions to these models against those using intact predators (for example, Hinde 1954; Curio 1975; Kerlinger and Lehrer 1982; Hanson and Coss 1997). For instance, free-ranging bonnet macaques make progressively fewer alarm calls and flight responses to an upright spotted model of a leopard than to an upside-down model, to a dark upright model, or to a dark upside-down model, indicating that both spots and configuration are involved in recognition (Coss and Ramakrishnan 2000). It is difficult to generalize about the salient features that prey use to make discriminations about predators, because experimental studies necessarily involve different prey species, different predators, and different model alterations; however, the presence of eyes or eyelike structures often triggers a response, and increasing realism of the model is usually important.

Studies investigating whether prey can recognize predators by the sounds that they make employ playbacks of predator vocalizations. In Kibale Forest, Uganda, when chimpanzee or crowned hawk eagle calls were played to red-tailed monkeys, blue monkeys, or red colobus monkeys, the monkeys looked in the direction of the speaker but not when calls of a black and white casqued hornbill or red colobus, neither of which are predators, were played to them (Hauser and Wrangham 1990). Similarly, American crows took flight on hearing playbacks of red-shouldered hawks and great horned owls, which prey on their nestlings and fledglings, but did not fly off on hearing a benign wood thrush, or a Madagascar harrier hawk that they could never have heard before (Hauser and Caffrey 1994). Even more impressive is when prey discriminate between dangerous and benign individuals of the same predator species. In the northeastern Pacific Ocean, resident killer whales live in large stable groups and eat fish exclusively, whereas transient killer whales live in smaller social groups and prey on marine mammals. Residents are vocal and their dialects differ between groups, but transients are usually silent. Off the British Columbian coast of Canada, harbor seals are an important prey of transients. Seals can distinguish between playbacks of residents and transients because, on hearing a transient call, they rapidly quit the water's surface if they are visible from below. Interestingly, seals showed a similar disappearing response to playbacks of unfamiliar Alaskan resident killer whales that eat fish as they did to transient Canadian whales. This demonstrates that lack of response to harmless British Columbian residents resulted from habituation rather than associating transients' calls with attacks, a far more dangerous mode of learning (Deecke, Slater, and Ford 2002). Similarly, California ground squirrels can distinguish large rattlesnakes from small ones, and warm ones from cold ones, on the basis of rattle amplitude, pitch, and click rates (Swaisgood, Rowe, and Owings 1999).

Finally, prey can recognize the odor of predators (chapter 2). For instance, moose show enhanced vigilance in response to smelling snowballs coated in grizzly, black bear, or tiger feces or snowballs soaked in wolf or coyote urine compared with control snowballs (Berger 1998; Berger, Swenson, and Persson 2001). Despite impressive recognition skills, there are limits to the recognition abilities of prey within a modality with prey, responding to some stimuli but not to others (see Curio 1975 for a detailed study), as well as differences in the extent to which prey use different modalities to recognize predators (Berger 1998, 1999).

1.5 Recognition by young animals

1.5.a Innate recognition
As young animals are smaller than adults, they are likely to suffer the attentions of a greater number of predator individuals, as well as a greater number of predator species (Kunkel et al. 1999; Pierce, Bleich, and Bowyer 2000). In addition, young are often poorly coordinated and can fly off or run less quickly and maneuver less well than adults. For all these reasons, young prey will be especially vulnerable to predators, and a great number of studies show that young suffer higher rates of predator-induced mortality than adults (for example, Ricklefs 1969b; Schaller 1972; FitzGibbon and Fanshawe 1989; McLeery and Perrins 1991). Thus, there should be strong selection pressure on predator-naïve young to recognize predators when they first encounter them. Innate recognition of predators, in the sense of no experience of an attack by the predator being required, has been documented in many species, such as those listed in table 1.4, using experiments in which models are presented to laboratory-reared young (called Kaspar-Hausers in old fashioned terminology: N. Tinbergen 1951; Curio 1975) or young that are naturally predator-naïve because they come from predator-free areas.

Table 1.4 Innate enemy recognition as demonstrated by deprivation experiments on vertebrates

"Kaspar Hauser" sp.	Predator stimuli and ranking where known[a]	Stimulus specificity controlled by A or I[b]	Comment
Atlantic salmon	Pike > burbot	—	Avoidance of pike only increases on perceiving it
Guppy	Giant rivulus, trahira, and cichlid piscivores	A	Adaptive geographic variation
Three-spined stickleback	Pike	—	Adaptive geographic variation
African jewel fish	2 staring eyes > 3 staring	A I	Serves avoidance of piscivores and dominant conspecifics
Hognose snake	Long-eared owl D; human with stare > human with eyes averted	A I	Recovery time from cataleptic response
Virginia opossum	Human (dog)	—	Feigning death when grabbed: dependent on rearing by mother and/or housing conditions
Wolf	Human	—	Only if not familiarized with humans in early life. Independent of parents' tameness
Fallow deer	Human	—	Only if not familiarized with humans during first 2 days of life. Independent of mother's tameness
California ground squirrel	Pacific rattlesnake > gopher snake	A	Adaptive (?) geographic variation
California ground squirrel	Pacific rattlesnake same as gopher snake	A	"Kaspar Hausers" both snake naïve + burrow-naïve dealt appropriately with snake in burrow but did not discriminate the 2 species as they do above ground
Golden hamster	Dog, odor of dog, +D	A	Flight into burrow; freezing rarely
Golden hamster	Odor of European polecat +D; European polecat +D	A	Freezing, threat, attack, occlusion of burrow
Norway rat 5 laboratory strains	Domestic cat	A I	Response most pronounced in least selected strain. Strain differences
White-throated wood rat	Gopher snake	A	
Deer mouse	Weasel sp., cat, 2 snake sp.	A	Adaptive geographic variation
Red-legged partridge	Human	—	Only if not familiarized with man within 48 hrs. of hatching

Table 1.4 (continued)

"Kaspar Hauser" sp.	Predator stimuli and ranking where known[a]	Stimulus specificity controlled by A or I[b]	Comment
Domestic fowl	Human	—	Avoidance differs among two stocks; modifiable more by reward/habituation in birds of tame stock
Domestic fowl	Staring eyes D, human face	I	Recovery time from tonic immobility. Domestic fowl, human experienced
Domestic fowl	Eurasian kestrel D, staring eyes	A D A D	Fear increases on stare of following chick. Domestic fowl
Scrub jay, Mexican jay	Great horned owl	—	Response matures earlier in less social scrub jay
Pied flycatcher	Owl D, owlet sp., great gray shrike	A I	Shrike: adaptive geographic variation
Zebra finch	Ferruginous pigmy-owl	A	Presence of pair mate and/or open view of landscape required. Dummies fail
Galapagos finch sp.	Snake sp. +D	A	Natural "Kaspar Hausers" on snake-free islands. Adaptive geographic variation
Galapagos finch sp.	Owl D, hawk D	A(D) I	Adaptive geographic variation

Source: Curio 1993.
[a] Denotes response to dummy (D) only; +D denotes response to dummy and to live predator.
[b] A: animate, denotes either live, taxidermic, or other replica of control stimulus. I: inanimate, denotes man-made control stimulus of about predator size, or simply test environment (the latter when not specified).

In birds, an illustration of the former type of study is where hand-reared turquoise-browed motmots or great kiskadees were found to avoid cylindrical models of venomous coral snakes (*Micrurus* sp.) painted with alternating vertical bands of wide yellow and narrow red color; they then generalized this aversion to horizontally painted stripes (S. M. Smith 1975, 1977). An example of the latter kind of study is the Seychelles warbler alarm calling at and attacking a mounted model of the Seychelles fody, an egg predator of this species. Warblers attacked the model whether they came from Cousin Island, inhabited by fodies, or Aride Island, where fodies are absent (Veen et al. 2000; see also Maloney and McLean 1995). The idea that predator recognition is innate in birds has a deep history in animal behavior, going back to studies of Lorenz (1939), Kratzig (1940), Tinbergen (1948), and Schleidt (1961), who flew model silhouettes in one direction to represent a hawk, and in the other direction to represent a goose, over recently hatched chicks, such as domestic turkeys, and noted differences in their fear responses.

Similarly, in mammals, naïve young show innate fear responses to predators. Inexperienced young California ground squirrels recognize snakes and respond to them using adultlike behavior patterns, including sniffing and tail flagging when they are aboveground (Owings and Coss 1977) and, when they are belowground, kicking sand and plugging the burrow (Coss and Owings 1978); zoo-bred black-tailed prairie dogs respond to snakes by jump-yipping, foot thumping, and tail flagging (Owings and Owings 1979); and laboratory-born subspecies of deer mice discriminate between predators and nonpredators as measured by alterations in time budgets, provided the predators are found in that subspecies' habitat (Hirsch and Bolles 1980). Therefore, in both mammals and birds there is compelling evidence that certain species of prey can react to predators appropriately on their first encounter. Innate antipredator responses are likely to be manifested where predation pressure is high straight after birth or hatching or where opportunities to learn about predators are limited, whereas learning to recognize predators is thought to be beneficial in situations in which the probability of meeting a particular predator species varies between generations or where individuals of a given predator species sound or look different from one another (Curio 1993).

1.5.b Learning to recognize predators   Many studies have also shown that predator recognition and responses change with age, indicating that maturation and the role of experience and learning are important (for example, Kramer and von St Paul 1951; Hinde 1954; Schaller and Emlen 1961; Curio 1975; McLean and Rhodes 1991; Hanson and Coss 1997). In some species, learning occurs extremely rapidly. In the New Zealand robin, adult birds from an island where no stoats occur treated a taxidermic mount of a stoat in the same way as a cardboard box when either model was presented at the nest containing their chicks, whereas adults from mainland New Zealand showed a far greater intensity of response to the stoat (fig. 1.3a). Yet, after being exposed for only 5 minutes to a moving model stoat with a dead robin in its mouth together with playbacks of alarm or distress calls at a distance of 10–15 m from the nest, robins reacted with strong intensity to a stoat the next day (fig. 1.3b). More commonly, however, antipredator responses have both genetic and learned underlying components. For example, in an early experiment, Hinde (1954) showed that innate recognition of an owl in the young chaffinch precedes the development of a mobbing response by 2 weeks, but mobbing then continues to develop over another 2-month period.

Learning can proceed in two ways: (i) by starting out with a specific predator image and expanding it by adding cues based on personal attack, or noting conspecifics' responses, both of which require risky experience with predators. Often young are predisposed to learn about the characteristics of predators

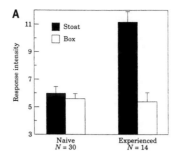

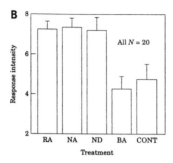

Figure 1.3 *A*, index of response by predator-naïve (island) and experienced (mainland) New Zealand robins to a standard presentation of a predator (stoat) and a control (box). *B*, index of response to a stuffed stoat by incubating female robins 1 day after exposure to a stoat for 4 training treatments and a control (CONT) which received no training. Treatments were RA, a mounted robin in a mobbing posture placed 2 m from the stoat and beside a speaker that played robin alarm/mobbing calls; NA, the same calls but no mounted robin; ND, the same robin and speaker but with playback of robin distress calls; and BA, a mounted blackbird and playback of blackbird alarm/mobbing ("tix") calls. Response intensity is presented as a mean (+ SE) calculated across nests for responses by the first bird to approach the model (the average response for a pair was used if they arrived together), and was determined for each nest by combining all responses as a simple arithmetic summary of behavior patterns scored on a scale from 0 to 2. Behavior patterns scored were number of flights, wing flicks, hops, alarm calls, head-feather and wing-droop displays, minimum approach to the model, and delay by the bird approaching the nest; minimum and maximum scores = 0, 16 (Maloney and McLean 1995).

rather than innocuous objects. Thus, juvenile rhesus macaques become fearful of a snake after seeing video sequences of adults behaving fearfully toward it, but they fail to develop a fear response to plastic flowers when these were paired with identical video footage (Mineka and Cook 1988; see also Curio 1988a; Magurran 1989; Griffin, Evans, and Blumstein 2002). Guided learning could result from selective attention, preferential associations made between particular stimuli and fear, require less experience or reinforcement, or rely on strong arousal in the observer, but the importance of these are unknown at present. Guided learning possibly serves to reduce the time and energy costs of responding inappropriately to harmless objects.

(ii) Learning can also proceed by starting out with a generalized response to a wide variety of stimuli, not all of them threatening, and developing into a

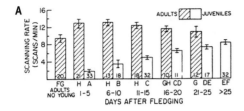

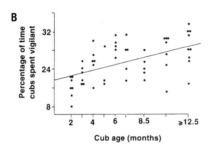

Figure 1.4 *A*, scanning rates (X + SE) for adult (*hatched bars*) and juvenile (*open bars*) yellow-eyed juncos plotted against time in days. Numbers denote sample sizes. Groups that do not have the same letter under their histogram bar differ significantly (K. A. Sullivan 1988). *B*, percentage of 5-minute scans during the midday rest period that cheetah cubs had their eyes open and were observing their surroundings plotted against cub age in months; each point represents an average for all cubs in a litter (Caro 1994a).

response to a specific set of dangerous objects (Ivins and Smith 1983; Pongracz and Altbacker 2000). Both phenomena carry costs of responding to false alarms but require no experience of predators (and the associated risks) in order to proceed. Lack of discrimination between the dangerous and the benign in young animals may be an adaptive response to their poorer physical capabilities and heightened predation pressure than adults (in essence, they are erring on the side of caution), or alternatively could constitute a developmental constraint requiring experience or maturation to become more specific, or both.

There are several potential mechanisms by which prey come to recognize predators as they grow older (Curio 1993). First, and most obviously, animals may put themselves in a situation in which they are likely to learn about a predator. Young birds and mammals show increasing levels of vigilance with age (fig. 1.4) and are therefore more likely to see a predator. Second, young animals may approach and inspect a predator when they see it (section 7.6.c) and perhaps learn about its characteristics, including its motivation and behavior (Magurran and Pitcher 1987). In a well-known experiment, Kruuk (1964) presented members of a gull colony with a stuffed stoat, next to which was placed a dead stuffed gull. Herring gulls and lesser black-backed gulls on the wing responded by flocking and alighting near the models. Gulls were more attracted to the predator than to a nonpredatory hedgehog and even more attracted to the stoat when a dead gull was placed next to it, and they alighted at a farther distance, indicating greater fear (fig. 1.5). Kruuk interpreted attraction to the

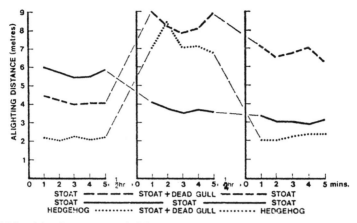

Figure 1.5 Mean alighting distance (in meters) of gulls from predator models plotted against time in minutes (Kruuk 1964).

predator as bringing birds into a position where they could collect information about a potential source of danger, and he thought that birds were more wary of a predator that had apparently killed a colony member because it might kill again (see also Conover 1987). In some species, such as Thomson's gazelles, inspection behavior is more prevalent in immature age classes, supporting this learning hypothesis (FitzGibbon 1994).

Third, early development of antipredator responses may be contingent on experience not associated with predatory attack. Thus, in fish, social encounters between fry or exposure to water movement in European minnows (Magurran 1990a) and fathers' attempts to chase fry back into the nest in three-spined sticklebacks (Tulley and Huntingford 1987) may enhance fry's subsequent ability to evade predators. Similarly, parents' alarm calls may enhance antipredator responses in offspring (Kullberg and Lind 2002). Fourth, animals may come to recognize and respond to predators after being pursued by a predator (McLean, Holzer, and Strudholme 1999). Tammar wallabies, for example, can be trained to become fearful (less relaxed) of stimuli that are paired with an adverse event, a simulated capture procedure by a person (Griffin, Evans, and Blumstein 2001). Here, captive-bred wallabies were exposed to a model red fox, and a person carrying a net pretended to capture them 3–5 seconds later (paired-experimental group). In an unpaired control group, presentation and capture were separated by a gap of 25–90 minutes and occurred in no particular order. Later the same day, experimental wallabies showed less relaxed behavior than controls when shown a fox or a cat, a transient fearful response to a wallaby (possibly for social reasons), but no fear of a young goat, cart, or blank stimulus (fig. 1.6). Thus, an aversive stimulus paired with the presentation of a predator changes subsequent antipredator behavior to that predator and generalizes to other predators but not all mammals.

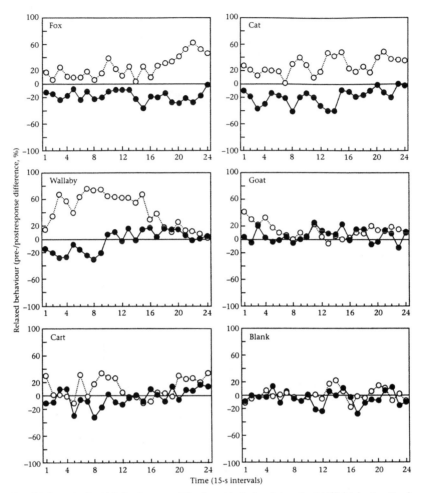

Figure 1.6 Changes in relaxed behavior of tammar wallabies after training for the paired-experimental (*filled circles, N* = 8) and unpaired-control (*open circles, N* = 8) groups. The mean pre/postresponse difference is plotted for 24, 15-second time intervals from stimulus onset, 1 minute during stimulus presentation and 5 minutes after the stimulus had disappeared from the stage. Note that enhanced responses to the predator will be reflected in a reduced proportion of relaxed behavior (Griffin, Evans, and Blumstein 2001).

Fifth, prey may use cues from conspecifics such as fear responses of mothers to learn to recognize sources of danger (Mineka and Cook 1988). In a classic series of experiments, Curio and colleagues demonstrated that mobbing by a conspecific can teach a bird to mob even an innocuous object (Curio, Ernst, and Vieth 1978a, 1978b; Curio 1988b, 1993). Their experimental apparatus consisted of a teacher European blackbird that could see a stuffed little owl from its aviary, next to which was a second "observer" blackbird in a parallel aviary that could see only a stuffed male noisy friarbird that was novel but not a predator of blackbirds. The observer showed a dramatic increase in standardized

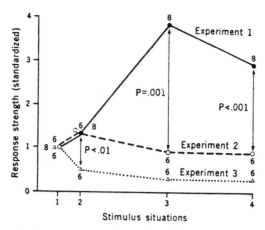

Figure 1.7 Strength of mobbing a honeyeater (noisy friarbird) by an observer blackbird. In experiment 1, the observer was shown a honeyeater while a conspecific mobbed a little owl in situation 3. In experiment 2, the honeyeater was shown without conspecific mobbing. In experiment 3, the honeyeater was removed from the box and the empty box was shown alone. Response strength was standardized with reference to the empty box control (situation 1). On the *x* axis, 1 denotes 09.30 hours: rotation of empty box as a control; 2, 10.00 hours: novel honeyeater presented alone to observer; 3, 12.00 hours: novel honeyeater presented to observer paired with teacher mobbing little owl; and 4, 14.00 hours: novel honeyeater presented to observer alone (Curio, Ernst, and Vieth 1978a).

response strength in the form of mobbing over and above the fear it had of novelty (situation 2 in experiments 1 and 2, fig. 1.7). Moreover, an observer took off faster if another bird was mobbing (Frankenberg 1981). Extraordinarily, observer blackbirds could be conditioned to avoid a multicolored plastic bottle using the same experimental design with a mobbing teacher, although the effect was not so marked, indicating that there are constraints on learning about sources of danger. Finally, by repeatedly placing the observer blackbird in the role of teacher, information about the aversive nature of the friarbird could be transmitted through a total of six different birds with no discernible decrement of information transfer ("cultural transmission"). Similar findings have been made in tammar wallabies responding to teachers that had been made fearful of a model red fox (Griffin and Evans 2003).

Prey may also come to discriminate sources of danger by cueing into alarm calls of conspecifics (Herzog and Hopf 1984; Cheney and Seyfarth 1985a) or even sympatric heterospecific prey (Hauser 1988b; chapter 6). In these circumstances, monitoring conspecifics' behavior in the presence of a dangerous animal and hearing bouts of alarm calling may condition young animals to danger. Conversely, lack of response of mothers or conspecifics to harmless objects may help to reduce fear in young animals that would otherwise start or flee from sources of novelty such as falling leaves ("conditioned inhibition": Seyfarth and Cheney 1980). In cheetahs, cubs show a progressive reduction in reacting to harmless objects (table 1.5) as they grow older, probably because of

Table 1.5 Instances of 1 or more cheetah cubs being fearful of relatively harmless animals. None were fearful of inappropriate animals after 4 months of age.

2 months of age
    Hisses and runs away from black-bellied bustard 7 m away
    Run away from 2 crowned cranes 25 m away
    Run off 10 m when giraffe 200 m away
    Slink away from Grant's gazelle
    Slinks away from hartebeest
    Slink away from kori bustard
    Back away from an adult male warthog 10 m away[a]
    Hisses at secretary bird 8 m away[b]

3 months of age
    Slink off when 2 subadult Grant's gazelles spar 35 m away
    Run when neonate Grant's gazelle bleats while dying
    Run off when vulture hisses at them 20 m away[a]
    Slink toward mother when 2 adult warthogs pass by at 30 m[a]
    Flee from a warthog 50 m away[a]
    Moves around to back of mother when zebra 40 m away[a]
    Run to mother when golden jackal howls at them 20 m away[b]

4 months of age
    Move away from giraffe 120 m away
    Slink off 2 m when 2 giraffes 200 m away
    Slinks off to family when bustard 800 m away
    Moves to mother when 12 zebras 20 m away[a]

Source: Caro 1994a.
[a] These species might be harmful to young cubs.
[b] Species known to pose a threat to young cubs.

an absence of reaction by mothers and her calling the cubs back after they have scattered. Carefully designed experiments to elucidate these possibilities have yet to be carried out, however.

1.6 Relaxed selection

Innate responses to predators in inexperienced animals suggest that there are genetic predispositions to respond to predators, whereas the role of subsequent learning raises the question of whether there are genetic predispositions to learn (Magurran 1990a). If prey move into areas where certain predators are absent, or live in areas where predators have been extirpated very recently or for several generations, there may be relaxed selection on innate responses and predispositions to learn about predators, as well as an absence of opportunity to learn at all. This is strictly an empirical issue, since there are no a priori hypotheses as to how rapidly predator recognition and antipredator responses should be lost from populations. A handful of studies have documented effects of relaxed selection on antipredator behavior in mammals but at present show no consistency regarding the speed at which prey-discriminative abilities are lost.

California ground squirrels co-occur with Pacific rattlesnakes and Pacific gopher snakes across most of their range in and around the Central Valley of California, USA, except in wetter areas and at higher elevations. Based on genetic analyses, ground squirrels are thought to have colonized snake-free ar-

eas between 70,000 and 300,000 years ago, and levels of resistance to rattlesnake venom broadly match the extent of snake predation across sites.
When ground squirrels from several populations were brought into the laboratory and observed in a room with a caged snake, all individuals recognized
snakes. Surprisingly, ground squirrels from snake-free populations exhibited
higher levels of physiological arousal and vigilance as measured by greater levels of tail piloerection, amount of time that they faced the snake, and percentage of time near the snake than their counterparts that were used to snakes,
whereas substrate throwing and tail flagging did not differ between relaxed
and snake-selected populations (fig. 1.8). In contrast, laboratory-born arctic
ground squirrels, a species that experiences no snakes in the wild and that diverged from California ground squirrels 3–5 million years ago, showed less
caution when investigating snakes in the laboratory, were less attentive to
them, and did not throw substrate at the snakes, compared with laboratory-
raised California ground squirrels (Goldthwaite, Coss, and Owings 1990).
Thus, antipredator recognition and responses do not disintegrate after
300,000 years in this taxonomic group but have disappeared by 3 million years
(Coss and Owings 1985; Coss 1991; Coss and Goldthwaite 1995; Coss 1999).

Similarly, tammar wallabies living on predator-free Kangaroo Island off
the coast of Australia that has been isolated since the last ice age, 9500 years
ago, still retain their ability to recognize a model of an extinct predator, the
thylacine, as measured by reduced foraging, and they show the same reaction
to a stuffed fox or cat. Nonetheless, they do not react to calls of these or similar species (Blumstein et al. 2000). Another, New Zealand population, isolated
from predators for 130 years, similarly retained its ability to recognize a thylacine (Blumstein, Daniel, and Springett 2004). Other marsupials, including
Western quokkas and Western gray kangaroos, isolated from predators for
7000 and 9500 years, retain a suite of antipredator defenses that appear sensitive to a now-absent predation risk (Blumstein, Daniel, and McLean 2001;
Blumstein and Daniel 2002). Also, Byers (1997) noted that antipredator behavior of pronghorn, apparently designed to evade capture by extinct predators, has persisted since the last ice age.

In contrast, Berger (1998) uncovered a rapid reduction in predator recognition abilities in North American ungulates isolated from predators for just
tens of years. Specifically, predator-naïve moose from Wyoming and parts of
Alaska, USA, that had not experienced grizzly bears or wolves for 40–75 years
(8–10 generations) showed lower levels of vigilance in response to wolf howls
or to calls of ravens that are associated with carcasses left by predators, and to
wolf urine or bear feces (Berger, Swenson and Persson 2001). Other studies
have found odor to be important, too. Across marsupials, ontogenetic experience of predators is important in altering the way that herbivore prey re-

A. Tail Piloerection

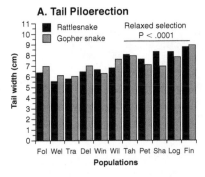

B. Facing Snake

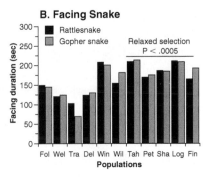

C. Proximity

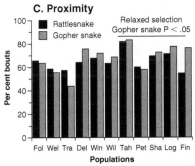

D. Substrate Throwing

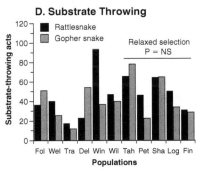

E. Tail Flagging

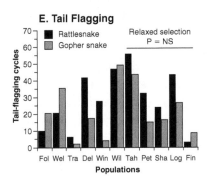

Figure 1.8 Comparisons of antisnake behavior of 11 populations of adult wild-caught California ground squirrels during alternate 5-minute trials with a caged rattlesnake and gopher snake ($N = 8$ squirrels/population). Solid lines above average values span 5 populations from habitats where rattlesnake and gopher snakes are rare or absent. Abbreviated names refer to sites in California (Coss and Goldthwaite 1995).

spond to the odors of predators (table 1.6). The tammar wallaby studies show that predator recognition abilities are lost at different rates depending on the modality, and the pademelon study shows that loss can occur within a generation or so of captivity.

Predator recognition capabilities might persist over time because there is little genetic variance on which selection can act to disintegrate innate recognition systems if selection has been strong in the past (Coss 1999). Alternatively, recognition systems may be physiologically or pleiotropically linked to other behavioral systems, such as recognizing conspecifics (Coss 1999). Another suggestion is that the presence of any predator, such as a raptor, may be sufficient

Table 1.6 Summary of studies on olfactory predator recognition in herbivorous marsupials

Species	Predator experience	Responds to scent of predators?
Tammar wallaby	Evolutionary, not ontogenetic[a]	No
Red-necked pademelon	Evolutionary, not ontogenetic[b]	No
Swamp wallaby	Evolutionary and ontogenetic[c]	Yes
Brushtail possum	Evolutionary and ontogenetic[d]	Yes

Source: Blumstein et al. 2002.
[a] Kangaroo island tammar wallabies were last exposed to natural mammalian predators about 9500 years ago.
[b] Red-necked pademelons were last exposed to mammalian predators before being brought into captivity, one or two generations ago.
[c] Field study on predator-experienced wallabies.
[d] Two independent studies, one on wild-caught, predator-experienced possums, the other a field study of predator-experienced possums.

to maintain rudimentary aspects of recognition of other predators (Blumstein et al. 2000; Blumstein, Daniel, and Springett 2004). Finally, there may be little cost to predator recognition, so that prey in predator-free populations will not be selected to lose responsiveness (Magurran 1999). Interspecific variation in loss of discrimination abilities might be expected if the degree to which genetic factors are involved varies across modalities and different prey give prominence to different modalities in recognizing predators. Alternatively, interspecific variation in loss might be contingent on danger posed by former predators. All of these ideas concerning persistence and loss of defenses are no more than possibilities at present.

Despite poor recognition skills of predator-naïve prey, wild animals learn about predators rapidly, just as they do in captivity (Curio 1988a; Mineka and Cook 1988). Individual moose quickly learned to respond to predators again: previously predator-naïve mothers whose calves were killed by wolves (that were in the process of recolonizing Yellowstone National Park, USA, at the time) elevated their post-playback vigilance in response to wolf calls by 500% and took 12 times longer to resume feeding than in years prior to wolf arrival. Additionally, mothers with wolf experience showed greater levels of vigilance than mothers who had lost calves to starvation. The same sort of rapid learning is seen in primates. In Tai National Park, Côte d'Ivoire, where poaching for monkeys is rife in certain areas, people imitate calls of crowned hawk eagles and distress calls produced by duikers when captured by a leopard. Both eagles and leopards prey on monkeys. Diana monkeys react to eagles and leopards by approaching and vocalizing (chapter 11), which gives away their location to humans hunting on foot, whereas they would otherwise remain silent and slip away if they encounter people. Monkeys in the poached part of the park have learned to discriminate between real eagle and duiker calls and imitations, usually reacting to the latter by silence. A small area of the park around the research site is avoided by hunters, however, and here Diana monkeys do not distinguish between imitations and real calls. Calculations show that discrimination abilities were acquired or lost within a span of just 4 years, or only one monkey generation (Bshary 2001). Similarly, mantled howler monkeys on Barro Colorado Island, Panama, learned to recognize the cries of newly reintroduced harpy eagles within 12 months (Gil-da-Costa et al. 2003). These three studies show that predator discrimination abilities can be acquired extremely rapidly in the wild, although they do not indicate the means by which this occurs.

1.7 Observer bias

Many studies of antipredator defenses use tethered or dead predators, predator mounts or models, domestic dogs or other carnivores, or even people to elicit antipredator responses in prey instead of waiting to observe infrequent natural predation events that are often confounded by several factors. Unfortunately, lack of movement or different types of movement from that shown by natural predators may result in quantitative or qualitative differences in antipredator responses, although it is commonly assumed that experimental and natural encounters produce the same results. For example, the same black-tailed prairie dog individuals barked more, and maintained a greater distance from western diamondback rattlesnakes and bull snakes when the snakes were tethered than when they were encountered naturally. This was probably because tethered snakes were unable to retreat and thus attempted to strike the prairie dogs more often. In any body of work that investigates antipredator defenses, it therefore important to witness antipredator responses in a natural context where predators can react to defensive actions; this makes research findings more credible (Loughry 1987b; Hennessy and Owings 1988).

Given that humans are important predators of larger species of birds and mammals (Robinson and Bennett 2000), there is a potential danger that the presence of an observer may affect routine antipredator activities in certain prey, interactions between prey and nonhuman predators, and the magnitude of predation attempts. All of these problems have been noted, but their relative importance across taxa is not well understood.

To illustrate, tamarins are surreptitious when entering their roost sites in the wild; in captivity they are slower to do this by about 15 minutes in the presence of an observer (Caine 1990). Red colobus monkeys in Gombe Stream National Park, Tanzania, move away from human observers immediately by fleeing to adjacent trees, thereby disrupting group cohesion and opportunities for group defense against predatory chimpanzees. Chimpanzees capitalize on this disruption and are more likely to hunt colobus when two human observers are present (table 1.7). Vervet monkeys in Amboseli National Park, Kenya, suffer higher predation from leopards when observers are away from

Table 1.7 Behavioral changes in the hunting tendencies of the Gombe chimpanzees reacting to red colobus when two Tanzanian observers were present or absent

Chimpanzees' reactions	Present (%)	Absent (%)
No interest	42	57
Detour	19	21
Test	3	11
Hunt	36	11
Total	72	31

Source: Boesch 1994.

their study site than when researchers are working in the field (Isbell and Young 1993a). In the first instance, observers affected prey's perception of risk; in the second, they facilitated predator opportunities for hunting successfully; in the third, they reduced the predator's opportunities for hunting.

It is time consuming and difficult to witness predation attempts in many circumstances, so observers often volunteer to be a proxy for a predator themselves (for example, table 1.4). This raises a number of problems, including the extent to which prey perceive humans as perilous, failure of humans to understand and respond to antipredator signals directed at them, and alterations in prey's defensive behavior as it becomes habituated to repeated predator simulation events or becomes bolder as a consequence of perceiving that its efforts "drove the predator away" (R. L. Knight and Temple 1986a, 1986b; Caro 1995; Gunness and Weatherhead 2002). In addition, even the behavior of humans is at issue, because direct staring is known to affect prey differently from an averted gaze (Scaife 1976a, 1976b; Coss and Goldthwaite 1995), and carrying a stick or wearing different clothing can alter antipredator responses (Kruuk 1964; Slobodchikoff et al. 1991). Potential problems associated with humans taking the place of natural predators have received relatively little research attention and are all too frequently dismissed as being unimportant.

More subtly, the way in which prey animals experience predation threat in experimental studies may affect their antipredator behavior. Thus, when animals are maintained under low risk but are then exposed to brief and infrequent high-risk situations, modeling demonstrates that prey are likely to exhibit a greater intensity of defenses than under field conditions where the contrast between high- and low-risk situations is not as marked. If high-risk situations become more frequent or lengthy, as, for example, when researchers repeatedly subject prey to predators, prey will be forced to lower antipredator effort. Thus, temporal variation in risk will alter the allocation of defense and feeding effort across different risk states (Lima and Bednekoff 1999a).

This chapter raises issues about the application of frequently used terms in the study of antipredator defenses. It concludes that it is most cautious and appropriate to avoid functional definitions but adhere to descriptions. This is because, unlike other subdisciplines in behavioral ecology, such as foraging theory or territoriality, the adaptive significance of many behavior patterns exhibited in the presence of predators is not yet known, and there are strong indications that functions vary across species. In discussing predator recognition, that some species recognize certain predators innately whereas others require learning is not at issue; most recognition systems develop as an interplay between both causal components. More contentious are the ways in which learning about specific predators proceeds either from the general to specific or vice versa. Furthermore, there is nothing approaching consensus on whether there are commonalities in certain learning mechanisms being used to learn about particular classes of predators, or being employed by certain prey taxa, or being employed by prey species living in particular ecological contexts. We simply have a checklist of the ways in which species can learn to recognize and respond to predators, but no underlying patterns.

1.8 Summary

The study of antipredator defenses began over a century ago, and the recent growth in information necessitates that any synthesis be carefully organized. Flow diagrams of prey-predator interactions and catalogs of stage-specific antipredator responses to predation threat have both been employed in the literature; the second method is utilized in this book.

The book centers on the functional and evolutionary significance of antipredator defenses, with a trait being termed adaptive if it confers reproductive benefits on its bearer, not in the sense of having done so in the past or because genetic differences are associated with differences in defenses. Evolution is used in its comparative sense rather than as demonstrating historical change or alterations in gene frequencies. *Avoidance of predators, alarm calls, warning signals, deterrence, distress calls, defense calls, group defense,* and *alarm responses* are common terms in the study of antipredator defenses, but they are sometimes used opaquely and carry double meanings. I make recommendations for how they might be applied in the future.

Bird and mammal prey are sophisticated at distinguishing between dangerous and harmless animals through visual, auditory, and olfactory sensory modalities. Young animals are subject to particularly strong predation pressure, and while some are capable of innate predator recognition, others have to learn about predator characteristics, albeit rapidly. Learning can occur by building out from an initially limited predator image or by narrowing a generalized response. Young may place themselves in a position where they are likely to learn about predators; inspect predators; acquire experience in other, nonpredatory contexts; learn as a consequence of having been pursued by predators; or learn from behavior of conspecifics or heterospecifics. In places where predators are now absent, prey can lose their ability to discriminate predators, but rates of loss vary greatly across species and between sensory modalities. Rapid learning about ontogenetically novel predators occurs within a few years in most cases.

Humans can affect predator-prey interactions if either prey or predator is differentially fearful of people. Use of humans as surrogate predators in anti-predator studies is risky but expedient.

2 Morphological Traits to Avoid Detection

2.1 Introduction

Prey can avoid detection if a predator fails to distinguish it from the background. The most obvious way in which an animal avoids being detected is through concealing coloration (Poulton 1890; Beddard 1892; G. H. Thayer 1909), but remaining stationary and choosing to settle in a particular aspect are important, too (Baker and Parker 1979; Endler 1981, 1984). Focusing on color, Endler (1978) defined a color pattern as cryptic if it resembles a random sample of the background perceived by predators at the time and age and in the microhabitat where the prey is most vulnerable to visually hunting predators. Crypsis is probably the most widespread antipredator defense in homeotherms and is often thought of as the default against which more flamboyant traits can be compared, although it is not necessarily the ancestral condition.

Crypsis and its converse, conspicuousness, are difficult to measure operationally. First, animals that are conspicuous at a short distance may be difficult to see a long way off (Endler 1978). Bright black and white bands of color, so distinctive close to or in a field guide, may break up the animal's outline at a distance (Gotmark and Hohlfalt 1995). Second, because contrast between an animal's color and the background depends on ambient illumination and spectral reflectance of the background (Hailman 1977; Lythgoe 1979; Burtt 1981), an animal may be cryptic or conspicuous at one time of the day but not at another, or against one type of background but not others (Endler 1978, 1990). Third, animals that are cryptic to humans may not be so to predators (Endler 1978, 1983, 1991b; Gotmark and Unger 1994). To take an extreme, birds can see ultra-

---

Figure 2.0 (*Facing page*) Red river hog piglets have blotched and striped coats like many other young suids and bovids. Their pelage almost certainly allows them to blend in with dappled light filtering in through the lightly wooded forests in which they live (reproduced by kind permission of Sheila Girling).

violet wavelengths, but humans cannot (A. T. D. Bennett et al. 1997). In addition, these factors may interact (Endler 1991b; Endler and Thery 1996).

Setting these issues aside, however, crypsis in relation to color can be achieved through at least three mechanisms. First, background matching or general color resemblance (Cott 1940) occurs when the animal matches the background on which it is normally found or when it resembles a compromise between two different habitats (Endler 1984; Merilaita, Tuomi, and Jormalainen 1999). Uniform pelage that matches the background, or spots or stripes whose pattern blends in with that of the background, are both forms of background matching (section 2.2); similar principles may apply to birds' eggs. An alternative form of matching occurs when prey resembles a conspicuous object in the environment that is avoided by predators, such as a plant part or bird dropping (Endler 1981; Starrett 1993; Hasson 1994). This form of concealment, termed masquerade, appears the least in birds and mammals, although certain birds' nests appear to show special resemblance to inanimate objects. Second, animals may avoid detection by minimizing shadow, possibly through countershading (A. H. Thayer 1896). Countershaded organisms have dark pigmentation on the dorsal surface and light pigmentation on the ventral surface that, when light comes from above, may disguise shadow on the underside of the body, and thus conceal the animal's shape (Cott 1940; Kiltie 1988) (section 2.3). Third, animals can be cryptic through disruptive coloration: contrasting colors here present a series of different objects to the predator's eye and thereby mask the animal's form (Cott 1940) (section 2.4). These three types of crypsis are not necessarily exclusive in the sense that the same part of the body could be disruptively colored at one distance but match the background at another, or cause different species of predators to miss seeing prey in different ways.

Some prey individuals may be able to avoid detection if the population is composed of individuals that are polymorphic. Where predators are presented with two or more color morphs, they may overlook one at the expense of the other, because they develop a "search image" for one morph (L. Tinbergen 1960) or search for one morph inappropriately (Endler 1991a). Where predators overlook the rarer morph and concentrate on the common, termed apostatic selection, they may allow morphs to avoid capture simply because they have a different appearance without being especially cryptic (B. C. Clarke 1969) (section 2.5).

Finally, there are cases that cannot be placed easily into any category. How does one classify the nest of the cape penduline tit, an oval domed nest made of cottony fibers of the seeds of the Kapok tree (*Ceiba pentandra*)? The entrance can be closed along its horizontal axis, but beneath the true entrance is a deep, rounded depression that looks exactly like another entrance (Skead

1959). Presumably, this blocked false entrance deceives snakes and other predators as to how to enter the nest. Examples such as this remind us that our attempts to categorize adaptations to avoid detection are still crude and somewhat arbitrary.

2.2 Background matching

There are many fascinating early accounts of mammals' overall coloration patterns matching their background (for example, Selous 1908; Buxton 1923). For instance, Hingston (1932) wrote of a pale-throated three-toed sloth in Guiana that had a greenish tinge to its black and gray hair produced by algae, which, he argued, made it difficult for raptors and jaguars to see the animal either from above or below the canopy, and of an almost pure white bat (*Mesophylla*) that spent the day hanging on the underside of a bright and silvery palm leaf instead of roosting in a dark tree hollow. As a consequence of these and many other anecdotal observations, many coloration patterns in birds and mammals are nowadays unequivocally accepted as being antipredator adaptations for camouflage on the basis of rather little or even no evidence. For example, the willow ptarmigan's white feathers and the mountain goat's white fur are thought to act as camouflage in snow, and the giraffe's reticulated coat pattern is assumed to hide it in the woodlands of Africa. Nonetheless, in addition to avoiding detection by predators, there are several other adaptive explanations for coloration patterns in animals, such as finding food, regulating temperature, or attracting mates (Cott 1940; Edmunds 1974; Burtt 1981; Butcher and Rohwer 1989; Booth 1990; Savalli 1995), so compromises between these selection pressures can be expected.

In mammals, research on background matching falls into two categories: early studies that mapped intraspecific variation onto environmental variables, and recent studies that relate interspecific variation to ecological and social variables, controlling for phylogeny. In birds, many experimental studies have addressed antipredator benefits of cryptic plumage in adults, although in recent years these have been overshadowed by studies of ornamentation and sexual dichromatism in the context of sexual selection (M. Andersson 1994).

2.2.a Color resemblance in mammals   For years it has been known that in some North American mammals, variation in uniform pelage coloration matches variation in soil coloration (Allen 1874; table 2.1); however, there was debate as to whether variation in coat color was due to local differences in temperature and humidity or whether it helped in concealment. For instance, Sumner (1921) thought that concealment from visually hunting predators was unimportant because desert rodents are nocturnal, differences in pigmentation ex-

Table 2.1 Early studies of variation in coloration patterns of some North American rodents

Species	Phenomenon	Reference[a]
Apache pocket mouse	Nearly white on white sands in New Mexico	1
Valley pocket gopher	Dark on humus-filled soils of Santa Catalina mountains	2
White-throated wood rat	Dark on lava beds in New Mexico	3
Grasshopper mouse	Dark on lava beds in Arizona	4
Deer mouse	Dark in montane forest belt of Santa Catalina mountains	1
Oldfield mouse	Pale on pale sands of Florida beaches	5

[a]1. Dice and Blossom 1937; 2. Goldman 1935; 3. S. B. Benson 1936; 4. Merriam 1890; 5. Osgood 1909.

tend to parts of the body not exposed to view, and some species, such as pocket gophers, live underground (see also Buxton 1923; Sumner and Swarth 1924). There are problems with this reasoning, however, because it fails to acknowledge that deserts are rarely completely dark at night, that tone rather than coloration may act in concealment, and that vision of nocturnal predators may be more acute than our own (Cott 1940). Indeed, subsequent quantitative work has shown that there are strong positive correlations between pelage brightness and soil brightness (for example, Heth, Beiles, and Nevo 1988). In oldfield mice, for instance, coat brightness matches soil brightness at different locations in the southeastern United States (Belk and Smith 1996), and brightness measurements of pale-agouti-colored Eastern fox squirrels are consistently and positively correlated with brightness levels from five different habitat types in the Mississippi River delta as determined from photography (Kiltie 1992a). Both of these studies support an antipredator function, because brightness per se is unlikely to affect thermoregulation.

Stronger support for rodent coat color acting as a concealment device comes from studies of predators that choose to attack conspicuously colored individuals. In an experiment in which light- and dark-brown oldfield mice were released onto light- and dark-colored soils, owls caught more of the conspicuous than the matching phenotype (Kaufman 1974; see also Dice 1947; but see Mueller 1974). Similarly, domestic cats in a barnyard in Missouri, USA, took more house mice with pale-yellow pelage in preference to normal agouti-colored mice, eventually eliminating them from the population (L. N. Brown 1965).

In other orders, the relationship between coat color and broad habitat categories appears to hold, at least in some instances. Ortolani and Caro (1996) divided carnivores' pelage into 19 different regions and classified them according to color or contrast, and then scored species according to environmental variables (including habitat type), geographic region, and behavior such as activity pattern. Coloration patterns were then matched to particular variables using the phylogenetic comparative method (Harvey and Pagel 1991) both within families and across the whole order (Ortolani 1999) to control for shared ancestry. They found that pale fur color was associated with desert or semidesert environments, although this depended on how discrepancies in the phylogenetic trees were resolved. They also found that spotted coats were associated with an arboreal lifestyle, which supports background matching against dappled light passing through leaves. Indirectly, the carnivore data lent some credence to Gloger's (1833) rule that species living in warm, humid climates have dark coats, because the data showed that dark fur was significantly associated with tropical forests, a potentially humid environment. Unfortunately, the adaptive significance of these associations is improperly understood. For example, dark fur in carnivores may be related to heat exchange rather than to concealment in tropical forests, and even if it was for concealment, it is difficult to disentangle the extent to which carnivores are camouflaged for protective concealment, that is, to escape predation from heterospecific predators (Palomares and Caro 1999; Creel, Spong, and Creel 2001) or for aggressive concealment, that is, to approach prey without being detected (Cott 1940).

Clearly the argument for coloration providing protective camouflage is more persuasive if it can be discerned in herbivores rather than carnivores. Using coloration and environmental data coded in a similar fashion as for carnivores, Stoner, Bininda-Emonds, and Caro (2003) found that there are associations between lagomorph species showing pale overall body coloration and living in open habitats, particularly in deserts. In adult bovids, spotted and striped coat colors are associated with living in lightly forested areas, although there are no associations between overall coat coloration and habitat variables after controlling for phylogeny (Stoner, Caro, and Graham 2003). Thus, there is comparative support from a handful of mammalian orders that pelage coloration aids in concealment from predators (Caro 2005).

Camouflage is more effective if an animal remains immobile, a behavior often termed "freezing." It is common to see rodents, lagomorphs, and galliformes crouch and remain still when they are approached closely by people or run a short distance and then crouch. When motionless, some species will allow predators to approach extremely closely before fleeing. Bitterns are notorious for assuming an upright, stiff, and immovable posture that allows them to blend in with the reeds in which they live. The upright bill can be touched and the head pushed back without the bird taking flight, and, incredibly, the little bittern will quickly turn the orientation of its body so that the dull yellow ventrum that resembles reeds is always displayed toward the observer (Cott 1940).

2.2.b Color resemblance in birds   In birds, apparent examples of camouflage are just as striking. The early literature contain reports of species, such as desert larks, resembling the color of the ground on which they nest (Cheesman 1926), and there are early experimental studies on domestic chicks that lend support to antipredator benefits of coloration (Davenport 1908; but see Pearl 1911). Cott (1940) suggested that cryptic coloration is required by certain species more than others. He remarked that nocturnal or crepuscular feeders that need to remain still and sleep during daylight, such as owls, frogmouths, nightjars, and woodcocks, are extraordinarily well camouflaged and remain immobile until approached closely. Also, species that do not nest in concealed places need to be cryptic. In these species the sex that incubates eggs has upperparts that are camouflaged; for example, plovers or the dotterel. On the other hand, he noted that large birds such as pelicans, birds of prey such as kites, or birds with formidable beaks and legs such as cockatoos have no need to be cryptic. In contrast with mammals, studies of the adaptive significance of cryptic and conspicuous coloration in birds have been far more extensive and thorough since Cott's (1940) great synthesis. Yet despite this, findings still offer no clear resolution of which pelage characteristics serve as camouflage devices, nor do they shed much light on the evolutionary causes of crypticity in birds.

Baker and Parker (1979; see also Baker and Hounsome 1983; Baker 1985) put forward an idea that species' differences in vulnerability were responsible for differences in crypticity. They suggested that bright coloration in prey was associated with being unprofitable, lumping many different aspects of behavior as being unprofitable, whereas cryptic coloration was connected with vulnerability (see also Cott 1946/47; Cott and Benson 1970). Using interspecific comparisons of European, North African, and Middle Eastern birds, whose different body regions were scored for levels of conspicuousness, Baker and Parker found that species that live on the ground or on water are less colorful and show more plumage dimorphism with one sex being dowdy; that adults of smaller species are less colorful and more dimorphic than adults of larger species; that species with exposed nests are less colorful than those with concealed nests; and that birds providing more parental care are less colorful than those that give less care. Others had already noted that in monomorphic species of waterfowl, where both sexes are tawdry, males are more likely to incubate eggs (Verner and Willson 1969) and care for young (Kear 1970) compared with dimorphic species, where dull-colored females are more likely to perform these duties. Thus, birds that live in vulnerable conditions or perform duties that make them vulnerable benefit from being inconspicuous, whereas birds that are unprofitable advertise the fact.

None of these studies controlled for shared ancestry, but more damaging,

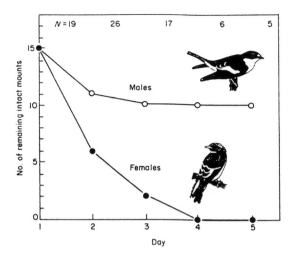

Figure 2.1 Attacks by avian predators on mounted male and female pied flycatchers. Mounts were set out on day 1 and checked on mornings 2–5. The y axis shows the number of mounts that had not been attacked; N = number of observed sparrow hawks; actual numbers were probably larger (Gotmark 1992).

across species there is no a priori prediction that plumage brightness will be associated with greater predation, although within species we might expect this (Lyon and Montgomerie 1985). This is because for each prey species, severity of predation is likely to vary according to many factors other than crypticity, including type of predator, behavior of prey, and availability of alternative prey species (Krebs 1979). Worse still, studies of predation in fish (Endler 1980) and mammals (Kaufmann 1974) have shown that conspicuous individuals suffer higher rates of predation. Indeed, rather than advertising unprofitability, conspicuousness might only appear in those species in which predation costs are low. Since Baker and Parker's study was correlative, one might believe that the key to solving some of these problems would be a series of experimental studies determining whether predators choose to capture or avoid cryptic birds (Butcher and Rohwer 1989).

Just such studies have been provided by Gotmark and his coworkers, although this research explored intraspecific rather than interspecific differences in coloration. Using remotely triggered cameras and stuffed mounts of male and female pied flycatchers set out in the open, these researchers could determine which sex was attacked first and how many of each sex were attacked by migrating raptors (Gotmark and Unger 1994). Gotmark (1992a, 1993, 1995) found that conspicuous black and white male mounts were attacked less often by sparrow hawks than drab female mounts (fig. 2.1), supporting Baker and Parker's hypothesis that conspicuous coloration signals some (unknown) aspect of unprofitability or that drabness was associated with vulnerability. Unfortunately, the reasons for raptors spurning male mounts are opaque. First, black and white pied flycatchers may have been novel prey to young sparrow hawks; indeed, European blackbird mounts whose wings had been painted red were avoided by raptors (mostly goshawks: Gotmark 1994a, 1996). Alternatively, black and white plumage may actually have been inconspicuous! In a test with humans, Gotmark and Hohlfalt (1995) found that although male pied flycatchers were easier to detect than females when seen on the ground, they were equally difficult to spot in trees. In contrast, male chaffinches that have a pink-brown, slate-blue, and chestnut body with black and white wings were as difficult for people to detect as females with gray-green bodies and smaller white wing patches when they were on the ground; however, males were easier to detect in trees. Thus, plumage that is conspicuous at close range may be cryptic at a distance when birds are in trees, perhaps through disruptive coloration (see below), and its effectiveness may vary according to background (see also Metz and Ankney 1991).

In addition, further studies have provided conflicting results to those using pied flycatcher mounts: male chaffinch mounts were attacked more often than female mounts (Gotmark 1993); black and white black-billed magpies were attacked more than magpie mounts painted brown (Gotmark 1997); great tits painted and ringed as fledglings were killed more often than controls left unpainted, as determined from remains in sparrow hawk nests and at plucking sites (Gotmark and Olsson 1997); and adult black and white male pied flycatchers suffered higher predation by sparrow hawks than dull brown males in the same breeding season (Slagsvold, Dale, and Kruszewicz 1995).

One important key to these conflicting results may lie in the way in which behavior modifies vulnerability to predation. In pied flycatchers, where brown females construct nests alone, rates of female disappearance in a woodland area in Norway were 0.42%/day during nest building, 0.53% during egg laying, but 0.05% during incubation, returning to 0.36% during the nestling period. These figures highlight the importance of inactivity in reducing predation. Furthermore, females were more likely to disappear (2.8%) than males (0%) during the period of nest building despite the males' brighter coloration (Slagsvold and Dale 1996). As much of the nest material is found on the ground, activity at ground level apparently increases risk of predation (see also A. Lindstrom 1989; Cresswell 1993). (High rates of nesting failure so commonly reported in the literature encompass many parts of the reproductive cycle and do not, in principle, conflict with Slagsvold and Dale's data showing that females suffer least predation during incubation.)

In a study of chaffinches in Sweden where females were killed by sparrow hawks consistently more than were males, females spent more time foraging than males and were therefore probably less vigilant; females were more active and may thus have been easier to detect; and females foraged closer to the ground, again increasing predation risk (Gotmark et al. 1997). In short, Got-mark's original flycatcher findings with mounts did not consider how sex differences in detectability and ability to escape might influence predation rates, factors that could easily override predators' decision to attack prey on the basis of coloration. It is still unclear whether in species in which females build nests, males can afford to be brightly colored; or, alternatively, in species in which males benefit from being brightly colored in male-male competition or through female choice, females are forced to build nests alone because predation costs prevent males from helping them.

An alternative hypothesis for crypticity in female birds is that females need to be cryptic because bright coloration can attract predators to the nest (Wallace 1889). In general, it is believed that open nesters suffer greater predation than cavity nesters, and that ground nesters suffer more predation than those that nest off the ground (but see T. E. Martin 1993a). Using more detailed data on rates of predation that vary according to canopy height (T. E. Martin 1993b, 1995), it is possible to match female plumage to nest height across species. In finches and warblers, where only females incubate eggs and brood young, nesting failure, often resulting from predation, is low on the ground and in the canopy compared with predation in shrubs (fig. 2.2). Plumage brightness patterns of females and males of these species also differ by nest height, with female brightness being most drab in shrub nesters, where rates of nesting failure are greatest. Male brightness, on the other hand, increases monotonically with canopy height (T. E. Martin and Badyaev 1996). As these findings control for phylogenetic effects, the results compel us to believe that nest predation places sharper controls on female than on male plumage brightness (but see Promislow, Montgomerie, and Martin 1992). Comparative and phylogenetic studies demonstrate that female plumage is more labile over evolutionary time than that of males (as found in nine-primaried oscines, Bjorklund 1991; New World blackbirds, Irwin 1994; tanagers, K. J. Burns 1998), but it is not clear whether these evolutionary changes are responses to predation threat while nesting.

In short, activities associated with nest construction and location are likely to drive the evolution of cryptic coloration in female birds, although other, as-yet-undiscovered factors associated with predatory behavior and escape strategies are bound to be involved. In species in which males are involved in nest building, they are likely to suffer similar constraints on gaudiness. Where males gain sexually selected benefits obtained from bright plumage, these are likely to be limited by predation, too, at least in a general sense, although these trade-offs may be solved using disruptive coloration and changes in behavior to reduce detectability.

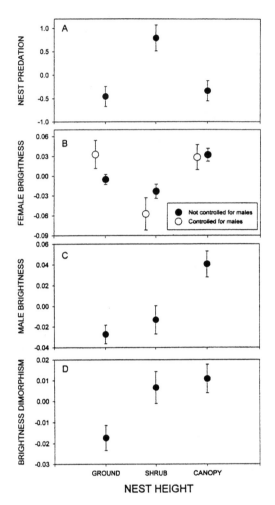

Figure 2.2 Variation in means and 1 standard error for standardized independent contrasts among 3 nest heights. *A*, nest predation; *B*, female brightness controlled and not controlled for effects of male brightness; *C*, male brightness, and *D*, dimorphism in plumage brightness between females and males (T. E. Martin and Badyaev 1996).

2.2.c Color resemblance in birds' eggs   In birds, eggshell coloration is a maternal rather than an offspring trait, and females of many species lay cryptically colored eggs consisting of a light background color broken up by dark speckles. At first sight, it seems quite obvious that eggshells are adaptively colored to resemble their background (Harrison 1985). Indeed, one of the most widely cited studies in all of animal behavior consists of a series of simple experiments using birds' eggs that demonstrated that removal of broken eggshells away from the nest by common black-headed gulls serves an antipredator function (Tinbergen et al. 1962). Black-headed gull eggs were more likely to be predated by herring gulls and carrion crows if eggs were painted white rather than their natural camouflaged color; nests were more vulnerable to predation if eggshells, colored white inside, were present and principally if they were close to the nest. As wet, newly hatched chicks are particularly vulnerable to predators, the survival value of eggshell removal is to reduce the nest's conspicuousness. More generally, however, it is substantially more difficult to explain patterns of egg coloration across species of birds. Specifically, it is not clear why only some species apparently lay camouflaged eggs, whereas others lay uniformly colored white or blue eggs that seem so conspicuous to us (Lack 1958; Montevecchi 1976; Y. Oniki 1985).

A partial answer to this question is that there may be a physiological cost to laying speckled eggs. For example, the mourning dove that incubates its eggs almost constantly after laying is a species that produces white eggs (Westmoreland and Best 1986), suggesting that once the probability of detection is circumvented, it is cheaper for females to avoid pigmenting their eggshells (see also Y. Oniki 1985). Another idea, that camouflaged eggs incur a heat cost (Montevecchi 1976; Bertram and Burger 1981), is erroneous, because naturally cryptically colored eggs are not necessarily strong absorbers of heat. Studies of the spectral reflectance of 25 species of birds' eggs show that they reflect nearly 90% of radiation in the near-infrared wavebands. In contrast with melanin found in dappled pelage of birds and mammals, cryptic coloration in birds' eggs is derived from protoporphyrin and biliverdin pigments that are highly reflective in near-infrared wavebands (fig. 2.3). Blue and blue-green eggs laid by herons and ibises have similar thermal properties.

Alternatively, white or uniform eggs may actually be camouflaged and not conspicuous. Across 60 species of Amazonian birds, eggs tend to be white (often with some spotting) if the nest in which they are laid is thin and can be viewed from below, suggesting that they blend in with patches of sky (Y. Oniki 1985). Similarly, uniformly colored eggs appear cryptic in other environments: blue eggs in Amazonian birds are associated with nests placed in leafy green sites and isolated sunny bushes, and buff (pink) eggs are often found in nests placed among dull-colored dead leaves on the ground or tree limbs (Y. Oniki 1985).

Yet another possibility is that that egg coloration is not the primary means by which predators detect eggs; instead, predators may focus their attention on nests. In a series of experiments using artificial song thrush nests containing quail eggs painted white, blue (the egg color of thrushes), or cryptic, Gotmark (1992b) discovered that predation, principally by jays and hooded crows, on the three types of eggs was similar whether nests were exposed or concealed. In contrast, when eggs were placed in exposed and concealed spots in the absence of nests, dark-spotted eggs survived for considerably longer than the uniformly colored eggs. These results indicate that corvids located eggs

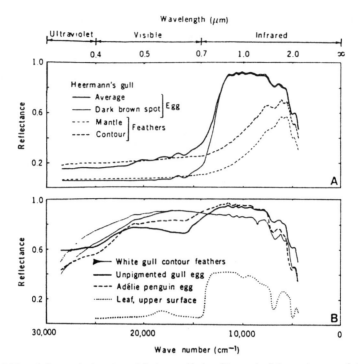

Figure 2.3 Spectral reflectance of various avian eggshells. *A*, eggs and feathers of Heermann's gull; *B*, comparison spectra of white gull feathers, pale-blue gull egg, pure-white egg (Adelie penguin), and *Populus deltoides* leaf (Bakken et al. 1978).

using nests rather than focusing on nests themselves. Parallel findings were made in a study of yellow-hammer, blackcap, and song thrush nests where white, blue, and brown artificial eggs suffered similar rates of attack but this time by mammalian predators (Weidinger 2001; see also Mason and Rothstein 1987). While there can be little doubt that cryptic egg coloration is a means to avoid detection in ground-nesting birds or birds that lay directly on the ground (Solis and Lope 1995; Yahner and Mahan 1996; Lloyd et al. 2000), egg color may have little influence on the probability of predators locating eggs in shrub-nesting birds.

As a matter of fact, crypticity and conspicuousness are complicated concepts that need to be measured operationally. Westmoreland and Kiltie (1996) photographed clutches of three species of blackbird eggs, scanned them into a computer, and extracted samples of pixels. They then used three measures to quantify crypsis: pattern disparity, which compares subsamples of pixels within the nest or in vegetation around the nest; mean brightness, which measures the difference between egg brightness and brightness of surrounding vegetation; and brightness disparity, the patchiness of brightness within the nest. Whereas pattern disparity did not differ among species, the other measures did, indicating that the measures represent different aspects of color resemblance. Interestingly, none of these measures reached their possible maxima, suggesting that selection for different aspects of crypsis was not all that strong. Nonetheless, there was a correlation between mean clutch pattern disparity and mean background pattern disparity. Thus, the disparity among blackbird eggs matches disparity in nest lining even though eggs are not particularly cryptic. Perhaps the extent to which the whole nest blends into the larger background of the nesting area is most important, but it is difficult to judge, because channels visible to bird and snake predators differ from ours. It is remarkable that the adaptive significance of egg coloration is still so unresolved, given opportunities for simple experimentation as well as enormous collections of eggs in museums, several egg field guides, information on nest sites, and robust phylogenies for avian taxa that would allow phylogenetically controlled comparisons to be conducted.

2.2.d Special resemblance in birds' nests   Aspects of nest morphology are also affected by predation (Collias and Collias 1984). Considering the form of the nest, enclosed nests (S. Oniki 1979) and nests constructed in the holes of trees (Skutch 1966) are more successful than open nests in South and Central American forests. In temperate regions, loss of whole nests occurs at a higher daily rate for bird species that nest in the open (0.028/day, $N = 12$ species) than in species that lay in cavities. Among cavity nesters, species that excavate their nest have a lower rate of nest loss (0.004/day, $N = 8$) than species that do not excavate (0.016/day, $N = 5$) (T. E. Martin and Li 1992). Nonexcavators may fare poorly because they tend to nest lower down than excavators or, by reusing nests, may suffer from ectoparasites or the unwanted attention of predators revisiting a known nest site (P. Li and Martin 1991). It is conjectured that oven- or pouch-shaped nests, nests that are domed, or nests constructed in burrows are all adaptations to reduce predation (Y. Oniki 1985; Hansell 2000).

Hornbills lay their eggs and incubate and feed their young in nests into which the female has sealed herself with the help of the male or other group members. After selecting a suitable hole in a tree or cliff face, the pair builds up the rim using mud, sticky food, feces, or even partially digested mud but leaves a small hole through which the female can enter. Once inside, the female usually seals herself in, leaving only a vertical slit through which the male feeds her throughout the rest of the reproductive period (Kemp 1995). The whole procedure is thought to be an antipredator defense mechanism preventing predators from gaining access to the nest contents and permitting a longer incubation and nestling period than that predicted by body size.

Considering nest size, some evidence suggests that small nests, such as those found in cotingas or tree swifts and many tropical species, are less likely to be detected by predators (Snow 1981). Indeed, when European blackbird nests were manipulated in size, nests exchanged for larger ones suffered an increase in predation (Moller 1990). Some birds, such as whip-poor-wills or night-hawks, have dispensed with constructing nests altogether and simply lay on the ground (although factors other than avoiding detection could be involved).

Considering nest design, it is clear that predation can select for nest appearance; for instance, nest coloration often matches the environment and coloration of supporting structure. Across Amazonian birds, Y. Oniki (1985) found that dark nests were found in closed habitats and pale nests in open habitats, and that dark nests were located on dark supports but light-colored nests on pale leaves, limbs, or grass. Having dismissed alternative hypotheses about heat exchange or availability of nest material, his findings suggest that birds tailor the color of their nest to match the background. T. E. Martin (1987) compared artificial wicker nests lined only with leaves, wicker nests lined inside and outside with mosses to resemble natural hermit thrush and MacGillivray's warbler nests, and real nests of those two species in small white fir trees in central Arizona, USA. Surprisingly, he found that leaf-lined artificial nests suffered lower rates of predation from a variety of mammalian predators. This implies that local predators had search images for the color and form of natural nests. Other studies show the opposite result, however: addition of feathers to nests set on the ground increased predation by corvids in Denmark (Moller 1987a). These discrepancies suggest that there may be pointed differences in the way that avian and mammalian predators detect nests (Soderstrom, Part, and Ryden 1998; Rangen, Clark, and Hobson 1999, 2000).

Nest design can also impede access for predators. Yellow-rumped caciques build tightly interwoven nests, many of which are inactive. Great black hawks shake nests to check for large cacique nestlings within, but they cannot check nests individually when they are in an interwoven mass; and hawks soon abandon a colony if they cannot find eggs or nestlings quickly (S. K. Robinson 1985). Other structural modifications also exist: blackstarts build substantial stone ramparts around their nest, possibly to slow predator entry, giving the adult time to escape (Leader and Yom-Tov 1998); and red-cockaded woodpeckers excavate resin wells below their nest, impeding access by snakes (Rudolph, Kyle, and Conner 1990).

Some species alter the design of their nest if they leave it, as, for example, when waterfowl cover their eggs with down, consequently hiding the shape and color of their eggs. Other species alter the area surrounding their nest and thereby camouflage their eggs by matching the background to their eggs.

Figure 2.4 Examples of fox squirrel color morphs. The three on the left are from the southeastern coastal plain of the USA, varying from 100% dorsal black (10% of the population), to an intermediate value (40%), to 0% dorsal black (50%); dorsocranial coloration does not vary appreciably. The two on the right illustrate completely light and completely dark agouti morphs from the lower Mississippi drainage (Kiltie 1989a).

Sandwich terns, whose eggs resemble the color of splattered pebbles, deposit large quantities of feces within a close vicinity of their nest. At Ravenglass, UK, a small sandwich tern colony was situated in the middle of a large black-headed gull colony that itself was devoid of feces. Predatory crows restricted their search to the profitable clean area, where they could find black-headed gull chicks. To examine the effects of this camouflage, Croze (1970) laid out on a pebble beach red mussel shells that hid meat and compared their "survival" to black and white painted shells that he laid out on a portion of the pebble beach painted white to mimic a ternery. Crows had developed a search image for red mussels before the black and white ones were laid out. Findings showed that the black and white mussels escaped attention for several days, and crows appeared to avoid the white area. This may explain why terns started to lay after the gulls have laid and why they splatter shingle with feces.

2.2.e Melanism   Various species show discontinuous variation in coloration, most notably melanism, in which some individuals have black pelage (Majerus 1996). Certain gray seals living in the North Sea are black, for example. In arctic ground squirrels, melanism can reach 20% in areas that are particularly susceptible to burning (Guthrie 1967). In a series of papers, Kiltie (1989a, 1992a, 1992b) investigated the incidence of color polymorphisms, including melanism, in fox squirrels in relation to the incidence of fire. Fox squirrels inhabit fire-climax pine savannas in Louisiana and Mississippi, USA, and the percentage of black hair on the dorsum is positively correlated with frequency of lightning-caused wildfires, as well as with climatic factors that influence fires (fig. 2.4). In contrast, gray squirrels that exhibit very low incidences of melanism concentrate their activities in moist bottomland hardwood stands that burn far less (Kiltie 1989a). Kiltie first argued that individuals with a variable amount of black on them might be cryptic in habitats with light and dark areas that were recovering from burning. When, however, he matched background brightness to squirrel pelage brightness by digitizing photographs, he found that intermediate and black-backed morphs matched their background better than light morphs only for a brief period, in just the first 2 weeks after an area had burned (Kiltie 1992b). Benefits accruing over such a short time frame make it difficult to argue for an antipredatory function for melanism. Photographs necessarily exclude any influence of movement, so it is interesting that in an experiment in which squirrel models were pulled in front of red-tailed hawks, birds responded more slowly to intermediate-colored morphs. Such an advantage would allow genes for black dorsal pelage to be retained in the population. In this reasonably well-studied example, then, partially black coloration appears to benefit squirrels when they are mobile (see also Kiltie and Laine 1992).

More famous mammalian examples of melanism are found in jaguars and leopards, where black individuals reputedly reach very high frequencies in some wet tropical forests (Majerus 1996). It is assumed, though not shown, that melanistic individuals are less likely to be detected by prey. The argument is flimsy, because it rests so heavily on these species being top predators but avoids considering the importance of predatory or intraspecific attack (Majerus 1996), physiological benefits, or even past trophic assemblages of predators that used to prey on these species. Melanism has also been reported in black-bellied hamsters (Gershenson 1945), silver-gray brushtail possums (Guiler 1953), and European rabbits (Barber 1954), where it is associated with high rainfall, and in gray squirrels, where it is associated with urbanization (Gustafson and Van-Druff 1990). While predation pressure on dark morphs might be reduced in either overcast or urban environments, and hence provide selection against wild-type morphs, the case for melanism being an antipredator device is weak at best in these cases.

In birds such as gray partridges, melanic morphs are found on black peaty soils (Sage 1962), pointing perhaps to a cryptic advantage, but other examples are not so clear. Whereas arctic skuas and snowgeese show increasing melanism at southerly latitudes which might be associated with protective or aggressive camouflage (O'Donald 1983; Cooke, Rockwell, and Lank 1995), increasing melanism in New Zealand bellbirds as one goes southward might be associated with humidity (Bartle and Sagar 1987), supporting Gloger's rule. Melanism in male European blackbirds may be a consequence of sexual selection, whereas black coloration in corvids may signal unpalatability (Cott 1946/47). This short list hazards that there are many evolutionary causes of melanism to which should be added resistance to abrasion, a barrier to ultraviolet light, and thermoregulation, all of which need to be considered in any systematic analysis (Burtt 1979; Booth 1990; Majerus 1996). At present, a convincing case for melanism being an adaptation for background matching in homeotherms is tenuous and restricted to a very limited number of species.

2.2.f Changes in coloration with changing environments   In some species, individuals change color in circumstances related to risk of predation (variable color resemblance: Cott 1940), either because they grow older and less vulnerable or because the color of their surroundings changes. Both phenomena provide additional, though inconclusive evidence to support the idea that pelage color reduces the likelihood of being detected by predators.

Across a wide span of taxa, neonates have pelage characteristics that differ from their parents but that take on adult coloration later. Booth (1990) defined ontogenetic color change as a nonreversible color change associated with normal progressive development of an individual of a species. There are numerous proximate mechanisms involved in ontogenetic color changes, and many ultimate reasons it should occur. In homeotherms, evidence points to heightened vulnerability among defenseless young as being a force selecting for concealed pelage in immatures. In nidifugous birds, where young accompany their parents soon after hatching, nestlings have a characteristic mottled, streaked, or striped down that makes them difficult to see. Quail, partridge, gull, grebe, and emu hatchlings are all colored in this way. In contrast, nidiculous young, which remain and are fed in the nest, hatch naked and do not develop downy coats of variable color in the nest (Cott 1940).

Cott (1940) also noted that many cervids, suids, and felids give birth to spotted or striped offspring, but he was cautious in suggesting that these coats necessarily provide concealment from predators, as he believed that parents defend their young (as in suids), and offspring are born in sheltered retreats (as in felids), so that young are protected in other ways. We now accept that young may still benefit from camouflage despite parental solicitude. In young ungulates, striped and especially spotted coats are strongly associated with inhabiting forested areas (Stoner, Caro, and Graham 2003). Ungulate species whose young are spotted always sequester them in hidden locations in the weeks after birth (fig. 2.5; section 3.5), suggesting spotted coats aid in concealment. There have been few systematic examinations of the distribution of neonate coloration in other taxa and none as to why some adults of some species lose the dappled pelage of their young whereas others retain it.

It is common knowledge that many species of birds wear a cryptic winter plumage owing to the presence of light feather tips but conceal conspicuous coloration beneath; these tips either wear away in spring or can be erected as necessary (Veiga 1996). Moreover, some birds and mammals living in the Northern Hemisphere change their pelage to white or yellow-white in win-

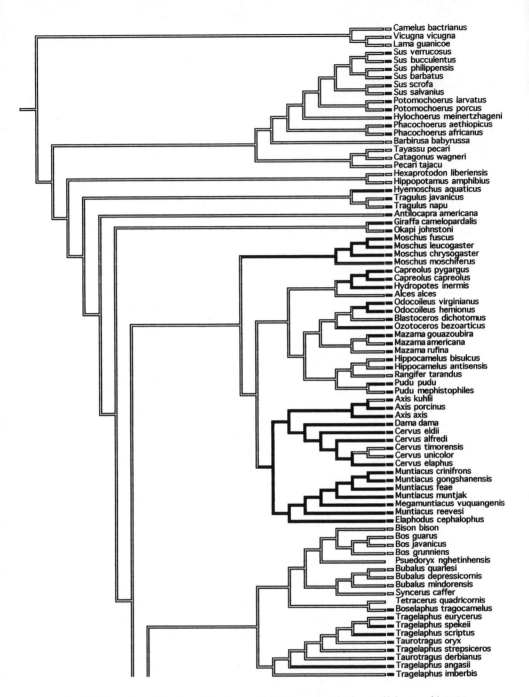

Figure 2.5 Phylogenetic tree showing the association between spotted coats in young artiodactyl species (*black sections of the tree*) and the hiding strategy (*black boxes*, to the right of the tree) in artiodactyls; gaps indicate no data available (Stoner, Caro, and Graham 2003).

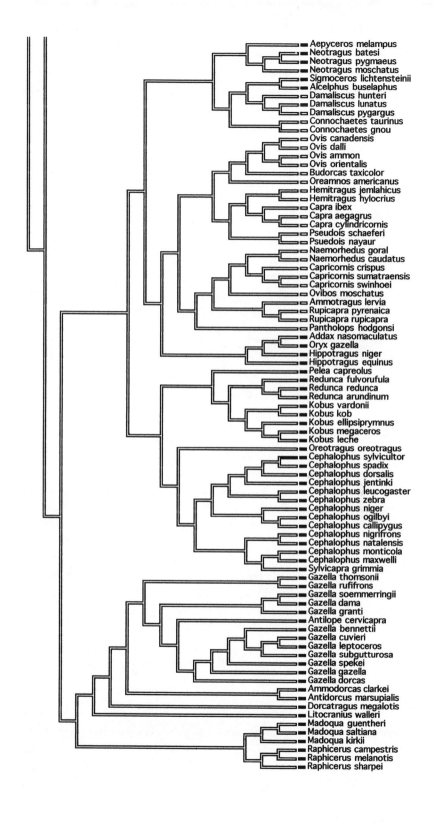

Aepyceros melampus
Neotragus batesi
Neotragus pygmaeus
Neotragus moschatus
Sigmoceros lichtensteinii
Alcelphus buselaphus
Damaliscus hunteri
Damaliscus lunatus
Damaliscus pygargus
Connochaetes taurinus
Connochaetes gnou
Ovis canadensis
Ovis dalli
Ovis ammon
Ovis orientalis
Budorcas taxicolor
Oreamnos americanus
Hemitragus jemlahicus
Hemitragus hylocrius
Capra ibex
Capra aegagrus
Capra cylindricornis
Pseudois schaeferi
Psuedois nayaur
Naemorhedus goral
Naemorhedus caudatus
Capricornis crispus
Capricornis sumatraensis
Capricornis swinhoei
Ovibos moschatus
Ammotragus lervia
Rupicapra pyrenaica
Rupicapra rupicapra
Pantholops hodgonsi
Addax nasomaculatus
Oryx gazella
Hippotragus niger
Hippotragus equinus
Pelea capreolus
Redunca fulvorufula
Redunca redunca
Redunca arundinum
Kobus vardonii
Kobus kob
Kobus ellipsiprymnus
Kobus megaceros
Kobus leche
Oreotragus oreotragus
Cephalophus sylvicultor
Cephalophus spadix
Cephalophus dorsalis
Cephalophus jentinki
Cephalophus leucogaster
Cephalophus zebra
Cephalophus niger
Cephalophus ogilbyi
Cephalophus callipygus
Cephalophus nigrifrons
Cephalophus natalensis
Cephalophus monticola
Cephalophus maxwelli
Sylvicapra grimmia
Gazella thomsonii
Gazella rufifrons
Gazella soemmerringii
Gazella dama
Gazella granti
Antilope cervicapra
Gazella bennettii
Gazella cuvieri
Gazella leptoceros
Gazella subgutturosa
Gazella spekei
Gazella gazella
Gazella dorcas
Ammodorcas clarkei
Antidorcus marsupialis
Dorcatragus megalotis
Litocranius walleri
Madoqua guentheri
Madoqua saltiana
Madoqua kirkii
Raphicerus campestris
Raphicerus melanotis
Raphicerus sharpei

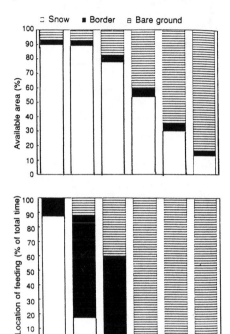

Figure 2.6 Proportion of snow-covered ground, border zone, and bare ground during the period of snow melt (*upper panel*) and the corresponding use for feeding by hen ptarmigan (*lower panel*). Note that the border zone in the upper panel remained roughly constant in size. Mean percentage pigmentation on the backs of hens is given at the very bottom of the figure. Data from all 5 hens studied (Steen, Erikstad, and Hoidal 1992).

ter. Prey species include rock ptarmigan, arctic lemmings, and mountain hares, and predators such as arctic foxes, weasels, and stoats. The probability of seasonal change is influenced by latitude: for example, within a species, stoats regularly turn white in the north of Scotland but less completely and less frequently in England (Cott 1940). Across species, the arctic hare remains white year-round, whereas northern populations of the varying hare turn white in winter, but those that range further south retain their summer coat all year (Cott 1940; see also Stoner, Bininda-Emonds, and Caro 2003). In the same way, permanent and seasonal white coats in carnivores are associated with living in the Arctic (Ortolani and Caro 1996), and in ungulates, species that turn white in winter live in Arctic and tundra habitats (Stoner, Caro, and Graham 2003).

In common with other aspects of crypsis, camouflage is enhanced by changes in behavior. In a detailed study of willow ptarmigan, researchers followed the behavior of five birds as they lost their white winter plumage during the month when snow melted in spring. While still white, the birds sat on snow, feeding on partly protruding dwarf birch and willow; when their back became partially pigmented, they preferred the border between snow and bare ground, whereas they fed on bare ground when pigmentation was nearly complete (Steen, Erikstad, and Hoidal 1992). Nevertheless, their feeding location was not entirely driven by camouflage considerations, as they were found feeding on willow in the border zone and on bare ground more than would be expected simply from the availability of these zones (fig. 2.6). While the association between microhabitat and plumage coloration does not signify that camouflage is the selective force driving coloration (white plumage may be sexually selected in males), it does indicate that the birds take account of their conspicuousness and attempt to reduce it. Moreover, before birds molt into summer plumage, males actively dirty their white plumage after snow melt if their mate is incubating (Montgomerie, Lyon, and Holder 2001). Nonetheless, it is still not clear why only certain species undergo seasonal pelage change. Cott (1940) and W. J. Hamilton (1973) suggested that species that are active predators or that suffer heavy predation tend to be white. More specifically, Cott argued that small prey that rely on crypsis, such as the varying hare, turn white in winter, but that large prey that can defend itself using force, such as the musk ox, do not; and that obligate predators, such as the ermine, turn white but that other predators that scavenge and are nocturnal, such as the wolverine, do not. These observations cry out for systematic investigation.

It is less appreciated that some dappled mammals wear uniform coloration in winter. Fallow deer living in deciduous forests in Europe have spotted coats, presumably to blend in with dappled light, but they lose their spots and take on a uniform gray or brown hue in winter when leaves fall. Japanese sika deer are also spotted and show the same pattern throughout most of their range, but in the evergreen forests of Taiwan, they retain their white spots year-round (Cott 1940). These are compelling anecdotes for background matching, but again they fail to address why changes in pelage coloration are so idiosyncratic.

2.2.g Masquerade   Some insects, amphibians, and fish closely approximate inedible objects, such as parts of plants or stones (Wickler 1968; D. Owen 1980; Endler 1981), but these forms of special resemblance, or masquerade, are rare in birds and mammals. Cott (1940) describes Australian frogmouths (Podargidae) that sit motionless on tree stumps or boughs during the day. Their lichen-gray and freckled brown coloration makes them almost impossible to see, since they look like a log or stump and will even remain motionless if a hand is placed on them! G. H. Thayer (1909) mentions a South American species of bat that is

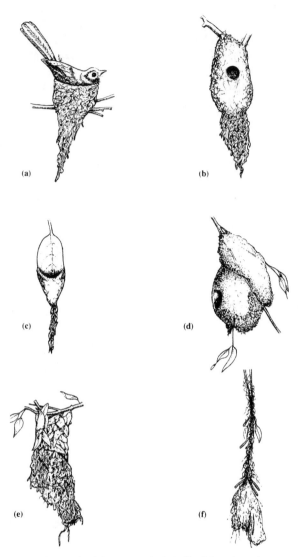

Figure 2.7 Possible examples of masquerade in birds' nests. (*a*), rufous-backed fantail; (*b*), scale-crested pigmy-tyrant; (*c*), long-tailed hermit; (*d*), red-faced spinetail; (*e*), eye-ringed flatbill: (*f*), black-tailed flycatcher (Hansell 2000).

marked and colored to look like a woody knot or other excrescence on the underside of a mangrove branch, whereto it clings, by day; not hanging downward, but pressed close against the bark, holding on with both feet and finger-hooks. (p. 120)

Hingston (1932) suggests that a three-toed sloth suspended from a branch looks like a termite nest or bulbous ephiphytic growth when its legs are bunched to-

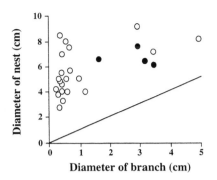

Figure 2.8 Diameter of 25 lichen-bearing nests, each of a different species, plotted against the diameter of the branch on which each was supported. *Open circles,* branches that bore no lichen; *solid circles,* branches bearing lichen. Line shows diameter of nests that have the same diameter as branches, so most nests were larger than their branches (Hansell 1996).

gether. Currently, no systematic investigation has been made of the distribution or ecological correlates of masquerade in birds or mammals.

The best examples of masquerade pertaining to homeotherms may actually be in nests constructed by birds (fig. 2.7). Hansell (2000) documents several miscellaneous examples of birds extending the top or bottom of their nest that may serve to masquerade as parts of a plant on which the nest is built (see also Collias and Collias 1984). More systematically, however, Hansell (1996) documented the number of species that applies lichen flakes (28) or white patches of spider cocoon silk (19) to the exterior of their nest. Hansell reasoned that if such ornamentation was a form of masquerade, nests should be smaller or of equivalent size to the branch supporting them; but if they served simply to reflect patches of light and hence be cryptic, nests might be of any size. Across 25 species that build nests using lichen, the latter hypothesis was supported (fig. 2.8). Nonetheless, both the form and application of additional material to the outside of the nest deserve much greater scrutiny, and as Hansell (2000) remarks,

> the application of lichen flakes, white spider cocoons and occasional other material to the outer surface of the nest may act to conceal the nest through crypsis, through masquerade, through disruptive camouflage, and possibly combinations of all three. (p. 101)

2.3 Concealing shadow

A second way in which an animal can achieve crypsis is by reducing the amount of shadow on its own body, because shadow highlights an animal's outline. Possibly, a common means of accomplishing this is through countershading, in which an animal's contour is partially concealed by dorsal pigmentary darkening that effectively reduces the amount of shadow falling on the ventro-lateral region when lit from above (A. H. Thayer 1896, 1902; Poulton 1902; fig. 2.9). Sloths that spend a proportion of time hanging upside down even have inverted countershading (darker bellies and lighter backs)

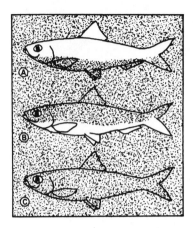

Figure 2.9 Classic depiction of self-shadow concealment. *A*, uniformly pigmented specimen with dorsal illumination; *B*, dorsomelanic specimen evenly illuminated; *C*, dorsomelanic specimen illuminated from above (Kiltie 1988).

(G. H. Thayer 1909). Countershading is a priori likely to be prevalent in aquatic birds and mammals, since light always emanates from a vertical source and there is less likelihood of casting an additional shadow on the substrate which would reveal the animal (Kiltie 1988). At first glance, penguins, whales, dolphins, and porpoises living in marine environments provide the best examples of countershading helping in concealment, because scattered light results in an invariate direction of illumination and because their predators can approach from any angle, although there is a suggestion that some of these color patterns cause pelagic fish school prey to break up and become susceptible to capture (R. P. Wilson et al. 1987; see also Gotmark 1987).

On land, many orders of mammals exhibit dark dorsal pigmentation: numerous species of marsupial, rodent, lagomorph, and ungulate have light or white ventral surfaces and uniform brown or gray dorsal surfaces. Associations were found between light ventral surfaces and both diurnal activity and living in deserts in bovids and in other ungulates (Stoner, Caro, and Graham 2003), although this was not replicated in lagomorphs (Stoner, Bininda-Emonds, and Caro 2003). Bright and very open habitats are where we would expect selection to act to minimize shadow (as well as minimize ultraviolet radiation).

Countershading is thus an extremely common pattern of coloration in numerous aquatic and terrestrial groups, but it could arise for reasons other than self-shadow concealment (Kiltie 1988), including background matching when viewed from above (few predators observe terrestrial prey from below) protection against ultraviolet radiation, as a thermoregulatory device, or even the counteraction of dorsal abrasion in some species (Burtt 1981; Walsberg 1988).

Recently the extent to which countershading helps to reduce shadow has come under close scrutiny (Ruxton, Speed, and Kelly 2004), but the sad fact is that there have been remarkably few tests of whether countershading helps to

Table 2.2 Numbers of different types of pastry taken by birds out of 25 of each type in 9 experiments on different days

Experiment	Dark	Light	Counter-shaded	Reverse-shaded	Total
1	7	12	3	11	33
2	3	21	1	10	35
3	4	12	1	8	25
4	3	4	3	9	19
5	4	16	5	14	39
6	5	2	1	7	15
7	3	8	1	8	20
8	5	8	1	6	20
9	7	13	2	9	31
Totals	41	96	18	82	237

Source: Edmunds and Dewhirst 1994.

conceal prey against predators, let alone as a consequence of self-shadow concealment (but see L. de Ruiter 1956; E. R. A. Turner 1961). Kiltie (1989b) assessed the shadow-reducing effect of countershading by photographing either the sides or the back of stuffed gray squirrel skins placed both vertically (as if climbing trees) or horizontally (as if running on the ground), in winter and summer, and in full sun and partial shade. "Transects" were then carried out along portions of the photographs, and the brightness of pixels was determined using densiometry. Kiltie discovered that when specimens were set horizontally, side views nearly always produced lower correlations between brightness and pixel position on a transect, as would be expected if countershading was working, than did transects across views of the dark back, where countershading would not be expected to be operating; the effect, however, was less marked for specimens placed vertically. On horizontal substrates, therefore, these results partially support Thayer's hypothesis, but whether the hypothesis can explain countershading more generally in this species depends on the extent of time that squirrels stay on flat ground or climb vertical trunks, and predation risk in each place. Choice of a correct orientation and substrate is always a key component of concealment (Edmunds 1974).

Edmunds and Dewhirst (1994) presented light, dark, countershaded dark and light, and reverse-shaded light and dark pieces of pastry to birds on a garden lawn in England and thus explored the putative antipredator benefits of countershading more directly. They found that light "prey" was taken significantly more than dark, and countershaded prey less than reverse shaded (table 2.2). Countershaded pastry was taken less than dark, which suggests

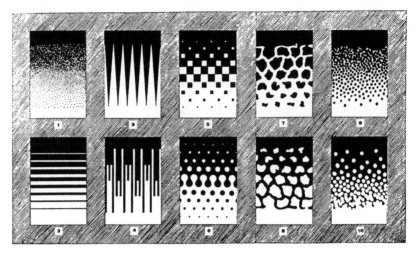

Figure 2.10 Types of patterns that at a distance produce graded tones similar to countershading (Cott 1940).

that simple background matching of countershaded pastry against dark-green grass cannot be an adequate explanation for their lower rates of predation. There must be something additionally beneficial to countershading, but this was unlikely to be due to any aversion to two-tone prey, because there was no significant difference in the extent to which reverse-countershaded pastries were preyed upon compared with light-green pastries. This important experimental study indicates that countershading can provide additional benefits over and above that of background matching, but it would be unwise to attribute all instances of countershading to self-shadow concealment. A recent study could only replicate these findings for European blackbirds, but not for robins or blue tits (Speed et al. 2005). In another counterexample, some naked mole rats have dark dorsal pigmentation but pink ventral surfaces. They are fossorial, but younger workers occasionally disperse aboveground at night. Braude and others (2001) argued that their dark backs must serve to match the background when viewed by aerial predators, rather than minimizing shadow, because mole rat legs are so short (heat exchange or ultraviolet protection were ruled out given their nocturnal disposition, as was protection from abrasion, because feet have no pigmentation). If pigmentation carries a cost, then background matching rather than self-shadow concealment is a sufficient explanation for countershading being so widespread, because we would expect animals to be darker above but refrain from producing melanin below simply in order to minimize production costs (Ruxton, Sherratt, and Speed 2004).

Mottram (1915) drew attention to the fact that different types of patterns on an animal's surface can produce graded tones similar to that seen in countershading when they are viewed from a distance (fig. 2.10). Seen in this way, many species may use countershading in addition to other forms of crypsis simultaneously. Thus Grevy's zebra has thicker black stripes on its dorsum than on its ventrum, while some felids have thick blotches on the dorsum that grade into spots on the flanks (for example, leopard cat) (Cott 1940). Those mammalian species without countershading may be nocturnal, such as roof rats; be diurnal but sit vertically, exposing their underside, such as chimpanzees; or perhaps have additional means of defense, such as size (rhinoceroses: section 9.3) or weaponry (porcupines: section 9.4) (G. H. Thayer 1909).

Birds and mammals can also reduce the probability of detection by crouching, which eliminates shadow on the substrate and reduces the height of their profile (Cott 1940; Edmunds 1974). In ungulates this behavior is called the prone response (Walther 1964; Lent 1974) and is found predominantly in young animals. For example, Thomson's gazelle fawns that drop down are less likely to be captured by cheetahs than those that flee; this response declines with age (FitzGibbon 1990a). Spotted thick-knee nestlings depress their bodies against the substrate when danger threatens, and their prone response becomes increasingly flattened from the first to second week after hatching before it disappears in development (Farren 1908). Lapwing and little tern chicks will remain in the prone position on carpets when turned upside down or even when washed in water (Russell 1938)! Adults also employ this behavior but to a lesser degree: African antelopes, steinboks, duikers, oribis, and reedbuck sometimes crouch slowly down and extend their neck on the ground, thereby reducing shadow (Selous 1908), and birds as large as turkeys will take on this posture when pursued by hunters and their dogs (Palmer 1909; section 12.10).

2.4 Disruptive coloration

A third method by which mammals might avoid being detected by predators is through disruptive coloration (also called obliterative coloration), in which the form of the animal is broken up by a pattern of contrasting tones and colors (Cott 1940; D. Owen 1980). Here, some components that match the background are overlaid with strongly contrasting tones, thereby creating abutting contrasting colors where no real border occurs; the lateral black stripe of a Thomson's gazelle or the black stripes of nestling Eurasian woodcocks or plovers are possible examples. Alternatively, real boundaries are broken up, as when contrasting patterns run against the body's contour, thereby concealing its outline, as in Burchell's zebra perhaps (but see Waage 1981). There are many plausible (but unproven) examples of mammals exhibiting disruptive coloration patterns but little systematic data. Anecdotes include the vertical stripes of the greater kudu, horizontal stripes of the golden-mantled ground squirrel or Brazilian anteater, and the bright-orange color patch bordered by black on the dorsum of the pale-throated three-toed sloth. More systematic data come from ungulates, where species with bold lateral side bands are found in open, desert, and tundra environments, habitats where alternative opportunities for pattern blending or hiding may be restricted (Stoner, Caro, and Graham 2003). These anecdotes and associations provide only weak support for disruptive coloration.

The problem with the topic of obliterative coloration is that it is very well accepted despite there being so few empirical tests of the phenomenon. One difficulty is identifying disruptive coloration in the first place. Until Gotmark (1992a) performed his experiments with pied flycatcher mounts and followed it up with experiments on human detection, no one ever considered that this black and white plumage might be disruptive. A second difficulty, as with countershading, is separating disruptive coloration from background matching, pattern blending in particular. Cott (1940) proposed certain elements of disruptive coloration, including patches of color that occur at the body's margin, thereby breaking up its silhouette; patches that stand out from other elements that themselves match the background; complex patches to give the impression of separate objects and strong contrast between elements so that the animal's real outline merges into the background. Merilaita (1998) reformulated these ideas, juxtaposed them with predictions from background matching, and then tested them quantitatively by digitizing photographs of a marine isopod *Idotea baltica*. This species has five heritable morphs, all of which have white spots superimposed on a background color that can be altered by chromatophore response to match the color of the brown alga on which it lives; the white spots look like white epizoites that also live on the alga. Isopod spots are variable in number, vary in size and shape, often touch the body's margin, and sometimes reach laterally across the individual. Analyses of pixels taken from the isopods and the background repeatedly supported predictions derived from disruptive coloration as opposed to background matching (table 2.3). Nonetheless, we must be very careful in assuming apparent examples of disruptive coloration necessarily provide an antipredator benefit. Silberglied, Aiello, and Windsor (1980) used the butterfly *Anartia fatima* to test whether the white band that runs along its wings acts to obliterate the butterfly's outline and so reduce predation. They found no difference in survival between butterflies whose white stripes they had colored black and control individuals.

Another problem is that in certain species, it is difficult to distinguish disruptive coloration from those that warn predators that the prey is dangerous or toxic (aposematism: sections 7.1–7.5, 9.4, 9.8). G. H. Thayer (1909) proposed that skunks and teledus (stink badgers) have obliterative coloration rather than warning coloration, as their outline is difficult to discern when

Table 2.3 Predictions for different traits of pattern elements from crypsis through background matching and through disruptive coloration, and observed traits for the white-spotted color morphs of *Idotea baltica*

Trait of pattern elements	Prediction from background matching	Prediction from disruptive coloration	Observed
Distribution	Same as in the background	More marginal spots than expected by chance	More marginal spots than expected by chance
Area	Same as in the background	—	Smaller than in the background
Shape	Same as in the background	Complex	More complex than in the background
Variation in area	Same as in the background	High	Higher than in the background
Variation in shape	Same as in the background	High	Higher than in the background
Colors	Same as in the background	Same as in the background, contrasted	Same as in the background, contrasted

Source: Merilaita 1998.

viewed from ground level. The whole topic of disruptive coloration needs systematic analysis (see also Cuthill et al. 2005).

Cott (1940) also called attention to the importance of using contrasting bands of color apparently to join separate parts of the body and thereby disguise their characteristic shape. A prominent structure requiring concealment is the vertebrate eye, as it stands out so well from the background (see also Barlow 1972; Gavish and Gavish 1981). One method is for the iris to match the tint of the rest of the head; another is for the black pupil to be subsumed within a black color patch, thereby masking its round shape (G. H. Thayer 1909). There are a plethora of mammals with black stripes or patches that run through the eye such as gemsbok and plains viscacha. No systematic tests of this idea are available, but dark patches around the eyes rather than through them (which may draw attention to the eye rather than away from it) are associated with crepuscular activity, riparian habitat, and grassland and terrestrial locomotion in carnivores (Ortolani 1999), results that speak to dark eye patches being antiglare devices (Ficken, Matthiae, and Horwich 1971) rather than obliterative coloration, although aposematism is an additional possibility (Newman, Buesching, and Wolff 2005).

2.5 Apostatic selection

Animals can avoid detection by predators by virtue of being dissimilar from conspecifics, provided morphs are somewhat cryptic (Allen 1988; Ruxton, Sherratt, and Speed 2004). Although predators might be expected to remove prey items of varying appearance according to their abundance in the population, L. Tinbergen (1960) noticed that tits (*Parus* sp.) concentrated their foraging efforts on palatable cryptic prey that were abundant and ignored other types of prey at very low densities. He argued that, given sufficient encounters, birds learn the cues, or selectively attend to cues (Langley 1996), that allow them to distinguish prey from the background; he called these search images (Dawkins 1971; Lawrence and Allen 1983), defined as transitory enhancement of detection ability for particular cryptic prey types or characteristics (Ruxton, Sherratt, and Speed 2004). It is unlikely that search images would be formed if prey are scarce, because the rate of encounters would be too low for learning to occur, and the search image requires additional encounters to be maintained; thus, rarer prey morphs should be disregarded and hence survive. This mechanism provides a proximate cause for frequency-dependent selection that acts against the common form of prey, also called apostatic selection (B. C. Clarke 1962, 1969; Greenwood 1984). There are other mechanisms that can lead to apostatic selection, however, including aversion to rare morphs, foraging in patches preferred by one morph, or differential handling skills on different forms (Endler 1988, 1991a).

Several experimental studies have shown that birds exhibit an increased ability to detect cryptic prey with successive encounters with one prey type (for example, blue jays: Pietrewicz and Kamil 1979; Kono, Reid, and Kamil 1998; blackbirds: Lawrence 1985a; bobwhite quail: Gendron 1986) and that they exhibit no such enhancement following encounters with two prey types; these results support the proposal for search image formation. Guilford and Dawkins (1987) put forward an alternative method by which predators might hunt. In situations in which one morph was more cryptic than another, predators might search at a slower rate in order to redress the balance of seeing and encountering inconspicuous prey (Gendron and Staddon 1983). They suggested that previous studies showing improvements in finding cryptic prey could simply result from a slower rate of searching, with the consequence that there would be no apostatic selection because predators would increase the chance of noticing both morphs (see also Endler 1991a; Ruxton, Sherratt, and Speed 2004). Recent work with prey that are equally cryptic (Plaistead and Mackintosh 1995) and careful analyses of time spent searching for prey (Lawrence 1988; Tucker and Allen 1993) have reaffirmed the existence of search images (but see Guilford and Dawkins 1988; Guilford 1992). Perhaps the best demonstrations of predators maintaining cryptic polymorphism in a prey population through the use of search images comes from the "virtual ecology" work of Bond and Kamil (2002). They train blue jays to hunt for an artificial digital moth on a computer screen. Moths can be made more or less cryptic compared to a granular background, and moth phenotypes are derived from algorithms that code for individual patch elements, overall changes in brightness and contrast, and even linkage mechanisms that protect favorable genotypes from recombination! Accuracy and latency of pecking can be entered into a selection algorithm and then reproduction can be allowed to occur, with the result that a new generation of moths can be presented to jays the next day. Bond and Kamil found that pecking accuracy decreased with moth crypticity as expected and that jay predation resulted in directional selection for increased crypticity across moth generations. For highly cryptic models, accuracy declined as dissimilarity between the target and the last previous correctly detected moth increased, as would be expected if jays were using search images to detect prey. This opens the door to frequency-dependent selection. Finally, phenotypic diversity of artificial moths (that is, polymorphism) increased over generations, as would be expected following frequency-dependent selection. In nature, the importance of search images is likely to be additionally influenced by palatability, choice of background matching selected by each morph, numbers of morphs, and prey behavior (Clarke 1962; Cooper and Allen 1994; Allen 1988; Forsman and Appelqvist 1998; Kono, Reid, and Kamil 1998).

Whether apostatic selection operates on birds and mammals in the wild is an open question, because most experimental studies of search images use arthropods, seeds, artificial food, or computer-generated images as prey rather than bird or mammal prey. The handful of studies on homeothermic prey are little more than interesting observations, and even these provide conflicting results. Mueller (1971, 1974) demonstrated that American kestrels form specific search images for either albino laboratory mice or mice dyed gray, although kestrels additionally focused on conspicuous mice when mouse and background colors were altered systematically. Against apostatic selection, Pielowski (1959) noted that goshawks selected white wood pigeons over dark ones when dark pigeons predominated in the population and the converse when white were prevalent. Among herons, seven species show color dimorphism, with some individuals being white and others gray or blue-gray. In the little blue heron, for instance, immatures are white and adults are always slate blue. Observations of predation attempts on both individuals and models by free-living hawks indicate that white morphs are attacked far more than blue ones (G. S. Caldwell 1986), which appears at odds with apostatic selection; the significance of such dimorphism is not understood. Nonetheless, there are so few other examples of color polymorphisms in bird and mammal prey populations that it suggests that apostatic selection may be relatively unimportant in these taxa. Perhaps predators of birds and mammals rely less on color when searching for prey but use cues such as movement, odor, and areas where prey aggregate in order to locate prey.

2.6 Summary

Animals can avoid detection by predators through crypsis: by matching the color of their background or resembling objects in their environment, countershading, and displaying contrasting colors that obliterate their outline. Surprisingly, however, even after a century of interest, there is little systematic evidence that avian or mammalian coloration patterns serve to avoid detection.

Background matching is extremely common in homeotherms. In some mammals, intraspecific variation in pelage color, or more specifically brightness, is known to match the brightness of soils on which individuals are found, and in experimental tests predators select against prey colored differently from their background, both of which support an antipredator function of background matching. In those species of birds where females are cryptic, females conduct nest-building activities without the help of the male. Females are also cryptic in avian species that suffer heavy nest predation. In experiments using stuffed birds, however, predation falls disproportionately on conspicuous males. Variation in behavior is probably an important mediating factor in making coloration effective and in reconciling these disparate results.

The extent to which the color of eggshells helps eggs in being overlooked by predators is still poorly understood. While speckled eggs appear cryptic, white eggs and blue eggs may also be cryptic under some circumstances. Moreover, certain predators may locate eggs using nests rather than eggs themselves. There is a stronger case for the form, size, and design of some birds' nests being mechanisms for concealment.

There is some evidence that partial melanism allows animals to avoid detection by predators in areas subject to regular wildfires; the adaptive significance of melanism and dark coat coloration in tropical forests is less understood. Seasonal changes in coat color, prey turning white at the onset of winter snow, and loss of dappled coats as vulnerable neonates grow older suggest that there has been selection for coat color acting on concealment. Very few homeotherms show a special resemblance to objects in their environment, but there are some anecdotes regarding birds' nests.

While many birds and mammals sport a dark dorsal and light ventral surface that is thought to reduce the amount of shadow falling on the animal's ventrum, there is presently very little evidence to show that countershading helps to conceal an animal by minimizing shadow on its body. Disruptive coloration, in which the outline of an animal is broken up by contrasting color patches, is also common in birds and mammals, but again firm evidence for its role in concealment is weak.

In theory, color polymorphism can benefit individuals of the rarer morph if predators develop a search image for the commoner form and a reduced ability to detect alternative prey. The extent to which color polymorphisms help individuals to avoid detection is unknown but may be of limited significance in birds and mammals. On the basis of the extent to which different types of crypsis are found in birds and mammals, background matching seems by far the most prevalent form of avoiding detection by predators.

3   Behavioral Mechanisms to Avoid Detection

3.1 Introduction

Whereas some invertebrates and vertebrates possess morphological traits to avoid being detected by predators throughout much or all of their life (chapter 2), individuals can additionally use more sensitive behavioral mechanisms to minimize exposure to threat. These include changing the timing or extent of their activities, selecting different habitats from predators, or altering patterns of feeding and reproduction (Lima and Dill 1990; Lima 1998 for important and wide-ranging reviews). Some of these changes in behavior occur as a response to indirect environmental cues associated with predation risk, such as changing light levels, whereas other changes are triggered by direct cues recently left by predators themselves, such as scent (Thorson et al. 1998). Moreover, behavioral responses to predation risk can occur over multiple timescales. In the short term, they include moving to a different habitat and modifying food intake or reproductive activities (Milinski and Heller 1978; Sih 1980; E. E. Werner et al. 1983; Lima, Valone, and Caraco 1985; Gilliam and Fraser 1987; Magnhagen 1991; Boyko, Gibson, and Lucas 2004), whereas over an animal's lifetime, individual growth rates may hasten (Werner 1986; Sih 1987), and over the course of evolution, a species may become arboreal or nocturnal (see also Charnov, Orians, and Hyatt 1976; Dill 1987 for effects on community structure), to give but two examples.

In this chapter, I consider the broad array of behavioral adaptations to avoid being detected by predators. Section 3.2 examines how birds choose

---

Figure 3.0 (*Facing page*) A white-tailed deer mother approaches her dappled fawn that has been left in hiding while she was away. The fawn stands unsteadily as she approaches; its flight from predators is slow and ungainly in the first weeks of life (reproduced by kind permission of Sheila Girling).

where to nest in relation to predation risk, whereas section 3.3 discusses parents' covert behavior around nest sites. Section 3.4 reviews behavioral changes in mammals involving use of refuges, alterations in activity, and developmental adaptations to avoid encountering predators. Section 3.5 focuses on shifts in habitat use under threat of predation and maternal attempts to diminish the probability that offspring will be detected by predators. Section 3.6 discusses decisions as to when, what, where, and how much to forage as well as food carrying. Section 3.7 examines changes in reproductive behavior under risk of predation. As a shorthand for behaviors that reduce the probability of encountering predators, one can call these predator avoidance mechanisms. Strictly speaking, any trait that lessens the probability of being consumed, such as flight or self-defense, could be termed a predator avoidance mechanism, but for convenience I have restricted the application of the phrase to situations in which the predator has not detected the prey, as suggested by Endler (1991a).

I have also limited prey's behavioral mechanisms to avoid encountering a predator to those circumstances in which the predator is not necessarily present. Thus, the cues that prey employ to detect predation risk must inevitably be indirect: either environmental cues often associated with predatory success or scent that could have been left by the predator earlier. They do not include seeing the predator, as this would mean that the predator was definitely close by. Treating predation risk as a dichotomous variable in this way (either the predator is in the immediate environment or it is not) raises at least three problems. First, how do we define immediate environment? Zones of detection will differ between prey species and between predator species. Second, the full span of predation risk lies on a continuum from a predator being entirely absent to a predator killing the prey. A predator having been in the area in the past, and being present and detected, are only two somewhat arbitrary points along this continuum of risk. Third, taking absence of visual cues as a way to define a subset of antipredator strategies does not help us understand the difference between predator avoidance mechanisms and flight in those species that rely heavily on olfaction to detect predators (for example, blind mammals). While somewhat arbitrary and biased toward the visual modality, this categorization nevertheless reflects current literature on predator avoidance mechanisms.

3.2 Nest site selection in birds

Where a bird constructs its nest is constrained not only by physical factors and availability of nest material but by predators (Collias and Collias 1984; Y. Oniki 1985; Hansell 2000) that can impose an enormous drain on survivorship for many bird populations (Nolan 1963; Ricklefs 1969b; Best and Stauffer 1980), accounting for as much as 76% of nest failures (Best 1978; see also Nice 1957). At

Table 3.1 Percentage of Passeriform and Piciform nests that were lost to predators (% nest predation) or that fledged at least one young (% nest success) among nest sites, and life history data gathered from the literature (number of species in brackets)

	Nest Predation (%)	Nest success (%)	Broods per year	Clutch size (eggs/nest)	Adult survival (%)	Annual fecundity
Forest habitat (N = 73)						
Excavator	11.0 (7)	85.3 (11)	1.09	4.56	0.67 (8)	5.0 (11)
Ground	30.7 (10)	63.3 (11)	1.19	4.44	0.63 (8)	5.1 (13)
Canopy	36.5 (17)	47.0 (18)	1.42	4.14	0.61 (10)	5.9 (18)
Shrub	48.5 (13)	45.5 (14)	1.83	3.66	0.53 (12)	6.7 (15)
Nonexcavator	31.4 (10)	62.9 (10)	1.63	5.71	0.44 (14)	9.1 (16)
Shrub/grassland habitat (N = 50)						
Ground	38.9 (15)	44.4 (14)	1.63	4.37	0.57 (8)	6.9 (15)
Shrub	43.7 (26)	40.1 (30)	2.07	3.72	0.53 (20)	7.8 (28)
Nonexcavator	37.5 (3)	54.8 (5)	2.00	4.54	0.44 (3)	9.0 (5)

Source: T. E. Martin 1995.
Note: Data are as follows: number of broods (number of successful broods attempted per year), clutch size, annual adult survival (percentage of individuals summing between years), and annual fecundity (product of numbers of broods and clutch size). Passeriform and Piciform species had nested in different sites and in forest and shrub/grassland habitats.

a coarse level, nest site location is related to different degrees of nest predation that are associated with different habitat types. Species that excavate nest sites have the lowest levels of nest predation, followed by ground and hole nesters that do not excavate, then canopy nesters, and finally shrub nesters (table 3.1). Nest site location is, in turn, associated with the number of broods that birds produce each season and with clutch size (which are negatively associated with each other): there is a pattern of nest site excavators having few broods and large clutches and shrub nesters having the most broods and the smallest clutches (table 3.1). In addition, nest site location explains length of nesting period, with shrub nesters having the shortest and excavators the longest periods, and it accounts for variation in adult survival and fecundity, with shrub nesters having the lowest survival and high fecundity but excavators showing the reverse. In short, key life history variables are related to nest site location and predation risk, although cavity-nesting birds that must find existing holes have larger clutches and annual fecundity for their level of nest predation, suggesting that limited availability of these sites may be the actual driving force behind their increased reproductive effort (T. E. Martin 1995).

Behavioral observations and comparisons between nest sites and the surrounding habitat at several spatial scales show that parents actively choose nest sites to reduce predation (for example, P. Lloyd et al. 2000; Doligez, Danchin, and Clobert 2002). Unfortunately, despite over 200 studies documenting variables that influence nest predation, there is still little consensus as to which

Table 3.2 Single illustrations of nest site selection that reduce the probability of detection by predators

Factor	Nest	Eggs[a]	Predator	Finding	Reference[b]
Habitat type	Dickissel	N	Snakes	Nests on prairies suffer greater predation than in old fields	1
Distance from edges	Artificial	Clay	Hooded crow, magpie, jay	Predation on open-ground nests lower far from edge, but not for partially covered nests	2
Habitat patch size	Artificial	Quail	Raccoon, red fox, gray fox, skunk, eastern chipmunk, gray squirrel, blue jay	Predation higher in small woodlots than large tracts	3
Vegetation around nest site	Hermit thrush	N	Red squirrel, gray-neck chipmunk, long-tailed weasel, house wren, Steller's jay	Reduced predation if surrounded by small white firs	4
Nest height	Various Arizona species	N	Red squirrel, gray-neck chipmunk, long-tailed weasel, house wren, Steller's jay	Predation greater in shrub than ground or canopy	5
Proximity to other nests	Artificial	Quail	Red fox, arctic ground squirrel, least weasel	Predation higher if adjacent to nest taken previously	6
Distribution of nests	Artificial	Hen	Raccoon, fox, skunk, blue jay, herring gull, American crow	In upland thicket, lower predation with uniform distribution than random or clumped	7

Source: Based on Collias and Collias 1984; Matessi and Bogliani 1999.
[a]N denotes natural.
[b]1. Zimmerman 1984; 2. Moller 1989a; 3. Wilcove 1985; 4. T. E. Martin and Roper 1988; 5. T. E. Martin 1993a; 6. O'Reilly and Hannon 1989; 7. Picman 1988.

factors are consistently important (Filliater, Breitwisch, and Nealen 1994). Results diverge in part because of differences in predator search techniques (for example, the relative import of vision or olfaction in finding nests and differences in abilities of predators to remember locations of nest sites) and the composition of predator communities that affects risks of, say, nesting on or aboveground, in simple or complex vegetation, or near habitat edges (Soderstrom, Part, and Ryden 1998; Matessi and Bogliani 1999). Findings also vary by landscape, with predation, particularly by avian predators, increasing in landscapes fragmented by agriculture (Bayne and Hobson 1997), and by geographic region (Tewksbury, Hejl, and Martin 1998). Nonetheless, it is relatively easy to distill a checklist of factors known to influence nest predation, each of which is likely to reduce the probability of detecting the nest or eggs inside it (table 3.2), bearing in mind that their relative import cannot be assessed at this stage. In what follows, I have taken examples from studies using real nests and artificial ones despite such studies sometimes yielding different results (see section 10.2.a). While artificial nests allow researchers to promulgate explicit experimental designs, they should be used in conjunction with real nests wherever possible.

3.2.a Habitat type   Complexity of the habitat in the general vicinity of a nest site is important in affecting predation risk (Partridge 1978), and individuals may sample several habitats before starting to nest (for example, Badyaev, Martin, and Etges 1996). Most obviously, ducks and waders reduce the impact of terrestrial predators by nesting on islands (for example, Duebbert, Lokemoen, and Sharp 1983; Frederick and Collopy 1989), and waterfowl show increased nesting success when foliage is thicker (for example, Bengston 1972; Duebbert and Kantrud 1974; Crabtree, Broome, and Wolfe 1989). In more terrestrial habitats, raccoons, one of North America's most vociferous mammalian predators on nests, exhibit reduced foraging efficiency and increased search times and locate fewer clutches of eggs as understory foliage density is increased artificially (Bowman and Harris 1980). Nonetheless, habitat complexity is not always beneficial: open-ground nests suffer greater predation in field-woodland boundaries than in open fields, at least in Denmark, because predatory magpies restrict their activities to woodlots (Moller 1989a).

Some studies have considered habitat characteristics at different scales around the nest site. Examining habitat selection by dusky flycatchers in limber pine-mountain juniper woodland in Wyoming, USA, Kelly (1993) measured vegetation characteristics at the nest site; within the nest patch, 0.04 ha around the nest site; and in the breeding territory. He found that flycatchers decided to nest in places with more foliage and greater tree densities than were generally available but that there was rather little selectivity at the scale of the territory. Successful nests were characterized by concealment from below, short distances from the nest tree to the nearest tree, and greater densities of small trees in the nest patch, factors that might conceal movements of parents on and off the nest. Different habitat characteristics may therefore be paramount at levels of the nest site, patch, and territory (see also Murphy, Cummings, and Palmer 1997).

3.2.b Distance from edges   As one moves from unsuitable to more suitable nesting habitat, nest density will increase; and if predation is density dependent, we would expect increased rates of predation along habitat edges (Fretwell 1972; Gates and Gysel 1978; Andren and Angelstam 1988; Marini,

Robinson, and Heske 1995). A great number of studies have documented increased predation (and cowbird parasitism) within 50 m of edges in North America and Eurasia (for example, Donovan et al. 1997). For instance, Soderstrom, Part, and Ryden (1998) attempted to integrate the influence of different predator species on both ground and shrub nests placed at varying proximity from a forest edge in different habitats in south-central Sweden. In a grassland habitat, they placed artificial nests containing two quail and one plasticine egg on the ground or in shrubs at 0, 15, and 30 m from a forest edge. (Plasticine eggs allow determination of predators' identities using bill or bite marks [but see Lariviere 1999; G. E. Williams and Wood 2002].) Results showed higher predation on shrub than on ground nests at all distances from the forest edge and strong differences in predator identity, with corvids targeting shrub nests and mammals seeking out ground nests (table 3.3), a common pattern. In addition, although there was little effect of forest edge, nests closer to a tree were more likely to be predated than those farther away ($>$ 30 m). This latter result is instructive, because it shows that nest predation in open habitats increases close to lookouts for avian nest predators. As avian predators are often the principal source of danger, one would expect no edge effects within forests at "soft edges," where there are scattered lookouts, but classic edge effects should be found at sharp borders between clear-cuts and grasslands (for example, Ratti and Reese 1988; Yahner and Scott 1988).

Early, influential studies (for example, Wilcove, McLellan, and Dobson 1986) in conjunction with two reviews (of 21 studies, Paton 1994; 27 studies, Andren 1995) of nest predation and edge effects concluded that nests were more likely to suffer attack along edges or gradients in productivity. Given that edges harbor greater predator species diversity than forest interiors (Vander Haegen and Degraaf 1996; Chalfoun, Ratnaswamy, and Thompson 2002), edges are often associated with anthropogenically deforested landscapes or agricultural practices (Rudnicky and Hunter 1993; M. J. Hartley and Hunter 1998), and human-dominated landscapes contain many generalist mammalian predators, such as domestic cats, opossums, and raccoons (Small and Hunter 1988; Dijak and Thompson 2000), edge effects have attracted a great deal of attention from conservation biologists (Murcia 1995; R. E. Major and Kendal 1996; Chalfoun, Ratnaswamy, and Thompson 2002). Recently, the link between edge effects and nest predation has been reexamined, now using a larger sample of 54 study sites (Lahti 2001). Of these, 13 exhibited an edge effect, 31 did not, and 10 showed an edge effect in at least one treatment, but not in at least one other. While this meta-analysis failed to consider habitat patch area and extent of habitat fragmentation, it does suggest that the existence of edge effects may have been accepted too readily. In addition, this analysis found no support for the ideas that edge effects are more prevalent along some

Table 3.3 Total number and percentage of depredated artificial ground and shrub nests at 3 distances from forest edges, and number of nests depredated by different predators

Distance from edge (m)	Ground nests predated		Artificial nest predators							
	n	%	Jay	Magpie	Crow hooded	Corvid	Fox Badger	Rodent	Not known	Egg gone
Ground nests										
0	18	47.4	1	1	0	0	5	3	5	3
15	14	36.8	0	1	0	1	3	4	0	5
30	13	34.2	0	0	0	0	5	3	2	3
Total	45	39.5	1	2	0	1	13	10	7	11
Shrub nests										
0	27	71.1	7	2	2	8	0	0	7	1
15	27	71.1	1	6	1	4	1	0	4	10
30	23	60.5	4	2	0	5	0	0	4	8
Total	77	67.5	12	10	3	17	1	0	15	19

Source: Soderstrom, Part, and Ryden 1998.

types of interface than along others or are more likely to occur with increasing fragmentation.

In another recent meta-analysis that considered 64 experiments from 32 studies, Batary and Baldi (2004) found that edge effects were more pronounced in North America and northwestern Europe than in Central America and central Europe; that nests in marshes and deciduous forests suffered predation-related edge effects but not those in coniferous or tropical forests or fields; and that edge effects were particularly evident for nests found less than 24 m from an edge but disappeared if they were more than 50 m from the edge.

Undoubtedly, certain bird populations suffer heightened nest losses from particular predators along habitat discontinuities, but in general, edge effects seem to be more varied and complex than originally thought and deserve continued but more skeptical study that incorporates confounding factors at multiple spatial scales and, best of all, behavioral studies of predators themselves.

3.2.c Habitat patch size   Conservation biologists have long been concerned about effects of habitat fragmentation on nesting success of migratory songbirds (Wilcove 1985; S. K. Robinson et al. 1995), but it is often difficult to separate effects of patch size from patch composition and the matrix surrounding patches. Studies that have examined habitat area per se find that predation on nests increases with declining area and that presence of known (Moller 1988a) or assumed predators (Small and Hunter 1988) exacerbates this effect. The most instructive studies attempt to integrate covarying habitat variables while separating the impact of different predators. For example, Donovan and colleagues (1997) explored patterns of nest predation on artificial nests using clay eggs in three types of fragmented landscape in Illinois, Indiana, and Missouri, USA, employing a randomized split-plot experimental design. Nests were placed at 50 m from forest edges and at 250 m inside forest cores within each of the three types of fragmented landscape, and microhabitats around nest sites were described. Of 1064 nests, 30.6% showed signs of predation, with raccoons, opossums, canids, and birds accounting for most of the losses. Predation by mammals but not by birds was higher in fragmented landscapes, but the key difference between edge and core was found in intermediately fragmented landscapes (contradicting Soderstrom, Part, and Ryden's 1998 study in Sweden; fig. 3.1). Thus, landscape context interacts with edge effects, and it is rash to consider one without the other. A meta-analysis of 120 and another of 117 tests of the effects of habitat fragmentation on nesting success both showed that effects were most evident at the landscape scale, followed by the patch scale and then the edge scale (Chalfoun, Ratnaswamy, and Thompson 2002; Stephens et al. 2003; Huhta et al. 2004).

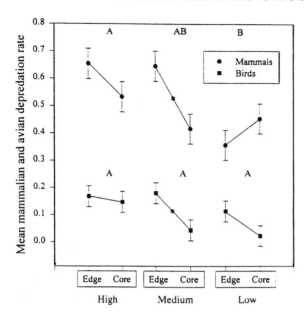

Figure 3.1 Mean mammalian and avian predation rates in edge and core habitats in highly fragmented (*left*), moderately fragmented (*center*), and unfragmented landscapes (*right*). Fragmentation classes with the same letter were not significantly different. Within landscapes, habitat means separated by an asterisk in the center of a line were significantly different (Donovan et al. 1997).

Fragmentation and habitat classes

3.2.d Vegetation around the nest site   In addition to choosing an appropriate habitat in which to nest, birds are selective about the choice of vegetation around their nest. For example, nest flora, cover, and vegetation structure all differ among eight species of shorebirds nesting in the same national park in southwest Saskatchewan, Canada (Colwell and Oring 1990), and between seven ground and shrub nesters in snowmelt drainage in Arizona, USA (T. E. Martin 1988a). In the latter study, nests placed in the preferred microhabitat, as defined by the vegetation type used most often, exhibited higher daily survival rates than those in the nonpreferred habitat. In theory, vegetation surrounding the nest could influence rate of nest predation by physically impeding access to the nest, increasing the number of potential nest sites in which a predator must search, or screening activities of parents from predators (Holway 1991); and by whatever means, foliage generally lowers rates of nest predation in both grassland/marsh and shrub/woodland habitats (Albrecht and Klvana 2004; table 3.4). To take one specific example, in mixed wood forests of Alberta, Canada, Rangen, Clark, and Hobson (1999) found that artificial wicker nests baited with Japanese quail eggs were most likely to survive if they were on the ground and well camouflaged by dense shrubs (see also Jackson, Hik, and Rockwell 1988). On the other hand, Seitz and Zegers (1993) found no effect of over eight measures of vegetation on nest survivorship in deciduous, coniferous, and successional habitats in Pennsylvania, USA, again using arti-

Table 3.4 Number of cases in which frequency of predation was or was not reduced by nest concealment from foliage cover

Taxonomic group	Habitat	Reduced	Not reduced
Artificial nests	Grassland/Marsh	6	0
	Shrub/Woodland	3	1
Anatidae	Grassland/Marsh	8	1
Phasianidae and Columidae	Grassland/Marsh	1	0
	Shrub/Woodland	4	1
Passeriformes	Grassland/Marsh	1	2
	Shrub/Woodland	6	2
Totals	Grassland/Marsh	16	3
	Shrub/Woodland	13	4

Source: T. E. Martin 1992a.

ficial nests and quail eggs (see also Howlett and Stutchbury 1996). Given that there are many additional examples of vegetation having positive effects on nest survival and many others showing no effect, researchers have outlined several possible reasons for such discrepancies (Major and Kendall 1996): (i) when generalist predators search for food items such as invertebrates and nest predation is incidental, concealment may be of little selective advantage (P. D. Vickery, Hunter, and Wells 1992); (ii) when the principal predators use olfaction rather than visual cues to locate nests, visual concealment may be of minor import (Davison and Bollinger 2000); (iii) when nests are predated by a variety of predators, birds may detect nests in the open, medium-sized mammals spot them in intermediate cover, and rodents find them in thick vegetation, thereby masking beneficial effects of each type of vegetation around the nest (Clark and Nudds 1991; Dion, Hobson and Lariviere 2000); and (iv) when parents visit the nest site to incubate or feed nestlings, predation risk against some predators may be mollified through nest defense by parental birds (T. E. Martin 1992b; Cresswell 1997a; chapter 10). Parental solicitude has a strong bearing on results based on experiments using natural rather than artificial nests, as the latter do not benefit from it (section 3.3).

Another possible factor responsible for discrepancies between studies using natural nests is that some species may not wish to construct nests completely concealed by vegetation. Adults may want to scan their surroundings for predators that prey on nestlings or themselves from the vantage point of the nest and thereby give themselves sufficient time to take flight or mount a counterattack. As illustrations, in response to gull-billed terns, a nest predator, Kentish plovers departed sooner from ground nests that had an unrestricted view than those with cover (Amat and Masero 2004); and when song thrush nests were grouped under five categories of crypsis, it was found that birds principally use nest sites of intermediate concealment despite the fact that artificial nests had a higher probability of survival the more that they were

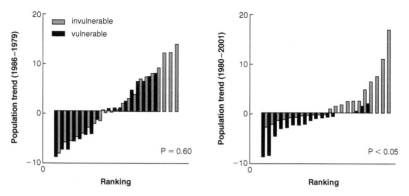

Figure 3.2 Breeding-bird-survey population trends (ranked left to right, from most negative to most positive) for raccoon-vulnerable (low-nesting) and invulnerable (high-nesting) species during the periods 1966–79 (*left panel*) and 1980–2001 (*right panel*) (K. A. Schmidt 2003).

hidden. Nonetheless, natural nest site survival was unrelated to concealment or surrounding tree density, implying that other factors, perhaps nest defense, might be involved (Gotmark et al. 1995).

3.2.e Nest height   Classically, predation risk has been assumed to decrease as nest height increases, with ground-nesting birds suffering greater predation than aboveground nesting species (Best and Stauffer 1980; Slagsvold 1982; Collias and Collias 1984). Worrisome data, because they were collected across a large spatial scale and are therefore of general significance, show that populations of species that nest closer to the ground (and are therefore vulnerable to raccoons and possibly opossums) have exhibited steady declines in Illinois, USA, since 1980, whereas species that nest higher up (and are therefore invulnerable to these predators) have shown population increases. In this study, bird species were categorized solely according to whether they nested above or below 2.5 m off the ground, and population trends were based on breeding bird surveys; the data were split before and after 1980, when the raccoon harvest reached its peak (fig. 3.2). While the data do not demonstrate that nest height is the definitive factor affecting population trends, they are suggestive (K. A. Schmidt 2003).

In a reappraisal of the importance of nest height, T. E. Martin (1992a, 1993a) controlled for habitat before examining nest predation among vegetation layers based on a long-term study of several species in mixed-conifer forest in Arizona; a 1-year study in deciduous forest in Arkansas, USA; and a review of the literature. Results showed that shrub-nesting species in Arizona suffered higher rates of predation (64.3%) than ground-nesting (33.8%) or subcanopy/canopy-nesting bird (39.9%) species; in Arkansas, ground nesters suffered lower rates (25.4%) than shrub- (45.2%) or subcanopy/

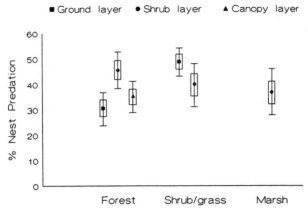

■ Ground layer   ● Shrub layer   ▲ Canopy layer

Figure 3.3 Mean (*symbol*), standard error (*box*) and 95% confidence interval (*error bars*) for percentage of nest predation found for ground-, shrub-, and canopy-nesting birds in 3 habitat types (*shown below x-axis*), based on data taken from the North American literature. Forest ground, $N = 3$ studies; forest shrub, $N = 11$; forest subcanopy, $N = 17$; shrub/grassland ground, $N = 24$; shrub/grassland shrub, $N = 22$; marsh shrub, $N = 12$ (T. E. Martin 1993a).

canopy-nesting (41.5%) species; and although studies from the literature showed greatest predation on the ground, this result disappeared once habitats were separated into forest, shrub/grass, and marsh (fig. 3.3). In forests, shrub-nesting species fared poorly. Evolutionarily, the height at which species nest may be relatively fixed, since it varies rather little across geographic regions (T. E. Martin 1988a), although variance in this trait within and between species is still poorly documented for most species. If there is little individual variation in nest height preference within species but greater predation risk at particular heights, other factors must be involved. While these may include competition (Schoener 1974), T. E. Martin (1988b) has argued that predation itself may partition nest sites vertically. Using artificial wicker nests baited with quail eggs all placed at the same height or alternatively at 1 and 3 m off the ground, he found that the percentage of nests remaining without signs of predation was greater when nests were stratified according to height (fig. 3.4). In a parallel experiment, predators were found to increase their searching intensity when artificial eggs were clumped, so one can argue that predators develop search images for nests set at particular heights. It therefore behooves species to place their nests at heights different from other species despite the fact that some heights are more dangerous than others. How interesting it would be to watch birds select or eschew nest sites according to the local layout of other birds' nests.

3.2.f Proximity to nests   A closely related argument states that because predators acquire search images for particular sorts of nests or nests placed at sim-

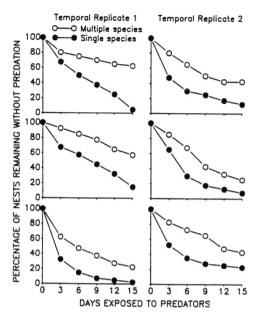

Figure 3.4 Percentage of nests that lose no eggs to predators. Each of the 2 columns is a temporal replicate, and each of the 3 rows is a spatial replicate. The "multiple-species" treatment (*solid circles*) represents the combined predation rate on 4 "species" (that is, height habitats); the "single-species" treatment (*open circles*) combines 2 sets of "single species" (same heights) in each temporal replicate; 2 sets were used so as to match the total number of nests (40) in the "multiple-species" treatment (T. E. Martin 1988b).

ilar sites (T. E. Martin 1988b), nests placed in close proximity may be subject to a higher rate of predation than those placed far apart if a predator locates similar nests close to one another. In a classic study, where N. Tinbergen, Impekoven, and Franck (1967) painted eggs and laid them out in scattered or in clumped formations, hooded crows took a greater number of eggs from the crowded plots. Such selection pressure should lead to overdispersion, where nests are placed farther away from one another than would be expected from the availability of nest sites. In an impressive study in which the effects of nest proximity were investigated at three spatial scales, duck nesting success was found to be unaffected by natural nest densities across fields, at three experimental densities inside 1 ha plots within fields, or in relation to the fates of nearest-neighbor nests, be they natural or artificial (Ackerman, Blackmer, and Eadie 2004). Damning as this extensive study might seem to be for effects of proximity on nesting success, other investigations show far more varied results even with Anatidae (table 3.5). Thus, the prospect that all predators show restricted searching may be too simplistic (Andren 1991; Schiek and Hannon 1993). Other factors may be involved in predation, including the presence of alternative prey (O'Reilly and Hannon 1989), the extent to which predators are egg specialists (Boag, Reebs, and Schroeder 1984), and the distance at which predators search elsewhere. For example, in a study involving crows, cats, and snakes as predators, Wada (1994) discovered that nest predation peaked between 50 and 70 m from other nests rather than at shorter distances. Furthermore, the presence of one type of nest (domed nests) can increase

Table 3.5 Summary of studies that have investigated the effects of nearest-neighbor distance and nest density on predation rates of duck nests

Scale	Nest type	Density of nests/ha or inter-nest distance	Nest spacing	Higher predation when nests closer together?	Major nest predators	Reference[a]
Small						
NN[b]	artificial	11–38 (range means)	random	yes (high density) no (low density)	striped skunks	1
NN	artificial	1000 m	linear transect	4 tests, no; 2, yes	not reported	2
NN	mallard	51 m (mean)	natural	no	hooded crows[c]	3
NN	artificial	10–80 (range)	uniform	yes	American crows	4
NN	artificial	12–27 m (range means)	random	no	striped skunks[c]	5
NN	*Anas* spp.	29 m (mean)	natural	no	striped skunks	5
Intermediate						
2–4 ha plots	artificial	2.5, 10, 25	random	no (early season) yes (late season)	striped skunks	1
10 ha plots	artificial	2, 10	uniform	yes	red foxes[c] mew gulls[c]	6
0.8–10.2 ha plots	artificial	1.5–102 (range)	uniform	yes	American crows	4
1 ha plots	artificial	5, 10, 20	random	no	striped skunks[c]	5
Large						
12–54 ha fields	*Anas* spp.	5.0–10.9 (yearly means)[d]	natural	no[e]	red foxes raccoons	7
5–33 ha fields	*Anas* spp.	2.4–11.1 (yearly means)[f]	natural	no	striped skunks[c]	5
among years	blue-winged teal	not reported	natural	yes[e,h]	striped skunks[c]	8
among years	mallard	not reported	natural	yes[e,g]	carrion crows[c] magpies[c]	9

Source: Ackerman, Blackmer, and Eadie 2004.

[a] 1. Larivière and Messier 1998; 2. Clark and Wobeser 1997; 3. Andren 1991; 4. Sugden and Beyersbergen 1996; 5. Ackerman, Blackmer, and Eadie 2004; 6. Esler and Grand 1993; 7. Duebbert and Lokemoen 1976; 8. Weller 1979; 9. Hill 1984.

[b] Nearest neighbor.

[c] Suspected.

[d] Range of (apparent) duck nest density among fields by year: 1971: 1.2–10.4 nests/ha; 1972: 3.0–21.7 nests/ha; 1973: 0.5–11.9 nests/ha.

[e] These studies used apparent nest density rather than the Mayfield estimate of nest density and are therefore considered weaker tests.

[f] Range of (Mayfield) duck nest density among fields by year: 1998: 0.1–16.6 nests/ha; 1999: 0.1–31.3 nests/ha; 2000: 0.1–8.0 nests/ha; 2002: 1.1–5.6 nests/ha.

[g] Negative relationship between yearly nest success and the total number of nests found each year.

[h] It is unknown whether this relationship was caused by density-dependent predation across years because nest success also correlated with alternate prey densities.

Case a

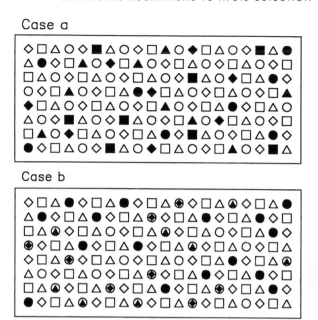

Case b

Figure 3.5 Schematic drawing of substrate and prey distribution for 2 experimental nest treatments. In both Cases a and b, 4 types of nest sites are represented by 4 open symbols. Four bird species are represented by 4 solid symbols. In the first treatment (*Case a*), each species occupies a different nest site. In the second treatment (*Case b*), density is the same as in Case a, but all species occupy the same nest site. The result is that cumulative density in a nest site (for example, *circles*) is 4 times greater when species use the same nest site (*Case b*) than when each species uses a different nest site (*Case a*). Note that almost all circles are occupied in Case b, rewarding predators to continue searching the substrate type but to ignore other substrate types because they are empty. This situation increases predation rates on all species using a given substrate type. In contrast, in Case a, the predator must search all substrates to encounter the same number of prey, and the increased encounters with unoccupied sites may cause the predator to give up before finding a prey, thereby reducing the cost of coexistence (T. E. Martin 1993b).

predation rate on other types (open nests) (Hoi and Winkler 1994), potentially confounding single-species studies.

3.2.g Distribution of nests   That predation rate sometimes increases as a function of nest proximity suggests that nests placed at high local densities will be subject to elevated rates of predation (Holt and Kotler 1987; Schmidt and Whelan 1998). Experimental work using artificial nests containing chicken eggs, for example, showed that nest survivorship declined with increasing density, although some of these densities were artificially high ($> 100/\text{ha}$) (Sugden and Beyersbergen 1996; see also Goransson et al. 1975). If predation risk increases with cumulative prey density, then predation risk should increase for all species using similar nest sites and mitigate against species coexistence (fig. 3.5). In a series of papers, Martin and his coworkers showed this to be the case. First, where artificial nests baited with real eggs were placed in locations that simulated

Table 3.6 Studies that have measured the effect of either nest density or inter-nest distance on predation rate on both real and dummy nests

Dispersion pattern	Range of density nests/km²	Range of inter-nest distance (m)	Effect of inter-nest distance at a given density[a]	Effect of density[a]	Reference[b]
Regular	12,500–4 × 10⁶	0.5–8	—	Yes	1
Overdispersed	250–400	5–90	Yes	—	2
Regular	100–400	50–100	—	Yes	3
Regular	2500	20	—	Yes	4
Random	150	40	No	—	5
Regular	17–667	40–250	—	Yes	6
Regular	25–100	100–200	—	No	7
Overdispersed	20–30	6–500	No	—	8
Regular	20–40,000	5–225	—	Yes	9
Overdispersed	10	—	No	—	10
Regular	13,000–89,000	—	—	Yes	11
Regular	27–100	100–190	Yes/No	No	12
Random	90	5–170	No	—	13

Source: Andren 1991.
[a] *Yes* indicates that it affects predation rate.
[b] 1. Tinbergen, Impekoven, and Franck 1967; 2. Krebs 1971; 3. Goransson et al. 1975; 4. Andersson and Wiklund 1978; 5. Loman and Goransson 1978; 6. Page et al. 1983; 7. Boag, Reebs, and Schroeder 1984; 8. Blancher and Robertson 1985; 9. Sugden and Beyersbergen 1986; 10. Galbraith 1988; 11. T. E. Martin 1988b; 12. O'Reilly and Hannon 1989; 13. Andren 1991.

those of two actual bird species, daily predation rate probabilities were higher when species coexisted than when they were separated. Among natural nests, individuals that showed greater microhabitat overlap with other species experienced elevated rates of predation in six out of seven species (Martin 1996). Second, removal of Virginia's warblers resulted in lower predation on orange-crowned warblers' nests and vice versa; these species use similar nest sites and food resources and share the same nest and adult predators (Martin and Martin 2001). Third, daily mortality rate on experimental nests was substantially reduced when T. E. Martin (1988b) compared artificial nests set at several different sites (on the ground, 1 m off the ground in either fir or maple, and at 3 m in maple, the "multiple-species" treatment in figure 3.4) with artificial nests set at the same site (that is, all on the ground) but at the same overall density (fig. 3.6). In summary, nest predation may be a sufficient selective pressure to force coexisting species to partition microhabitats among themselves.

Nevertheless, the relative importance of both inter-nest distance and nest density varies across studies (table 3.6) and probably depends on the predators involved. For predators such as ground squirrels, with 1- to 2-ha home ranges, individuals are unlikely to encounter sufficient nests to be able to respond by increased predation. Predators with larger ranges, such as gulls or

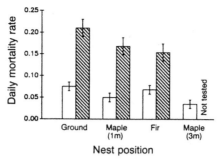

Figure 3.6 Mean (and standard error) in daily mortality rates (rate at which nests are lost to predators/day) for artificial nests when nests were placed among 4 different types of sites (*open bars*, Case a of fig. 3.5) and when nests were all placed in the same type of sites (*hatched bars*, Case b of fig. 3.5). Density of nests was held constant between the 2 treatments. A total of 480 nests were used among 3 replicate sites and over 2 temporal replicates. The situation of nests placed in the same type of sites was not tested for the 3 m maple (T. E. Martin 1993b).

raccoons, will be able to search for many nests and could exert a pattern of density-dependent predation on nests at least when nests are at intermediate or high densities (Andren 1991; Sugden and Beyersbergen 1996; K. A. Schmidt and Whelan 1999). Moreover, the extent to which a predator is satiated after raiding a nest will affect whether it searches for additional nests, a factor highly dependent on the ratio of clutch and egg size to predator body mass (Ackerman, Blackmer, and Eadie 2004).

3.2.h Proximity to social insects   Ever since Gosse (1847) observed that some birds build nests close to colonies of social hymenoptera to make them unattractive to predators, at least 100 species of birds, principally tropical passerines, have been reported as nesting within 1.5 m or less of nests of wasps, bees, or ants (Myers 1935; MacLaren 1950; Hindwood 1959; S. K. Robinson 1985; Wunderle and Pollock 1985; Brightsmith 2000). For instance, rufous-naped wrens in Costa Rica nest in *Acacia collinsi* trees with particularly active ant species that sting and bite. Moreover, they peck holes in artificial eggs placed in these acacias, suggesting that they try to inhibit other birds from nesting in these trees as well (Young, Kaspari, and Martin 1990).

To examine the consequences of choosing to nest near social hymenoptera, Joyce (1993) moved occupied *Polybia rejecta* wasp nests next to rufous-naped wren nests. Wrens whose nests were near experimentally relocated wasp nests were more likely to fledge young (6/16 in 1 year and 9/12 in another) than were wrens whose nests had no wasp nests placed near them (0/16 and 3/15, respectively). Even within the same tree, nests associated with wasp nests were more likely to fledge young than nests without wasp nests nearby. In this study, unsuccessful nesting attempts far from wasp nests were more likely to succumb to predation from white-faced capuchin monkeys (78.5%, $N = 28$)

Table 3.7 Proportions of yellow-rumped caciques' nests preyed on and number of colonies attacked by predators in different kinds of sites by different predators in Cocha Casu, 1979–83, and Cocha Totora, 1981–83, Peru

Colony type	Number of nests[a] (number of colony years)	Proportion preyed upon (number of colonies attacked)			
		Mammals	Snakes or mammals[b]	Birds	Total
Island	503 (5)	0.00 (0)	0.03 (1)	0.05 (4)	0.08
Wasp nest	125 (7)	0.21 (1)	0.06 (1)	0.09 (3)	0.35
Marsh shrub	138 (13)	0.00 (0)	0.57 (10)	0.04 (5)	0.61
Edge tree	230 (14)	0.20 (4)	0.07 (2)	0.27 (10)	0.54
Overhanging branch	133 (14)	0.65 (9)	0.12 (4)	0.11 (8)	0.88
Total	1129 (66)	0.14 (14)	0.12 (18)	0.11 (30)	0.36

Source: S. K. Robinson 1985.
Note: All marsh shrub colonies are in Cocha Totora. Edge trees are in figs isolated from the surrounding forest by low (<4 m tall) shrubby vegetation. Overhanging-branch colonies are on branches of canopy trees over water. Marsh shrubs are completely surrounded by low (<2 m) marsh grasses. Sites are given in order of decreasing safety.
[a] Sum over all years for the occupied colony sites of each type.
[b] Nocturnal attacks in which the predator entered through the opening at the top. Some nests may have been taken by an unidentified mammal.

than those near wasp nests (31%, $N = 13$). These monkeys passed by wasp nests and were obstructed by wasps if they tugged branches supporting wren nests.

It is important to come away with the idea that different aspects of nest site selection confer antipredator benefits as a consequence of predators hunting in different ways. In a detailed study of colonial nesting in yellow-rumped caciques, S. K. Robinson (1985) found that colonies surrounded by water or marsh vegetation could not be reached by primates, and those placed next to polistine wasp nests or at the edges of forest patches were attacked rarely by these predators (table 3.7). Colonies on islands were also protected from snakes that were themselves eaten by black caimans and giant otters in the water, but colonies were vulnerable to snakes if placed in marshes. Nest site selection had no strong influence on protecting caciques from avian predators (although those along edges were more susceptible), but colony size had a strong effect, because larger colonies mobbed black caracaras and Cuvier's toucans in great numbers which drove these predators off (chapter 11). Thus, a single prey species may use many antipredator ploys to thwart predation when subject to diverse predatory threats (chapter 13).

3.3 Behavior reducing the probability of predators detecting nests

Aside from judicious choice of nest site, birds can reduce the probability of their eggs and nestlings being detected through covert behavior (I cover active defense in chapter 10). For example, the prairie warbler builds its nest for only brief sessions and for a limited part of the day, thereby reducing likelihood of detection. Also, birds may desert their nest if discovered building (Nolan 1978). Ducks and geese cover their eggs with down or vegetation before they leave to forage (Collias and Collias 1984). Once eggs have hatched, broken eggshells are removed from the vicinity of the nest to reduce attracting attention to hatchlings (N. Tinbergen et al. 1963). After nestlings start feeding, adults remove their fecal sacs from the nest, presumably to reduce odor around the nest (Petit, Petit, and Petit 1989). Nonetheless, perhaps the most dangerous activity performed by parents is simply repeatedly arriving at and leaving the nest in order to incubate the eggs or (more frequently) to feed nestlings, for this can attract watchful predators to an otherwise well-concealed nest, as outlined by Skutch (1949; see also Cody 1973). For example, Siberian jays nesting in areas of high predation risk reduce parental visits as corvid predator activity increases during the day (Egges, Griesser, and Ekman 2005).

Recently, Skutch's hypothesis has come under scrutiny. Some studies of single species uncover no association between parental feeding rate and rate of nest predation, for example, in situations in which the principal predator is nocturnal (J. J. Roper and Goldstein 1997). In a suite of characteristically thorough studies, however, Martin and his coworkers (T. E. Martin and Ghalambor 1999; T. E. Martin et al. 2000; T. E. Martin, Scott, and Menge 2000) have shown that nest predation and parental visits to the nest are positively associated within species but are associated negatively across species, because species suffering high predation risk have apparently reduced parental activity over an evolutionary timescale. In their Arizona study site, Martin and colleagues videotaped parental activity at the nests of 10 species and regularly monitored loss of eggs and nestlings. Although there was no greater predation during the nestling stage across species, as had been predicted by Skutch, these results failed to control for predation differences among nest sites. Species with larger increases in parental activity during the nestling versus incubation stages (to control for differences in nest site susceptibility) suffered greater nest predation even after controlling for phylogeny. Then, across 19 species in Arizona and Montana, USA, Martin and colleagues found that incubation feeding rates were negatively associated with the percentage of nests lost (fig. 3.7), confirming Skutch's hypothesis. They also found that within three Arizona species, visitation trips during incubation were lower in successful than predated nests; and then, extending the study to include Argentinean species, nestling visitation rates declined with increasing predation rates, food delivery rate increased with visitation rate, and, crucially, food delivery rate decreased with increases in daily nest predation rates, confirming the North American study across a wider geographic area. Reductions in parental activity, specifically delivery of food to nestlings, is clearly related to predation and probably serves to lower it by making detection more difficult.

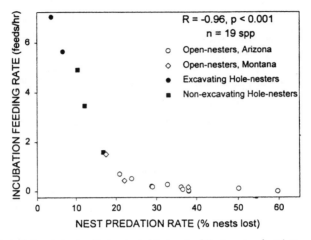

Figure 3.7 Incubation feeding rates (number of feeds per hour) relative to nest predation (percentage of nests lost to predators) across species. The relationship between incubation feeding and nest predation within each nest type is significant: open-nesting species, $r = -0.91$, $P < 0.0001$; cavity-nesting species, $r = -0.98$, $P = 0.001$ (T. E. Martin and Ghalambor 1999).

Birds may also attempt to reduce the probability of predators locating nests during a renesting attempt that follows the loss of a brood, because certain predators return to the site at which they earlier killed a brood. A model shows that dispersing birds should have higher success than nondispersing birds following predation especially if dispersers move beyond the home range of the predator (L. A. Powell and Frasch 2000). In southern Norway, predation on Tengmalm's owl clutches is heavy, 48%, and at least 70% of this is due to pine martens. Frequency of predation increased with nest box age and was higher in locations where the previous clutch was preyed upon than where the clutch had been successful. Female owls preferred to use new nest boxes in areas where boxes were added rather than ones that had been used in previous years (Sonerud 1985b). Moreover, when nest boxes were relocated by the researchers, the proportion of nesting attempts that suffered predation declined, but it increased in boxes that were not moved (Sonerud 1989). This indicated that it was the same location rather than the nature of the old nest box itself that resulted in elevated predation risk. Similar arguments have been put forward for the reasons that woodpeckers excavate new nest cavities each year (Nilsson, Johnsson, and Tjernberg 1991). Moreover, it is possible that birds benefit by the presence of old nests nearby: if predators find nests that are empty, they may move on (S. K. Robinson 1985; Watts 1987). This opens up the possibility that some species construct nests of materials that will last a long time in order to increase subsequent nesting success, and that supernumerary nests may be an avenue of avoiding detection by predators rather than a male courtship strategy (Collias and Collias 1978).

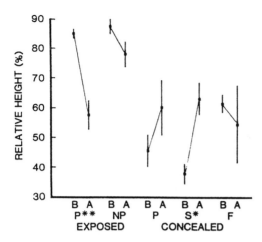

Figure 3.8 Relative nest height (mean and SE) before (*B*) and after (*A*) various stimuli (*P*, predation; *NP*, no predation; *S*, snowfall; *F*, successful fledging) at exposed and concealed nest sites. Median heights before stimuli at exposed sites were equal, but at concealed sites, snowed-upon nests were significantly lower than nests fledging young. Asterisks below *x*-axis indicate significant changes in nest height after a stimulus (*$P < 0.01$; **$P < 0.001$, using paired *t*-tests). Sample sizes at concealed nests were as follows: P, 20; NP, 7; and at concealed nests: P, 9; S, 16; F, 5 (Marzluff 1988).

Some birds show sophisticated behavior by changing the location of subsequent nesting attempts according to the reasons that their previous clutch failed (Dow and Fredga 1983; S. K. Robinson 1985; Amat, Fraga, and Arroyo 1999). Around Flagstaff, Arizona, the nests of the pinyon jay, a bird that lives for over 16 years, are subject to predation from ravens and American crows. They are also adversely affected by cold weather and snow in microhabitats where solar radiation is low (Marzluff 1988). Whereas nests that are higher and farther away from the trunk are more likely to be preyed upon because they are more visible from the air, concealed nests are sometimes abandoned following heavy snowfall, presumably because the energetic cost of incubation is too severe. Jays renest following failure and occasionally following success, but new placement differs according to the fate of earlier nests. Following predation on exposed nests, subsequent nests are placed lower in the canopy, but following failure for other reasons, nests are again placed in exposed places (fig. 3.8). When a concealed nest is preyed upon, height of the new nest does not differ from the old one, but following cold, snowy weather, the next nest is built in a less concealed location, higher up. Extraordinarily, following cat predation specifically, nests are built much higher off the ground. This long-lived bird appears to accumulate experience of predators and uses it to select nest sites. Nonetheless, each season breeders renest at high locations, perhaps because of poor memory, because height is an imperfect predictor of predation risk, or because of overcompensation against ground predators. Behavioral mechanisms to avoid nest detection are impressive but not perfect.

3.4 Refuges

Natural history studies of many animals indicate that they use refuges to reduce predation risk. Refuges can be physical structures, such as trees or cliff faces, or localized areas, such as thick vegetation, that are not specifically tied to short-term heightened risk of predation. Such refuges may be used principally during certain portions of the day; over longer periods of time, as during reproduction; or during most of an animal's life. In essence they are habitats or areas of greater safety than others (Krivan 1998; Ylonen, Pech, and Davis 2003). Refuges are also used in a separate sense, however, that is directly linked to predation, namely as safe areas to which an animal restricts its activity in the presence of predators. Here predators may simply be in the area but not present at the time (for example, Banks 2001), or refuges may be a bolt hole to which prey run when they see a predator or are pursued by one (for example, Holmes 1984; Sundell and Ylonen 2004). These different meanings of the word lead to long- and short-term changes in behavior being conflated, and to degrees of predation risk being poorly specified.

3.4.a Physical structures   Regarding the first meaning of *refuges*, the most obvious example during periods of the 24-hour cycle is of birds roosting in trees at night, presumably to avoid encountering predators (termed anachoresis), but mammals show parallel behavior. For example, North American porcupines sleep in buffalo-berry and willow groves but forage in the open (Sweitzer and Berger 1992), and Pacific harbor seals haul out far from shore to reduce risk of terrestrial predation (Nordstrom 2002). At night, diurnal terrestrial primate species move to trees and sleeping ledges, where they may be less likely to encounter predators (J. R. Anderson 1984). Monkeys and apes seem to choose sleeping sites on the basis of comfort (warmth, in places where the temperature drops sharply at night) and safety from predators. Focusing on the latter, primates prefer tall trees with branches that start high, trees with few branches, and, for smaller species, concealed places such as tangles of lianas (Garcia and Braza 1993) where they can rest (J. R. Anderson 1984). Sites with such characteristics become traditional sleeping sites. In addition, monkeys seek out terminal branches, which cannot support the weight of a predator and where they may feel vibrations of an approaching predator, and sometimes sleep over water, perhaps to reduce avenues of approach from other trees (Ramakrishnan and Coss 2002). Construction of sleeping refuges is common also. For instance, chimpanzees always fabricate night nests out of vegetation, but where large carnivores are present they build these off the ground (Whiten et al. 1999).

Mammals such as aardvarks and rodents construct subterranean burrows, appearing on the surface only at night. The extent to which these refuges mollify predation risk or influence thermoregulation is relatively unexplored (J. Martin and Lopez 1999), but burrow construction is certainly influenced by predation risk in some species. For example, prairie voles and meadow voles build more burrow entrances and construct simpler burrow systems in experimental pens that are open to terrestrial predators (weasels) that can enter burrows (Harper and Batzli 1996).

Many female mammals build rudimentary or sophisticated refuges prior to giving birth, a period when mothers and offspring are particularly vulnerable to predation and sometimes infanticide. It seems probable that nests, lairs, dens, and burrows so commonly enjoyed by juveniles during their mothers' temporary absences are chiefly antipredator devices; spotted hyena cub and warthog sounder burrows being cases in point (East, Hofer, and Turk 1989; Estes 1991). Particular species of birds such as storm petrels also nest in burrows, where they are protected from avian predators. Offspring of many nonprimate mammals are unable to urinate or defecate without being licked by their mother. After a period of absence, she licks their genital area and consumes their feces and urine. While mothers may recoup nutrients, restricting offspring elimination to the time that mothers can devour waste products must also reduce odor around the lair. Female mammals frequently move nest sites (for example, cheetahs: Laurenson 1993) especially at time of parturition and during lactation (for example, deer mice: Sharpe and Millar 1990). All these strands of evidence point to offspring refuges having an antipredator function in mammals, although the extent to which these serve to prevent predators detecting offspring or entering the refuge must vary enormously.

Some mammals are entirely fossorial. The degree to which the evolution of a fossorial lifestyle shown, for example, by moles and mole rats has been promoted by antipredator concerns or other factors, such as exploiting subterranean food sources, is unlikely to be resolved without greater knowledge of predation pressures on these species.

3.4.b Habitat shifts in rodents   Turning to the second meaning of refuges, that is, limiting activities to a particular locality less frequented by predators, several studies have focused on the way that rodents, ungulates, and other species (for example, red-bellied moustached tamarins: Caine and Weldon 1989; cheetahs: Durant 1998; arboreal primates: Sterck 2002) shift habitats under predation risk. Many demonstrated that rodents use illumination and cover as indirect cues of predation risk and shy away from using open areas when the moon is full, moving to microhabitats with cover (for example, J. A. Clarke 1983; Travers, Kaufman, and Kaufman 1988; Simonetti 1989; Daly et al. 1992; Diaz 1992; Dickman 1992; Price, Waser, and Bass 1994). Time spent in cover lowers time exposed to aerial predators and increases likelihood of escape if a bird attacks (Kotler 1984a; Kotler and Brown 1988; Longland and Price 1991; but see Schooley, Sharpe, and van Horne 1996). Indeed, both large body size and familiarity with an area both increase the probability that rodents will be

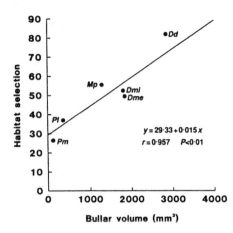

Figure 3.9 Relationship between mean habitat selection and mean bullar volume of rodents at Tonopah Junction, Nevada, USA. The habitat selection axis represents the percentage of captures occurring in the open microhabitat (arcsine transformed). A score of 0 indicates that all captures occurred in the bush microhabitat; 45, that equal numbers of captures occurred in bush and open microhabitats; and 90, that all captures occurred in the open. Dm: *Dipodomys merriami;* Dmi: *D. microps;* Dd: *D. deserti;* Mp: *Microdipodops pallidus;* Pl: *Perognathus longimembris;* Pm: *Peromyscus maniculatus* (Kotler 1984a).

found in cover, implying that there could be competition for safe areas (for example, voles: Koivunen, Korpimaki, and Hakkarainen 1998). Comparative evidence from deserts has shown that some rodent species are sensitive to moonlight, while others are not (Kotler 1984a; Bowers 1988). Species such as Merriam's kangaroo rat, with its enlarged hind limbs for bipedal locomotion that aid in erratic leaping, and inflated auditory bullae for increased hearing acuity and sensitivity to low-frequency sounds, remain in more-open areas in response to increased illumination than do pocket mice, such as the Arizona pocket mouse, that are quadrupedal and have no auditory bullae. Pocket mice move into cover under increased illumination (J. S. Brown et al. 1988). More generally, species with the greatest auditory bullar volume use open habitats the most (fig. 3.9). There is some evidence that bipedal species suffer lower rates of predation by avian predators than do quadrupeds: bipeds have a lower probability of being captured by great horned owls in the open part of an aviary (Longland and Price 1991), and in the field are caught less frequently by long-eared owls than are quadrupeds (Kotler 1985; but see Kotler et al. 1988).

In contrast with owls, snakes hide and hunt in cover. Bipedal locomotion is not supposed to help escape from the rapid strike of a snake, and acute hearing probably helps little in detecting relatively silent predators. Thus, it is no surprise that in laboratory experiments, rattlesnakes captured bipedals and quadrupedals equally often (Pierce, Longland, and Jenkins 1992). Differences in species' abilities to detect and avoid avian predators are therefore thought to influence trade-offs between predation risk and foraging efficiency and to sway habitat selection and thus opportunities to forage on different types of seeds. Mechanisms used to avoid detection are therefore believed to promote species coexistence (Rosenzweig 1973; Kotler and Brown 1988; Kotler and Holt 1989).

Kotler, Brown, and Mitchell (1994) have argued that predation contributes to species coexistence in some deserts but not in others. In the Sonoran Desert,

Table 3.8 Studies of ungulates shifting habitats at parturition

Species	Area where young are born	Reference[a]
Moose	High-elevation islands	1
Caribou	High elevation	2
Mountain sheep	Steeper slopes	3
Dall's sheep	Steep slopes	4
Bighorn sheep	High-elevation bajadas	5

[a] 1. Bowyer et al. 1999; Edwards 1983; 2. Bergerud, Butler, and Miller 1984; Bergerud and Page 1987; 3. Bleich, Bowyer, and Wehausen 1997; 4. Rachlow and Bowyer 1998; 5. Festa-Bianchet 1988; Berger 1991.

USA, seven species of rodent select different microhabitats as a consequence of differential trade-offs in the cost of predation and energetic cost of foraging, as outlined above; and in the Great Basin Desert, USA, four species coexist, and three of these suffer seasonally distinct reductions in foraging efficiency due to the necessity of shifting habitats away from a suite of predators. In the Negev Desert, Israel, however, five species coexist not because of predation risk but because seeds dispersed by intermittent wind action create trade-offs between foraging efficiency and resource abundance, at least between two of the gerbil species (Kotler, Brown, and Hasson 1991). Thus, both predation risk on the one hand, and temporal and spatial variations in resource abundance on the other, are able to promote species coexistence (Brown 1989).

A related body of literature has called attention to the way in which different suites of predators can facilitate predation. Here, risk of attack by terrestrial predators is thought to assist predation by avian predators and vice versa. For example, under experimental conditions, field voles move to open microhabitat in the presence of a least weasel but occupy cover in the presence of a European kestrel (Korpimaki, Koivunen, and Hakkarainen 1996). Allenby's gerbil and the greater Egyptian gerbil perceive snakes as a greater threat in bush microhabitats but owls as greater in the open (Kotler, Blaustein, and Brown 1992; Kotler et al. 1993). Although terrestrial predators may have the potential to facilitate predation by birds, the extent to which they do so over the course of a year in the wild is unclear: desert and Merriam's kangaroo rats only stayed in open habitat and hence under threat of owl predation during summer months, when snakes were active in bushes; in winter, snakes became quiescent and the kangaroo rats quit the open microhabitat (Bouskila 1995, 2001).

3.4.c Habitat shifts in ungulates Observational studies of North American ungulates have generated a small but consistent picture of expectant mothers moving to habitats with lower predation risk (table 3.8). Carnivores such as wolves and coyotes are apparently encountered less often on steep slopes, possibly because of low hunting success as a consequence of prey seeing them

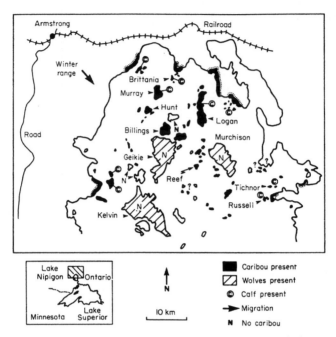

Figure 3.10 Distribution of caribou and wolves in summer 1975 on islands in Lake Nipigon, Ontario, Canada. The majority of the central and lower portions of the figure is water (Bergerud, Ferguson, and Butler 1990).

from vantage points and then moving to higher ground (Bergerud, Butler, and Miller 1984; Bowyer et al. 1999). In several species, mothers suffer reduced food intake, because forage quality is lower in these safer areas (see also Kohlmann, Muller, and Alkon 1996). For instance, each spring a small herd of caribou migrate to small islands in Lake Nipigon, Ontario, Canada, where they stay for the summer (Bergerud, Ferguson, and Butler 1990). Despite the fact that small islands are overgrazed and have poor forage, caribou refrain from moving to larger islands or the mainland, where wolf populations are very high (fig. 3.10). In some locations prey populations live in relatively safe havens year-round. For instance, remnant caribou populations persist on offshore islands in Lake Superior never visited by wolves, while populations on the adjacent mainland have gone extinct (Bergerud 1985; Ferguson, Bergerud, and Ferguson 1988); white-tailed deer survive for longer periods in the buffer zones between wolf territories, which are shunned by wolves for fear of intraspecific strife (Mech 1977; Lewis and Murray 1993); and in the 1800s, large-game animals were found in great numbers in zones contested by warring Native American peoples (Hickerson 1965; P. S. Martin and Szuter 1999). In most of these examples, however, we do not have the behavioral data to know whether prey actively sought out low-risk habitats or were selectively predated.

Table 3.9 Height on slopes (in meters) of deer before and after coyote hunts at the study site in southern Alberta, Canada (1995–96 data), and change in height for groups that moved when they had the option of moving in any direction

Time	Mule deer		White-tailed deer		
	n	median height	n	median height	p-value[a]
Before[b]	81	61 m	48	15 m	<0.0001
After[b]	71	69 m	38	8 m	<0.0001
Change	31	16 m	19	−8 m	<0.0001

Source: Lingle 2002.
Note: Data are from 3 winters.
[a] p-value on Mann-Whitney U test.
[b] There was a significant difference between height of groups before and after hunts for both species.

A breakthrough in teasing these factors apart is a study of habitat segregation in mule deer and white-tailed deer living sympatrically in southern Alberta, Canada (Lingle 2002), where the topography was varied. Mule deer occupied the slopes (98% of 83 groups) and were high up, whereas white-tailed deer inhabited the gently rolling or flat (51% of 143 groups) or else steep rolling terrain (19%), but in general shunned the slopes (30%). Two factors contributed to habitat segregation. First, mule deer were less likely to be encountered or attacked by coyotes, the main predator, the higher up the slope that their group was, and none were attacked or killed above 61 m, whereas there was no effect of height on predation risk for whitetails. Second and crucially, mule deer were more likely to move up slopes and white-tailed deer were more likely to move down slopes during a coyote hunt (table 3.9).

Avoidance of predation risk has also been imputed to address discrepancies in population sizes of sympatric sedentary and migratory ungulate prey. Fryxell, Greever, and Sinclair (1988) suggested that migratory populations, such as wildebeest, tiang, and white-eared kob, are numerically larger than sedentary populations because they can avoid predators that are restricted to denning sites. The idea is that resident ungulate herds are subject to resident carnivore predation throughout the year and are perhaps limited by them, whereas migratory herds can move beyond the range of these predators; instead, their population may be limited by food availability, especially in the dry season (Fryxell and Sinclair 1988). This hypothesis has been challenged by Hofer and East (1995), who maintain that large African predators, namely wild dogs, cheetahs, leopards, spotted hyenas, and lions, have nomadic segments of their population that can move with migratory herds (see also Nicholson, Bowyer, and Kie 1997). In addition, these carnivores show extended offspring dependence, during which time young accompany their parents on hunts (excepting spotted hyenas); and, furthermore, predators do not necessarily focus on resident as opposed to migratory prey species at least in the three African habitats examined by Hofer and East. These behavioral observations are persuasive, but the nub of Fryxell and colleagues' argument centers on which factors limit migratory and resident populations, and neither camp directly addresses this question, leaving the matter unresolved.

3.5 Reduced activity

Prey can alter their activities irrespective of whether they are or are not in a refuge, usually reducing activity when predators are active or in circumstances in which predators find it easy to locate prey. Sometimes reduced activity is tied to a diurnal cycle, but otherwise it may be contingent upon indirect cues associated with predator activity. Changes in prey activity may also be age dependent.

An obvious method of avoiding predation is to time activities to when predators find it most difficult to locate prey. Most small mammals, including rodents, lagomorphs, and bushbabies, are crepuscular or nocturnal. For example, spring hares and bats emerge only at night (Butynski 1984; Speakman 1991). Crepuscularity may be adaptive, because rods and cones in the vertebrate eye operate poorly in twilight (Lythgoe 1979). Nevertheless, there have been surprisingly few studies explicitly tying nocturnality to predation risk. Fenn and Macdonald (1995) discovered an unusual population of Norway rats that exhibited diurnal activity living at a red fox midden; red foxes are nocturnal. Not only did seasonal variation in the rats' diurnal activity patterns covary negatively with seasonal use of the midden by foxes, but rats released into a fox-proof enclosure reverted to normal nocturnal behavior within 4 weeks.

Many studies have linked nocturnal activity to predator avoidance indirectly by demonstrating that activity declines with increasing moonlight (for example, deer mice: O'Farrell 1974; Brillhart and Kaufman 1991; voles: Doucet and Bider 1969; shrews, Vickery and Bider 1978; African brush-tailed porcupines: Emmons 1983; Indian crested porcupines: Alkon and Saltz 1988; bats: Morrison 1978; Fenton et al. 1977). For example, early research showed that bannertail kangaroo rats increased surface activity almost threefold when the moon was down (Lockard and Owings 1974). This and subsequent studies (for example, Wolfe and Summerlin 1989) have disclosed that changes in rodent activity can be quite sensitive to different lunar phases. Indeed, the conventional wisdom among mammalogists is that small-mammal trapping success is greatly influenced by moon phase in temperate regions (Price, Waser, and Bass 1984; J. S. Brown et al. 1988). Seasonal differences in nocturnal activity have also been uncovered, as when mammals abandon their adherence to the lunar cycle on short summer nights when time available for foraging is curtailed (Lockard and Owings 1974; Alkon and Saltz 1988) or when snakes are active (Bouskila 2001).

Evidence that small nocturnal mammals suffer higher rates of predation under illumination is reasonably strong. J. A. Clarke (1983) showed experimentally that first, captive deer mice restrict their activity to cover under full moonlight, and second, short-eared owls are better able to search for, chase, and capture deer mice under brighter moonlight, mainly due to shorter search times. The importance of increased activity in augmenting predation has also been demonstrated in field situations: in Merriam's kangaroo rats, more-mobile individuals (as determined from radio fixes) were more likely to be killed by predators than their less-mobile counterparts (Daly et al. 1990; see also Smallwood 1989).

A surprising number of bird species are nocturnal; for instance, members of eight orders (27 families) of colonial waterbirds are strictly or regularly active at night (McNeil, Drapeau, and Pierotti 1993). While some of these orders feed at night, such as the Strigiformes, the case for nocturnality being a predator avoidance mechanism is forcible for many nocturnal seabirds. Anecdotal, correlational, and experimental data all support this view. For example, all storm petrels are nocturnal, except for the wedge-rumped storm petrel that breeds on the Galapagos, where the only predator, the Galapagos owl, is crepuscular and nocturnal (Harris 1969). Although Leach's storm petrel nests in burrows that are immune from gull predation, the adult petrels are subject to intense attacks from slaty-backed gulls when they leave land to forage at sea. In contrast with gulls, adult petrels are strictly nocturnal, even reducing their activity on moonlit nights at those times of the year when predation rate is high (fig. 3.11). The link between activity pattern and variation in light intensity, and the negative association with gull activity is highly suggestive of predator avoidance (see also Daan and Tinbergen 1979). Finally, predatory brown skuas cue in on playback calls of male blue petrels, while nocturnal blue petrels, thin-billed prions, Antarctic prions, and common diving petrels all reduce their vocal activity in response to the playback of a brown skua territorial call (Mougeot and Bretagnolle 2000).

Certain species of mammals also modify their activity in response to predator odor (Kats and Dill 1998). Experimental studies using urine as an indicator of predation risk have established that rodents avoid areas or traps tainted with predator odor (for example, Stoddart 1976; Dickman and Doncaster 1984; Gorman 1984; Jedrzejewski and Jedrzewska 1990; I. Robinson 1990; Dickman 1992; Sweitzer and Berger 1992; Lagos et al. 1995; Barreto and Macdonald 1999; see also Thorson et al. 1998; Blumstein et al. 2002), although there is an emerging debate on this point hinging on positive results being found under laboratory but negative results under field conditions (Wolff and Davis-Born 1997; Jonsson, Koskela, and Mappes 2000; F. Powell and Banks 2004; section 3.7). Rodents also shun odors of species other than carnivores: meadow voles avoid

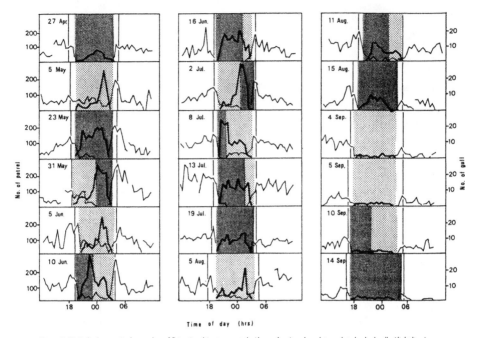

Figure 3.11 Daily changes in the number of flying Leach's storm petrels (*heavy lines*) and predatory slaty-backed gulls (*light lines*) per 5 minutes. Heavily shaded areas represent dark periods, and lightly shaded areas represent moonlit periods; vertical lines indicate sunrise and sunset (Watanuki 1986).

short-tailed shrew odor (Fulk 1972), and the water vole eschews Norway rat odor (Barreto and Macdonald 1999); shrews and rats eat young voles. Avoidance of predator odor may be specific to particular predators (for example, Swihart 1991) or general (for example, Nolte et al. 1994). Predator odor has demonstrable effects on foraging (for example, mountain beavers: Epple et al. 1993) and even on scent-marking behavior (for example, house mice: Roberts et al. 2001; but see Wolff 2004). In a stimulating study, antipredator responses of bank voles to odors of seven predator species were investigated (Jedrzejewski, Rychlik, and Jedrzejewska 1993). Scents of European weasels and red foxes caused a decrease in mobility (see also Borowski 1998) as well as changes in other behavior patterns: avoidance of pens with scent of mammalian predators, quitting artificial burrows, and climbing twigs (table 3.10). Immobility in response to a predator is a way of remaining silent, and weasels and red foxes are known to hunt by sound (Osterholm 1964; Jedrzejewski, Rychlik, and Jedrzejewski 1993). Most interestingly, the number of antipredator responses shown by bank voles increased in proportion to the degree to which each predator specialized on bank voles (fig. 3.12), possibly suggesting an evolutionary arms race between prey and various predator species (chapter 13).

In passing, it may be worth mentioning that avoidance of predator odor is

Table 3.10 Antipredator responses of bank voles to odors of 7 species of predators inhabiting central Europe—summary of experiments

| Predator | Bank vole response to predator odor | | | |
	Reduced mobility	Escape out of borrows	Avoidance of pen visited by predator	Climbing twigs
Weasel	+	+	+	+
Stoat	No	+	+	+
Stone marten	No	+	+	+
Polecat	No	No	+	+
Red fox	+	No	+	No
Raccoon dog	No	No	+	No
Tawny owl	No	No	No	No
Control (rabbit)	No	No	No	No

Source: Jedrzejewski, Rychlik, and Jedrzejewska 1993.
Note: No = no clear response; +, observed strong response.

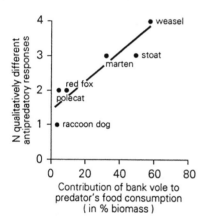

Figure 3.12 Number of qualitatively different antipredator behavior patterns of bank voles to 6 species of mammalian predators plotted against the extent of bank voles in autumn-winter food of these predators in Bialowieza National Park, Poland (Jedrzejewski, Rychlik, and Jedrzejewska 1993).

an area where antipredator behavior has applied significance. Concoctions of predator scents are used to keep wild herbivores such as snowshoe hares, black-tailed deer, and gray squirrels from commercially important crops (T. P. Sullivan and Crump 1984; T. P. Sullivan 1985; F. Rosell 2001).

In contrast with these observations, some species are attracted to predator odor: rodents to the scent of snakes, for example (for example, California ground squirrels: Coss and Owings 1978; Hennessey and Owings 1978; and Formosan squirrels: Tamura 1989; section 9.4). Many snakes are ambush predators, waiting motionless in rodent burrows or behind bushes, and scent may be the only effective means of detecting the threat and allowing prey to monitor the serpent's behavior and movements from then on (Randall, Hatch, and Hekkala 1995).

Finally, some mammals modify activity after hearing a predator. Owls hunt rodents silently, but they do give territorial calls (hoots), often at some distance from where they will hunt; and this may be the only cue by which rodents can detect danger. In response to an owl call, social voles crouch and retreat to experimental burrows, dormice reduce the distance they travel and stay close to the burrow, and lesser Egyptian gerbils crouch and freeze. Cairo spiny mice ignore owl calls—perhaps because they live in rocky habitats and are safer from aerial predation? Interestingly, some individual voles reduce activity, but others increase it in response to owl calls despite exhibiting equivalent rises in corticosterone concentrations. Classically, freezing or fleeing is thought to be intimately tied to predator proximity, with individuals freezing when the predator is a long way off, fleeing when it is close by, and actively defending itself at very close quarters. In this experiment, call proximity was the same for all the voles, but freezing versus fleeing could be respectively predicted from individuals' lowered versus raised activity levels prior to stimulus presentation. These results suggest that individual differences may be an important factor in reducing activity in response to predation risk (Hendrie, Weiss, and Eilam 1998; Eilam et al. 1999).

Cheetahs modify activity in response to the sound of lions and spotted hyenas, both of which prey on cheetah cubs. Cheetahs move away from playbacks of hyena whoops and lion roars and are unlikely to instigate hunting (Durant 2000a, 2000b). As cheetah females age the reproductively successful ones alter their responses to playbacks of spotted hyenas, although not to lions (the hyena is a smaller carnivore than a lion and a less dangerous threat). Thus, young cheetah females (< 4 years) that had raised a greater number of surviving cubs per year spent a greater proportion of time moving and moved a greater distance following the playback of a lion or spotted hyena. Of older females (> 8 years), however, the ones that had been more successful in reproductive terms spent less time moving and moved only a small distance in response to hyena playbacks. In contrast, they moved for longer and greater distances when lion roars were played. Durant argues that females switch from avoidance of hyenas to direct confrontation as they grow older and more successful but cannot afford to switch with more dangerous lions.

As described in chapter 1, there is some evidence that recognition of indirect cues provided by predators is innate. Field voles from the Isles of Orkney avoid the odor of the red fox, a predator that has never occurred on the archipelago (Calder and Gorman 1991). Similarly, short-tailed voles and black-tailed deer avoid odors of a bizarre collection of large, exotic carnivores that they could never have personally encountered (Muller-Schwarze 1972; Stoddart 1982; but see Epple et al. 1993). Nonetheless, the cheetah data indicate that avoidance of indirect (nonvisual) cues provided by predators additionally alters with both age and reproductive experience.

3.5.a Hiding in ungulates   One trait that has attracted particular attention in relation to reducing the probability of being detected by predators is hiding in neonate ungulates. Early work suggested that ungulates could be dichotomized into species in which neonates initially stay close to their mother ("follower species"), such as caribou and wildebeest, and those that spend the first part of their lives some distance from their mother and bedded down ("hider species"), such as Reeves's muntjac and waterbuck (Lent 1974; Leuthold 1977; Walther 1984). Behavioral differences are reflected in differences in musculature, with follower species showing greater muscle mass at birth (Grand 1991). Subsequent behavioral observations of species in captivity have shown that there are intermediates between these extremes (Ralls, Kranz, and Lundrigan 1986; Ralls, Lundrigan, and Kranz 1987; Green and Rothstein 1993; Bowyer, Kie, and Van Ballenberghe 1998).

Although there are several hypotheses regarding the adaptive significance of hiding, including protection from intraspecific aggression and energy conservation (see Carl and Robbins 1988), reduction of predation risk has received most support. First, whereas follower species generally have short birth seasons when the sudden appearance of many neonates swamps predation attempts (section 8.5), birth seasons in hider species are more variable and prolonged, as would be expected if young are avoiding encountering predators through crypsis, a strategy that would be enhanced at low neonate densities (Rutberg 1987). Second, there is a very strong association between ungulates with spotted or striped young and species exhibiting a hiding strategy after controlling for phylogeny. Indeed, 40 out of 40 species with spotted young hide them (fig. 2.5). Given support for predictions about dappled coats in adult mammals acting as camouflage, it seems probable that hiding in combination with being spotted aids in concealment (Stoner, Caro, and Graham 2003). Third, hiding is associated with alarm bradycardia, a condition in which heart rate declines by an average of 38%, at least in white-tailed deer fawns (Jacobsen 1979); and alarm bradycardia is a very predictable response to wolf howls in these fawns under 45 days of age (Moen et al. 1978; see also Gabrielsen, Blix, and Ursin 1985). Fourth, hider mothers exhibit behavior that suggests they are attempting to minimize predation risk. Pronghorn mothers maintain distances from their fawns at which it would be more profitable (based on energy calculations) for a coyote to search for other prey than systematically search around the mother. Also, mothers are no more likely to perform particular actions near their fawn than far from it, a practice that might reveal the where-

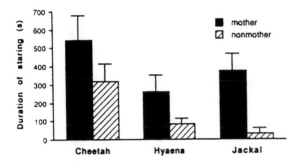

Figure 3.13 Mean (and SE) duration for which mother Thomson's gazelles with hidden fawns stared at 3 different predator species on the Serengeti Plains, Tanzania, compared with nonmothers in the same group (from FitzGibbon 1993).

Table 3.11 Number of Thomson's gazelle fawns found by cheetahs when in hiding and when active compared with the number predicted from the proportion of time fawns of each age class spent in hiding

Age		In hiding	Active	Total
1–2 weeks	Observed	9 (36%)	16 (64%)	25
	Predicted	19 (76%)	6 (24%)	25
3–8 weeks	Observed	5 (15%)	29 (85%)	25
	Predicted	19 (56%)	15 (44%)	34

Source: FitzGibbon 1990a.
$X^2 = 22.00$, df $= 1$, $P < 0.001$ for 1–2 weeks and $X^2 = 23.42$, df $= 1$, $P < 0.001$ for 3–8 weeks.

abouts of their offspring (Byers and Byers 1983). In addition, Thomson's gazelle mothers with hidden fawns spend more time being vigilant when they detect a predator than do nonmothers, and they delay returning to a hidden fawn (fig. 3.13), thereby reducing the advantages of a predator waiting for a fawn to stand to nurse (FitzGibbon 1993). Nonetheless, all these maternal behaviors might be expected whatever the function of hiding. Crucially, therefore, the most important piece of evidence is that Thomson's gazelle fawns that lay hidden were hunted by cheetahs less often than expected based on the proportion of time that fawns spent in hiding (table 3.11). Moreover, this resulted in an estimated 2.6% increase in fawn survivorship per year (FitzGibbon and Lazarus 1995).

3.6 Changes in foraging under risk of predation

In species in which animals flee to a refuge when alarmed but feed elsewhere, foraging is a perilous business. To estimate relative exposure to predation in a mammal, Blumstein (1998) multiplied time allocated to vigilance, foraging, self-grooming, and play by the distances from a refuge at which golden marmots performed these activities; by the relative differences in response time to the playback of an alarm call when performing each behavior; and finally, by the behavior-specific time it takes to return to a physical refuge if alarmed. He calculated that foraging scored 1.20 on an index of estimated risk and was far riskier than vigilance (0.70), grooming (0.01), or play (0.01), chiefly due to the time actually spent foraging. We might therefore expect ground-dwelling mammals and other species that use refuges to have sophisticated ways of ameliorating threat of predation while feeding, including reducing risk of detection.

The way in which mammals and other animals alter their feeding behavior under risk of predation constitutes a large and well-understood area of behavioral ecology both theoretically and empirically (for example, Dill and Fraser 1984; Houston, McNamara, and Hutchinson 1993; Sih and McCarthy 2002) and can, for convenience, be placed under four headings of When, Where, What, and How Much to Eat (Valone and Lima 1987; Lima and Dill 1990). An important methodological breakthrough in these studies is the measure of giving-up density (GUD) (J. S. Brown 1988; Valone and Brown 1989). The GUD is the resource density at which a forager ceases to feed in a resource patch. Its central tenet is that the forager stops feeding in a patch when the marginal benefit of harvesting no longer outweighs the marginal cost of foraging. Thus, the GUD reflects the harvest rate (H) that the forager enjoyed at the point when costs of foraging outweighed the benefits. Foraging costs include the energetic cost of foraging (C), predation risk (P), and missed-opportunity costs, (MOC) such that

$$H = C + P + MOC$$

If two patches share the same harvest-rate function and same energetic cost of foraging, then differences in quitting-harvest rates (measured as GUDs) should reflect differences in patch-specific risks of predation. GUDs in rodents can be measured in the field or laboratory using trays filled with sand into which seeds of different densities or size are mixed, and can be used in models of decision rules as to when to give up foraging under different sorts of predation risk (J. S. Brown 2000). GUDs are sensitive indicators of predation risk and foraging efficiency that vary according to type of predator, sort of food, and type of prey species, and the interactions between these factors (fig. 3.14).

The way in which animals perceive risk of predation can also be judged by the way in which they alter their feeding behavior at a fine-grained level. Operationally, seed-eating rodents can be offered seeds mixed throughout a tray of sand, half the number of seeds mixed throughout the central half of the tray only, or half the number of seeds mixed in the bottom half of the tray only. As apprehension increases, attention will be redirected from exploiting the seed trays to listening and watching for predators (Dall, Kotler, and Bouskila 2001; Abramsky, Rosenzweig, and Subach 2002). Therefore, the subject should be

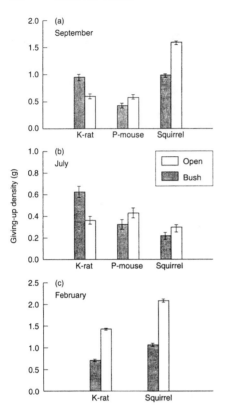

Figure 3.14 Giving-up densities (GUDs) of Merriam's kangaroo rat (*K-rat*), Arizona pocket mouse (*P-mouse*), and round-tailed ground squirrel (*Squirrel*) during different seasons of the year and by microhabitat (bush and open). Food patches consisted of metal trays filled with 3 g of millet seeds mixed into 3 L of sifted soil. GUDs were generally higher in the open microhabitat. In July, however, kangaroo rats indicated a higher perceived predation risk in the bush microhabitat, probably as a result of predation risk from rattlesnakes. GUDs are also a measure of foraging efficiency. Also, a low GUD provides one measure of interspecific competitive ability. There is a seasonal rotation in which species has the lowest GUD: pocket mice in September, squirrels in July, and kangaroo rats in February (J. S. Brown 1989).

less able to distinguish boundaries of micropatches and less able to take advantage of seeds concentrated in the bottom of the patch. An apprehensive animal should thus exhibit less selectivity with the half-filled trays compared with the full tray (Schmidt and Brown 1996).

3.6.a When to eat   Nocturnal foraging activity declines with moonlight, as well as activity and use of open habitat, in many rodents (for example, Kotler 1984b; Brown and Alkon 1990; section 3.5). For example, dune hairy-footed gerbils in the Namib Desert have higher GUDs at full moon than at new moon (that is, trays were depleted less in moonlight) and have higher GUDs on seed trays far from, as opposed to near, cover (J. J. Hughes and Ward 1993). Thus, a fewer number of distant trays are visited at full moon than at new moon (Kotler et al. 2002). Since the moon rises and sets at different times each night, one might expect the timing of foraging to alter accordingly. When the moon rises late, over the new moon or when it is waning, Allenby's gerbil and the greater Egyptian sand gerbil forage most intently during the first five hours of the night, and are most apprehensive then. During full moon or nights of waxing, foraging effort occurs throughout the night and at low intensities, as measured by GUDs or by levels of apprehension (Kotler, Ayal, and Subach 1994; Kotler et al. 2002).

In interspecific comparisons, species that are subject to high predation risk begin to forage late in the evening unless their food source forces them to start early. Bat species that feed on relatively immobile prey such as flightless arthropods or fruit emerge later in the evening than other chiroptera that feed on small aerial insects that come out at dusk. While cause and effect are difficult to untangle, aerial feeding chiropterans have high wing loadings and are therefore fast fliers, enabling them to catch insects on the wing and escape predation (G. Jones and Rydell 1994).

Similar effects of predation risk in choosing when to forage are seen in birds. Lima (1986) put forward the optimal body-mass hypothesis, which showed, through modeling, that the body mass maintained by wintering birds reflects a trade-off between the probability of surviving periods when food is unavailable and predation risk, because foraging exposes birds to predators and heavier birds are less able to escape predatory attack than lighter birds (section 12.7). The hypothesis predicts that with increasing resource predictability or increasing predation risk, fat reserves should decline (McNamara and Houston 1990). The strength of Lima's model is that it shows that body mass is the outcome of many factors acting simultaneously: mass decreasing with increasing predation risk and increasing temperature (because less energy is required to keep warm), and mass increasing in bad weather as food becomes less predictable, and also increasing with greater food abundance.

Energy reserves in birds show great variation, being depleted overnight and replenished during the day, so birds are hungriest in the early hours of the morning. We would therefore expect birds to feed a great deal early in the day, as their probability of surviving any shortage of food is lowest them. Unfortunately, dawn is also a time when nocturnal and diurnal predators could both be hunting, and predatory attacks may be difficult to detect in the gloom. What should a bird do? Several factors affect the onset of feeding. Most obviously, birds exposed to predation risk postpone regaining weight losses until the end of the day (van der Veen and Sivars 2000). Other detailed studies show that dark-eyed juncos arrive and initiate feeding at lower light intensities when food is available in safe cover than away from cover, and they start feeding earlier on clear days than on cloudy days, both of which speak to the importance of being able to detect and evade predators (Lima 1988a). In great tits, dominant individuals that are heavier than subordinates start to feed later than subordinates in habitats where nocturnal pigmy owls and diurnal sparrow hawks are found, and they stop feeding earlier (Krams 2000). In a habitat where the owls were absent and nighttime predation risk was therefore low, dominants started to feed earlier (actually in darkness) and stopped feeding later than subordinates. Predator avoidance in early light allowed dominants to restore energy reserves before the sparrow hawks began to hunt. As Lima's optimal body-mass model predicts, risk of predation affects individuals' patterns of feeding differentially according to their body mass. Finally, dark-eyed juncos stop feeding at much greater light intensities than those at which they started to feed, relinquishing up to an hour of potential feeding in the evening, presumably because they are now well satiated and are not prepared to incur predation risk associated with the onset of twilight (Lima 1988a).

3.6.b Where to eat   Predation risk also influences where animals are prepared to eat. To illustrate, gray squirrels forage less in patches far from cover (Newman and Caraco 1987); European hedgehogs avoid feeding at sites tainted with feces of European badgers, which prey on them (Ward, Macdonald, and Doncaster 1997); and willow and crested tits forage in safer innermost parts of trees in years when predation risk from pigmy owls is high (Suhonen 1993). Cowlishaw (1997a) investigated how populations of yellow baboons traded off foraging against predation risk. He quantified food availability in four habitats by recording stem density, plant part availability and plant protein and energy per unit area. Attack risk was determined on the basis of habitat visibility and predator ambush distances, whereas capture risk was derived from predator and prey travel velocities, habitat visibility, and nearest refuge distance. Baboon behavior was assessed through all-day follows. Baboons spent less time feeding in high-risk, food-rich habitats but more time feeding in low-risk, relatively food-poor habitats, and they restricted resting and grooming activities to the safest area, where key foods were absent. The arresting point about this study is Cowlishaw's attempt to stratify predation risk by habitat. He recognized two subcomponents, risk of attack and risk of capture, but dismissed risk of detection as being unimportant, since terrestrial baboons do so little to disguise their whereabouts. He argued that because attacks by lions and leopards were more successful at short ambush distances, the important variable for risk of attack was the frequency with which visibility fell below a predator's critical ambush distance (10 m for leopards and 25 m for lions) in different habitats. He calculated capture risk by measuring median distance to a refuge but included relative flight velocities of both prey (v_{prey}) and different predators (v_{pred}) and the maximum distance at which a predator could be detected (d_{vis}). Thus, prey must remain within a maximum distance from a refuge, R_{max}, where

$$R_{max} < d_{vis}([v_{pred}/\{v_{pred}-v_{prey}\}] - 1)$$

Risk of capture by a given predator can therefore be estimated for each habitat by the average maximum visibility in that habitat. Thus, risk of attack and capture can be calculated separately using explicit assumptions.

Predation risk also influences the extent to which animals will carry food items to protective cover. Lima (1985) and Lima, Valone, and Caraco (1985) developed a model predicting that the tendency to carry a food item to cover should decrease with the distance to cover (travel time) but increase with item size (handling time). In the absence of predation risk, energy intake is maximized by eating food items in the patch in which they are found, but if predators abound, eating food in situ may expose individuals to predation. When handling time (h) is less or equal to twice the travel time to cover (t), there is no conflict about where to consume food, and no carrying should occur. When $h > 2t$, however, a conflict develops, because carrying food items minimizes exposure time but minimizes intake rate, whereas staying to eat at the patch maximizes intake rate but also exposure time. Two predictions therefore arise: (i) holding food item size constant, the proportion of items carried should decline as distance to cover (travel time) increases, and (ii) holding distance constant, the proportion carried should increase as item size (handling time) increases. Experiments with both black-capped chickadees and gray squirrels, in circumstances in which $h > 2t$, supported these predictions. In a situation in which chocolate-chip cookies were cut into 1, 2, or 3 g pieces and placed at varying distances from cover (a stand of mature oaks, where predation risk was assumed to be low), gray squirrels spent a greater amount of time handling cookies the larger the size of the item, and less time handling items the further they were to cover. These results are not surprising. Crucially, however, a greater proportion of cookies were carried back to a tree to be eaten the larger they were and the closer they were to cover (fig. 3.15), confirming the model's predictions (see also Newman et al. 1988). Many other mammal species carry food items to cover, including rodents (for example, Kimberly rock rats: Begg and Dunlop 1980; Darwin's leaf-eared mouse: Vasquez 1994) and carnivores (for example, cheetahs: Caro 1994a), but the trade-off between foraging efficiency and predation risk is yet to be quantified in these species.

Valone and Lima (1987) tried to replicate their gray squirrel study with 10 species of birds, using small and large seeds placed at three distances from cover. For four species, the proportion of seeds carried decreased with increasing distance from cover and increased with increasing item size. For four other species, though, the proportion of large food items carried did not increase with increasing distance from cover. For cardinals and house sparrows, very few items were carried to cover at all. In short, the trade-off between foraging efficiency and predation risk failed to predict adequately patterns of feeding in all the species of birds. While the tendency to carry items increased with item size in 9 out of 10 species, the tendency to carry declined with increasing distance to cover in only 4 out of 10. One possible explanation may be that a minority of species were trying to minimize time spent vulnerable to

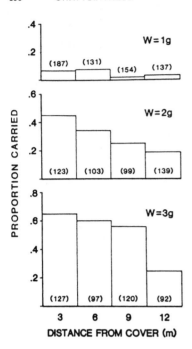

Figure 3.15 Proportion of food items carried by solitarily foraging gray squirrels for 3 weight and 4 distance combinations (Lima, Valone, and Caraco 1985).

attack, whereas the majority were relying on escaping should an attack occur and accepted that any time spent in the open was dangerous. If this is true, then a given amount of time spent handling an item cannot be equated with the same amount of time spent transporting a food item. Another idea is that attack risk does not increase linearly with distance from cover for every species, so different species perceive a given distance from safety as differentially risky. Yet another possibility is that cover is a mixed blessing, with some species or individuals perceiving it as safe, whereas other individuals see it as a source of attack (section 5.5). Thus, birds that passed through cover before feeding foraged closer to it, having inspected it for predators (Lima, Wiebe, and Dill 1987). Whatever the case, future models need to incorporate energy gained from foraging and risk of predation into a single currency of fitness.

Rodents pay greater attention to indirect cues from the environment than to direct cues left by a predator when deciding when and where to forage. When direct cues, namely urine of carnivore species, are pitted against indirect cues, that is, microhabitat differences such as feeding far from vegetative cover, Eastern fox squirrels, thirteen-lined ground squirrels, and oldfield mice each increase their GUD in response to habitat but not olfactory cues (Thorson et al. 1998; Orrock, Danielson, and Brinkerhoff 2004). The study of old-field mice is particularly instructive, because mice removed less seeds from trays that were exposed, when it was not raining, and on moonlit nights than

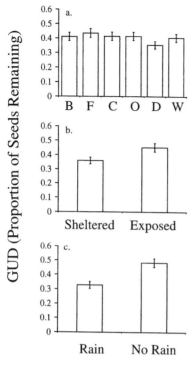

Figure 3.16 The mean proportion of seeds (+ SE) left in foraging trays by oldfield mice at Savannah River, South Carolina, USA. (*a*) There was no significant difference in the proportion of seeds removed in response to the scent of native predators (*B*, bobcat; *F*, fox), recently introduced predators (*C*, coyote), nonnative predators (*O*, ocelot), a native herbivore (*D*, white-tailed deer), and a control (*W*, water). (*b*) Oldfield mice removed significantly more seeds from trays located in microhabitats sheltered beneath vegetative cover (*Sheltered*) than adjacent trays less than 1 m away in microhabitats that were not under cover (*Exposed*). (*c*) More seeds were removed from trays during nights with rain compared with nights when no precipitation occurred (Orrock, Danielson, and Brinkerhoff 2004).

GUD (Proportion of Seeds Remaining)

in other circumstances, but they did not alter their behavior in response to evolutionarily novel or familiar carnivore odor (fig. 3.16). Perhaps indirect cues reflect predation risk from many potential predators, whereas direct cues only provide information about one or a few individuals of a single species.

3.6.c What to eat   Predation can alter diet selectivity, forcing individuals under risk to pick a narrower array of more-nutritious food items. In experiments involving different sorts of food, rodents become more choosy in situations of heightened predation risk brought on by increasing distance from cover (Hay and Fuller 1981) or full moonlight (Bowers 1988). Lima and Valone (1986) predicted that in circumstances in which carrying food items to cover is possible, small items may be rejected in favor of large, less easily handled items, because increasing item size tends to enhance the benefits of carrying. Thus, the relative value of a food item is a function of its distance from cover, energy content, and handling time. In quantitative support of the hypothesis, gray squirrels never carry small cookie items to cover, and they are more likely to reject small items and carry large items as item size increases and distance to cover decreases. There are few parallel data on diet selectivity in ungulates, but field observations of moose reveal that cows with calves ate significantly more herbs of low preference and fewer shrubs of high preference when they confined their movements to islands on Lake Superior that were free from wolves (Edwards 1983).

3.6.d How much to eat   Heightened predation risk reduces the amount of food consumed, a predictable consequence of curtailing time available for eating (Abrahams and Dill 1989). In Darwin's leaf-eared mouse, heightened predation risk in the laboratory brought about by increased nighttime illumination (to mimic a full moon) resulted in fewer sunflower seeds being consumed. Compared with dark nights, mice had 50% higher GUDs and a 14.5% reduction in total seed consumption, corresponding to a sacrifice of 17.0 kJ of potential energy. Although they lost weight during both dark and bright nights, they lost 5% more weight on "full-moon" than moonless nights (Vasquez 1994).

That predation influences foraging efficiency is without question, but the way in which it is predicted to do so is often open to post-hoc justification. If it is paramount that prey detect a predator early on, then prey must maintain a given level of vigilance when foraging in the open; here, heightened predation risk will increase handling time and thereby lower foraging efficiency (Krebs 1980). Similarly, a forager may select less profitable food items that require reduced handling time under these circumstances (Lima 1988b). If, however, a prey animal wants to minimize time spent foraging in an open area, perhaps because it has little chance of escaping if a predator does arrive, it will increase feeding rates or possibly food carrying under heightened predation risk (Newman et al. 1988). Without reasonable knowledge of other antipredator defenses in a prey species' repertoire, it is difficult to make foraging predictions with confidence.

3.6.e Effects of age and reproductive condition on risk-sensitive foraging   Some juvenile mammals are more risk sensitive than adults (as found in many other taxa: Stein and Magnuson 1976; Sih 1982; K. A. Sullivan 1988). Juvenile North American porcupines forage primarily in relatively safe buffalo-berry and willow groves, whereas 2-year-olds use risky areas more often, and adults forage most frequently in open habitats (fig. 3.17). In both hoary marmots (Holmes 1984) and collared pikas (Holmes 1991), juveniles forage closer to a talus, piles of rock where they can seek refuge, than do adults. Young marmots appear more vulnerable than adults because they are slower to detect a threat and take longer to reach safety, forcing them to remain close to a talus, where food is often depleted due to intense grazing (Huntly 1987). Although differential predation risk seems key to explaining juvenile hoary marmot patch use, age-specific variation in the closely related yellow-bellied marmot can be explained by nutrient demands of the growing animal and the small window of time for weight gain before hibernation (Carey 1985).

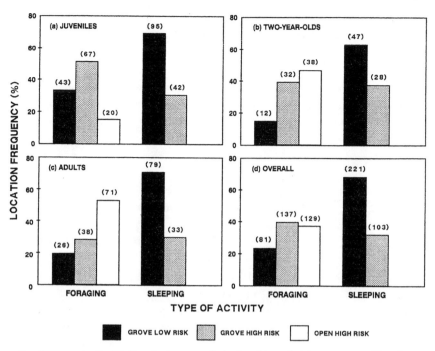

Figure 3.17 Frequencies with which different locations were used by foraging and sleeping North American porcupines of 3 age classes and all individuals together in the Granite Range, Nevada, USA (Sweitzer and Berger 1992).

Female mammals accept greater risks of predation when they are reproductive. Compared with nonparous females, Columbian ground squirrels spend more time feeding aboveground during lactation (MacWhirter 1991), collared pikas forage farther from the talus when pregnant (Holmes 1991), and bighorn sheep trade safety for access to more nutritious food during the last month of pregnancy (Berger 1991). These trade-offs are seen as a means to increase food intake to supply the growing fetus or for milk production.

For mammals in general, studies of foraging have advanced further than any other area of risk sensitivity, and researchers have investigated the perception of risk using indirect cues such as illumination and cover, species' differences in their perception of risk, and even the effects of different avian and terrestrial predator species on feeding (Kotler, Brown, and Hasson 1991). Using GUDs, it is possible to quantify how risk influences classical optimal foraging models (J. S. Brown 1992) and to determine the extra resources needed to produce a switch to a more hazardous food site (Gilliam and Fraser 1987). Last, rodent species' differences in risk sensitivity are thought to be an important influence on community structure. For mammals, there are at least three lacunae in this active field: an inability to translate foraging decisions into fitness payoffs in mammals (as attempted for other groups: Abrahams and

Dill 1989; Nonacs and Dill 1990; but see Kotler 1997), a delay in replicating rodent findings to areas outside desert communities, and difficulties in transferring methods to other orders, such as ungulates.

3.7 Changes in reproduction under risk of predation

Many species show changes in reproductive behavior under risk of predation (Sih 1995), but evidence for this in mammals is under active debate. The argument runs as follows. In northern Fennoscandinavia, where densities of microtine rodents and their main predators, small mustelids, fluctuate synchronously in 3- to 5-year cycles, the population crash is usually shorter than the maximum reproductive life of an individual field vole. In southern Scandinavia and central Europe, microtine population dynamics are less cyclic. It might therefore pay individuals from cyclic populations to limit their reproduction, and there is a certain amount of evidence for this. For instance, bank vole males from cyclic populations grow more slowly than males from noncyclic areas (Heikkila et al. 1993). Ylonen (1994; Ylonen and Ronkainen 1994) has argued that it will pay microtines to delay reproducing in cyclic populations until after the crash, when specialist predator numbers and intraspecific competition are reduced. In support of this, research has shown that when field voles are exposed to predator odor, no copulatory behavior is seen, and there is a tendency for males to be more aggressive and avoid females when compared with controls (Koskela and Ylonen 1995). In bank voles, where much of the work has been conducted, females refuse to copulate with males (whose behavior does not change) compared with controls (Ronkainen and Ylonen 1994). Furthermore, females under risk of predation have abnormally long and fewer estrus cycles (Koskela et al. 1996), possibly induced by lowered food intake (Ylonen and Ronkainen 1994). As nonreceptive females will not permit males to mount, absence of copulation could be due to anestrus.

Additional mechanisms to limit reproduction apparently exist: sexual maturation of young red-backed voles is delayed rather than hastened in the presence of mustelid odor, suggesting that rapid sexual maturation has predation costs (Ylonen et al. 1992; Heikkila et al. 1993). Certainly, estrus deer mice are more vulnerable to predation than nonestrus females, probably because estrus females are more active and also because least weasels can discriminate and select estrus over diestrus rodent scents (Cushing 1985). Finally, under risk of predation, litter weights are reduced in red-backed voles (Heikkila et al. 1993), and litter sizes are smaller in bank voles (Korpimaki, Norrdahl, and Valkama 1994; see also Slagsvold 1984a).

Unfortunately, almost all the empirical studies in support of suppressed reproduction come from laboratory studies and have not been replicated in the field (Wolff and Davis-Born 1997; T. Mappes, Koskela, and Ylonen 1998; Jonsson, Koskela, and Mappes 2000; see also Parsons and Bondrup-Nielsen 1996; Herman and Valone 2000 for parallel concerns about foraging). Laudatory re-examination of the laboratory data by some of the very authors who generated them has raised the possibility that laboratory animals might be extremely neophobic to any strange odor and that they consequently lower their activity and feeding; this, in turn, reduces energy reserves and fecundity of females. Moreover, the whole edifice may be suspect: T. Mappes, Koskela, and Ylonen (1998) suggest that the underlying assumptions of the breeding suppression hypothesis, namely reproduction having survival costs in female voles which can be reduced through suppression, high probability of survival to the next breeding season, and high predation risk that will decline before the next breeding attempt, are assumptions that need verification.

In short, while reproductive suppression under risk of predation remains a growth area in antipredator research (for example, Ruxton and Lima 1997), laboratory studies need revisiting, and explanations as to the distribution of different types of predator-sensitive reproductive suppression across species, the presence of more than one mechanism in some species, and sensitivity to particular predators but not others (Klemola, Korpimaki, and Norrdahl 1998) all require attention.

In conclusion, since the indirect fitness costs of predation risk principally mediated by limited opportunities for feeding but also for reproduction were made explicit by Lima and Dill (1990), the explosion in interest in this topic has centered on the extent to which predation alters aspects of feeding and re-production, and consequently fitness and community composition. Rather less attention has been paid to the antipredator benefits of altered behavior under threat of predation, the tack chosen in this chapter; but, that said, pred-ator avoidance must be the most common form of antipredator defense in an-imals. While such a view stems from logic rather than empirical data, it is based on the supposition that animals encounter dangerous situations almost daily but encounter predators far more rarely (Thorson et al. 1998).

Instead of trying to synthesize such a wide body of literature (see Lima 1998), I have highlighted areas where predator avoidance might influence fitness, focusing first on young (nest site selection) and then on adult homeo-therms. In altricial birds, nest placement must be under acute selection pres-sure, because a nest and its contents cannot be moved in response to subse-quent predator arrival. In most birds, judicious nest site selection, surreptitious behavior at nests, and nest defense (chapter 10) are the only antipredator de-fense options available to parents in keeping predators from their young. The constraint applies less forcefully to precocial birds and to mammals, some of which can move after birth, and others of which can carry neonates limited distances.

In contrast, adult birds may need to rely less on behavioral mechanisms to avoid being detected by predators than adult mammals, because birds can fly away from terrestrial mammals and snakes. Only in relation to aerial predators does it seem that birds show risk-averse foraging behavior. Rather, it is among mammals where studies have generated so many empirical examples of predator-sensitive behaviors that include refuge use, changes in activity, and foraging behavior. As demonstrated in the rodent studies of Brown and Kotler and others, rodents living in open, two-dimensional habitats are subject to predation from many predators, including owls, carnivores, and snakes.

Risk-sensitive behavior can be observed best in open desert rather than in closed forests, so these are the environments from which we have good data. Thus, we cannot yet hazard guesses as to whether behavioral mechanisms to avoid detection, such as nest site selection or changes in reproductive activity, are more or less restricted to certain ecological contexts than to others. Thus, we do not understand the environmental variables (habitat complexity and predator community) that promote the evolution of sophisticated and keen predator avoidance mechanisms in homeotherms or other animals.

3.8 Summary

Animals try to avoid encountering predators by responding to indirect cues associated with predation or to direct chemical or auditory cues given by predators themselves. Principally, prey may shift microhabitats, reduce activity, curtail foraging, and suppress reproduction.

Predation on eggs and nestlings is an extremely important source of avian mortality, and birds attempt to conceal their nest by choosing a particular habitat, avoiding forest edges or small habitat patches, by building the nest in thick vegetation, constructing it at a certain height, taking account of the proximity of other nests, and avoiding placing the nests where nest densities are high. Some birds even nest close to biting and stinging insect colonies. The relative importance of these nesting decisions differs according to how predators in the local community search for and remember location of nest sites. Birds lower the likelihood of visual predators locating their nest by reducing the number of visits that they make to the nest. By nesting at a new location following the loss of a clutch to predators, they can increase their chances of remaining undiscovered.

By retreating to physical structures or constructing their own refuge, animals can reduce predation risk when predators are active. Thus, some mammals sleep in trees, others raise offspring in dens, and others spend much of their life in refuges such as burrows, all of which likely help reduce probability of detection by predators. Animals can also limit their activities to certain areas where predation risk is low. Desert rodents move to cover in response to nighttime illumination, but quadrupedal rodents do this to a greater extent than bipedal species that are better at detecting and escaping avian predators. Differential response to predation risk is thought to foster species coexistence in some desert ecosystems. Ungulate mothers seek out safe areas at parturition which may be a form of avoiding detection. Arguments that ungulate populations survive in areas that are free from predators, and that ungulate migrations incur antipredator benefits are controversial.

Animals can also reduce predation risk by reducing activity. Rodents lower activity in response to increased illumination at night, change behavior when exposed to predator odors in laboratory settings, and exhibit conservative behavior on hearing predator calls. Young of certain ungulate species remain quiet and hidden during periods when their mother is absent.

Predation risk alters the timing and location of feeding, and type and amount of food that animals will eat, with individuals often forced to trade off between minimizing predation risk and maximizing food intake. Giving-up densities at seed trays are used as a tool to quantify perceived risk while feeding, and level of apprehension can be measured as the extent to which animals assess the profitability of food in microhabitats. Different species appear to perceive risk of predation in the same habitat in different ways.

Juvenile mammals show a greater willingness to avoid encountering predators than adults, whereas pregnant females will take greater risks to obtain nutritious foods than will nonparous females. Some evidence suggests that animals employ behavioral and physiological mechanisms to forgo reproduction when exposed to predator odors, but this is controversial. In rodents these include anestrus, reticence to copulate, and delayed maturation.

Options to avoid detection by predators are limited to nest site selection in altricial birds, whereas precocial birds and mammals can move young. In contrast, adult birds can fly from nonavian predators, whereas adult mammals are confronted by many predator guilds, which may explain why so many behavioral mechanisms to avoid detection have been documented for mammals.

4 Vigilance and Group Size

4.1 Introduction

If a predator approaches and comes within the prey's visual field, the prey animal may be able to observe the predator at a distance, thereby giving itself the opportunity to move away or hide and even alert others before the predator draws too close. Vigilance is usually defined operationally as raising the head above a certain level off the ground. Antipredator vigilance, often also referred to as scanning for predators, sometimes precludes other activities such as foraging, however. A key way to recoup this lost opportunity for foraging is to be in a group and rely on other group members to take on some of the burden of scanning for predators. In addition, many pairs of eyes may improve predator detection. Unfortunately, relying on others to spot predators opens up the opportunity for relaxing personal watchfulness to a point where no group member maintains vigilance. That this does not occur in groups in nature demands innovative explanation. This chapter examines these issues. First, different measures of vigilance (section 4.2), antipredator benefits (section 4.3), and costs (section 4.4) of vigilance for individual prey animals are described. Next, an influential model of how grouping modifies both individual and group vigilance is presented, and empirical evidence in support of this idea is tabulated (section 4.5). These data confirm that individuals gain antipredator benefits through group vigilance and that individual vigilance declines with increasing group size. Nevertheless, the reasons that individuals relax but do not curtail personal vigilance as group size increases are still not well understood. Five current hypotheses to explain why individuals forgo the temptation to curtail personal vigilance in groups are discussed (section 4.6). Finally, individuals can gain group-related vigilance benefits by joining up with het-

Figure 4.0 (*Facing page*) Yellow-throated sand grouse drinking from temporary rainwater pools. Birds living in or aggregating in groups accrue antipredator benefits from detecting predators at greater distances and, as a result of sharing vigilance, obtain greater opportunities for other activities, such as drinking and feeding (reproduced by kind permission of Sheila Girling).

erospecifics. Despite these potential benefits, patterns of vigilance in mixed-species groups are complex and are influenced by species' asymmetries in opportunities for vigilance and predation risk (section 4.7). Environmental and demographic factors that affect vigilance are addressed in chapter 5.

4.2 Measures of vigilance

Head position is central to most definitions of vigilance. As a case in point, FitzGibbon (1990b) defines a visual scan as starting

> when the focal gazelle lifted its head above shoulder height and to have finished when the gazelle lowered its head and returned to feeding. (p. 1117)

Similarly, gramnivorous birds scan for very short periods of time using "head-cocks" between bouts of pecking at seeds. Vigilance in small birds is often measured using number or duration of interscan intervals, that is, the periods of time between scanning events (K. A. Sullivan 1984). Animals can alter either the duration or the rate of scanning behavior. For example, in response to increased cover that may conceal a predator, ruddy turnstones increase the duration of scans, whereas purple sandpipers increase the frequency of scans (Metcalfe 1984a). G. Roberts (1995) found that great crested terns increased their interscan intervals but not the duration of their scans as flock size increased, whereas Studd, Montgomerie, and Robertson (1983) discerned that house sparrows reduced their scan bouts but not scan rate in response to increases in group size. Through modeling, the extent to which animals alter their vigilance in response to social and environmental variables can be shown to be contingent on the predator's behavior. If predators attack at random, constant interscan intervals will minimize predation risk. If, however, predators modulate their attacks to start when interscan intervals begin (Hart and Lendrem 1984), prey should choose variable interscan intervals; but now they are vulnerable to predators that wait for long intervals before attacking (Scannell, Roberts, and Lazarus 2001). In prey species that need a long handling time for food items, increasing the length of scans might be necessary, because frequent scanning could impede food processing (Metcalfe 1984a). Other, more mechanistic interpretations of vigilance are occasionally used (for example, Lawrence 1985a).

Prey animals can assume different positional levels of vigilance. For instance, squirrels can remain on all fours but lift their head from the ground, sit upright on their haunches, or climb a rock and look around. Bipedal vigilance probably enables them to spot both terrestrial and avian predators at a greater distance than when they are in a quadrupedal stance (Schooley, Sharpe, and van Horne 1996). This idea is supported by data obtained by observing that Columbian ground squirrels climb rocks more frequently after an alarm call has been given, although alternative explanations are possible (MacHutchon and

Harestad 1990). Prey might also use vigilance to scan for different types of predators; perhaps they have something analogous to a search image (section 2.5) for a particular type or species of predator, although our knowledge of this possibility is currently limited to gross categories of aerial or terrestrial threats (Arenz and Leger 1999a).

For the sake of convenience, it is assumed that birds and mammals are not vigilant when involved in activities such as feeding, preening, or moving (Coolen and Giraldeau 2003). For example, gray squirrels and eastern chipmunks both spend more time pausing when moving away from safe forest cover than when returning to it (McAdam and Kramer 1998; Trouilloud, Delisle, and Kramer 2004), suggesting that their ability to detect predators is impeded during locomotion. Nonetheless, this must be an unrealistic assumption, most obviously because animals can observe their surroundings when they fly or chew cud (Frid 1997), or when birds husk seeds with their bill (Gluck 1987). Waders that pause and travel while foraging show lower levels of vigilance than species continuously searching for food with their head down, presumably because the former can scan as they move forward (Barbosa 1995). Indeed, quite recently, the whole premise that vigilance and other activities are incompatible has received repeated questioning.

First, it is now known that the visual field of many bird species extends a long way around the back of the head, enabling them to see in several directions while feeding. Thus, northern shovelers that have a greater binocular field width than wigeon spend less time in a head-up posture while dabbling than do wigeon while grazing (Guillemain, Martin, and Fritz 2002; Fernandez-Juricic, Erichsen, and Kacelnik 2004), and European starlings spend more time scanning if their vision is occluded when they have their head down foraging (Fernandez-Juricic, Smith, and Kacelnik 2005). Second, careful field observations indicate that there is a tradeoff between time spent feeding and the ratio of food handling that is compatible with vigilance (H_c) to time spent searching for food (S). For foraging bouts of samango monkeys feeding on different types of food, Cowlishaw and colleagues (2004) recorded the time spent feeding, search time between bites, and proportion of handling time that was compatible and incompatible with vigilance by noting whether monkeys were observing their surroundings while chewing. This allowed them to calculate different H_c/S ratios for feeding on berries and flowers of different plant species. Now they found that the relationship between feeding rate and proportion of time spent scanning or median scan duration was far shallower when $H_c/S > 1$ than when $H_c/S < 1$. In short, the classic feeding/vigilance tradeoff is modified by the extent to which food handling requires visual attention, raising the possibility that animals seek out particular sorts of food that are compatible with simultaneously checking their surroundings. Third, in experiments in which mounted hawks were "flown" toward feeding dark-eyed juncos or blue tits, birds with their head down could still detect the model (Lima and Bednekoff 1999b; Kaby and Lind 2003). With chaffinches tested using the same protocol, individuals that were good foragers were also good at detecting an approaching model sparrow hawk (Cresswell et al. 2003). This was because every time that a bird found a seed it raised its head to handle it, allowing it to detect approaching danger rapidly. In all three species under experimental conditions, feeding and vigilance were not mutually exclusive. Moreover, the type of feeding task affected detection and takeoff latencies in blue tits, with foraging on cut-up mealworms not affecting these responses but foraging on whole mealworms slowing them considerably (Kaby and Lind 2003; see also Lawrence 1985b; Dukas and Clark 1995; Dukas and Kamil 2000). Given the extent to which vigilance and feeding are compatible varies according to food type and presumably mode of predator approach and habitat, interpretation of several models of vigilance (Beauchamp 2001 and below) needs urgent revisiting (see Lima 1988b), and this is already starting (Fernandez-Juricic et al. 2004). For instance, it would be instructive to match the shape of well-established vigilance-feeding relationships to categories of food type across species.

Conventionally, decisions that animals make regarding patch use (chapter 3) and vigilance (chapters 4 and 5) are treated separately, in part for historical reasons, because harvesting rules are an ecological concept and vigilance is a behavioral measure. Nonetheless, both are complementary and alternative ways of managing predation risk, and each should be influenced by changes in threat, although in general, changes in time allocated to foraging will be negatively associated with vigilance. A few attempts have been made to link aspects of patch use and vigilance (for example, Repasky 1996; Ranta et al. 1998). For example, J. S. Brown (1999) modeled links between both time devoted to harvesting food patches differing in predation risk and feeding rates there, and vigilance while feeding in patches. He found that if he increased the encounter rate with predators or increased predator lethality, it caused GUDs (giving-up densities; see section 3.6), harvest quitting rates, and vigilance rates to rise, whereas if he increased the marginal energetic value of food in patches, these measures decreased. When the effectiveness of vigilance for spotting predators was increased, vigilance was found to rise, but GUDs and harvest quitting rates declined. Such models explore the relationships between vigilance and habitat or patch use rather than between vigilance and time spent foraging, as is more common.

4.3 Benefits of individual vigilance

Although animals look at their surroundings for many different reasons, the principal function of vigilance is assumed to be enhanced predator detection.

Evidence that vigilance increases the probability that an individual will detect a predator comes from the laboratory, from fieldwork, from observations, and from experiments. (i) Mallard ducks open one eye periodically during sleep. Individuals with one eye open initiated escape behavior within only 0.17 seconds after an expanding video image that simulated predatory attack was presented. Moreover, ducks on the edge of a sleeping group kept the outward-facing eye open (Rattenborg, Lima, and Amlaner 1999a,b). (ii) J. G. Robinson (1981) showed that female wedge-capped capuchins that were more vigilant were more likely to detect a ground predator than less vigilant individuals (see also Baldellou and Henzi 1992; Cowlishaw 1997b). In addition, van Schaik and van Noordwijk (1989) found that male white-fronted and brown capuchin monkeys, which are more vigilant than females, were more likely to detect model predators than females. (iii) In a laboratory experiment, chaffinches responded earlier to an approaching model sparrow hawk the higher the rate at which the birds raised their head (Cresswell et al. 2003). (iv) In a field experiment, Hunter and Skinner (1998) were able to compare the rate and duration of vigilance in impala and wildebeest in Phinda Resource Reserve, South Africa, before and after historical predators, lions and cheetahs, were introduced. Findings signified that vigilance was uniformly greater under high predation risk. Frequency and duration of vigilance increased over months following predator reintroductions for all measures except wildebeest scanning rates (fig. 4.1; see also Laundre, Hernandez, and Altendorf 2001). (v) Anecdotal comparative evidence supports this same conclusion: Weddell seals basking on fast ice in the Antarctic (where there are no terrestrial predators) sleep so soundly that they can be approached to point-blank range by humans on foot. In contrast, ringed seals living sympatrically with polar bears, arctic foxes, and wolves in the Arctic are so vigilant that they retreat into water at the slightest disturbance (Stirling 1977). (vi) Some studies even suggest that more-vigilant individuals have greater reproductive success than less-vigilant individuals, a key to demonstrating an antipredator benefit of vigilance. Durant (2000b) found that free-living female cheetahs that were more vigilant, as measured by looking at the speaker following the playback of a lion roar, were somewhat more likely to be those females that had achieved greater reproductive success to that point in their life.

Specific age-sex classes of birds and mammals obtain benefits from vigilance that are alternative or additional to detecting predators. Some researchers have attempted to formulate different predictions for different functions of vigilance, suggesting, for example, that males but not females will be more vigilant in areas of territory overlap if males are looking out for competitors rather than predators (for example, Steenbeek et al. 1999). Others have scored glances directed at conspecifics as opposed to the environment

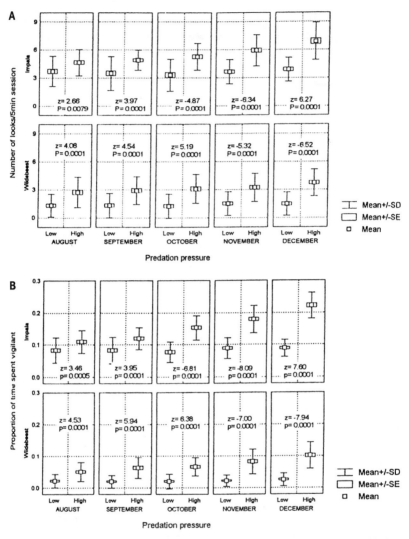

Figure 4.1 A, rate of looking and B, proportion of time spent looking for free-living impala and wildebeest in low- and high-predation conditions before and after predators had been introduced. z-statistics and p-values result from 2 sample Wilcoxon comparisons between predation conditions within months (Hunter and Skinnner 1998).

(Caine and Marra 1988; Treves 1999a; Slotow and Coumi 2000). Both types of studies show that non-antipredator functions of vigilance principally fall under three categories (table 4.1): (i) male mating effort where males attempt to enhance paternity by being vigilant for competitors (for example, white-faced capuchins: Rose and Fedigan 1995) or mates (for example, harbor seals: Renouf and Lawson 1986); (ii) where subordinates attempt to avoid aggression from dominants through heightened vigilance (for example, squirrel monkeys:

Table 4.1 Studies of birds and mammals showing that vigilance is used for reasons other than predator detection

Species	Laboratory (L) or field (F)	Evidence	Reference[a]
Mating effort			
White-faced capuchin	F	(a) α males most vigilant (b) Males scan more than females and looked most at other males	1
Vervet monkey	F	Males more vigilant than females but did not alarm call more at predators	2
Thomas's langur	F	(a) Species shows infanticide and males more vigilant in areas of home range overlap (b) Females with infant more vigilant in areas of home range overlap	3
Cheetah	F	By elimination, males look out for females	4
Harbor seal	F	Males scan more at time of mating	5
Social factors			
Tasmanian devil	F	Vigilance at carcasses to avoid intraspecific competition	6
Brown capuchin	F	Vigilance increases as number of neighbors increases	7
Squirrel monkey	L	Hierarchical squirrel monkeys look at social targets more than nonhierarchical red-chested moustached tamarins	8
Talapoin monkey	L	(a) Dominants monitor opposite sex more than subordinates (b) Subordinates monitor other monkeys more than dominants	9
Yellow baboon	F	(a) High-ranking daughters glance less than low-ranking daughters (b) High-ranking sons glance more than low-ranking sons	10
European rabbit	F	More vigilant when stuffed rabbit is present	11
Foraging factors			
Cheetah	F	Mothers' vigilance during the day is associated with measures of hunting, not antipredator behavior	12
Northwestern crow	F	(a) Proportion of time spent scanning and length of scan bout increased with increasing prey quality and decreasing prey quantity (b) Time spent scanning higher when bird scrounged than found food (c) Scanning bout length increased with group size	13
Nutmeg manikin	L	Hopping with head up increased when opportunities for scrounging increased	14

[a] 1. Rose and Fedigan 1995; 2. Baldellou and Henzi 1992; 3. Steenbeck et al. 1999; 4. Caro 1994a; 5. Renouf and Lawson 1986; 6. M. E. Jones 1988; 7. Hirsch 2002; 8. Caine and Marra 1988; 9. Keverne et al. 1978; 10. Alberts 1994; 11. S. C. Roberts 1988; 12. Caro 1987; 13. Robinette and Ha 2001; 14. Coolen and Giraldeau 2003.

Caine and Marra 1988); and (iii) where individuals look for opportunities to steal food from other group members (for example, northwestern crow: Robinette and Ha 2001). While these studies show that vigilance serves social and foraging as well as antipredator functions in a qualitative sense, the extent to which a given bout of scanning serves to detect predators is difficult to decipher. Nonetheless, inroads are being made. In a study of nutmeg manikins where birds keep their head up while stationary, while hopping, and while eating, the use of stationary and eating head-up postures increases with distance from cover. In contrast, use of head-up while hopping is not influenced by distance to cover but does increase when opportunities for scrounging food are enhanced through changing the spatial distribution of food. In short, different aspects of vigilance in nutmeg manikins serve differing functions (Coolen and Giraldeau 2003).

4.4 Costs of individual vigilance

Classically, the principal cost of vigilance is thought to be a time cost, where opportunities for alternative behaviors are forfeited. The most common trade-off between vigilance and other behaviors is with foraging (Godin and Smith 1988; J. S. Brown 1999). Most models assume that vigilance is incompatible with foraging and that, across species, time spent vigilant is usually, but not always, correlated negatively with time spent feeding (see Lima 1988b). Conversely, difficulties in searching for food result in reduced antipredator vigilance (Lawrence 1985b; Dukas and Kamil 2000). Nonetheless, as reported in section 4.2, this assumption has come under scrutiny.

Vigilance can also compete with other activities, such as sleep (Gauthier-Clerc, Tamisier, and Cezilly 1998). For example, barbary doves spent more time with their eyes open and less time with their eyes closed in Lendrem's (1984b) experiments where he patrolled their aviary with a tame ferret, compared with controls that were left alone! In several species, vigilance and grooming seem incompatible: an impala grooming another animal took longer to spot a person walking toward it than did the groomee or nongroomers (Mooring and Hart 1995); territorial adult male antelope from four species spent half as much time grooming as did bachelor males or females, owing to their heightened vigilance necessary to defend territories and females (although vigilance was not actually measured in this study: Hart et al. 1992); grooming blue monkeys have lower rates of vigilance than those feeding or resting (Cords 1995); and when laboratory rhesus macaque mothers groomed other individuals and their infants were away, maternal glance rate toward infants was reduced and infants suffered higher rates of harassment and aggression (Maestripieri 1993).

Energetic costs of vigilance have never been examined, but it seems reasonable to argue that standing or sitting upright and head movements require relatively little energy used over the course of a day. Survivorship costs of vigilance are unknown for most species, but in dwarf mongooses vigilant individuals suffer disproportionate predation costs. This social carnivore flushes and catches arthropods as the group moves through the undergrowth. During foraging, one individual runs ahead of the group, climbs to a vantage point, a tree branch or termite mound, and scans for predators. Such sentinels or guards (section 5.4) give a repetitive alarm call on sighting a raptor. As guards are often left 60–80 m behind the group as it moves past them, they are forced to make a solitary dash to rejoin the group. Over a period of 2½ years, Rasa (1986) noted that guards constituted 67% of mongooses that were predated, although this was not actually while on guard duty and must therefore constitute an ancillary cost of vigilance. Sentinels are usually found in permanent groups of closely related animals (Bednekoff 1997) and may therefore be willing to incur greater survivorship costs than vigilant individuals in temporary aggregations. In contrast, in a closely related species, the meerkat, where there also are sentinels, no guards were attacked or killed during 2000 hours of observation, because they were the first to see predators and were close to burrows (Clutton-Brock et al. 1999).

4.5 Effects of group size on vigilance

For over a century, biologists have realized that a flock or herd will have a greater chance of detecting an approaching predator than a solitary individual (Galton 1871; Belt 1874; Darwin 1871; R. C. Miller 1922). The idea was formalized by Pulliam (1973), who considered a flock of birds whose scans were of negligible duration. It was assumed that scans occurred randomly following a Poisson distribution, flock members scanned independently, and if one member detected a predator it would alert all the others. If t is the time a predator requires to make its final exposed approach from a hidden position to within striking range of the flock, n is group size, and λ is the mean rate of scans per individual, then the probability, P_n, that any individual scans in a time less or equal to t is

$$P_n = 1 - e^{-n\lambda t}$$

Some of this model's assumptions were tested and supported early on (see Bednekoff and Lima 1998a). These were that scans are highly variable in length and therefore approximate a Poisson distribution, and that individuals scan independently, making it difficult for a predator to predict when a group member will be looking up (Bertram 1980; Elgar and Catterall 1981; Elcavage and Caraco 1983; Studd, Montgomerie, and Robertson 1983; Scannell, Roberts, and Lazarus 2001); but other assumptions, such as the duration of a scan not being influenced by the duration of previous scans and, in particular, in-

(a)

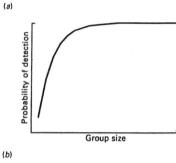

Figure 4.2 Relationships between group size, detection, and scanning rates of prey. In (a) the probability that a group detects the predator before its final uncovered approach increases with group size. In (b) an individual may decrease its scanning rate with increased group size, for any fixed probability of detecting the predator (Elgar 1989).

(b)

dividuals alerting others, were examined much later. Two predictions follow from Pulliam's model: (i) if λ and t are held constant, the probability that a predator is detected increases with flock size, and (ii) if t is held constant, a bird that joins the flock should be able to reduce its scanning rate without lowering P_n (fig. 4.2). Using biologically sensible figures for n, λ, and t, most attempts to test the model have found that P_n reaches asymptote at relatively small group sizes, sometimes as low as between 5 and 10 individuals, although this is species dependent (for example, Elgar and Catterall 1981). Thus, predator detection improves rapidly at small group sizes, whereas increased time devoted to other activities such as foraging are thought to be the principal benefit from being in large groups (Lima and Dill 1990).

4.5.a Increased probability of predator detection The first prediction from Pulliam's model is that individuals in groups have a greater probability of detecting danger. Early tests of Pulliam's model were conducted on birds and showed that larger flocks detected predators at greater distances or earlier than smaller flocks (table 4.2). For example, Kenward (1978) trained a male goshawk to attack woodpigeons encountered alone or in flocks when they were foraging at brassica feeding sites, in grasslands, and in suburban gardens. Attacks on single pigeons and those in small flocks were more successful than those on flocks comprising more than 10 birds. Birds that detected the goshawk at some distance, as determined by their alighting response, were more likely to escape, and their response distance increased with flock size (fig. 4.3). Later studies of

Table 4.2 Studies showing how increasing group size influences the ability of birds to detect predators

Species	Effect of group size	Reference[a]
Emu	Larger groups quicker to detect approaching person	1
Common redshank	Probability of a redshank being captured in an attack declined with increasing flock size	2
Wood pigeon	Attacks on single/small flocks by goshawks were more successful than those on flocks of >10 birds	3
Laughing dove	The larger the flock, the quicker the response to a model of a life-sized hawk	4
European starling	Responded more quickly to flying model hawk	5
Cliff swallow	Small colonies never detected a snake; large colonies (>500 nests) detected a snake at greater distances than middle-sized colonies	6
Barn swallow	Larger colonies responded to a stuffed little owl earlier	7
House sparrow	Larger flocks detect an approaching person farther away	8
Red-billed quelea	Captive flocks detected a goshawk flying over their cage with greater probability than single birds. Probability of detection of a brief, artificial alarm stimulus increased with flock size	9

[a] 1. Boland 2003; 2. Cresswell 1994a; 3. Kenward 1978; 4. Siegfried and Underhill 1975; 5. Powell 1974; 6. C. R. Brown and Brown 1987; 7. Moller 1987b; 8. Harkin et al. 2000; 9. Lazarus 1979.

mammals have shown the same (table 4.3). For example, harbor seals on haul-out sites in the Bay of Fundy, Canada, detected an approaching canoe at greater distances as group size increased, as measured from direct and prolonged staring at the boat (fig. 4.4).

Larger groups detect predators at greater distances or earlier because there are more individuals watching their surroundings at any given time, irrespective of reductions in personal vigilance as proposed by Pulliam's model (Bertram 1980; Lendrem 1984a). Thus, when measures of corporate vigilance (that is, any group member being vigilant) are plotted against group size in a variety of mammals (fig. 4.5), they always rise with increasing group size. In short, a simple model predicts and ample experimental and observational data confirm that potential prey are more likely to notice a predator as group size increases, although the rate at which this diminishes with group size varies across species.

Indirect evidence that animals benefit from conspecifics' vigilance comes from a study of house sparrows foraging at experimental food trays set out on the roof of the University of Cambridge's Zoology Department, UK. There, Elgar (1986) found that before starting to feed, sparrows made a "chirrup" call that recruited other sparrows; pioneers that chirruped more frequently were joined by conspecifics more rapidly. Chirrup rate declined rapidly with flock size (fig. 4.6), mimicking the scanning rate group-size relationship (fig. 4.2.b). This suggests that birds benefited from the presence of additional flock members because it enabled the pioneer to scan less without sacrificing the group's ability to detect danger.

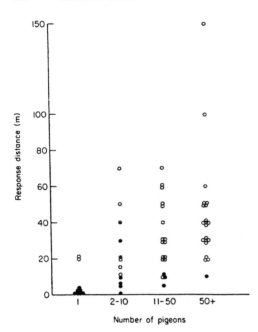

Figure 4.3 Response distance in meters of pigeons in different flock sizes attacked by a trained goshawk. Open circles denote unsuccessful attacks; filled circles, when a kill took place (Kenward 1978).

Table 4.3 Studies showing how increased group size influences the ability of mammals to detect predators

Species	Effect of increasing group size	Reference[a]
Gray kangaroo	Larger groups detect approaching dingoes at greater distances than smaller groups	1
Long-tailed macaque	Detect an approaching human at greater distance	2
Pig-tailed macaque	Detect an approaching human at greater distance	2
Harbor seal	Detect approaching canoe at greater distance	3
Thomson's gazelle	Detect an approaching cheetah at greater distance	4
Black-tailed prairie dog	Detect a stuffed American badger earlier under experimental conditions	5

[a] 1. Jarman and Wright 1993; 2. van Schaik et al. 1983; 3. da Silva and Terhune 1988; 4. FitzGibbon 1990b; 5. Hoogland 1981.

4.5.b Reduced individual vigilance   The second prediction to come out of Pulliam's model, that individual vigilance declines with group size, has now been demonstrated in numerous birds and mammals. In studies of vigilance, group size normally refers to group members that are visible to one another rather than the number of members that are actually in the group (Metcalfe 1984b). Considering just mammals, there are at least 36 species showing a significant decline in individual vigilance with group size (table 4.4). In the large majority of studies in which vigilance declines, however, evidence that there is a leveling off of individual vigilance at low group sizes is decidedly mixed or frequently goes without mention (but see Elgar and Catterall 1981).

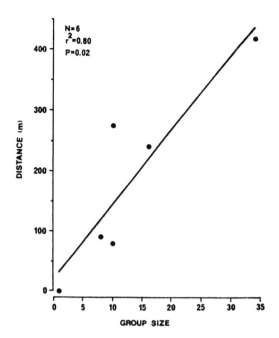

Figure 4.4 Distance at which harbor seals detected an approaching canoe as a function of group size (da Silva and Terhune 1988).

Nevertheless, there are a minority but increasing number of studies that have found no effect of group size on vigilance, most of these being primates (table 4.5) (see also Colagross and Cockburn 1993; Slotow and Coumi 2000; Laundre, Hernandez, and Altendorf 2001; Robinette and Ha 2001). These exceptions cannot be ascribed to methodology or sampling large groups where vigilance might decline little, and their common taxonomic affiliation suggests that vigilance in primates may differ from other mammals (Treves 2000; Treves, Drescher, and Ingrisano 2001). One possible idea is that primates, in contrast with most other mammals, can simultaneously feed and be vigilant while processing food items, is poorly supported; detailed data on time spent scanning in four species show that it is significantly lower when foraging than when resting. A second possibility, that within-group vigilance increases with group size, thereby washing out the effect of reduced scanning for predators, is of little help either. The argument runs that in contrast with those ungulate species on which there are vigilance data, many primates are subject to infanticide by new males that have already entered and are living in the group, which might force females and infants to be vigilant for conspecifics (Steenbeek et al. 1999). Currently, however, limited data on within-group vigilance in red colobus and redtail monkeys show no increase in glances at conspecifics as group size increases (Treves 1999a). In contrast, in some nonprimate species, in which group members share and compete over food items or carrion, individuals do look at one another or at potential food sources more as group

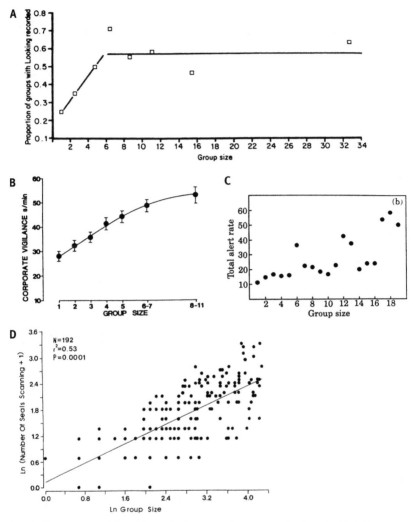

Figure 4.5 Changes in corporate vigilance with group size. *A*, eastern grey kangaroos: proportion of group in which any kangaroo was looking up (Jarman 1987). *B*, brown hares: mean proportion of time that at least 1 hare was vigilant (Monaghan and Metcalfe 1985). *C*, capybaras: mean number of alert events per individual per hour multiplied by the average number of individuals for each group size (Yaber and Herrera 1994). *D*, harbor seals: log number of seals scanning simultaneously (da Silva and Terhune 1988), all plotted against group size.

size increases (S. K. Knight and Knight 1986; M. E. Jones 1998; Beauchamp 2001; Coolen, Giraldeau, and Lavoie 2001). At present, the puzzle as to why vigilance fails to decline with group size in these species is not solved; more generally, however, vigilance is unlikely to decline where (i) animals routinely face predators that do not rely on surprise, (ii) conspicuousness increases as group sizes increase, (iii) certain age or sex classes are preferentially targeted

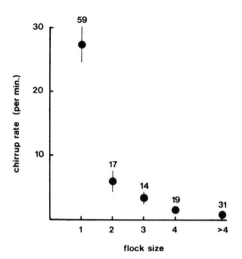

Figure 4.6 Mean (and SE) chirrup rate per house sparrow for each flock size of sparrows, both foraging and sitting on a wall adjacent to food trays (Elgar 1986).

by predators irrespective of group size, (iv) individuals limit the ability of others to scan their surroundings effectively, and (v) predation pressure is extremely low (see Childress and Lung 2003). The importance of these possibilities depends critically on quantitative assumptions; nonetheless, they remain unexplored, and the accent of research is still on reaffirming the group size effect instead of investigating exceptions and their causes.

Elgar (1989) has criticized empirical data purporting to show a decline in individual vigilance with increasing group size, because the majority of studies of birds and mammals failed to control for key confounding variables. These were (i) that larger groups might congregate in high-density food patches and thus increase their time spent feeding, thereby lowering time spent vigilant (Beauchamp and Livoreil 1997); (ii) individuals in larger groups might compete more for a given quantity of food and thus devote less time to vigilance (Clark and Mangel 1986); (iii) larger groups have relatively more individuals in the group's center, and central individuals are less vigilant than peripheral animals (chapter 5); and (iv) individuals of poor foraging ability may choose to forage away from competitors, resulting in solitary animals feeding less and consequently being more vigilant. Moreover, (v) males or subordinates, both of which have higher levels of vigilance (chapter 5), might predominate in small groups, whereas juveniles, which have lower degrees of vigilance, might predominate in large groups; and (vi) distance from cover, habitat, obstructions, and time of day might confound results, as might breeding status and the presence of predators or observers (chapter 5). Elgar provided no solutions to these valid concerns, but since 1989, studies have chipped away at these confounds through carefully designed experiments, through statistical controls, or by recording vigilance in situations in which

Table 4.4 Studies of mammals showing that individual vigilance declines with group size

Species	Group size range[a]	Effects on vigilance	Effects on feeding	Reference[b]
Tammar wallaby	1–17	Proportion of time looking declines	Proportion of time foraging increases	1, 2
Eastern gray kangaroo	1–33	Proportion animals looking and in upright posture declines	Proportion animals feeding increases	3
Quokka	1–27	Percentage of time vigilant decreases with number of individuals within 10 m	Percentage of time foraging increases with number of individuals within 10 m	4
Red-necked pademelon	1–24	Mean scan index declines for adult females with young and other age-sex classes	Feeding rate and scan index negatively correlated	5
Cheetah	1–4	Percentage of rest period females vigilant declines with with more old cubs	Belly size smaller with more old cubs	6
White-nosed coati	1–13	Percentage of drinking bout vigilant declines	Percentage of drinking bout drinking increases	7
Harbor seal	1–40+	Rate of scanning declines and length of time scanning declines	Not measured	8–10
Burchell's zebra	37 ± 4! ng	Percentage of time vigilant and rate of scanning decline	Not measured	11, 12
Warthog	ng	Rate of scanning declines	Not measured	11
Red deer	5–7*	Rate of scanning declines	Not measured	13
White-tailed deer	1–>14	Percentage of time alert declines	Percentage of time feeding increases	14
Mule deer	1–11+	Percentage of time vigilant declines	Not measured	15
Pronghorn	1–12	Proportion of group with head up and percentage of time vigilant decline	Proportion of time foraging increases	15, 16
Impala	36 ± 5! ng	Proportion of individuals with head up, rate of scanning, and percentage of time vigilant decline	Time spent looking and feeding negatively correlated	12, 17, 18
Coke's hartebeest	ng	Rate of scanning declines	Not measured	19
Wildebeest	33 ± 8! ng	Proportion of individuals with head up, rate of scanning, and percentage of time vigilant decline	Time spent looking and feeding negatively correlated	11, 12, 17, 18
Topi	ng	Rate of scanning declines	Not measured	11
Tsessbe	ng	Proportion of individuals with head up declines	Time spent looking and feeding negatively correlated	17
Springbok	3–536	Proportion of individuals vigilant and percentage of time vigilant decline	Proportion of individuals feeding does not increase	19, 20
Grant's gazelle	ng	Rate of scanning declines	Not measured	11
Thomson's gazelle	1–>20 ng	Percentage of time vigilant and rate of scanning decline	Not measured	11, 21

Table 4.4 *(continued)*

Species	Group size range[a]	Effects on vigilance	Effects on feeding	Reference[b]
Bison	1–14	Percentage of time vigilant declines	Not measured	15
African buffalo	78 ± 28! ng	Proportion of individuals with with head up, rate of scanning, and percentage of time vigilant decline	Time spent looking and feeding negatively correlated	11, 12, 17
Spanish ibex	2–9	Rate of scans declines	Not measured	22
Bighorn sheep	1–36	Rate of scanning and percentage of time vigilant decline	Proportion of time foraging increases	15, 23, 24
Waterbuck	11 ± 5!	Percentage of time vigilant declines	Not measured	12
Reedbuck	ng	Proportion of individuals with head up declines	Time spent looking and feeding negatively correlated	17
White-tailed prairie dog	<8 and >15	Large wards spend a greater percentage of time scanning than small wards	Not measured	25
Black-tailed prairie dog	<30 and >80	Large wards spend a greater percentage of time scanning than small wards	Not measured	25
Hoary marmot	1 and >1	Rate of looking up lower in a group	Amount of time feeding higher in a group	26
Golden marmot	1–10	Proportion of time vigilant declines marginally	Proportion of time foraging increases marginally	27
Yellow-bellied marmot	1–26	Percentage of time juveniles and yearlings look up declines	Not measured	28
Great gerbil and fat sand rat	1–2	Percentage of time in low-cost vigilance lower in great gerbil pairs than solitary great gerbils or fat sand rats	Percentage of time foraging greatest in great gerbil pairs	29
Capybara	1–19	Rate of head raising declines	Not measured	30
Brown hare	1–11	Proportion of time scanning and scan duration declines	Proportion of time feeding increases	31
European rabbit	1–12	Rate of looking declines	Not measured	32

[a] Maximum range given if more than one study; * median values; ! mean and standard error; ng not given.

[b] 1. Blumstein, Evans, and Daniel 1999; 2. Blumstein and Daniel 2002; 3. Jarman 1987; 4. Blumstein, Daniel, and McLean 2001; 5. Wahungu, Catterall, and Olsen 2001; 6. Caro 1994a; 7. Burger and Gochfeld 1992; 8. Krieber and Barrette 1984; 9. Terhune and Brilliant 1996; 10. da Silva and Terhune 1988; 11. Scheel 1993a; 12. Burger and Gochfeld 1994; 13. Clutton-Brock, Guinness, and Albon 1982; 14. LaGory 1986; 15. Berger and Cunningham 1988; 16. Lipetz and Bekoff 1982; 17. Underwood 1982; 18. Hunter and Skinner 1998; 19. Siegfried 1980; 20. Bednekoff and Ritter 1994; 21. Fitzgibbon 1990b; 22. Alados 1985; 23. Risenhoover and Bailey 1985; 24. Berger 1978a; 25. Hoogland 1979; 26. Holmes 1984; 27. Blumstein 1996; 28. Carey and Moore 1986; 29. Tchabovsky, Popov, and Krasnov 2001; 30. Yaber and Herrera 1994; 31. Monaghan and Metcalfe 1985; 32. Burnett and Hosey 1987.

most of these variables do not apply (for example, house sparrows and European starlings: Lazarus and Symonds 1992; silver-eyes: Catterall, Elgar, and Kikkawa 1992; white-nosed coatis: Burger and Gochfeld 1992; common redshanks: Cresswell 1994a; green-winged teal: Poysa 1994; cheetahs: Caro 1994a;

Table 4.5 Primate studies that tested for group size effect on time spent vigilant

Species	Group-size effect	No. of groups	Group size range	Notes on methods	Reference[a]
Black howler monkey	None	6	2–10		1
White-faced capuchin	None	4	12–24	No glances	2
Weeping capuchin	Negative	2	8–25	No glances	3
Vervet monkey	[b]	6	2–19	No glances	4
Redtail monkey	None	6	1–29	[c]	5, 6
Blue monkey	None	5	1–30+	Only males[c]	7
Blue monkey	None	2	20–36	Only females	8
Blue monkey	None	3	1–12	Only males[c]	5
Yellow baboon	None	4	22–55		9
Red colobus	None	5	20–76		6
Human	Negative	394	1–5	No WGV[d]	10
Chimpanzee	None	9	1–13	Only fems, juv	5

Source: Treves 2000.

[a] 1. Treves, Drescher, and Ingrisano 2001; 2. Rose and Fedigan 1995; 3. J. R. de Ruiter 1986; 4. Isbell and Young 1993b; 5. Treves 1997; 6. Treves 1998; 7. Tsingalia and Rowell 1984; 8. Cords 1990; 9. Cowlishaw 1998; 10. Wirtz and Wawra 1986.
[b] Internal contradiction: Isbell and Young (1993b) reported a significant negative relationship across groups but showed no effect of group size when sex and dominance rank were controlled for in a factorial ANOVA.
[c] Comparisons of the same males in versus out of groups (Tsingalia and Rowell 1984) or as group residents versus solitaries (Treves 1997).
[d] Students at a refectory: group size = number of people at a table; WGV = within-group vigilance.

great crested terns: G. Roberts 1995; red-necked pademelons: Wahungu, Catterall, and Olsen 2001). For example, in a field experiment where dark-eyed juncos were given continuous access to unlimited millet seeds and cornmeal for 4 months, birds showed a consistent decline in scan durations from group sizes 1 to 6. As food was always present and no aggressive behavior was observed, it is difficult to argue that vigilance declines because birds spent more time in scramble competition (Lima, Zollner, and Bednekoff 1999). In another observational study, Burger and Gochfeld (1992) watched white-nosed coatis drinking at a waterhole, thereby avoiding the problems of food abundance or type, competition over food, and where they could standardize for proximity to safety, presence of observer, and possible predators, as well as habitat visibility across group sizes. They found that both the length of drinking bouts and the percentage of time spent drinking increased with group size. Indeed, of these more sophisticated studies, only the one on silver-eyes far from predators has failed to find the predicted negative correlation with group size. In short, a few exceptions notwithstanding, the association between reduced vigilance, as far as it relates to looking out for predators, and increased group size appears robust, although Pulliam's model has rarely been tested quantitatively because it is so difficult to attach numerical values to parameters such as predatory risk (Lima 1990a; Quenette 1990).

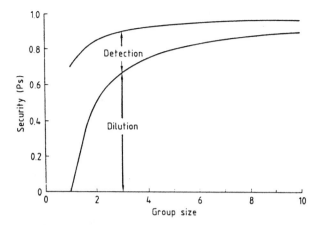

Figure 4.7 Security as a function of group size based on a model by Dehn (1990):

$$P_S = P_{GD} \times P_P + \frac{(1 - P_{GD}) \times (N - 1) \times P_P}{N} + 1 \times (1 - P_P),$$

where P_S is the probability that an individual survives, P_{GD} the probability of the group detecting a predator, P_P the probability that a predator is present, and N is group size. The lower curve shows the dilution effect alone, where vigilance is set to zero. The upper curve shows dilution and detection, where vigilance is set to 0.7; for both curves $P_P = 1$, i.e., a predator is present.

W. D. Hamilton (1971) argued that being in a group confers benefits on an individual if a predator attacks and kills only one individual per group. Assuming that predators attack different-sized groups with equal probability, an individual has a $1/n$ chance of dying in a group of n individuals, and thus the chance of being killed declines with group size (chapter 8). Termed "the dilution effect," this phenomenon provides an alternative or additional explanation for the decline in individual vigilance with increasing group size. Simply put, it is less dangerous being in a larger group, so there is less need to maintain high levels of personal vigilance (see Lazarus 1979). Logically, groups containing more individuals will better detect predators; additionally, individuals in those groups will experience a reduced chance of being killed (fig. 4.7), although the shape and height of the two curves plotted against group size are unknown except for isolated examples relating to single specific predators (Dehn 1990; chapter 13). Given that the effectiveness of corporate vigilance must reach asymptote fairly early on with increasing group size, vigilance becomes relatively less important and dilution more important in large groups, although both will show diminishing returns as group size continues to increase. Experiments in which hungry, and therefore effectively nonvigilant, birds are introduced into a flock of well-fed birds (Lima 1995a) have found that individual vigilance of nondeprived birds declines with group size, which either supports the idea that well-fed individuals are responding to the dilution effect since the additional birds were hardly vigilant at all (fig. 4.8), or that

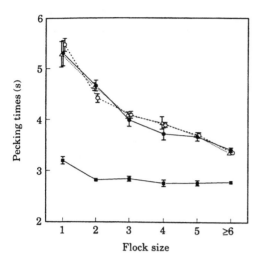

Figure 4.8 Mean (and SE) pecking times of juncos as a function of flock size. Longer pecking times indicate greater vigilance. Solid squares are for food-deprived birds; solid circles are nondeprived birds feeding in the presence of food-deprived birds. Open triangles are control data when no deprived birds whatsoever were introduced into flocks. Open circles are another sort of control, when no deprived birds were present at the feeding site on days when deprived birds had been released. Vigilance of nondeprived birds was indistinguishable in the presence and absence of nonvigilant deprived flockmates (Lima 1995a).

they were erroneously assuming that extra birds were being vigilant. In contrast, work on white-capped capuchin monkeys demonstrates that individual vigilance is inversely related to the group's percentage of males, which are particularly vigilant in this species, rather than to the total size of the group (Rose and Fedigan 1995). This indicates that detection rather than dilution benefits are paramount. Attempts to prize apart these processes underlying the group size–vigilance relationship are just starting (for example, G. Roberts 1996; Cresswell, Lind, et al. 2003; Fernandez, Capurro, and Reboreda 2003).

One promising modeling exercise shows that detection and dilution are intertwined once the point at which a predator targets a particular prey individual is considered (Bednekoff and Lima 1998b). The exercise employed a series of models in which the advancing predator may choose which of two prey animals to target at any one of three decision points as it advances toward prey (fig. 4.9). The first model proposes perfect collective detection such that if one prey individual detects the predator before τ seconds have elapsed, it will transfer this information to the other member of the pair without delay (D_p). Here, the probability that a prey individual fails to look up before the predator reaches the critical point, C, is a negative exponential of its scanning rate—that is, $e^{-\lambda\tau}$ or $e^{-\lambda^*\tau}$—for the other prey individual (Pulliam 1973). As each member of the prey group scans independently of one another, the probability that one prey animal will be caught is this term multiplied by itself but divided by 2, as the prey captures only one prey animal (Bednekoff and Lima 1998b). Thus,

$$D_p = (e^{-\lambda\tau} e^{-\lambda^*\tau})/2$$

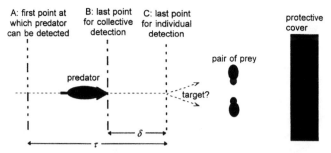

Figure 4.9 The attack scenario. A predator is approaching from the left. It can first be detected at point A. If a prey individual detects the predator before point C, that individual can escape to protective cover. If a prey individual detects the predator before point B, not only does it escape, but also the other prey has time to react and escape. The predator takes τ seconds to move from A to C and δ to move from B to C. At some point, the predator targets one of the prey for attack. The figure shows targeting at point C, assumed by the third model (see text) (Bednekoff and Lima 1998b).

In this equation, the numerator describes detection and the denominator describes dilution.

In a second model, prey must still detect the threat before τ seconds have elapsed, and it can either do this itself or else by reacting to its partner but after a delay of δ seconds (section 4.6.a). Assuming the predator commits to pursuing one of the two prey animals before it is $\tau - \delta$ seconds into the attack, the danger to the individual under early targeting is

$$D_e = (e^{-\lambda\tau} e^{-\lambda^*(\tau-\delta)})/2$$

In this scenario, collective detection becomes less effective as the delay, δ, increases; but again, the numerator describes detection and the denominator describes dilution, and they are still separate entities.

In a third model, the predator targets the prey individual at τ, the last point for effective detection. Here the situation changes radically, because any prey individual leaving before τ alters the risk of dilution for those left behind. Now the danger to the focal prey animal is the probability of not looking up by τ multiplied by [(the probability that the other animal will look up between $\tau - \delta$) plus one half of the probability that the other animal does not look up in time to escape].

$$D_\tau = e^{-\lambda\tau}[(e^{-\lambda^*(\tau-\delta)} - e^{-\lambda^*\tau}) + e^{-\lambda^*\tau}/2] = e^{-\lambda\tau}(e^{\lambda^*(\tau-\delta)} - e^{-\lambda^*\tau}/2)$$

Here the numeral 2, signifying dilution, appears in only one term, because the probability of the focal animal being targeted depends on whether and when the other animal detects the predator. Thus, both detection and dilution change with vigilance rates in late targeting scenarios. As delay, δ, increases, collective detection becomes less important, but so does dilution as the focal prey becomes increasingly likely to be the sole target. In sum, Bednekoff and Lima show that detection and dilution combine in different ways depending on when and how the predator targets its prey, with the distinction diminishing the later that the predator chooses its victim.

4.5.c Increased foraging   As vigilance competes with all other activities, it is an empirical issue as to whether a group-related decline in vigilance results in an increase in time spent feeding. Generally, studies show that time spent feeding does increase with group size. In table 4.4, for instance, 10 out of 11 species for which data are available show increased individual feeding as group size increases; another study demonstrated a greater proportion of time devoted to drinking in larger groups. Across birds, a decline in vigilance is also often associated with an increase in proportion of time spent feeding or in feeding rate (table 4.6). Using a far more extensive list of examples than in table 4.6, Beauchamp (1998) found that vigilance declined in 31 out of 32 studies in which feeding rate increased with group size, although many of these studies had confounding factors. The number of such associations in both mammals and birds is persuasive, but most do not show the direction of the causal arrow. There may be other reasons that food intake rises with group size, such as local enhancement, that may then reduce time available for vigilance (Beauchamp and Livoreil 1997). Experimental work in which colony sizes of black-tailed prairie dogs were manipulated show that the proportion of time that prairie dogs fed declined when individuals were removed from the group for 48 hours. Group sizes were reduced by over half, from 10–58 adults down to 8–15 adults. Concomitantly, the time for which they kept their head up or assumed a bipedal posture increased after conspecifics were removed. Replacing the animals produced roughly opposite effects. This experiment demonstrates that feeding and alertness are tightly linked and that both are driven by changes in group size (Kildaw 1995). Nonetheless, for many species, an increased food intake resulting from reduced vigilance is often assumed rather than demonstrated. For example, Cresswell (1994a) found that while common redshanks increased pecking rates as vigilance declined, these were unsuccessful pecks and intake did not increase. In addition, in situations in which animals are hungry and move to more promising food patches, vigilance can increase because gains in food intake eventually diminish in these patches, so there is more time available for vigilance (McNamara and Houston 1992; Repasky 1996).

4.6 Why don't individuals cheat?

Pulliam's model demands that each group member take account of others' vigilance or at least an approximation of their vigilance and adjust its own level accordingly. But what is to stop an individual from cheating—lowering its vig-

Table 4.6 Effects of group size on individual vigilance in birds

Species	Group size range	Effects on vigilance	Effects on feeding	Reference[a]
Greater rhea	1–12	Proportion of time vigilant declines with group size	Proportion of time foraging increases with group size	1
Ostrich	1–3 or 4	Overall decline in percentage of time heads up as the number of birds in the group increased	Not measured	2
Emu	1–20	(a) No categories of vigilance affected by group size	(a) Not measured	3
		(b) Proportion of time vigilant declined with group size	(b) Proportion of time foraging increased with group size	4
Pink-footed goose	N/A	Time spent head up decreased while time with head on back (eyes closed) increased with with flock size	Time spent grazing decreased with increased group size	5
Common teal	1–70	Peeking (vigilance while sleeping) decreased with increasing flock size in females	Not measured	6
	3–15	Vigilance not related to group size	Not measured	7
Bald eagle	1–42	Percentage of time scanning increases with group size	Piracy increases as group size increases	8
Common redshank	1–41+	Interscan interval increased with flock size	Feeding rates remained fairly constant over all flock sizes	9
Lapwing	1–100+	Scan less often in large flocks	Birds in large flocks take a more profitable mixture of worm sizes	10
Spotted dove	2–10	Proportion of time individual spent scanning decreased as flock size increased	Not measured	11
Barbary dove	1–12	As flock size increased so did time spent with eyes closed	Not measured	12
Laughing dove	2–21	Birds in large flocks appeared less fearful	Singletons did not feed or fed in a desultory manner	13
White-winged chough	4–33	Proportion of time vigilant declines between group sizes 4 and 20	Proportion of time foraging increases from group size 4 to 20 but then declines slightly	14
Red-winged chough	1–55	Vigilance declines in larger groups	Feeding frequencies unrelated to flock size	15
Silver-eye	1–>8	No relationship between scanning and group size	Not measured	16
Willow tit	2–11	Negative correlation between proportion of birds scanning and the flock size	Time spent foraging increased with increased flock size	17
	1–6	Reduced frequency alert as flock size increases	Not measured	18
House sparrow	1–4	Scanning decreases with increase in flock size	Not measured	19
	1–7	Scanning rate declines with flock size	Not measured	20
White-browed sparrow-weaver	2–>4	No association between group size and percentage of time vigilant	Sparrow weavers in large groups do not feed more than those in small groups	21
White-crowned sparrow	2–35	No correlation between group size and vigilance rate	Pecking rate increases with group size	22

Note: N/A, not applicable.

[a] 1. Fernandez, Capurro, and Reboreda 2003; 2. Bertram 1980; 3. Hough et al. 1998; 4. Boland 2003; 5. Lazarus and Inglis 1978; 6. Gauthier-Clerc, Tamisier, and Cezilly 1998; 7. Poysa 1987b; 8. S. K. Knight and Knight 1986; 9. Cresswell 1994a; 10. Barnard and Stephens 1981; 11. Sadedin and Elgar 1998; 12. Lendrem 1984b; 13. Siegfried and Underhill 1975; 14. Heinsohn 1987; 15. Rolando et al. 2001; 16. Catterall, Elgar, and Kikkawa 1992; 17. Hogstad 1988; 18. Ekman 1987; 19. Lima 1987b; 20. Harkin et al. 2000; 21. Ferguson 1987; 22. Slotow and Rothstein 1995.

ilance relative to flockmates and thereby increasing its time spent feeding? Put another way, cooperating is not evolutionarily stable. Pulliam, Pyke, and Caraco (1982) modeled games between selfish and cooperative vigilant group members. A selfish individual always scanned at the rate that, if adopted by all flock members, had the property that any deviating individual had a lower probability of surviving. A cooperative individual always scanned at a cooperative equilibrium rate regardless of what its flockmates were doing. They found that individual scanning rates of selfish individuals rapidly dropped to zero as soon as group size reached only four or more (see also G. A. Parker and Hammerstein 1985; McNamara and Houston 1992; Rodriguez-Girones and Vasquez 2002). Surprisingly, observed rates of scanning in yellow-eyed juncos matched the cooperative but not the selfish model of vigilance (table 4.7). In order to explain this, they suggested that juncos use a "judge" strategy, whereby flock members remain cooperative as long as other members remain cooperative. Watching others would be costly, however, and evidence that group members monitor one anothers' vigilance is weak or indirect at best (Inglis and Isaacson 1978). In addition, subjects that were checking conspecifics might also be able to look out for predators as well (P. I. Ward 1985; Packer and Abrams 1990; but see Coolen and Giraldeau 2003)! Moreover, coordination might reduce unpredictability of scans, lowering the ability to outwit stalking predators (Rodriguez-Girones and Vasquez 2002). In a direct test, Lima (1995a) introduced first food-deprived and hence nonvigilant and, at another time, satiated yellow-eyed juncos to a flock of nondeprived birds, as described earlier. Nondeprived juncos showed a decreasing level of vigilance with increasing flock size, as measured by time necessary to ingest 10 pieces of cornmeal, irrespec-

Table 4.7 Observed individual scanning rates and predicted rates from cooperative and selfish models in different group sizes

N^a	Observed	Cooperative	Selfish
1	13.9	18.6	18.6
2	7.85	10.4	3.4
3	6.22	8.7	0.6
4	6.02	7.8	0
5	5.87	6.9	0
6	5.66	6.4	0
7	5.58	5.9	0
8	5.59	5.5	0
9	4.88	5.0	0
10	4.65	4.9	0

Source: Pulliam, Pyke, and Caraco 1982.
[a] Group size.

tive of whether these flockmates were deprived (nonvigilant) or nondeprived (vigilant) (fig. 4.8). This result demonstrates that juncos do not monitor the vigilance of their flockmates, although they do monitor group size. Thus, we are still left with the question of why group members do not cheat.

4.6.a Predator detection is not collective   Five main avenues of thought are being used to explain why individuals maintain personal vigilance in groups. The first questions the assumption that an individual that detects a predator will alert all other members to danger. Lima (1995a, 1995b; Lima and Zollner 1996) has spearheaded tests of this assumption and found it to be unjustified. First, using the same feeding pad as described above, Lima rolled a ball silently down a narrow ramp directly at a target bird, either a yellow-eyed junco or an American tree sparrow, so that only it could see the ball. The ball was subsequently shunted sideways, out of sight of the flock. Upon detecting the ball, the target adopted a stiff, upright posture before flushing for cover. Lima found that very few nontarget birds responded to this protocol: in only 20% of 69 ball attacks did any nontarget birds become alert, and only 4.1% of a prospective 485 nontarget birds took up this stance. Birds that did become alert were invariably close, within 2 m or so, of the target individual (see also Hilton, Cresswell, and Ruxton 1999; section 5.2). Similarly, in response to the target flushing for cover, nontarget birds flushed in only 40% of 64 cases, and this was usually only one or two birds. Judging from the birds' behavior, flushing for cover appears to be a signal of danger. In other studies the proportion of birds flushing also increased when feeding occurred far from protective cover or the target departing birds were on the perimeter of the flock (Lima 1995b), situations both associated with risk. In contrast, the proportion decreased when birds were visually separated by a wall (see also Elgar, Burren, and Posen 1984; Metcalfe 1984a) or were separated spatially (15 cm versus 4 cm) from detecting flockmates (Lima and Zollner 1996), situations in which risk might be lower.

In a subsequent experiment (Lima 1995b), the proportion of nondetectors flushing was found to increase when more than one bird was targeted, especially in the case of small flocks (fig. 4.10). This showed that nondetectors respond to the number of birds departing. The idea that multiple threat-induced departures over a short time has a markedly greater influence on escape behavior than single departures has since been modeled several times (for example, Lima 1994a; Ruxton 1996; Ruxton and Roberts 1999). If, for example, birds respond to single departures of flockmates, many of these could simply be false alarms, so flighty individuals will lose time feeding, incur unnecessary energetic costs of flight, and may even be subject to greater predation risk later when they are forced to reduce vigilance as they subsequently feed to recoup losses (Proctor, Broom, and Ruxton 2001). Observations confirm that false

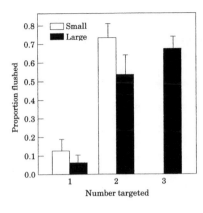

Figure 4.10 Average (and SE) proportion of nondetectors that flushed to cover as a function of group size (small groups are 5–7 birds; large groups are 9–15 birds) and the number of birds targeted for attack with a ball. $N = 15$ ball attacks in all cases. There were no data for 3 birds attacked in small groups (Lima 1995b).

alarms can potentially be a serious problem in the wild: on Tyninghame estuary in Scotland, flocks of common redshanks embarked on 170 alarm flights from raptors, but there were also 115 cases of mistaken identity in which redshanks flew from nonraptor species and 572 flights with no apparent cause (Cresswell, Hilton, and Ruxton 2000). Given that so many alarms are false, it is not surprising that birds use a rule of thumb to make an immediate escape with the simultaneous departure of two or more friends, but otherwise quickly assess the reason for a single departure before acting (Lima 1994a; Ruxton 1996), although more departees than two may be required when flocks are large (G. Roberts 1997). Postponements in responding to flockmates' departures support this contention, because delays until the next redshank or the next redshank but one flying off were shorter as the number of detectors increased from one to two to more than two (Cresswell, Hilton, and Ruxton 2000). Only in circumstances in which flock size is small and there is likely to be only one predator detector, the rate of false alarms is low, or the time taken to attack is extremely short is it best for birds to leave immediately regardless of the number of detectors (Proctor, Broom, and Ruxton 2001).

In summary, individual birds seem to respond simply to nearby flockmates' departures and are unable to distinguish departures induced by predation threat from departures for other reasons, such as moving to a new food patch. Lima (1995b) argues that

> this ambiguous, departure-based form of collective detection implies that individuals must rely considerably on personal vigilance to detect predatory attack. Such a reliance may leave animals more vigilant than suggested by models of vigilance based upon the conventional version of collective detection. (p. 1097)

Failure to alert others specifically of impending danger and having to rely on conspecifics' flight behavior may be the key selective force maintaining individual vigilance in groups.

4.6.b Vigilant nondetectors are at an advantage   In addition to differences in escape probabilities between detectors and nondetectors, there are differences between vigilant nondetectors and nonvigilant nondetectors. Using the same experimental ball-rolling technique, Lima (1994b) found that vigilant junco and American tree sparrow nontarget birds had shorter latencies to flushing than did birds that had their head down at the time the ball was rolled at a single target bird (table 4.8). Similarly, house sparrows that were vigilant at the initiation of a flush to safety departed more quickly than nonvigilant house sparrows (Elgar, McKay, and Woon 1986), and common redshanks that had their head up at the start of an attack by a sparrow hawk flew earlier than those with their head down (by about 0.1 seconds or 3 m of additional approach by the raptor: Hilton, Cresswell, and Ruxton 1999). Thus, personally vigilant birds are at an advantage to nonvigilant individuals, possibly because they are able to react more quickly to danger. Again, this would select for maintaining personal vigilance in groups.

4.6.c Predators select low-vigilance individuals   Where predators selectively target individuals exhibiting low levels of vigilance from a group of animals, there will be strong selection for maintaining personal vigilance. Stalking predators that employ a stealthy approach to get close to unwary prey might use such a rule. Thus, when FitzGibbon (1989) recorded the vigilance of two unwary Thomson's gazelles being stalked by a cheetah that were matched for distance to the cat, she found that the cheetah chose the least-vigilant individual in 14 out of 16 occasions. Similarly, under laboratory conditions, a blue acara cichlid preferred to attack foraging guppies over nonforaging ones, as well as nose-down foragers over horizontal foragers (Krause and Godin 1996). In both examples, foraging individuals are not only less vigilant, but their feeding activity could possibly be a cue to being in poor condition and hence the ease with which they could be captured. In contrast, an opportunistically hunting predator, the sparrow hawk, attacked birds mounted in feeding and vigilant pos-

Table 4.8 Average (*N*, SE) time delay in seconds in flushing by nondetecting yellow-eyed juncos and American tree sparrows following the departure of a detector in the ball-rolling experiment

Species	Nondetecting class	
	Vigilant	Nonvigilant
Junco	0.22	0.35
	(*17*, 0.021)	(*16*, 0.032)
Tree sparrow	0.20	0.29
	(*11*, 0.032)	(*11*, 0.041)

Source: Lima 1994b.

tures at similar frequencies (Cresswell, Lind, et al. 2003), implying that the importance of personal vigilance in reducing the likelihood of being targeted will be restricted to limited circumstances and that the benefits of vigilance could depend on the frequency of attack of different sorts of predators. More generally, the relationship between patterns of prey selectivity and mode of hunting is poorly understood: selectivity could be widespread in stalking predators but rare in coursing predators if they pick out a victim after prey have seen them, or it may be uncommon in stalkers if they target the nearest prey individual to them but common in coursers if they choose a victim on the basis of its speed once the prey are flushed (Caro and FitzGibbon 1992; Murray et al. 1995). Nevertheless, risk of attack by just one stalking predator might be sufficient selection pressure to at least contribute to personal vigilance in groups.

Packer and Abrams (1990) contrasted "cooperative" and "selfish" models of group vigilance, incorporating both the advantage that detectors have in escaping predation and predator selectivity for nonvigilant individuals into ESS (Evolutionarily Stable Strategy) models. (In the cooperative model, each individual is vigilant at a rate that gives the highest fitness if all group members play the same strategy [see also Lima 1987a]. In the selfish model, each individual is vigilant at a rate that gives a higher fitness when played against itself than any alternative vigilance rate would gain against it.) They found that individuals in selfish groups were likely to be more vigilant than those in cooperative groups when predators select an individual that was not vigilant at time of attack. These models indicate that the high levels of vigilance observed in flocks and herds may not be due to cooperation, as formerly supposed (Pulliam, Pyke, and Caraco 1982), but selfish attempts at minimizing predation risk (see McNamara and Houston 1992 for a review and general ESS model).

4.6.d Individuals maintain vigilance so as not to lose group members   It follows from the dilution effect that even unrelated individuals in small, permanent groups might be vigilant in order to lower the chance of losing a group member to a predator, since maintaining a large n will minimize $1/n$. ESS models suggest such a scenario is plausible (Kaitala, Lindstrom, and Ranta 1989; Lima 1989), but its applicability is restricted to fairly small groups and to groups that maintain permanent membership. The idea cannot explain vigilance in temporary aggregations of birds at feeding stations.

4.6.e Multiple attacks are possible   Most models assume that the predator makes but one attack. If, however, the group is subject to multiple attacks, a small increase in the probability of escaping as a result of personal vigilance will be amplified over time. Thus, when the probability of surviving each attack is 90%, there is only a 35% chance of surviving 10 attacks (0.9^{10}). If personal vigilance increases survival probability to 95%, survival chances rise to 60% (0.95^{10}) over the long run (Dehn 1990). In addition, modeling shows that vigilance against different types of threat, such as aerial and terrestrial predators, may have unexpected outcomes, with innocuous predators driving unexpectedly high levels of vigilance (Lima 1992a), although the impact of several predation threats on the vigilance–group size relationship is unknown.

In conclusion, the idea that individuals in groups cooperate in being vigilant for predators no longer stands up to scrutiny. High levels of personal vigilance in small groups and lower levels in larger groups can potentially be explained through the benefits accruing to the individual and not to group members, although the type and relative importance of these benefits will depend on the nature and extent of predatory attempts. More generally, the elegant theory proposed by Pulliam and the apparent supporting evidence seem to be mismatched. There are assumptions of the theory, such as randomness in sequential scans, predators remaining concealed until they rush from cover, and the group detecting danger if any of its members does, that do not stand up empirically; thus, it is extraordinary that we so often witness reduced individual vigilance as group size increases (Bednekoff and Lima 1998a). Questions as to why individuals scan independently of one another rather than on a rotational basis, or why they scan for different lengths of time are still open.

4.7 Vigilance in mixed-species groups
Given that increasing group size is associated with a reduction in levels of individual vigilance in so many species, it is likely that individuals might increase the probability of detecting predators as well as relax personal vigilance if they associate with members of other species, provided that these species overlap their geographic distribution and are similarly matched in size and vulnerability and therefore share common predators (Metcalfe 1984b), as well as perhaps respond to each others' alarm calls. Mixed-species associations are common in birds (for example, Moynihan 1962; Ulfstrand 1975; D. B. A. Thompson and Barnard 1983; Thiollay 1991), mammals (for example, Struhsaker 1981; Terborgh 1983; A. C. Smith, Kelez, and Buchanan-Smith 2004), and fish (for example, Ehrlich and Ehrlich 1973; Barlow 1974), but the potential antipredator advantages of polyspecific associations are not limited to vigilance. Aside from obtaining warning of predators from other flock members (section 6.5), they include confusion of predators (Morse 1977) or mounting a better defense against them (Struhsaker 1981). Alternatively, or additionally, individuals might gain foraging benefits, including enhanced rates of prey capture (Munn and Terborgh 1980), or be better able to find resources (Krebs 1973; Barnard and Stephens 1983). Finally, mixed-species associations might simply occur by chance (Waser 1984; Whitesides 1989). Therefore, it is perhaps not surprising that

individual vigilance levels have been shown to decline, remain the same, or even increase when individuals associate with heterospecifics (tables 4.9 and 4.10).

In the Kibale Forest in Uganda, for example, where five species of primate associate together, changes in vigilance in the presence of heterospecifics are difficult to predict (C. A. Chapman and Chapman 1996). Compared with associating with conspecifics, both red colobus and redtail monkeys show significantly lower amounts of vigilance when associating with black and white colobus and gray-cheeked mangabeys, as do blue monkeys with red colobus; but red colobus and redtails show greater amounts of vigilance with blue monkeys, and blue monkeys show more vigilance with mangabeys (table 4.11).

A number of confounding variables can help to explain inconsistent results such as these. (i) Each species may eat food that requires different handling times, leaving them with different opportunities for vigilance. (ii) Species may enter polyspecific associations at different-sized groups. As personal vigilance

Table 4.9 Some effects of mixed-species grouping on vigilance in birds

Species	Effects	Reference[a]
Ruddy turnstone and purple sandpiper with each other and other species	Turnstone vigilance reduced more in presence of turnstones and purple sandpipers than redshanks or oystercatchers. Turnstones have less effect on reducing purple sandpiper vigilance than other purple sandpipers	1
Lapwing and black-headed gull	Lapwings scanned more often with more gulls present	2
Wigeon and wader species	Wigeon males, but not females, vigilant for shorter periods	3
Downy woodpecker and black-capped chickadee, tufted titmice, white breasted nuthatch, hairy woodpecker	Downy woodpeckers foraging with three or more flock members showed low head-cocking rates and increased feeding rates compared with foraging alone. Woodpeckers benefit from signals other flock members are exchanging	4
Great tit and coal tit, blue tit, crested tit	Great tits showed no effect on vigilance with with conspecifics or other species	5
Great tit and yellowhammer	Handling time of food items but not scanning rate declines with more yellowhammers present	6
Willow tit and coal tit	Willow tits scanned less when they were in a flock with coal tits	7
Chickadee and tufted titmice	Chickadees reduced vigilance when foraging with titmice, and titmice also reduced their vigilance in two-bird heterospecific groups compared with when they were housed singly	8
American goldfinch, purple finch, and pine siskin	American goldfinches and purple finches reduce vigilance in each other's company but increase it with pine siskins; siskins maintain the same level of vigilance whatever group composition	9

[a]1. Metcalfe 1984b; 2. Barnard and Stephens 1981; 3. O. W. Jacobsen and Ugelvik 1994; 4. K. A. Sullivan 1984; 5. Carrascal and Moreno 1992; 6. Poysa 1985; 7. Hogstad 1988; 8. Pravosudov and Grubb 1999; 9. Popp 1988.

Table 4.10 Effects of mixed-species grouping on vigilance in mammals

Species	Effect on vigilance	Reference[a]
Redtail monkey and blue monkey	Decline when in association	1
Redtail and black and white colobus, red colobus, blue monkey, gray-cheeked mangabey	Increase when in associations[b]	2
Blue monkey and redtail monkey	Decline when in association	1
Red colobus and redtail monkey	No effect on either species	3
Red colobus and redtail, blue monkey	Increase in both associations	4
Red colobus and Diana monkey	More often below closed canopy, descend to feed on termite hills, more exposed from below, and look down less often when foraging	5
Red colobus and black and white colobus, redtail, blue monkey, mangabey	Decline when in associations[b]	2
Springbok and zebra, gemsbok, giraffe	No effect on springbok	6
Grant's gazelle and Thomson's gazelle	Decline with increasing number of Thomson's gazelle when number of conspecifics is low	7
Thomson's gazelle and Grant's gazelle	Decline with increasing number of Grant's gazelle when number of conspecifics is low	7
Thomson's gazelle, Grant's gazelle warthog, topi, hartebeest, Burchell's zebra, buffalo	Decline with increasing number of heterospecifics taking all species together	8

Note: Effect refers to the first species listed unless stated.
[a] 1. Cords 1990; 2. Chapman and Chapman 1996; 3. Treves 1999a; 4. Stanford 1998; 5. Bshary and Noe 1997a; 6. Bednekoff and Ritter 1994; 7. FitzGibbon 1990b; 8. Scheel 1993a.
[b] See Table 4.11 for details.

declines rapidly with increasing group size in small groups but levels off in large groups (fig. 4.2.b), individuals from species with a smaller initial group size will show greater reductions in vigilance. For example, in central Amazonia, red-cap moustached tamarins live in groups 1.6 times as large as saddle-back tamarins and also form mixed-species groups with them (Peres 1993). In heterospecific groups, moustached tamarins show higher average levels of vigilance than saddlebacks (9.7% versus 4.4% of the time). (iii) One prey species may be able to detect different predators better than the other prey species (Gautier-Hion, Quris, and Gautier 1983). Saddleback tamarins spend a greater percentage of time looking downward at the ground, and moustached tamarins spend more time looking sideways or up at the canopy (Peres 1993). As a result, saddleback tamarins see a greater number of terrestrial predators, and moustached tamarins a greater number of aerial predators (fig. 4.11). Differences in type of vigilance shown by each species might therefore diminish the benefits of reduced scanning for both. (iv) There may be an imbalance in the extent to which each species is subject to predation risk. In the Peres (1993) study, antipredator benefits of association were greater for saddleback than for mous-

Table 4.11 The percentage change in the number of times animals looked up per minute when in association with a specific monkey species, compared with when that species was alone ([in association − alone]/alone)

Species in association	Focal species				
	RC	BW	MG	RT	BL
Red colobus (RC)	—	−6.9	+2.3	+15.9	−26.6*
Black and white (BW)	−37.4*	—	NA	−83.6*	NA
Gray-cheeked mangabey (MG)	−49.5*	NA	—	−59.6*	+69.6*
Redtail (RT)	−2.8	+2.4	−27.4	—	+18.7
Blue monkey (BL)	+51.7*	+0.8	+11.3	+37.1*	—

Source: Chapman and Chapman 1996.
Note: Minus sign denotes that attention increased when the focal species was alone.
NA: not applicable.
*$P < 0.05$ using a two-tailed t-test.

tached tamarins, because aerial predators were more common than ground threats. (v) Predators common to both species may prefer one species over another; thus, members of one may continue to maintain high levels of vigilance in the presence of heterospecifics if that species is likely to be selectively targeted in mixed-species groups. For instance, smaller species will usually be more vulnerable than larger species, as when smaller Thomson's gazelles associate with Grant's gazelles (FitzGibbon 1990b). In this situation, one species will obtain benefits through corporate vigilance but not through dilution and may thus be reticent about relaxing vigilance. (This provides an opportunity to tease apart these two processes, but this has not yet been attempted using polyspecific associations.) Each of these possibilities could help explain the discrepancies in results for the same-species pairings in different study sites (table 4.10).

Species in association may both show reductions in vigilance with heterospecific group size but for different antipredator reasons. FitzGibbon (1990b) studied groups of Thomson's and Grant's gazelles in both conspecific and in heterospecific groups on the Serengeti Plains of Tanzania. Both showed declines in the percentage of time spent vigilant as the number of heterospecifics in the group increased (fig. 4.12). For reasons that are not clear, Grant's gazelles nonetheless spent a greater percentage of time being vigilant (8.7%) than Thomson's gazelles (6.1%) after controlling for several confounding factors such as total group size. Grant's gazelles are also the taller of the two species. Thus, it was not surprising that Grant's gazelles were more likely to detect a cheetah first when gazelles were in mixed-species groups and that such groups detected cheetahs at significantly greater distances than single-species Thomson's gazelle groups (Xs = 214 m versus 173 m, respectively). (There was no significant difference between detection distances, however, for Grant's gazelles in mixed-species versus single-species groups [X = 190 m]).

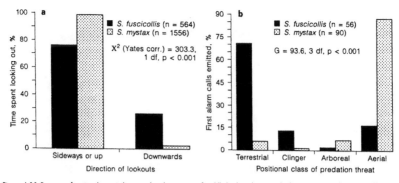

Figure 4.11 Patterns of antipredator vigilance and early warning of saddle-back and moustached tamarins according to (*a*) directions at which animals looked out from the group and (*b*) positional class of animate threatening stimuli toward which first alarm calls were emitted (Peres 1993).

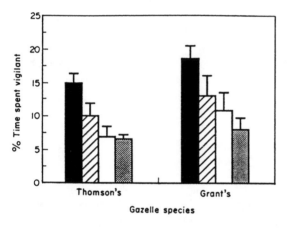

Figure 4.12 The effect of the number of heterospecific group members on the mean percentage (and SE) of time spent vigilant by Thomson's and Grant's gazelles in groups with 1 to 5 conspecifics. *Filled bars*, 0 heterospecifics; *hatched bars*, 1–5; *open bars*, 6–20; *gray bars*, > 20 (Fitz-Gibbon 1990b).

Cheetahs were less successful at hunting Thomson's gazelles found in mixed-species than single-species groups, in part because they abandoned a greater proportion of their stalking attempts toward mixed-species groups and even avoided hunting mixed-species groups altogether. Thus, Grant's gazelles provided Thomson's gazelles with an early warning of cheetahs, but the former derived few vigilance benefits from the mixed-species association. Despite this asymmetry, Grant's gazelles still lowered personal vigilance levels as heterospecific group size increased, presumably because of a greater number of eyes and cheetahs' preference for chasing smaller Thomson's gazelles if an attack was initiated on a mixed-species group.

A similar example comes from red colobus and Diana monkeys that form mixed-species associations in the Tai National Park, Côte d'Ivoire. Whereas red colobus looked down and sideways less often when they were with Diana monkeys than when they were in monospecific groups, Diana monkeys showed no alterations in their vigilance. Also, red colobus were more often found below closed canopy, came down to feed from termite mounds more frequently than expected, and were more exposed from below when they were with the other species. Diana monkeys were more likely to raise an alarm in response to a terrestrial predator (a person approaching) or an experimental presentation of a model crowned hawk eagle than were red colobus. Nonetheless, antipredator benefits were not entirely one-sided: Diana monkeys were found higher up and in more exposed situations when they were with red colobus (Bshary and Noe 1997a).

Mixed-species associations may also be beneficial because a decline in vigilance afforded by the association can lead to greater foraging efficiency (Barnard, Thompson, and Stevens 1982; D. B. A. Thompson and Barnard 1983; K. A. Sullivan 1984; Beveridge and Deag 1987). In a rather extraordinary experiment in which parids (tufted titmice) and either Carolina or black-capped chickadees (so-called nuclear species, because they facilitate flock formation and initiate movements) were completely removed from eight woodlots in Ohio, the response of satellite or follower species, such as downy woodpeckers and white-breasted nuthatches, was monitored (Dolby and Grubb 1998). Satellite species showed an almost doubling of vigilance (head cocks per minute) compared with control woodlots, and a decline in mass. While vigilance might have increased simply because group sizes declined, and body mass might have suffered because parids could no longer lead the satellite species to food or were not available for kleptoparasitism, the findings suggest that mixed-species associations allow for reduced personal vigilance and consequent enhanced foraging efficiency. Indeed, nuthatches became reticent about visiting an exposed bird feeder in woodlots where the nuclear species had been removed (Dolby and Grubb 2000).

Variation in vigilance responses between species involved in mixed-species associations and across different mixed-species associations is frustrating but hardly surprising. First, the principal advantage to polyspecific associations may be food related, making a-priori predictions about vigilance difficult. Second, different polyspecific associations will be subject to different predator species. Third, relative predation risk between members of different mixed-species pairs will differ because of differences in relative body weights and modes of escape. Fourth, within a single association, each of the species may be subject to different predators or magnitude of predation attempts from the same predator. Generalizations about vigilance in mixed-species associations therefore appear very difficult without further systematic data gathering, perhaps focusing on one species associating with a series of different heterospecifics but suffering the attentions of the same predator community.

4.8 Summary

Vigilance, or scanning the environment, allows potential prey to detect predators at some distance. Vigilance is usually assumed to be incompatible with other activities, such as sleeping. Several lines of evidence show that vigilance is used for predator detection, but vigilance is also used in social situations, as between competitors, when subordinates check the activities of dominants, and males monitor potential female suitors. The principal cost of vigilance is thought to be a lost opportunity for feeding; energetic and survivorship costs are assumed to be negligible.

An influential model by Pulliam showed that individuals would benefit from being in a group, because many pairs of eyes would be better able to detect a predator than a solitary animal, and individuals could reduce personal vigilance without compromising group vigilance and thereby obtain feeding benefits. Empirical studies show that groups detect predators at greater distances than individuals and that group (corporate) vigilance does rise with group size. The great majority of studies show that individual vigilance declines with group size. A small number of studies cannot find a vigilance–group size effect, however; reasons for these discrepancies are poorly understood. In general, individual foraging increases as vigilance declines in larger group sizes. Confounding factors that could account for the negative association between vigilance and group size, such as scramble competition for food or proximity to cover, have been investigated and in the main discounted. Nevertheless, the decline in vigilance in larger groups might be due to lowered individual risk of predation in large groups rather than benefits of corporate vigilance.

Pulliam's model of group vigilance is open to cheating by selfish nonvigilant individuals. Five explanations are currently on the table to account for individuals being vigilant in groups. These are (i) that individuals do not share information about threat; (ii) that vigilant individuals can escape more quickly than nonvigilant individuals; (iii) that predators select low-vigilance individuals; (iv) that individuals are vigilant so as not to lose group members who contribute to the dilution effect; and (v) that multiple attacks mean that some personal vigilance is important whatever the group size. The relative importance of these explanations is unknown.

Individuals can reap reduced vigilance and predator detection benefits from forming polyspecific as well as conspecific associations. While mixed-species associations are common in birds and mammals, vigilance does not always decline with heterospecific group size. Asymmetries in species' initial group size, differential susceptibility to predation, and abilities to detect different kinds of predators may help to explain patterns of vigilance in mixed-species groups.

5   Factors Affecting Vigilance

5.1 Introduction

Predation risk is a function of prey's vulnerability to and being attacked by a predator. Risk will therefore be heightened in circumstances in which individuals are forced to trade off antipredator defenses, such as being with conspecifics or using a refuge, for maintenance behaviors, usually feeding (Lima 1987a). Risk will also be raised in environments in which predators are numerous. Many social and environmental factors therefore affect risk, and these influence the amount of time and the rate at which animals scan their environment for predators (Quenette 1990). Thus, when individuals are found at some distance from the safety of other animals or from cover, or enter predator-rich environments, they increase vigilance in an attempt to counteract heightened risk. Mothers with relatively defenseless offspring show strikingly higher vigilance levels, too. The extent to which predation risk alters patterns of vigilance within and between species has been examined many times in birds and mammals, but in most cases quantitative differences in vigilance have only been explained as adaptive responses to qualitative differences in risk. Very few studies have been able to quantify predation risk; thus, detailed understanding of the relationship between vigilance and predation hazard still remains limited.

This chapter begins by continuing to address aspects of group-related vigilance, focusing on two factors that affect individual vigilance in groups: proximity to conspecifics and location within the group (sections 5.2 and 5.3). It then examines the factors that foster individuals taking on roles as sentinels, a form of vigilance that occurs occasionally in species that live in permanent

---

Figure 5.0 (*Facing page*) A meerkat assumes an elevated vantage point while taking on sentinel duty. Guards rotate within the group and take over when they are well fed principally for selfish reasons, to see predators as early as possible; group members additionally gain from the early warning (reproduced by kind permission of Sheila Girling).

groups (section 5.4). Next, the chapter reviews how closeness to cover can transform vigilance; findings here depend on whether individuals view cover as a refuge or a place from which predators can launch an ambush (section 5.5). Sections 5.6 and 5.7 examine the influence of age, mother's parity, sex, and dominance; and section 5.8 tackles miscellaneous factors that affect vigilance, such as time of day. Section 5.9 outlines how vigilance changes with predator abundance. Section 5.10 broadens the focus to include cross-species comparisons in vigilance as they relate to body size and antipredator defenses.

Vigilance increases considerably when a predator appears in view. Scanning rates of European starlings (G. V. N. Powell 1974), yellow-eyed juncos (Caraco, Martindale, and Pulliam 1980), downy woodpeckers (Sullivan 1984), European goldfinches (Glueck 1987), and common teal (Poysa 1987b) all increase after exposure to a real or model predator, and the same is true in mammals (for example, hoary marmots: Holmes 1984). Watching a predator is distinct from being vigilant before a predator is seen, however. The latter serves to detect a predator as early as possible; the former may serve to monitor the predator's movements, its distance, and even its demeanor, such as hunger level and condition, any of which may be indicative of the probability of attack. Moreover, a failure to be vigilant under these two circumstances may lead to very different consequences. As a result, factors affecting vigilance in these two contexts are likely to differ. Certainly, the predator's proximity and behavior will override many of the factors known to influence vigilance under less precarious circumstances; for example, group size is likely to have little effect on individual vigilance if a predator is nearby. Conflating vigilance in the two contexts is therefore of little merit, so the rest of this chapter focuses on vigilance before a predator appears.

5.2 Distance from conspecifics and perceived group size

Strong negative effects of group size on vigilance in mammals and birds (tables 4.4 and 4.6) suggest that individuals fashion their vigilance to take account of the number of animals in their group. Yet it seems implausible to believe that animals monitor the exact number of individuals in their group, especially if they are in large groups. Furthermore, Lima's (1995a) experiment showing that yellow-eyed juncos treat food-deprived (nonvigilant) birds as additional group members demonstrates that individuals are not monitoring conspecifics' vigilance. The proximate cues that individuals use to set their own level of vigilance when they are in a group might therefore be exclusive of counting exact number of group members or detailed behavioral monitoring. Instead, some evidence points to individuals using both nearest neighbor distance and number of individuals close by as proximate measures of personal risk. Elgar, Burren, and Posen (1984) tried to investigate whether house sparrows adjusted their

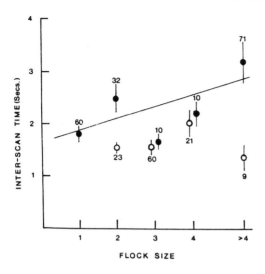

Figure 5.1 Mean interscan time (and SE) in seconds for house sparrows in different flock sizes when a barrier was placed across the feeder. Solid circles show data where birds were only on the observed side of the barrier; the line is a linear regression of these points. Open circles represent data where there was a solitary bird on the observed side of the barrier, but an increasing number of birds on the other side of the barrier; numbers refer to sample sizes (Elgar, Burren, and Posen 1984).

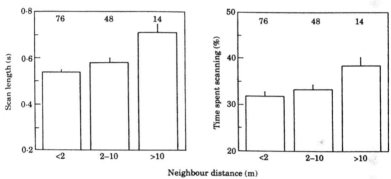

Figure 5.2 Mean (and SE) scan length (*left panel*) and percentage of time spent on vigilance (*right panel*) in common teal in relation to neighbor distance. Number of individuals are given above (Poysa 1994).

scanning rate according to the number of birds visible or the total number of birds foraging within a flock. Watching sparrows alighting on a parapet and then descending to a food source that was divided in two by a 12 cm high wall, they recorded interscan intervals of birds and plotted them against total flock size and separately against the size of the flock that the birds could watch on the side of the wall nearest to the observers. Sparrows clearly adjusted scanning rate according to the number of birds that they could see rather than the number that was present (fig. 5.1). In a second experiment, in which two feeders were placed either together or at different distances apart, sparrows greater than 1.2 m apart (not a great distance) scanned independently of one another.

Common teal adjust their scan length and time spent scanning according to nearest neighbor distance (fig. 5.2). When the usual effect of group size was held constant in these birds, nearest neighbor distance had a significant effect on scanning measures, but when the effect of nearest neighbor distance was held constant, group size did not affect measures of vigilance (see Rolando et al. 2001 for parallel findings in red-billed choughs). Bekoff's (1995) discovery that evening grosbeaks spend a greater proportion of time scanning when birds were arranged in a line while feeding as compared with being in a circle, where they could see one another more easily, corroborates these findings (see Caraco and Bayham 1982). Experiments with European starlings in which density was altered show that birds far from one another point their heads in random directions, and those at intermediate distances look at one another, whereas those close together point heads in opposite directions (Fernandez-Juricic, Siller, and Kacelnik 2004; Fernandez-Juricic, Smith, and Kacelnik 2005). All these studies suggest that birds are visually checking whether conspecifics are nearby. The picture for mammals is remarkably similar (table 5.1), although it must

Table 5.1 Studies of mammals demonstrating an effect of neighbor distance on vigilance

Species	Effect	Reference[a]
Black howler monkey	Less time scanning when 1 or more associates in the same tree	1
White-fronted capuchin	Less vigilant with neighbors present	2
Brown capuchin	Less vigilant with neighbors present	2
	More vigilant with increasing number of neighbors	3
White-faced capuchin	Time spent vigilant increases with the proportion of time in which there were no neighbors	4
Wedge-capped capuchin	Proportion of time spent vigilant increases with nearest neighbor distance	5
Redtail monkey	Time spent scanning by females decreases as number of neighbors in <2 m increases from 0 to 2	6
Yellow baboon	Proportion of time vigilant is greater for both males and females when dispersed rather than aggregated	7
Thomas's langur	Percentage of time vigilant was lower with increasing number of neighbors	8
Red colobus	Time spent scanning by males decreases as number of neighbors in <2 m increases from 0 to 2	6
Harbor seal	Increased proximity reduced individual scanning rate	9
Impala	Proportion of time spent looking declined with nearest neighbor distance	10
Buffalo	Proportion of time spent looking declined with nearest neighbor distance	10
Reedbuck	Proportion of time spent looking declined with nearest neighbor distance	10
Hoary marmot	Individuals feeding with >1 marmot <10 m away looked up less than those feeding alone	11
European rabbit	Rate of scanning decreased as proximity to consort increased	12

[a] 1. Treves, Drescher, and Ingrisano 2001; 2. van Schaik and van Noordwijk 1989; 3. B. T. Hirsch 2002; 4. Rose and Fedigan 1995; 5. S. R. Robinson 1981; 6. Treves 1998; 7. Cowlishaw 1998; 8. Steenbeck et al. 1999; 9. Krieber and Barrette 1984; 10. Underwood 1982; 11. Holmes 1984; 12. S. C. Roberts 1988.

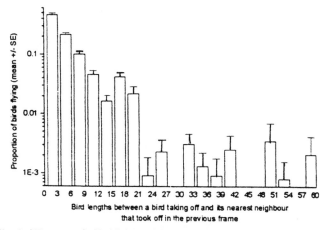

Figure 5.3 Mean (and SE) proportion of redshank flock that took flight with respect to the distance between the focal bird and its nearest neighbor to have taken flight within the previous 0.02 seconds ($N = 38$ attacks). The y axis has a log scale. Few birds took off immediately after distant birds took flight (Hilton, Cresswell, and Ruxton 1999).

be noted that Underwood (1982) found that the proportion of time reedbuck, impala, and buffalo looked up decreased rather than rose as nearest neighbor distance increased (see also B. T. Hirsch 2002). The influence of the number of close neighbors on reducing vigilance in red-tailed monkeys and red colobus but the lack of an effect of group size (Treves 1998) is particularly interesting, because it replicates findings in common teal and red-billed choughs.

While the importance of nearby neighbors is well supported by these studies, it is not clear whether it is the number of nearby neighbors, the proximity of the nearest neighbor, or a combination of the two that is preeminent. For species in which a rapid, coordinated departure is important because it might confuse the predator or because a tardy individual is subject to particular risk, it may be important to copy the behavior of neighbors; thus, close proximity might allow the behavior of flockmates to be monitored and copied easily. For common redshanks feeding on the Firth of Forth estuary in Scotland where attacks by sparrow hawks are common, coordinated departure of flock members is extremely impressive (Hilton, Cresswell, and Ruxton 1999). Videotapes of 38 attacks on flocks varying in size from 7 to 61 birds showed that all birds left within just 0.7 seconds of the first redshank to fly off! Redshanks that had their head up when the predator appeared flew earlier than those with head down, reiterating the importance of selfish vigilance (section 4.6.b). Birds were more likely to leave if a nearby bird had just departed than if a far-away conspecific had left. Indeed, the elapsed time taken for a bird to fly off after another departed was highly dependent on the distance of the departing bird (fig. 5.3). With sparrow hawks approaching at a speed of 25 m/sec, a delay of 0.1 sec would mean a 3 m reduction in distance. In this prey species, therefore, the behavior of nearby birds provides a cue for responding to danger and thereby allows birds to relax personal vigilance when conspecifics are very close but not when they are far away, irrespective of group size. In contrast, for species in which the predator takes a single prey individual from a subgroup of animals on the edge of the group closest to its approach, the number of nearby neighbors may be key to lowering vigilance through dilution of risk.

5.3 Position in the group

W. D. Hamilton's (1971) selfish-herd paper drew attention not only to the dilution of predation risk with increasing group size but also to a second important concept: that compared with individuals in the center of a group, those on the periphery would be more exposed to predation risk from predators that approached prey from the side. Empirical data from birds and mammals support the idea that individuals on the edge are more vulnerable (for example, Jakobsen and Johnsen 1988; FitzGibbon 1990c; see Stankowich 2003 for a review; section 8.7.b), although the best evidence comes from stationary groups, such as birds nesting in colonies (Kruuk 1964; Horn 1968; section 8.7.a). Thus, peripheral individuals should be more vigilant than those in the center of the group. Many studies have shown that peripheral individuals exhibit higher vigilance than those in the center (for example, birds: Lazarus 1978; Inglis and Lazarus 1981; Petit and Bildstein 1987; Keys and Dugatkin 1990). For instance, starlings feeding on Newcastle Town Moor in Britain spent fewer seconds per minute vigilant in central than in peripheral positions, and birds in midway positions fell between the two (fig. 5.4). This was reflected in more time spent pecking by central versus midway versus peripheral birds (Jennings and Evans 1980). Similarly, the number of scans made by different sex classes of capybaras was significantly greater than expected on the outside and less than expected on the inside of groups composed of up to 19 animals (table 5.2). Comparable results have been found in most other studies of mammals (table 5.3).

Unfortunately, many of these studies suffer from the problems outlined by Elgar (1989) that hound the relationship between group size and vigilance (Krause 1994). For example, individuals in the center of the group might be less vigilant because feeding competition is greater in central positions, because they are farther from cover that could hide a predator, because they are farther from the observer (Bednekoff and Ritter 1994), because they are monitoring neighbors rather than predators (Hoogland 1979; Fernandez-Juricic et al. 2004), because they simply cannot see their surroundings very well (Scheel 1993a), or simply because periphery is defined in such a way that it includes only the most vulnerable individuals (Stankowich 2003). In addition, in groups moving through the environment, some peripheral individuals will be at the rear and hence vulnerable to attack (Underwood 1982). Age, sex, and

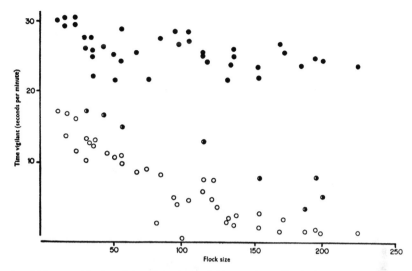

Figure 5.4 Time spent vigilant (seconds/minute) plotted against flock size in starlings. Solid circles show peripheral birds; half-filled circles, birds in the midway position; open circles, birds in central positions. Each point represents the mean for birds in that position in one of the flocks studied (Jennings and Evans 1980).

Table 5.2 Number of times (scans) that each class of individual capybara was observed in the alert posture in each of 4 imaginary concentric zones

		Observed frequency	Expected frequency
Females	Innermost	81	106
	Intermediate	177	152
	Peripheral	8	13
	Outside the group	54	49
Dominant males	Innermost	50	69
	Intermediate	176	156
	Peripheral	12	14
	Outside the group	44	42
Subordinate males	Innermost	6	24
	Intermediate	114	111
	Peripheral	23	20
	Outside the group	158	146

Source: Yaber and Herreva 1994.
Note: Expected values calculated based on the number of scans that each class was in each zone and the mean alert rate of the class, assuming equal probability of being alert in each zone.

dominance could also play a role if particular classes are found in certain locations (Black and Owen 1989). Among some mammals, for instance, highly vigilant mothers with young offspring move to the rear of the group when a predator is sighted (fig. 12.4). Each of these factors could drive elevated vigilance levels in peripheral individuals in addition to, or instead of, enhanced risk of predation. At present, studies of mammals, at least, frequently measure the effects of location separately for sex and sometimes age and parity but

Table 5.3 Studies of mammals showing an effect of increased vigilance between individuals on the periphery of the group compared with the center

Species	Notes on controlling factors	Reference[a]
White-fronted capuchin	Sexes examined separately: effect seen in both	1
Brown capuchin	Sexes examined separately: effect seen in both	1
Wedge-capped capuchin	Adult and subadult males and females, and juveniles examined separately: effect only found for adult and subadult females	2
Harbor seal	No factors considered	3
African elephant	No factors considered	4
Burchell's zebra	Examined by male, female with and without young, and neonate: effects for all except neonate but no statistics	4
Pronghorn antelope	Examined in relation to group size: no effect found	5
Impala	No factors considered	4
	Compared center, front, side, and rear in open and closed habitats: center least vigilant in closed habitat	6
	Compared center, front, side, and rear: center has lowest vigilance	7
	Examined by male, female, and mother in center, front, side, and rear: center and mothers most vigilant but no statistics	7
Wildebeest	Examined by male, female with and without young, and neonate: effects seen but no statistics	4
	Compared center, front, side, and rear in open and closed habitats: rear least vigilant in open, effect in closed not clear	6
	Compared center, front, side, and rear: center has lowest vigilance	7
	Examined by male, female, and mother in center, front, side, and rear: center and mothers most vigilant but no statistics	7
Tsessebe	Compared center, front, side, and rear in open and closed habitats: measures of vigilance equivocal	6
Springbok	Examined adult females and juveniles: effects seen for both	8
Buffalo	No factors considered	4
	Examined by night and day for adult males, subadult males, adult females, and subadult females: effects at night in peripheral adult and subadult males	9
	Compared center, front, side, and rear in open and closed habitats: no effect	6
Spanish ibex	Males, females, and juveniles examined separately: effect found in adult females and juveniles	10
Waterbuck	No factors considered	4
Uganda kob	Examined by male, female with and without young, and neonate: effects confirmed but no statistics	4
Reedbuck	Compared center, front, side, and rear in open and closed habitats: center least vigilant in closed habitats	6
Black-tailed prairie dog	No factors considered	11
Capybara	Females, dominant males, and subordinate males examined separately: effect seen in each of them (table 5.2)	12

[a] 1. van Schaik and van Noordwijk 1989; 2. J. G. Robinson 1981; 3. Terhune and Brilliant 1996; 4. Burger and Gochfeld 1994; 5. Lipetz and Bekoff 1982; 6. Underwood 1982; 7. Hunter and Skinner 1998; 8. Bednekoff and Ritter 1994; 9. Prins and Iason 1989; 10. Alados 1985; 11. Hoogland 1979; 12. Yaber and Herrera 1994.

make few attempts to control for other confounding variables. While peripheral individuals are more vigilant than central ones in most species (but see Black et al. 1992), it is not clear that this unambiguously reflects antipredator behavior associated with heightened predation risk.

5.4 Sentinels

In a small number of unrelated bird and mammal taxa, an individual acts as a sentry, sentinel, or guard while other members of the group engage in some other activity, usually foraging. Empirical data show that sentinels are more likely to spot predators at long distances than others (Manser 1999); sentinels then give alarm calls or depart and thus alert other group members. Small groups have one guard, whereas larger groups may have more; individuals rotate sentry duty. Sentinels have been reported in monogamous birds and mammals such as Carolina wrens (Morton and Shalter 1977), American crows (D'Agostino, Giovinazzo, and Eaton 1981), red-winged blackbirds (Beletsky, Higgins, and Orians 1986), and klipspringers (Tilson 1980; Dunbar and Dunbar 1974), but they have received more attention in group-living species, avian examples of which include jungle babblers (Gaston 1977) and Arabian babblers (Wright et al. 2001), white-browed sparrow weavers (J. W. H. Ferguson 1987), yellow-rumped and red-rumped caciques (Feekes 1981), pinion jays (Balda, Bateman, and Foster 1972), and Florida scrub jays (McGowan and Woolfenden 1989; Hailman, McGowan, and Woolfenden 1994); mammal examples embrace dwarf mongooses (Rasa 1986, 1989a, 1989b), meerkats (Moran 1984; Clutton-Brock, O'Riain, et al. 1999), and vervet monkeys (Horrocks and Hunte 1986; Baldellou and Henzi 1992). One might also think of a parent in charge of offspring as a sentinel, but as juveniles are less adept at spotting predators (section 5.6), the literature treats maternal vigilance as a form of investment.

Sentinels usually occupy exposed positions, such as termite mounds, boulders, or high branches, and consequently were assumed to suffer higher rates of predation than group members foraging below. In monogamous species, the concomitant benefits of sentinel behavior were sought in the context of parental care, because individuals were thought to call to warn their mate and possibly attendant offspring of danger. For instance, male red-winged blackbirds utter continuous repetitive calls while perched atop vegetation near the nest on which the female is incubating eggs. In response to the appearance of a predator, he switches to (one of a number of) other call types. When males are absent from the territory and calling has ceased, females spend more time being vigilant off the nest (Beletsky, Higgins, and Orians 1986; Beletsky 1989). Sentinels can thereby reduce their partner's energy expenditure or allow their partner more time for foraging. In group-living species, however, benefits of sentinel behavior were initially sought through kin selection or reciprocal altruism.

Supposed high survivorship costs of guard duty do not reflect empirical data, however (Bednekoff 1997). Sentinels are the first to detect danger in 92.3% of cases in dwarf mongooses (Rasa 1987) and 92.5% of cases in Florida scrub jays (McGowan and Woolfenden 1989), far more than would be expected given the ratio of sentinels to other group members. Furthermore, both dwarf mongoose and meerkat sentinels choose guard posts close to protective cover (Rasa 1989b; Clutton-Brock, O'Riain, et al. 1999), whereas the principal threats to scrub jays come from terrestrial predators far below the sentinel's perch, suggesting that sentinels are safer than other group members. Indeed, in 2000 hours of observations, no meerkat guard was ever attacked or killed by predators (Clutton-Brock, O'Riain, et al. 1999). In short, it is now thought that sentinels are safer than foragers, suggesting the behavior is a selfish means of reducing risk. Crucially, solitary individuals take on guard duty, in meerkats spending 12%–22% of their time as sentinels (Clutton-Brock, O'Riain, et al. 1999). How, then, do interchanges between sentinels appear so orderly, with groups rarely having either zero or many sentinels simultaneously on guard?

Using a dynamic-state model, Bednekoff (1997, 2001) ran simulations when the spread of information about predators is fully shared between both sentinels and foragers, when it passes only from sentinels to foragers, or when no information about threat is transferred between individuals. In his models, foragers could encounter 0, 1, or 2 units of food per unit time while sentinels encountered none, but sentinels always had a greater probability of detecting predators. Bednekoff found that both personal energetic reserves and extent of information sharing dramatically alters the relationship between the actions of others and the decision to take on guard duty. In both of the cases in which information is shared, the individual is far more likely to become a guard if no others are on duty but is much less likely if one or more sentries are there already (fig. 5.5). If information is never shared, predation risk increases as the number of sentinels remove themselves from danger, leaving foragers at greater risk. As a result, individuals would be willing to take on guard duty at increasingly lower energetic reserves as the number of sentinels rises (fig. 5.5). Thus, when information is shared, it pays to be a sentinel when no one else is, provided reserves are not dangerously low, and it pays to forage if a sentry is on duty. If information is not shared, every individual should alternate between guarding and foraging, probably in unison.

Empirical support for the model comes from free-living meerkats (Clutton-Brock, O'Riain, et al. 1999). Individuals fed with hard-boiled egg the day before increased their sentry duration by 30%, and the probability of an individual going on guard was twice as high when no other guard was on duty. If an individual tried to assume guard duty while another was posted, one of the

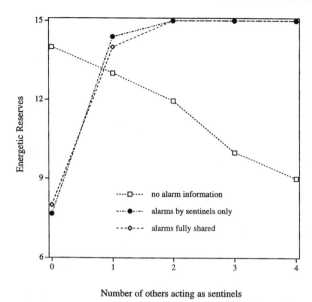

Number of others acting as sentinels

Figure 5.5 Optimal decision rules for a member of a group of 5 animals. The animal makes decisions based on its own energetic reserve levels and the actions of other group members. For each situation, it forages if its reserves fall below the line in question and acts as a sentinel if its reserves fall above the line. Energetic reserves cannot increase above 15 in this model (Bednekoff 1997).

two usually terminated its guarding. Although an individual meerkat seldom assumed responsibility for two consecutive bouts of guarding, there was no order to sentry duty. Indeed, latency to resume guarding was reduced after observers interrupted the animal's bout of guarding and was further reduced if it was additionally offered egg, indicating that energetic reserves were important in decision making. That unrelated meerkats guarded as much as relatives speaks against kin selection, and absence of cost speaks against reciprocal altruism, leaving self-interest as the most likely explanation for sentinel behavior, with by-product mutualism as the explanation for coordination between sentinels, because individuals and others both benefited. Similar state-dependent sentinel behavior has been recorded in Arabian babblers, where food-supplemented individuals increased sentinel effort (Wright, Maklakov, and Khazin 2001; Wright et al. 2001).

On a proximate level, sentinels are able to rotate guard duty more efficiently by communicating with foragers (the "watchman's song": Wickler 1985). Vocalizations have been reported in dwarf mongooses (Rasa 1986), white-browed sparrow weavers (J. W. H. Ferguson 1987), Florida scrub jays (McGowan and Woolfenden 1989), and meerkats (Manser 1999), which circumvents the considerable cost of foragers having to check visually whether anyone is on duty (P. I. Ward 1985). In meerkats, sentinels utter continuous soft vocalizations while on duty that decline in number just before the guard relinquishes responsibility (Manser 1999). The probability of another group member taking up duty is less likely when the guard is vocalizing. Moreover, the period of overlap, when two sentinels are on duty, was shorter when the sentinel was calling. Thus, the total time that the group was guarded rose and the overlap between sentinels declined as a result of vocalizations. While vocalizations are not essential for guard rotation, they help it run more smoothly.

Explanations for the taxonomic distribution of guarding behavior are less developed, but models that consider the evolution of coordinated vigilance bear on this issue. Realistic assumptions that individuals can sometimes detect a predator without raising their head from feeding and that information about predator presence does not always spread through a group each have the effect of decreasing the value of coordinated vigilance. When individuals have a low probability of detecting a predator without raising their head but a high probability of being warned of danger if another group member detects a predator, coordination, such as that seen among groups with sentinels, is favored (Fernandez-Juricic et al. 2004; see also Rodriguez-Girones and Vasquez 2002). Local ecology will alter the importance of these factors. Thus, sentinels are associated with moderately open habitats, where information can easily pass between group members; for example, vervet monkeys feeding in open maize fields but not in dense bush (Horrocks and Hunte 1986), meerkats in semidesert habitats but not banded mongooses in woodland (but see J. W. H. Ferguson 1987). In addition, moderately open but not completely open habitats may favor guarding, because sentinels can watch from the relative safety of trees (see also Yasukawa, Whittenberger, and Nielsen 1992), thereby reducing the cost of sentinel behavior. While these observations explain in a qualitative sense why sentinels are absent from very open or dense habitats, they do not properly explain why there are so few examples from species inhabiting semiopen habitats.

5.5 The influence of cover

In general, one might expect that small birds and mammals would be less likely to be killed by a predator in closed than in open habitats (Watts 1990) and would therefore relax vigilance when in cover. Cover affects vigilance in different ways (Lima 1987b), because some species are ambushed by predators from cover (for example, Amat and Masero 2004), whereas others use cover to hide from predators (for example, Watts 1990). Observing ruddy turnstones and purple sandpipers along a stretch of rocky outcrops and sandy mud in western Scotland, Metcalfe (1984a) saw that the species respectively increased scanning duration and scanning rates when their vision was restricted by rocks close by. In contrast, in another study, house sparrows showed greater vigilance with increased distance from cover (Barnard 1980). Metcalfe interpreted his findings in the light of the shorebirds' escape tactics: he surmised that as waders have no refuge and instead form maneuverable flocks when attacked, they are forced to detect predators as early as possible by using an unrestricted view. Visual obstruction may have also impaired their ability to observe other shorebirds, which could cause them to misjudge flock size and others' behavior (Harkin et al. 2000). Barnard construed the sparrows' behavior as an adaptive response to increased predation risk away from the protection afforded by vegetation. These two sorts of findings are reflected in studies of other bird and mammal species (tables 5.4 and 5.5).

To test this idea in a single species, Lazarus and Symonds (1992) constructed a frame that either was filled with loose plant cuttings that did not restrict visibility severely and thus acted as protective cover, or contained a solid wooden board that obscured the birds' view and that could not be used as a refuge (obstructive cover). They then fed house sparrows and European starlings near (0.5 m) or far (5 m) from the two sorts of frame. The researchers

Table 5.4 Studies of birds in which individual vigilance levels changed in relation to cover

Species	Measure	Reference[a]
Vigilance increases further from cover		
House sparrow	Birds look more often in open fields than in cattle sheds	1
Mixed flocks of house sparrows, chaffinches, European starlings, robins and blackbirds, and gray partridges	Vigilance increased with distance to protective cover	2
No clear effect		
Common teal	Scanning length was shortest at the intermediate distance to cover but increased again as distance increased	3
White-crowned sparrow	No change in scanning rate with distance from cover	4
Dark-eyed junco	Increasing distance to cover had no consistent effect on vigilance	5
Vigilance increases near to cover		
Bald eagle	Raised heads more frequently when feeding near cover than when feeding farther away	6
Mixed flocks of house sparrows, chaffinches, European starlings, robins and blackbirds, and gray partridges	Vigilance decreased with distance to obstructive cover	2
House sparrow	Birds spend less time scanning when feeding in a patch 5 m than in 0–5 m from cover	7
White-browed sparrow weaver	Sparrow weavers feeding in dense grassland had vigilant birds perched nearby during a greater proportion of time than did birds feeding in open grassland	8

[a] 1. Barnard 1980; 2. Lazarus and Symonds 1992; 3. Poysa 1994; 4. Slotow and Rothstein 1995; 5. Lima 1988b; 6. S. K. Knight and Knight 1986; 7. Lima 1987b; 8. J. W. H. Ferguson 1987.

Table 5.5 Studies of mammals in which individual vigilance levels changed in relation to cover

Species	Measure	Reference[a]
Vigilance increases further from cover		
Tammar wallaby	More vigilant further from cover at high predation risk site	1
Redtail monkey	Number of scans increases as density of foliage declines	2
Blue monkey	Number of scans increases as density of foliage declines	2
Yellow baboon	Proportion of time vigilant increases at 5 m and 50 m from tree or cliff	3
Warthog	Scan more in open woodland and short grass than in dense woodland or tall grass	4
Moose	Females show increased vigilance as move from 0 m to 20 m from cover	5
Wildebeest	Scan less when near thick bush or tall grass but more in short grass	4
Dall's sheep	Percentage of time that females were vigilant increased with distance to cliffs but was mollified by larger group size	6
Hoary marmot	Scan more often when >50 m from talus	7
Yellow-bellied marmot	Juvenile marmots spend more time looking up when density of safety burrows was low	8
Guinea pig	Scanning rate increases at 1–2 m from cover	9
Vigilance increases near to cover		
Western gray kangaroo	More vigilant nearer to cover at high predation risk site	1
Vervet monkey	Frequency and duration of vigilance increases in dense vegetation	10
Pronghorn	Duration and frequency of scans greater in shrub habitat than in meadows	11
Impala	Rate of looking and percentage of time looking greater in closed habitats	12
Wildebeest	Rate of looking and percentage of time looking greater in closed habitats	12
Tsessebe	Rate of looking and percentage of time looking greater in closed habitats	12
Springbok	More vigilant when approaching clumps of trees than moving in the open	13
Reedbuck	Rate of looking greater in closed habitats	12

[a] 1. Blumstein and Daniel 2002; 2. Cords 1990; 3. Cowlishaw 1997b, 1998; 4. Scheel 1993a; 5. White and Berger 2001; 6. Frid 1997; 7. Holmes 1984; 8. Carey and Moore 1986; 9. Cassini 1991; 10. C. Chapman 1985; 11. Goldsmith 1990; 12. Underwood 1982; 13. Bednekoff and Ritter 1994.

found that for both species the proportion of observations during which an individual bird was vigilant increased with distance to protective cover but decreased with distance from obstructive cover (table 5.6); thus, birds were attributing differential risk to the cover depending on its nature. Specifically, they were treating the first type of frame as a refuge but the second, solid frame as an impediment to vision. Nonetheless, the way in which cover affects vigilance is now believed to be more complex than this (Arenz and Leger 1999a, 1999b).

(i) Where species escape predation through flight and therefore require an unimpeded view of their surroundings to detect predators as soon as possible

Table 5.6 Mean proportion of observations in which house sparrows and European starlings
were vigilant at different distances from different types of cover

	Protective cover		Obstructive cover	
	Near	Far	Near	Far
Sparrows	0.40	0.62	0.68	0.42
Starlings	0.39	0.63	0.70	0.43

Source: Lazarus and Symonds 1992.

in order to take evasive action (Arenz and Leger 1997a; Amat and Masero
2004), they will scan more in habitats with cover because it limits clear views
and makes the probability of failing to detect a predator more likely (for example, shorebirds: Metcalfe 1984a; Cresswell 1994a; and impala: Underwood
1982). Treves (2002) championed this view, because he found that three species
of monkeys failed to demonstrate predictions about vigilance based on vulnerability. He predicted that foraging monkeys exposed near the ground, near
tips of branches, in sparse foliage, or while foraging upside down faced a high
risk of predation and would thus be more vigilant. Observations did not match
these expectations; rather, monkeys were most vigilant near the trunk of the
tree, where they were surrounded by large branches, or in dense foliage, where
cover obstructed their ability to see; they were least vigilant at branch tips,
where visibility was good. Specifically, scan rate was much higher when monkeys were in trees with large and many leaves (fig. 5.6). In fact, habitat visibility
is increasingly regarded as a critical variable in structuring vigilance in other
species that are accustomed to ambush predation (for example, vervet monkeys: C. Chapman 1985; yellow baboons: Cowlishaw 1998). Alternatively, prey
might scan more in habitats without cover if they can evaluate efficacy of scanning and find it to be of little use (for example, warthog and wildebeest: Scheel
1993a). (ii) Where species rely on refuges to escape predators, individuals far
from cover might scan more than those close to cover who can escape into a
bolt hole, such as hoary and yellow-bellied marmots (Holmes 1984; Carey and
Moore 1986), or who can hide from predators in dappled light that some types
of cover afford, such as moose perhaps (White and Berger 2001). Alternatively,
individuals in the open might feed more rapidly and hence reduce vigilance in
order to spend less time exposed (Lima 1987a, 1987b), such as gray squirrels
(Newman et al. 1988). (iii) Where species are subject to predation by concealed
predators, they will be more vigilant in or close to cover; possible examples include vervet monkeys (C. Chapman 1985) or springbok (Bednekoff and Ritter
1994). In short, the way in which prey detect predators and escape from them,
and the way in which predators approach prey, may all be involved in determining how cover influences vigilance.

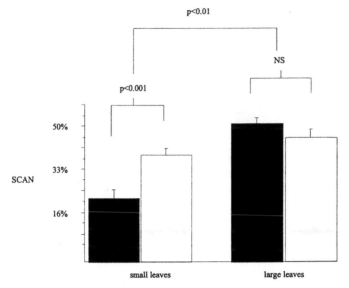

Figure 5.6 Mean (and SE) percentage of time monkeys spent scanning (SCAN) in relation to the size of leaves and number of leaves in selected food-tree species; black bars indicate few leaves; white bars, many leaves. Red colobus and redtail monkeys are pooled (Treves 2002).

5.6 Age and parity

In species in which neonates and juveniles are subject to greater predation risk as a result of their poor ability to flee, they are expected to be more vigilant than adults. There is abundant evidence that young mammals (Sibly et al. 1997; FitzGibbon and Lazarus 1995) and young birds (Solonen 1997) are less able to escape predators than adults, and there is some evidence that juvenile mammals (for example, hoary marmots: Holmes 1984; yellow-bellied marmots: Carey and Moore 1986) are more vigilant than adults. Nevertheless, the majority of studies show that juveniles are actually less vigilant than adults (for example, white-fronted geese: M. Owen 1972; yellow-eyed juncos: K. A. Sullivan 1988; white-winged choughs: Heinsohn 1987; table 5.7 for mammals). How can we explain this? One set of possibilities is that young animals are less competent than adults, making vigilance less profitable. (i) They might fail to notice dangerous predators. Cheetah cubs were less likely to attend to predators that passed by the family group than their mother (Caro 1994a). (ii) Young animals might be slower to notice predators than adults. The percentage of occasions when predators were spotted by cheetah cubs before they were spotted by their mother was low throughout most of the period of cub dependence (fig. 5.7). (iii) Young animals might observe predators but fail to recognize them as a threat. Vervet monkeys were less likely to give alarm calls to raptors than were juveniles or adults, despite looking up at the sky (Seyfarth

Table 5.7 Mammalian species examined for age differences in antipredator vigilance

Common name	Type of association with age	Reference[a]
Positive		
Wedge-capped capuchin	Positive	1
Burchell's zebra	Positive	2
Wildebeest	Positive	2
Klipspringer	Positive	3
Spanish ibex	Positive	4
Bighorn sheep	Positive	5
Uganda kob	Positive	2
Black-tailed prairie dog	Positive	6
California ground squirrel	Positive	7
Columbian ground squirrel	Positive	8
	Positive	9
Thirteen-lined ground squirrel	Positive	10
Negative		
Impala	Negative	2
Hoary marmot	Negative	11
No effect		
Nine-banded armadillo	No effect	12
African elephant	No effect	2
Fallow deer	No effect	13
Springbok	No effect	14
Golden marmot	No effect	15
Mixed results		
Eastern gray kangaroo	Positive	16
	No effect	17
Yellow baboon	Negative (females)	18
	No effect (males)	18
African buffalo	Negative	2
	No effect	19
Yellow-bellied marmot	Negative	20
	Positive	21

Source: Arenz and Leger 2000.

[a] 1. J. R. de Ruiter 1986; Fragaszy 1990; 2. Burger and Gochfeld 1994; 3. Tilson 1980; 4. Alados 1985; 5. Risenhoover and Bailey 1985; 6. Loughry 1993; 7. Loughry and McDonough 1989; 8. Betts 1976; 9. MacHutchon and Harestad 1990; 10. Arenz and Leger 1997b; 11. Holmes 1984; 12. McDonough and Loughry 1995; 13. Schall and Ropartz 1985; 14. Bednekoff and Ritter 1994; 15. Blumstein 1996; 16. Heathcote 1987; 17. Colagross and Cockburn 1993; 18. Alberts 1994; 19. Prins and Iason 1989; 20. Carey and Moore 1986; 21. Armitage and Chiesura 1994.

and Cheyney 1986). (iv) Young might become aware of the predator but orient in the wrong direction (for example, mule deer: Lingle and Wilson 2001). Acting alone or together, each factor would reduce the effectiveness of vigilance in juveniles and hence the selective pressures acting to maintain it.

A second possibility is that juveniles have greater nutritional needs than adults, are forced to feed more, and hence have less opportunity to be vigilant. Arenz and Leger (2000) videotaped free-ranging thirteen-lined juvenile ground squirrels that they supplemented with either high-energy peanut butter and oats or with low-energy food (lettuce). Over a period of 2½ weeks, the

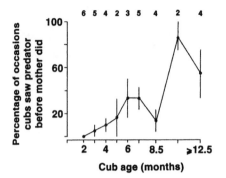

Figure 5.7 Mean (and SE) percentage of occasions when potential mammalian predators were seen by a family of cheetahs in which cubs saw the predator before their mother did, plotted against cub age in months. Number of litters are shown above (Caro 1994a).

juveniles fed on low-energy diets reduced their time spent vigilant, while those on the high-energy diet increased time spent vigilant on a par with adults. This was because low-energy juveniles spent an increasing percentage of time feeding, whereas their counterparts supplemented with oats and peanut butter did not. Juveniles must forage in order to grow and to deposit fat for hibernation, but adults only have to lay down fat for the winter, which may explain juvenile-adult differences in this and related species.

As very young animals are notoriously vulnerable to predation, it is not surprising that parents of neonates and hatchlings are far more vigilant than females without young. Among birds, the effect is seen across all types of breeding systems. For example, in cooperatively breeding Siberian jays, where young remain on their natal territory for up to 3 years, individuals with no close relatives show the predicted decline in vigilance with group size, but parents who have offspring feeding alongside them increase their vigilance, and their rate of vigilance is higher than breeders in other groups that have no offspring remaining with them (fig. 5.8). Similarly, monogamous barnacle geese show higher levels of vigilance when they overwinter in the company of offspring than when none are present (Black and Owen 1989).

Among mammals it is incontrovertible that mothers with young are far more vigilant than females without young (for example, Frid 1997; Swaisgood, Owings, and Rowe 1999; Toigo 1999). Most data come from open-country species, in which early predator detection might be the key that allows slow-moving young to escape, putting maternal vigilance at a premium. As examples, female elk with calves are more vigilant than any other age-sex class (Laundre, Hernandez, and Altendorf 2001; Childress and Lung 2003); female elephant, Burchell's zebra, impala, wildebeest, buffalo, waterbuck, and Uganda kob with young are all more vigilant than conspecific females without young (Burger and Gochfeld 1994; see also FitzGibbon 1993; Hunter and Skinner 1998); and reproductive female Eastern gray kangaroos are more vigilant than any other age-sex class (Colagross and Cockburn 1993). In contrast, moose

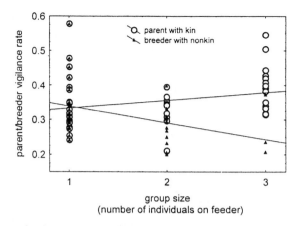

Figure 5.8 Nepotistic vigilance behavior of Siberian jay breeders and parents. Vigilance rates of parents/breeders are given for different group sizes when feeding together with retained offspring (*open circles*) or with unrelated immigrants (*filled triangles*). Data for parent/breeder alone is shared for both regressions. Parent/breeder alone: $N = 24$; parent with offspring: $N = 12$; breeder with unrelated immigrant: $N = 12$; parent with offspring and other individual: $N = 15$; breeder with 2 unrelated individuals: $N = 3$ (Griesser 2003).

mothers are no more vigilant than females without young, that spent 7.1% of their time vigilant, but they are more vigilant when their calves are active (12.6% of the time) than when they are recumbent (2.6% of the time). As active calves are more likely to be seen by predators than calves that are lying down, mothers are sensitive to offspring vulnerability (White and Berger 2001; Molvar and Bowyer 1994; see also Hauser 1988a; Swaisgood, Owings, and Rowe 1999).

Maternal vigilance is a type of parental care, and because it reduces maternal food intake (for example, Lipetz and Bekoff 1982), it may constitute a form of parental investment in the sense of lowering the mother's ability to invest in other offspring (Clutton-Brock 1991). Vigilance is an unusual type of parental investment, however, in that it may benefit all offspring instead of just one (section 10.11). Where maternal vigilance reduces the probability of all offspring being killed by a predator, mothers would be expected to show greater vigilance with larger litters, because a given amount may lead to greater reproductive success than for mothers with smaller litters (Lazarus and Inglis 1986). This situation pertains to cheetah mothers when cubs are less than 4 months old, a time when most cubs in a litter are killed in a given predatory attack by lions. Here mothers might be expected to show greater vigilance with larger litters, and this was confirmed. Later, however, when cubs are more mobile, only one cub is likely to be taken by a lion during an attack. Now mothers would be expected to be vigilant independent of litter size, a prediction that was again supported (Caro 1987). Nevertheless, in situations in which increasing brood size results in young making more noise or foraging farther away from their parents, the probability of predation may rise, so we would expect parental vigilance to increase whether or not vigilance benefits all offspring equally. Positive relationships between parental vigilance and brood size have been found in some studies of geese where nidifugous young do stray from their parents (Schindler and Lamprecht 1987; Sedinger and Raveling 1990; Forslund 1993), although findings are by no means consistent (Lessels 1987).

5.7 Sex differences and dominance

Patterns of vigilance vary between the sexes, with male mammals usually being more vigilant than females (for example, white-faced capuchin: Rose and Fedigan 1995; yellow baboon: Cowlishaw 1998; vervet monkey: Isbell and Young 1993b; Baldellou and Henzi 1992; white-fronted and brown capuchins: van Schaik and van Noordwijk 1989; harbor seal: Renouf and Lawson 1986; Burchell's zebra, waterbuck, and wildebeest: Burger and Gochfeld 1994) than vice versa (for example, Thomson's gazelle: FitzGibbon 1990c). Greater vigilance in males is often thought to stem directly from intrasexual competition rather than greater attention to predators, however (table 4.1). For example, in polygynous vervet monkeys and harbor seals, males become more vigilant as the breeding season approaches. Nonetheless, males in some polygynous species may be vigilant for antipredator reasons. Male capuchin monkeys are far better at detecting model eagles and snakes than are females (van Schaik and van Noordwijk 1989); female vigilance may be constrained by foraging activities in these species. In addition, male vigilance can also alert other members of the social group to the threat of predation.

In some monogamous species, such as klipspringers, males spend significantly more time standing and looking around than do females. Enhanced vigilance in males of these species may allow the female to spend more time feeding, with beneficial consequences for gestation and lactation (Dunbar and Dunbar 1974). In gallinaceous birds, there is stronger evidence that male vigilance substitutes for female vigilance and thereby grants her greater opportunities to forage. White-tailed ptarmigan males spend 22%–30% of their time vigilant while accompanying their mate before the onset of incubation (Artiss and Martin 1995) and are more likely to become vigilant in the presence of a foraging female (Artiss, Hochachka, and Martin 1999). Also, these males were more vigilant in a year when terrestrial carnivores and raptors were prevalent but were no more vigilant when females were fertile or when males were more common in the population. These latter results suggest males were on the lookout for predators rather than other males. Interestingly, the proportion of time females fed was positively related to male vigilance (fig. 5.9). In gray partridge, a related and again monogamous species, females spent more time near more-vigilant than less-vigilant males in a laboratory choice experiment

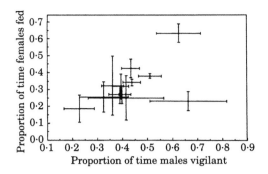

Figure 5.9 Mean (and SE) proportion of time that female white-tailed ptarmigan spent feeding relative to the mean (and SE) proportion of time their mates were vigilant during preincubation periods. Data points are for individual pairs (Artiss and Martin 1995).

(Dahlgren 1990; see also Beani and Dessi-Fulgheri 1995). If female preferences for more-vigilant males are widespread across monogamous species, male vigilance may be determined by both natural and sexual selection.

In certain species, male birds are more vigilant than females when approaching their nest. A northern mockingbird male spends more time and uses a greater number of perches when approaching and leaving the nest than does its mate. When a predator approached the vicinity of the nest, the male became silent until his mate arrived; but then he called, in response to which she interrupted her visit and perched until the predator had departed (Breitwisch, Gottlieb, and Zaias 1989). Sex differences in nest visitation are unknown for most species, however. Among nidifugous species where offspring quit the nest, male barnacle geese are also more vigilant than their partner in the company of their young (Forslund 1993).

In some species, vigilance is additionally influenced by dominance, although the function of such vigilance is almost certainly to monitor group members rather than detect predators (table 4.1). In multimale primate groups, dominant males may scan more than subordinate males because they are trying to detect other males attempting to join the group (Baldellou and Henzi 1992; Rose and Fedigan 1995) or because they are trying to prevent subordinate males within the group from mating. Having more relatives in their group than subordinates, they might scan more to protect a greater number of offspring either from predators or from other groups attempting to steal resources. These latter two possibilities also apply to species in which dominant females are more vigilant than other females, such as ring-tailed lemurs (Gould, Fedigan, and Rose 1997).

In various species, subordinates are more vigilant than dominants, however (table 4.1). Sometimes this difference can be attributed to social factors, at other times to antipredator factors. For example, that juvenile daughters of low-ranking female yellow baboons glance more frequently than daughters of high rankers, and that sons of high rankers glance more than sons of low rankers are difficult to explain as adaptations to predation risk (Alberts 1994).

Explanations that relate to antipredator behavior are nonetheless possible. Willow tits forage in small groups of one to six individuals, and there is a marked reduction in alertness as flock size increases (Ekman 1987). In addition, alertness declines when tits forage higher up in spruce and pine. Removal experiments show that dominants exclude subordinates from the upper sections of trees in winter, and as a consequence of foraging lower down, subordinates are forced to be vigilant and hence accrue reduced benefits of grouping compared to dominants. Although it is still worthwhile for subordinates to join dominants in order to reduce personal vigilance, the reduction in personal vigilance time for subordinates in flocks with four as opposed to two members was only half that for dominants. More generally, subordinates are often found on the periphery of the group where vigilance is expected to be greatest. Thus, capybara subordinates are more likely to be located on the periphery of the group where they are very vigilant (Herrera and Macdonald 1993), but in the center of the group they are no more vigilant than would be expected (table 5.2). Forcing subordinates into exposed positions is a common phenomenon in animal groups (see Hegner 1985).

5.8 Miscellaneous factors

Levels of individual vigilance may be altered by local habitat characteristics that animals perceive as safe or dangerous. California ground squirrels, for example, are more vigilant when close to rocks or burrows (Leger, Owings, and Coss 1983), and chaffinches are more vigilant when singing from higher perches in saplings (Krams 2001). Vigilance is also influenced by a raft of environmental factors, including time of day, temperature, level of starvation, body condition, and season (Pravosudov and Grubb 1998a).

Looking first at time of day, predation risk is usually greater at night, when many predators are active. Data from several studies of African ungulates show that vigilance is higher at night (for example, Prins and Iason 1989; Bednekoff and Ritter 1994). For example, scan rates of Burchell's zebras and wildebeests were higher on the Serengeti Plains on moonlit nights; lions are more likely to hunt at night than during the day (Schaller 1972; Scheel 1993a). Furthermore, zebras scanned significantly more on nights when the moon was clearly visible than on nights when the skies were cloudy which may contribute to reducing lion hunting success on bright nights (see van Orsdol 1984). Similarly, dark-eyed juncos showed high levels of vigilance in the dim light of very early morning, when both nocturnal and diurnal predators may be active (Lima 1988c). An additional factor promoting high levels of vigilance at night is that prey may find it difficult to detect predators under conditions of low light intensity. At dawn, under experimental conditions, European minnows demonstrated aversive behavior patterns to a model pike predator far later on its approach than during conditions of daylight, indicating that they found it more difficult to detect danger at low light intensity (Pitcher and Turner 1986).

As increasing amounts of vigilance are associated with reduced time spent feeding (tables 4.4 and 4.6), individuals that need to feed rapidly or for long periods of time will have less opportunity to be vigilant (Milinski and Heller 1978; McNamara and Houston 1986). This can be seen over the short, medium, and long term. In the short term, birds that are hungry or starving drastically reduce the amount of time devoted to vigilance. For instance, tufted titmice, which must feed after sunrise, show increasing levels of vigilance as the morning progresses (Pravosudov and Grubb 1998a). Birds are less vigilant at cooler temperatures or when their body mass is low because they must feed to obtain energy for maintaining metabolism (Pravosudov and Grubb 1998a, 1998b). Competition over food may also reduce personal vigilance (Caraco 1979; Barnard 1980; Elgar 1987) because individuals will attempt to increase the rate at which they process food (Milinski and Parker 1991).

Over a medium timescale, Bachman (1993) studied the way in which body condition affects the trade-off between vigilance and foraging in juvenile female Belding's ground squirrels. Squirrels were captured and fed either an energy-rich peanut butter mixture or energy-poor lettuce for 6 days, in addition to their ad-libitum foraging in the field. As a result, they gained or lost weight in comparison with each other. Bachman then played a squirrel alarm call under standardized conditions to squirrels eating from a tray of crushed peanuts and noted their responses. She found that poorly supplemented squirrels were much less vigilant than richly supplemented individuals. Her key findings were that squirrels' vigilance response and intensity of response were positively correlated with both body weight at the time and with weight increase over the previous week, but that the foraging response (measured as head down) was negatively correlated with these measures. This was the case even though none of the squirrels were starving at time of the test playback, showing that ground squirrels were sensitive to a longer-term mass set point that they wanted to reach.

Finally, over a longer timescale, seasonal limitations on food availability can put a premium on gaining access to food and hence reduce time available for vigilance. In temperate regions this can occur in winter; in the tropics, in the dry season. Thus, certain ungulates living in Kyle Recreational Park, Zimbabwe, showed reduced vigilance from May to October, the period when little rain falls, compared with the rest of the year (Underwood 1982).

For the sake of convenience, factors affecting levels of vigilance have been treated independently in the literature but are interrelated in nature (Underwood 1982; Lima 1987b; FitzGibbon 1990c; Burger and Gochfeld 1994; Poysa 1994; Hill and Cowlishaw 2002). Subordinate individuals may be found on the periphery of groups, peripheral individuals may be closer to cover, hungry individuals may be more willing to aggregate together at food sources, nearest neighbor distances may be reduced in cover or in large groups, and so on. In documenting the influence of one factor, it is important to take others into account statistically or through experimentation, as Elgar (1989) stressed in his influential review of the effects of group size on vigilance in birds and mammals. As yet, few observational studies have achieved this, owing to the need to collect large quantities of data. For example, casting behavioral observations into three categories of position in a herd and four age-sex classes can result in such small sample sizes that they defy statistical analyses (Burger and Gochfeld 1994). Experimental studies have been more successful in removing potential confounding variables, such as temperature or time of day, but many other confounds are often left out. One notable exception is a study of the way in which vigilance of elk in Yellowstone National Park, USA, alters according to group size and risk of encountering wolves for different age-sex classes (Childress and Lung 2003). For males, time spent scanning was little influenced by herd size or encounter risk, because they have to feed voraciously to put on fat after the rut and before winter. For yearlings, time spent scanning declined with increasing herd size but was unaffected by risk of encounter; they need to put on fat rapidly, too. For females without calves, time spent scanning declined with herd size and likelihood of encountering wolves, as expected from classic "dilution effect" (chapter 8) and risk-sensitive behavior. For mothers, herd size had no effect, and reduced encounter risk had only a marginal effect in lowering vigilance: calves are selectively targeted by wolves, so that early detection and dilution would have little influence on reducing the likelihood of predation. Differential responses of these age-sex classes show that these subtleties cannot be ignored. Rather than demonstrate that a particular factor exerts an influence in yet another species, it would be more profitable to reexamine its influence, taking many confounding variables into account.

5.9 Predator abundance

Where risk of predatory attack differs between two areas but other environmental factors such as cover are equivalent, vigilance can be expected to vary accordingly. As vigilance increases, costs accumulate because of reduced opportunities to feed, which may negatively affect chances of future reproduction. A cost curve might increase slowly at first but then accelerate as individuals or parents have to curtail feeding almost entirely (fig. 5.10). Benefits of vigilance as measured by spotting a predator and taking evasive action might increase rapidly from no vigilance but then decelerate as every additional minute spent vigilant decreases in importance. In habitats with a great number of predators, a given level of vigilance is less effective than in habitats with fewer predators, so the high-predation site benefit curve should be shifted to

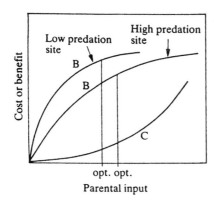

Figure 5.10 A model of cost (*C*) and benefit (*B*) functions related to parental input (that is, vigilance for predators) on high- and low-predation brood-rearing sites. The optimal amounts of parental input at the 2 sites are indicated (Forslund 1993).

Table 5.8 Effect of rearing site on intensity of vigilance in barnacle geese parents

	Predation		*p*-value on *t*-test
	Low site	High site	
Mortality of young	37% (*N* = 45)	81% (*N* = 69)	
Male vigilance	0.68	0.84	*p* = 0.0007
Female vigilance	0.51	0.55	*p* = 0.41
Pair vigilance	0.83	0.95	*p* = 0.014
Summed pair	0.61	0.69	*p* = 0.0071

Source: Forslund 1993.
Note: Least squares means when brood size held at its mean value.

the right; the cost curve, on the other hand, will not differ between sites, because costs are measured as lost opportunities for feeding. In these circumstances, the optimal amount of, say, parental vigilance should be greater at the high-predation site. Among barnacle geese, vigilance of male parents and two indices of pair vigilance were greater at a site suffering high predation from gulls (table 5.8), lending support to the model.

The model also can be used more generally, because an area can be dangerous not only because many predators lurk there but because a prey animal faces difficulties in detecting predators, detects more dangerous predators, or is far from a refuge. Heightened vigilance in such conditions may be pervasive among birds and mammals (Elgar 1989). A selection of examples includes greater vigilance among small birds foraging in early dawn, when predators are difficult to spot (Lima 1988c), ground squirrels spending a greater amount of time standing bipedally in response to seeing and hearing large and warm rattlesnakes (Swaisgood, Owings and Rowe 1999; Swaisgood, Rowe and Owings 1999), rodents being more vigilant in the open than among vegetation (for example, Vasquez, Ebensperger, and Bozinovic 2002), and terrestrial primates being more vigilant far from trees and cliff faces (Cowlishaw 1997b). Exceptions to this phenomenon can often be attributed to confounding factors, as when animals form larger groups living in more dangerous areas and thereby counter the necessity of increasing vigilance (Wahungu, Catterall, and Olsen 2001).

5.10 Interspecific differences in vigilance

There are several reasons that we might expect to observe interspecific differences in vigilance, assuming that it principally serves an antipredator role. At first blush, vigilance might be predicted to decline with increasing body size, at least in mammals, because larger prey are subject to predation from fewer species; moreover, larger predators that pose a danger to large prey live at low abundances (Geist 1974; Jarman 1974). Nonetheless, in birds, very small species might suffer fewer predatory attacks than medium-sized species because of their great maneuverability and skulking behavior. Second, larger species might be less vigilant if their larger eyes allowed them to detect predators at greater distances (Brooke, Hanley, and Laughlin 1999; Kiltie 2000). Having controlled for phylogeny, there is no evidence that birds with larger eyes do detect danger at greater distances, however (Blumstein et al. 2004). Third, larger animals might be less vigilant because their height allows them a better view of their surroundings, so that they need to scan less frequently or for shorter periods (Andersson 1981). Fourth, as smaller species are more selective feeders and therefore have higher intake rates, they might have less need for foraging time and more for vigilance (Illius and FitzGibbon 1994).

Empirical evidence from ungulates supports the proposition that larger species are less vigilant than smaller ones, although data are somewhat equivocal. Taking group size into account, because heavier species live in larger groups (Jarman 1974), two out of three studies found that larger ungulates were less vigilant than smaller species, and a third supported this without controlling for group size (table 5.9).

A second reason for expecting interspecific differences is that vigilance is correlated directly with frequency of predatory attack irrespective of body size. Scheel (1993a) used data on lion predation attempts to test this hypothesis directly, but surprisingly he found a significant negative correlation between the number of hunts suffered per individual and average scan rate (table 5.10). He suggested that species under high risk from lions cannot gain much benefit from being vigilant, and that it pays them to feed instead. There is some circularity to this argument, given that lions appear to hunt species that show least vigilance. A more engaging possibility raised by Scheel (1993a) is that the species showing high vigilance in his study, Thomson's and Grant's gazelles, hartebeest, and topi, have few antipredator tactics open to them other than early detection of predators. In contrast, warthogs can retreat into burrows, wildebeest may reduce the impact of predation by forming very large herds,

Table 5.9 Interspecific differences in patterns of vigilance in ungulates

Species	Effect	Confounding variables	Reference[a]
Impala, Uganda kob, waterbuck, wildebeest, Burchell's zebra, buffalo, African elephant	Proportion of time vigilant declines with body size. Not examined statistically	No factors taken into account	1
Thomson's gazelle, Grant's gazelle, warthog, topi, Coke's hartebeest, wildebeest, Burchell's zebra, buffalo	Scan rates not correlated with body size. Scan rates significantly negatively correlated with predation risk	Group size taken into account statistically	2
Pronghorn antelope, bighorn sheep, mule deer, bison	Time spent vigilant declines statistically with body size	Group size controlled for	3
Steenbok, oribi, impala, reedbuck, tsessebe, wildebeest, sable antelope, buffalo	Proportion of time spent looking significantly declines with group size	Group size taken into account statistically	4

Note: Species listed in ascending order of body weight in each entry.
[a] 1. Burger and Gochfeld 1994; 2. Scheel 1993a; 3. Berger and Cunningham 1988; 4. Underwood 1982.

Table 5.10 Comparisons of predation risk and vigilance in Serengeti ungulates

Species	Adult female body weight (kg)	Hunts/individual by lions	Average scan rate
Thomson's gazelle	16	0.0003	81.8
Grant's gazelle	42	0.001	68.4
Warthog	52	0.12	35.0
Topi	108	0.007	64.6
Coke's hartebeest	126	0.007	75.6
Wildebeest	203	0.00009	46.6
Burchell's zebra	219	0.0002	39.4
Buffalo	446	0.003	44.3

Source: Scheel 1993a.
Note: Species listed in ascending order of body weight.

and zebra and buffalo actively defend themselves. Scheel's study is one of a very few to relate vigilance to quantitative differences in attack rate, a point raised at the beginning of the chapter; but the discrepancy between Scheel's results and those of the other ungulate researchers requires explanation.

A third reason for interspecific differences in vigilance is that different species practice antipredator ploys that necessitate detecting predators at different distances. In comparing mule deer with white-tailed deer (an interesting juxtaposition, because they have similar body sizes), Lingle and Wilson (2001) reported that mule deer detected coyotes and approaching humans at greater distances than white-tailed deer. For a given distance that deer became alert to coyotes, however, encounters with mule deer progressed further along the approach, pursue, attack, and kill sequence than encounters with white-tailed deer. White-tailed deer can outdistance coyotes, but mule deer cannot. In other words, mule deer must become aware of coyotes at a greater distance if a given encounter is to end at the same stage of the hunt sequence. While mule deer actually demonstrate lower levels of scanning than white-tailed deer, they are more likely to orient directly at an approaching person; and in response to coyotes, they often leave the area entirely on seeing them, in contrast with white-tailed deer. Mule deer have far larger ears and more laterally placed eyes than white-tailed deer that may enable them to detect predators at longer distances (and their ears have interesting parallels with heteromyid rodent ear structure!). Prey differences in susceptibility to a particular predator and antipredator behavior following predator detection must have an enormous bearing on the patterns of vigilance across species, so it is rather surprising that so little attention has been paid to the way that comparative aspects of predator-prey interactions relate to vigilance.

Difficulties in making firm predictions about interspecific differences in vigilance resurrect the problems of assessing predation risk. Predation risk is a composite measure of prey vulnerability and predatory attempts on prey, both of which are influenced by many factors. Vulnerability is affected by prey coloration and morphology, by trade-offs in where and when to forage, by group size, by dominance, and by reproductive condition, to name but a few; predatory attempts are influenced by which predators are present, predator abundance, and preferences for different types of prey. In addition, however, predator preferences and even predator numbers are influenced by prey vulnerability and even vulnerability of alternative prey species (Sih and Christensen 2001). With predation risk being subject to so many influences, quantifying interspecific differences in risk is an intricate if not foolhardy endeavor, and this makes hazardous a-priori predictions about comparative levels of vigilance.

The multitude of factors that affect prey vulnerability also limits our ability to rank factors determining levels of vigilance, or other antipredator defenses for that matter. Consider a new parent with offspring sequestered in a lair: few changes in antipredator behavior are necessary; but for a parent with precocial young that are unable to fly, severe changes in antipredator behavior may be necessary, leading to reductions in feeding, limited movement, and as a consequence, heightened personal danger. Interactions between factors present additional complications, as, for example, when close proximity to group members facilitates quick escape but creates competition over food, forcing individuals into riskier habitats. It is difficult to see a way through this morass at present, so our principal achievement is still a checklist of factors affecting vigilance.

5.11 Summary

A wide variety of ecological and social factors influence the extent to which birds and mammals are vigilant. Most have been investigated in isolation, although many covary in nature. It is helpful to separate vigilance shown before a predator is seen from vigilance that monitors a predator after it has appeared, because they may have different causal and functional attributes. In establishing levels of personal vigilance, individuals take account of the number and proximity of nearest neighbors rather than group size per se or the level of vigilance exhibited by neighbors. Neighbors may provide cues to presence of danger and allow a coordinated response, or may dilute predation risk.

Theory and empirical evidence show that individuals on the periphery of groups are more vigilant than those in the center, because they are more prone to predatory attack. Nevertheless, confounding variables are often ignored in studies examining vigilance in relation to position in the group.

In a limited number of species, one individual acts as a sentinel and alerts other group members to danger while they forage. Guard duty is probably selfish behavior, as sentinels usually spot danger first and take up position after feeding; nonetheless, other group members benefit from sentinels in their group. While the circumstances under which sentinel behavior may be maintained have now been modeled, the taxonomic distribution of sentinel behavior is poorly understood.

Where cover impedes view, prey may increase scanning rates close to cover or, alternatively, lower them if prey assess vigilance as being ineffective. Where cover provides a refuge for escape, individuals may scan more far from cover in order to increase the probability of early detection, or scan less to reduce time spent exposed in the open. Where species are subject to predation by ambush, they will be more vigilant close to cover.

As young animals are subject to greater predation risk than adults, they might be expected to show higher levels of vigilance. Some evidence supports this contention, but juveniles are less vigilant than adults in many species. They may be less effective at recognizing and responding quickly to predators or, because of nutritional demands, have less time for vigilance. Mothers with offspring are extremely vigilant in most species. Male mammals are usually more vigilant than females, but this is often a consequence of male-male competition in polygynous species. Males in monogamous species may carry some of the burden of vigilance for females, giving the latter greater opportunities for feeding. Differential vigilance among dominant and subordinate group members can usually be attributed to social reasons.

Vigilance is also influenced by a variety of other factors that bear on predation risk and feeding demands over different timescales. Interspecific differences in vigilance are related to body size, rates of predatory attack, and antipredator options open to prey.

6 Conspecific Warning Signals

6.1 Introduction

Previous chapters have examined antipredator defenses that do not require prey to actually see a predator, but I now move to part of the predatory sequence in which prey first notices a predator. At this juncture, it may emit an auditory, chemical, or visual warning signal, also called an alarm signal (Klump and Shalter 1984). These are unfortunate terms, because in a colloquial sense *alarm signal* implies that the animal is behaviorally or physiologically stressed, which has rarely been verified; and *warning signal* implies that other animals are informed of danger (section 1.3.b). "Other animals" are most commonly assumed to be conspecifics that are believed to benefit from this information. In nature, however, signals could alternatively be directed at the predator or even at heterospecific prey species or any combination of these players, and could benefit the signaler at the expense of the receiver. As warning signals are assumed to be costly because they are thought to attract predators' attention to the signaler, a considerable amount of theoretical and empirical work has sought to identify the targets of alarm signals and the nature of benefits to both signaler and recipients. A different, more proximate area of investigation has examined the constraints on signaling, particularly the necessity of reducing the predator's ability to locate the signaler and maximizing the signal's detection by receivers; these can be thought of as ways in which the signaler respectively minimizes costs and maximizes benefits. Yet a third line of research has examined the nature of the information conveyed by signals that, if correctly interpreted, might facilitate adaptive responses in receivers.

Figure 6.0 (*Facing page*) A white-tailed prairie dog in bipedal posture uttering an alarm call. Many sciurids have a considerable suite of aerial and terrestrial predators, and their alarm calls to these 2 classes of threat are often different (reproduced by kind permission of Sheila Girling).

The scope of this chapter is to examine warning signals directed at conspecifics; chapter 7 discusses those targeted at predators. Classically, alarm signals directed at conspecifics are viewed as unmistakable vocalizations that in birds consist of a high-frequency sound with an onset and offset that are difficult to discern (Marler 1959; Latimer 1977), but in larger mammals consist of a broadband-frequency sound that is repeated several times (for example, Waser and Waser 1977; Atkeson, Marchinton, and Miller 1988). Auditory signals are often accompanied by an alarm movement, as illustrated by Canadian beavers slapping their tail on water (Hodgdon and Larson 1973) or species of West African tree squirrels showing tail piloerection and foot and tail movements prior to calling (Emmons 1978). Vigilant behavior, not normally regarded as a warning signal, may also serve to alert conspecifics. For example, prairie dogs run to their burrow and take up an alert bipedal posture (posting) when they first detect a ground predator. The principal function of posting is probably to get a better view of the surroundings, but it also alerts other prairie dogs to danger (Hoogland 1981; Schooley, Sharpe, and van Horne 1996). Similarly, escape behavior may inadvertently serve as a warning signal. Common voles either freeze or run to cover if voles in a neighboring experimental arena freeze or run away (Gerkema and Verhulst 1990). This occurs when voles are in both visual and nonvisual contact, so auditory cues must be involved. While ultrasonic sounds cannot be discounted, voles could simply be paying attention to increases in noise made by conspecifics or lack of noise as cues to danger. In general, however, warning signals are usually thought of as specific calls.

In section 6.2, I examine the ways in which birds reduce localizability of alarm calls, and the means by which they make calls detectable to conspecifics but minimize conspicuousness to predators. After reviewing these methods of reducing signal cost, I examine survivorship costs of alarm calling (section 6.3) that vary considerably depending on the type of predator and environmental context. Section 6.4 looks at the range of possible benefits of calling and data that support these hypotheses, the most well-worked of which is warning kin of danger. Alarm calling between species in mixed-species associations is an additional benefit subject to different sorts of selection (section 6.5). In section 6.6, the various types of information conveyed by sciurid, avian, and primate warning calls are reviewed. This includes knowledge about the extent of danger and about the class of the predator and constitutes a very active area of research effort on warning signals. Next, the development of alarm calls in general, and predator-specific alarm calls in particular, is documented (section 6.7). Finally, the use of alarm calls in deceiving conspecifics is described (section 6.8).

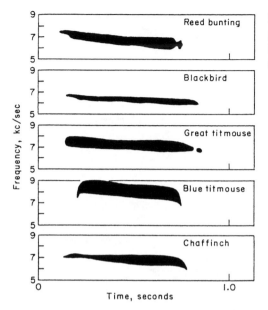

Figure 6.1 Sound spectrograms (frequency plotted against time) of aerial predator calls of 5 species of birds. Note the similarity in call structure (Marler 1959).

6.2 Acoustic constraints on alarm calls

6.2.a Localizability   Owls and other aerial predators use phasic differences to localize sounds; low-frequency sounds that have wavelengths less than twice the distance between their ears are easy for them to locate. They also pay attention to intensity differences in sounds, which makes discontinuous high-frequency sounds relatively easy for them to find (but see Konishi 1973). Therefore, moderately high, uniform call frequencies of standard intensity ought to be the sort of calls that are most difficult for them to localize (Marler 1955, 1959; Marler and Hamilton 1966; Brown 1982). Small birds do indeed produce a single, high-frequency, pure alarm tone in the 6- to 9-kHz range, with a narrow bandwidth and a gradual low-amplitude onset and offset (fig. 6.1). Similarly, small mammals, such as Columbian, arctic, and Richardson's ground squirrels (Melchior 1971; Koeppl, Hoffmann, and Nadler 1978; Owings and Virginia 1978) and marmosets (saddleback tamarins: Vencl 1977), which are also subject to predation by raptors, emit high-frequency alarm calls.

Empirical research investigating Marler's hypothesis has generally been positive. For example, in the laboratory, C. H. Brown (1982) demonstrated that red-tailed hawks and great horned owls were less able to orient their head correctly toward the source of a "seeet" aerial alarm call given by an American robin than to the mobbing call of a red-winged blackbird and that the error in orientation was greater at the end than at the beginning of the "seeet" call, suggesting that the call might be ventriloquial in character (Perrins 1968; see also

Konishi 1973). Nevertheless, there is also evidence to the contrary. In the laboratory, where the source of a sound is necessarily close by, goshawks and pigmy owls can localize the source of "seeet" alarm calls, as determined by rotation of their head (Shalter 1978; see also Shalter and Schleidt 1977), although they often fail to respond to calls. The extent to which predators localize warning calls under natural conditions is unknown, however. Given that high-frequency calls are directional and attenuate rapidly (Klump, Kretzschmar, and Curio 1986), they may alert conspecifics nearby without an approaching predator being able to hear them. Based on physical characteristics, selected experiments, and the difficulty that humans have in locating "seeet" alarm calls, it is generally held that they are somewhat difficult for raptors to locate (Klump and Shalter 1984), but the reasons for this are still opaque.

Interestingly, alarm calls of Australian passerines do not show the same characteristics as their Palearctic counterparts in that a number of species give low-frequency alarm calls sounding like a low-pitched whistle or "chit" (Jurisevic and Sanderson 1994). Furthermore, Australian raptors find these sounds relatively easy to locate (Jurisevic and Sanderson 1998), implying that passerines there may put less of a premium on remaining inconspicuous. More work needs to be carried out on how Australian raptors respond to alarm calls of prey.

Common selective pressures on reducing localizability seem to have resulted in structurally similar alarm calls arising in diverse taxa (Marler 1955, 1957; Vencl 1977; Emmons 1978; Maier, Rasa, and Scheich 1983; Jurisevic and Sanderson 1994; E. Greene and Meagher 1998). Most of these examples consist of prey species that share the same suite of predators, implying that selection has arrived at analogous solutions to avoid prey being found by the same predator. For example, in response to raptors, red squirrels and sympatric passerines in North America produce similar "seeet" alarm calls (fig. 6.2). Whether high-frequency calls characteristic of small-bodied species are due to structural constraints on the vocal apparatus or are tailored to auditory capabilities of similar predators is unclear. Common environmental pressures, such as attenuation by leaves and branches in tree canopies that affect sympatric species, may have also been important in shaping convergent calls (Vencl 1977).

6.2.b Detectability   From the standpoint of prey, it may be important that predators cannot locate the individual giving a signal, but it probably matters less as to whether conspecifics localize the source of a call, because they may simply need to know that a predator is in the vicinity. Therefore, selection can be expected to mold alarm call structure to reduce localizability, but it is less easy to formulate predictions regarding detectability of alarm calls. Much research has focused on the physical factors that attenuate and degrade sounds in the environment (see Wiley and Richards 1982; Klump and Shalter 1984;

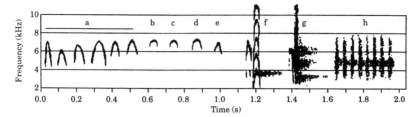

Figure 6.2 "Seeet" alarm calls (shown as frequency against time) produced in response to aerial predators by *a*, 6 red squirrels; *b*, a lazuli bunting; *c*, black-capped chickadee; *d*, northern junco; and *e*, ruby-crowned kinglet; *f*, seet-bark and *g*, bark, the last two of which are produced by red squirrels toward all and terrestrial predators, respectively; and *h*, the territorial rattle call of red squirrels. Note the similarity in call structure between the alarm calls of birds and squirrels to aerial predators (Greene and Meagher 1998).

Endler 1992; Bradbury and Vehrencamp 1998 for reviews), but most of this work has assumed that selection has acted to maximize sound transmission of, for example, birdsong (for example, Richards and Wiley 1980; Wiley 1991). Less attention has been paid to how alarm call structure might both maximize detectability by conspecifics and minimize detectability by predators.

A number of predictions arise from signal detection theory, but data indicating that warning calls meet these dual expectations are mixed. (i) Sound intensity declines at a rate of 6 dB for every doubling of distance between source and receiver (Schleidt 1973), so that alarm calls might be expected to be low in amplitude if there was selection against predators hearing the call. Alarm calls of some small birds and mammals are of low amplitude; for example, black-capped chickadee alarm calls have sound-pressure levels of only 55.6 dB measured at a distance of 1 m (Witkin and Ficken 1979). Nonetheless, alarm calls of some homeotherms are loud—forest-dwelling primates, for example (Waser and Waser 1977). Perhaps such calls are given at frequencies only weakly audible to predators, there is weak selection against predators hearing their alarm calls, or it is actually beneficial to the prey if predators hear conspecific warning calls, as they may abandon hunting when detected (section 7.6). In many instances, loudness of calls may simply be related to the need for transferring information to other group members. Species in widely dispersed feeding groups, such as the Eurasian curlew, give loud alarm calls, whereas those in tightly knit groups, such as dunlin, give soft ones (Owens and Goss-Custard 1976).

In an important study of auditory acuity of both prey and predator, Klump, Kretzschmar, and Curio (1986) found that European sparrow hawks were sensitive to sounds in the 1- to 4-kHz region but were insensitive to frequencies of 8 kHz and above, whereas great tits were sensitive to sounds in the 1- to 8-kHz region but insensitive at 10 kHz. Comparison of the two species shows that great tits' hearing is superior at 8 kHz (fig. 6.3), which is the dominant frequency of the "seeet" alarm call that tits use when confronted by a sparrow hawk flying at some distance away. For "seeets" given under environ-

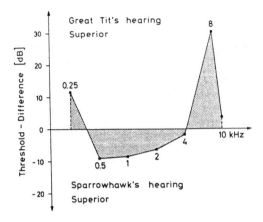

Figure 6.3 Differences in the absolute thresholds between the great tit and European sparrow hawk (calculated using signal detection theory, $d' = 1.5$) (Klump, Kretzchmar, and Curio 1986).

mental masking noise, detection distances are only about 10 m for the sparrow hawk but up to 40 m for the tits, which effectively solves the trade-off in informing the two parties.

(ii) At environmental boundaries, between, say, air and the ground, sound transmission suffers from interference due to reflection and additional waves propagating in and near the ground, principally below 1 m (Marten and Marler 1977; Marten, Quine, and Marler 1977). That sciurids such as thirteen-lined (Schwagmeyer and Brown 1981) and California ground squirrels (Owings and Virginia 1978) alarm call from promontories hints that they are attempting to increase transmission distance by ascending above this area of interference. These calls are often made from the safety of the burrow entrance, thereby possibly reducing the necessity of making the call inconspicuous to some predators.

(iii) Sounds are attenuated by atmospheric absorption and scattered by both vegetation and atmospheric turbulence as their frequency increases (Wiley and Richards 1978). In less-vegetated habitats, short pulses and high rates of repetition reduce distortion, but in thick vegetation lower frequencies reduce reverberation. Generally, however, avian and mammalian alarm calls are of high frequency, so it appears that selection has done little to reduce attenuation or scattering, either because it is not necessary to do so or because of the conflicting demand of reducing localizability using high-frequency sounds (C. C. Smith 1978; Perla and Slobodchikoff 2002; see section 6.2.a).

(iv) Predictions regarding signal structure in different environments, such as using lower frequency sounds in forests than in open habitats in order to reduce attenuation by vegetation (E. S. Morton 1975; M. J. Ryan and Brenowitz 1985; Waser and Brown 1986), are not generally met for alarm calls. For example, noting that the alarm calls of four species of marmot differed, Daniel and Blumstein (1998) broadcast and then rerecorded each species' alarm call in each of four different habitats where the species were found. Although they could show that habitats differed markedly in transmission properties and that calls differed in transmission properties, too, there was no evidence that the call of a given species transmitted better in its native habitat than in others. They concluded that factors other than maximizing long-distance transmission must be important in the evolution of species-specific alarm calls in marmots. On the other hand, other studies have shown that certain characteristics of calls do differ in different habitats. Intraspecific comparisons of alarm calls in Gunnison's prairie dogs found that the number of syllables and length of alarm call were greater in more complex habitats containing cover, rocks, and tree stumps (Slobodchikoff and Coast 1980), perhaps to increase transmission distance or to overcome greater levels of background noise (Tamura and Yong 1993).

(v) While it is known that factors such as time of day and season (associated with changes in temperature, wind, and vegetation thickness) that increase turbulence and background noise are likely to influence the time of day when animals communicate (Wiley and Richards 1982), these are unlikely considerations in regards to alarm calls given in response to the appearances of predators which may occur at any time, although usually during restricted periods of the day. Nevertheless, the effectiveness of alarm calls in warning conspecifics will be influenced by these factors.

In summary, low amplitude of warning calls and the absence of structural characteristics that serve to minimize attenuation suggest either that there has been selection against long-range transmission that a predator might cue in on, or that it is unnecessary to transmit long distances within most social groupings.

6.3 Costs of warning signals

The logic that underscores why selection has reduced localizability and detectability of conspecific warning calls to predators assumes that prey individuals make themselves more open to predatory attack if they utter an alarm call. While modeling shows that the equilibrium probability of an individual giving an alarm call is independent of the benefit-to-cost ratio, simply that the benefit B is greater than the cost C to the caller, at least when the benefit is not contingent on group size (R. J. Taylor, Balph, and Balph 1990), it is still important to measure costs of alarm calling. These have been demonstrated only rarely, however. Obviously, callers are likely to incur energetic costs from repetitive calling, although these have never been measured in birds and mammals; costs could mount if false alarms are common. Also, alarm callers may incur time costs, although these may be minor, as the caller is likely to be watching the predator anyway; again, time costs have not been measured. Nonetheless, it is potential survivorship costs that are at issue in regard to both call structure (section 6.2) and benefits of alarm calling (section 6.4) (but see R. J. Taylor, Balph, and Balph 1990). Unfortunately, there are conceptual difficulties in measuring survivorship costs, because if calls are costly, they may be given only in situations in which prey are exposed to little danger; thus, an observer may conclude that a caller incurs little risk of predation (Lima and Dill 1990). Where costs have been estimated, current evidence appears equivocal.

In examining the design features of alarm calls, authors have argued, often as an afterthought, both for and against survivorship costs. Observing that free-living Gunnison's prairie dogs closest to a stuffed model of an American badger or that had entered a burrow because of the badger's proximity were less likely to alarm call than prairie dogs that were farther away or that had not disappeared (fig. 6.4), Hoogland (1996) argued that alarm calls must be costly (see Ivins and Smith 1983 for another mammalian example). Similarly, Alatalo and Helle (1990) found that willow tits did not usually give alarm calls if an aerial predator was passing nearby (at 10 m), while more than half the individuals called in response to a distant predator 40 m away, again supporting the idea that calling is risky. Evidence that some species use ultrasonic alarm calls that may be inaudible to predators is strongly suggestive that calling imposes survivorship costs (Wilson and Hare 2004).

Against alarm calls being risky, Noyes and Holmes (1979) noted that hoary marmots in an open meadow call while running to a refuge. They suggested that if calling was costly, they might be expected to call only after reaching safety. MacWhirter (1992) argued that because Columbian ground squirrels were as likely to utter nonrepetitive calls when they were both close and far from a stuffed American badger, calling did not increase vulnerability (see also Barash 1975a; Dunford 1977). In regards to birds, authors have argued that the structure of passerine "seeet" alarm calls make small birds unlikely to be located or detected by predators, so that "seeet" calls are not costly (C. H. Brown 1982; Klump and Shalter 1984; Klump, Kretzschmar, and Curio 1986). Sadly, all of these studies rely only on verbal arguments to make their cases. A sole study, of Belding's ground squirrels in Tioga Pass, California, USA, has data on capture rates in relation to calling (Sherman 1977, 1985). In this 8-year program, ground squirrels that alarm called were more likely to be captured by terrestrial predators, such as long-tailed weasels, than those that did not call (table 6.1). In contrast, individuals voicing a different type of alarm call in response to aerial predators, such as goshawks, were less likely to be captured that those that did not call (table 6.1). In this one species, the costs of calling appear highly predator specific (MacWhirter 1992), mimicking perhaps the conflicting results from studies of other species. It seems likely that predation

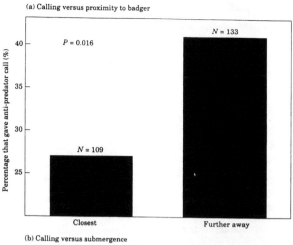

(a) Calling versus proximity to badger

P = 0.016

N = 109

N = 133

Closest

Further away

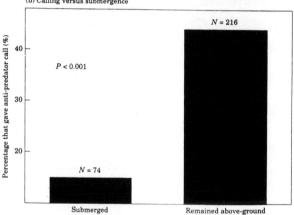

(b) Calling versus submergence

P < 0.001

N = 74

N = 216

Submerged

Remained above-ground

Figure 6.4 Antipredator calls of Gunnison's prairie dogs to a stuffed American badger. (*a*), Calling versus proximity to badger; (*b*), calling versus whether prairie dogs are submerged belowground. *N* refers to the number of experimental trials (Hoogland 1996).

Table 6.1 Alarm calling and survival in Belding's ground squirrels at Tioga Pass, California, USA

| Category | Number of ground squirrels | | | |
	Captured	Escaped	Percentage captured	P-value*
Aerial predators				
Callers	1	41	2%	
Noncallers	11	28	28%	<0.01
Total	12	69	15%	
Terrestrial predators				
Callers	12	141	8%	
Noncallers	6	143	4%	<0.05
Total	18	284	6%	

Source: Sherman 1985.
*Using Chi-square test.
Note: All data are from observations of attacks made by hawks (*n* = 58) and predatory mammals (*n* = 198) that occurred naturally from 1974 to 1982.

costs are contingent on the predator's hunting style, predator's speed of approach, distance between prey and predator, and prey's degree of exposure (R. J. Taylor, Balph, and Balph 1990). If costs are indeed this context dependent, it is ill-advised to make generalizations until far more comparative data have been amassed.

6.4 Benefits of warning signals

Based mostly on unmeasured assumptions that warning signals increase the probability of predation, several hypotheses have been advanced about the adaptive significance of alarm calls (Klump and Shalter 1984; Hauser 1996; Bradbury and Vehrencamp 1998). One class of hypothesis states that alarm calls are directed at the predator and that they inform the predator that it has been seen (perception advertisement) or that the prey individual is additionally in good enough condition to escape (condition advertisement) (section 7.6). Remaining hypotheses consider alarm signals as being directed at conspecifics and fall into three general categories: those that are selfish (benefit the caller but harm the receiver), those that are mutualistic (benefit both caller and receiver), and those that are altruistic (in the sense of harming the caller but benefiting the receiver now, but the caller receiving delayed or indirect benefits). Since costs and benefits to each participant are measured rarely, these are only qualitative categories at present. Further, it is now recognized that an alarm call can have multiple effects on conspecifics that may involve more than one type of payoff, and can additionally influence the predator's behavior (Zuberbuhler, Noe, and Seyfarth 1997; Zuberbuhler, Jenny, and Bshary 1999; Shelley and Blumstein 2005).

6.4.a Apparently selfish alarm calls   Charnov and Krebs (1975) derived a model to suggest that a calling individual might benefit if movements of alarmed receivers distracted the predator's attention away from the sender. Thus, by calling, a bird might cause its flockmates to dash for cover without regard to the predator's whereabouts, whereas the caller could protect itself by seeking cover on the far side or in the middle of the flock, or might benefit if the predator became confused by moving prey. Charnov and Krebs (1975) used the term *manipulation*, implying, unfortunately, that the caller caused conspecifics to behave maladaptively. Several authors have incorrectly interpreted the Charnov and Krebs argument as a caller reducing the receiver's fitness to $B/C < 1$, but such behavior is unlikely to be maintained in a population for long. Indeed, Eastern chipmunks increase their vigilance and load food items at a slower rate after hearing an alarm call, suggesting that instead of being gullible, they are attempting to assess the extent and location of danger for themselves (Baack and Switzer 2000). Actually, Charnov and Krebs (1975) also wrote

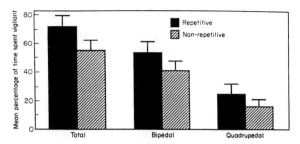

Figure 6.5 Effects of repetitive and nonrepetitive calls on the mean (and SE) percentage of time spent vigilant by California ground squirrel receivers. Values represent the percentage of time spent vigilant during the first 5 minutes of a repetitive calling bout and the 5 minutes after a nonrepetitive call. Only total time spent vigilant was significantly different, whereas percentage of time spent vigilant in bipedal and quadrupedal stances was not (Loughry and McDonough 1988).

(The recipients) use the information for their own benefit, but in doing so make it possible for the caller to benefit even more. (p. 110)

Thus, the argument really hinges on caller $B/C >$ responding receiver $B/C >$ nonresponding receiver B/C.

Empirical studies that address this hypothesis have not been supportive. For example, S. F. Smith (1978) could find no evidence that alarm-calling Sonoma chipmunks enjoyed improved survivorship over others as a result of the other chipmunks fleeing. In-depth analyses of California ground squirrel alarm calls have uncovered two classes of calls: nonrepetitive chatters and repetitive chatter-chats that are heard during and after encounters with mammalian predators (Owings and Virginia 1978; Leger, Owings, and Gelfand 1980; Loughry and McDonough 1988; Owings et al. 1986). Owings and Hennessey (1984) argued that repetitive calls might act tonically (see Schleidt 1973) to promote and maintain vigilance in receivers. Whereas receivers become more vigilant after both types of call, the percentage of time spent vigilant is greater following repetitive calls (fig. 6.5), although this effect eventually begins to wane as calling continues (see also Hare 1998). As adult males, who are less related to group members than females, also call and elicit increases in vigilance among group members (Loughry and McDonough 1989), the idea that the calling male directly benefits from everyone being vigilant seems reasonable. Nonetheless, as Owings and colleagues point out (for example, Owings and Leger 1980), this can hardly be construed as selfish behavior, as the receivers benefit from monitoring the whereabouts of the predator, too. Similar arguments have been advanced for alarm calling in spider monkeys (Chapman, Chapman, and Lefebvre 1990) and red colobus (Stanford 1998), and for tail flagging in California ground squirrels that occurs in response to snakes and attracts the attention of nearby squirrels (Hersek and Owings

1993). In all these examples, calling increases vigilance in group members and in all probability benefits both caller and receivers in a mutualistic fashion, because some group member is monitoring the predator.

Alarm calling can also benefit both participants in other ways. After a Belding's ground squirrel alarm calls in response to a raptor, other individuals flee for cover. Sherman (1985) has argued that if this individual does not call but flees, it would be picked out because of its singular movement (thus, $B > C$ for giving a call). If a conspecific hears the call but remains stationary, however, it would be singled out as the only individual not moving or the only one left (thus, $B > C$ for responding). Therefore, ground squirrels are forced to flee as a result of others' behavior. These speculations help explain why nomadic, nonparental males call as frequently as females. Again, this is an example of mutualism, because receivers are better off responding than ignoring the call. In short, Charnov and Krebs's hypothesis is about mutualistic benefits for caller and receiver, with the nature and magnitude of benefits being different for sender and receiver.

6.4.b Mutually beneficial alarm calls   Another category of hypotheses explicitly acknowledges that both caller and recipient accrue net benefits from alarm calls and includes a slew of mechanisms as to how altered behavior of conspecifics affects the predator's ability to make a successful attack. Thus, if, for sake of convenience, we begin at the start of the predatory sequence, an alarm call might cause conspecifics to act cryptically, thereby reducing the likelihood that the predator sees the group (R. Dawkins 1976). Informal observations of birds suggest this is likely, but experimental confirmation is lacking. Alarm calls might cause conspecifics to increase their vigilance, resulting in short-term benefits for both caller and receivers obtained from monitoring predators' movements (Owings and Leger 1980; section 6.4.a), or result in longer-term benefits for the caller that come from protecting a mate, with the consequence of furthering the caller's reproductive success (G. C. Williams 1966). Thus, female Carolina wrens may give "chirt" calls to inform the male of the danger imposed by a red-tailed hawk nearby. Alarm calls may reduce his immediate chance of mortality, but in addition, being relieved of vigilance, he is allowed more foraging time and energy for defense of their territory (E. S. Morton and Shalter 1977; see also Witkin and Ficken 1979). In perhaps a parallel instance of alarm calling by members of a pair, a male willow tit alarm calls more frequently if he can see his mate in an aviary, possibly because he assumes she is not in danger when she is out of sight (Hogstad 1995a).

Alarm calls might cause group members to bunch together and travel as a more cohesive group, thereby benefiting from a dilution effect or other group-related benefits (chapter 8). Long-distance calls of male Diana mon-

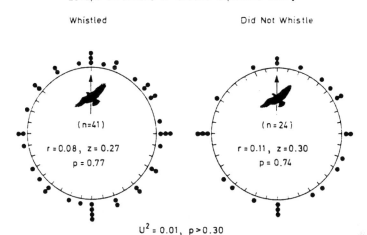

Escape Directions of Ground Squirrels that:

Whistled   Did Not Whistle

(n=41)   (n=24)

r = 0.08, z = 0.27   r = 0.11, z = 0.30

p = 0.77   p = 0.74

$U^2 = 0.01, \; p > 0.30$

Figure 6.6 Directionality of Belding's ground squirrels' escape movements relative to Harris hawks' flight paths (0°). Animals that did not alarm whistle (*right-hand side*) could not have seen the raptor, owing to an intervening hill or rock outcrop (Sherman 1985).

keys that are reliably given to certain predators result in nearest neighbor distances and group spread decreasing while travel and association rates of the group increase (Shultz, Faurie, and Noe 2003).

Alarm calls might precipitate prey animals fleeing en masse (Owens and Goss-Custard 1976). For example, alarm calls cause common redshank to fly off together, which might confuse a predator. That alarm calls occur more often on rocky shores, where vision is obstructed, than on flat salt marshes indirectly supports this idea, because redshanks can see one another fly off in the open habitat but can see one another less easily on rocky mussel beds. In response to attack by sparrow hawks, redshanks on the ground are more vulnerable than birds in flight; thus, both caller and recipients benefit from being in communication (Cresswell 1994b). Following an alarm call, Belding's ground squirrels flee at random with respect to the direction of the flight path of a trained Harris's hawk (fig. 6.6), thereby creating pandemonium, which may hinder the raptor in homing in on its quarry (Sherman 1985).

Alarm calls might also act to recruit conspecifics to defend themselves against attack, perhaps by mobbing the predator (Stone and Trost 1991; chapter 11).

Additional hypotheses do not consider the mechanism by which predators are thwarted by the behavior of caller and recruits, but argue that an unsuccessful attack has longer-term beneficial consequences for the group. Thus, R. J. F. Smith (1986) suggests that alarm calls will be under positive selection if they help to prevent the demise of group members, as it is important to maintain an optimum group size for foraging and other activities. There are no tests of this idea, and it will be difficult to measure short-term costs of giving an alarm call against the long-term and diffuse benefits of retaining group members. Others have argued that an unsuccessful attack will lower the probability of the predator visiting the area again (Curio 1976; Trivers 1971; Cresswell 1994b). Such benefits could pertain to relatively sedentary prey, to prey that return to an area to feed, and to predators that return to the same area repeatedly to hunt (for example, sparrow hawks: Wilson and Weir 1989). Whereas this idea seems intuitively appealing, there are precious few data to show that individual predators avoid places where they have earlier been unsuccessful, or return to sites of successful attacks (Lima 2002). Early and rare data comparing the number of times that coyotes and long-tailed weasels returned to sites where Belding's ground squirrels had or had not been caught by these species (20.9 versus 18.9 days return times, respectively) unfortunately says little about factors influencing predator hunting methods, because different individual predators may have been involved (Sherman 1977).

6.4.c Altruistic and kin-selected alarm calls   Two hypotheses suggest that signalers suffer a net cost as a result of alarm calling (usually assumed to be a reduction in survivorship), but that recipients gain a net benefit. In his original paper on reciprocal altruism, Trivers (1971) used alarm calls as a putative example in which an actor warned nonrelatives of danger and thereby increased its chances of being killed; but at a later date, the recipient returned the favor by alarm calling, again at a personal cost, but benefited the original actor. Reciprocal altruism demands that groups are stable, ensuring repeated interactions between prey individuals, which is the case for some groups of mammals but applies less to aggregations of birds. Very few tests have been made of this hypothesis to explain alarm calling, but Sherman (1977) reported that female Belding's ground squirrels refrained from calling when no kin were alive despite being surrounded by unrelated females, and that females did not discriminate against noncallers, both of which speak against reciprocal altruism. Alarm calling is an unusual act in that it benefits many recipients simultaneously, and models have shown that it is easy for noncalling individuals to be maintained in populations showing alarm-calling behavior (Tamachi 1987), making it problematic for alarm calling to evolve through reciprocal altruism (see also Hoogland 1981).

The second means by which costly alarm calling might be maintained in a population is if recipients are kin (Maynard Smith 1965), such that $C < rB$, where r is the coefficient of relatedness between actor and recipient (W. D. Hamilton 1964). There is strong evidence for alarm calls alerting both de-

Table 6.2 Important comparisons derived from studies of ground-dwelling sciurids supporting the idea that kin selection has been an important force in the evolution of alarm calling

Common name	A	B	C	D	Reference[‡]
Ground squirrels					
California	f = m*				1
Belding's	f > m	rf > if	Yes (f)	Yes (f)	2
Columbian			Yes (f)	Yes (f)	3
Arctic	m > f				4
Richardson's	m > f[†]		Yes (f,m)		5
Round-tailed	f > m				6
Thirteen-lined	f > m		Yes (f)		7
Chipmunks					
Sonoma	f > m				8
Prairie dogs					
Gunnison's			Yes (f)		9
Black-tailed		rm > im	Yes (f)	Yes (f,m)	10, 11
Marmots					
Yellow-bellied			Yes (f)		12

Note: Column headings are as follows: A. Do sexes differ significantly in rate of alarm calling (f, female; m, male)? B. Do resident individuals (r) alarm call more than immigrants (i)? C. Do females or males with emergent young alarm call more than females or males with no young? D. Do females or males with nondescendant kin alarm call more than females or males with no kin?

* Dominant and presumably breeding males call.

[†] Males defend territories throughout the breeding season.

[‡] 1. Owings and Leger 1980; 2. Sherman 1977; 3. MacWhirter 1992; 4. I. G. McLean 1983; 5. Davis 1984; 6. Dunford 1977; 7. Schwagmeyer 1980; 8. Smith 1978; 9. Hoogland 1996; 10. Hoogland 1983; 11. Hoogland 1995; 12. Blumstein et al. 1997.

scendant and nondescendant kin to danger in group-living mammals, less evidence for alerting descendant kin in birds, and a handful of examples of alerting nondescendant kin in birds.

In mammals, most research has been conducted on ground-dwelling sciurids (table 6.2) where different researchers have made a variety of comparisons to test predictions from kin selection theory. These are that females who are philopatric are more likely to call than males who leave their natal coterie, that residents who are more likely to have kin present call more than immigrants, that adults call more frequently when they have offspring than when they do not, and that adults with nondescendant kin call more than when they have none. For example, in an exemplary study, Sherman (1977) demonstrated that female Belding's ground squirrels were more likely to alarm call in response to a terrestrial predator if they were reproductive, had a living daughter or granddaughter with them, had a mother or sister still alive, or were residents (table 6.3). Moreover, he systematically dismissed alternative hypotheses for alarm calling, including reciprocity and pursuit deterrence. In addition, sciurid exceptions to expectations derived from kin selection theory can be explained by differences in breeding structure. For instance, male Richardson's ground squirrels defend territories during the breeding season and lac-

Table 6.3 Kinship and asymmetries in tendencies to give alarm calls to ground predators among female Belding's ground squirrels

Category of females	Observed	Expected	*P*-value
Reproductive with no known living relatives	14	9	<0.025
Nonreproductive with no known living relatives	2	7	
Reproductive with a living daughter or granddaughter but no other living relatives	18	12	<0.025
Reproductive with no known living relatives	5	11	
Reproductive with mother or at least one sister alive but no living descendants	13	8	<0.025
Reproductive with no known living relatives	3	8	
Reproductive residents: known to have lived in the same area the previous year or years	64	56	<0.025
Reproductive nonresidents: temporary invaders to the area	9	17	
Reproductive with either mother, sister, or descendant alive and present when predatory mammal appears	9	9	NS
Reproductive with either mother, sister, or descendant alive but not present when predatory mammal appears	6	6	
Reproductive without their mother or any sisters but with nursing young known to be alive	21	22	NS
Reproductive without their mother or any sisters and whose young were destroyed	4	3	

Source: Sherman 1977.
Note: Reproductive means pregnant, lactating, or with living postweaning young of the year. Expected frequencies were computed by assuming that animals call randomly, in proportion to the number of times they are present when a predatory mammal appears.
NS denotes nonsignificant.

Table 6.4 Evidence of kin-related benefits to alarm calling in nonsciurid mammals

Common	Findings	Reference[a]
Red colobus	a. Higher rates of calling with more vulnerable juveniles in group	1
Kloss's gibbon	a. Both sexes give siren calls which can be heard by neighboring groups that are composed of relatives	2
Spanish ibex	a. Females give more calls than males b. Females with kids give more calls than those without	3
Great gerbil	a. Higher frequency of vocalizing and foot drumming in family groups with newly emergent pups	4
America pika	a. Males call less often than females to pine martens b. Residents call more than nonresidents	5

[a] 1. Stanford 1998; 2. Teneza and Tilson 1977; 3. Alados and Escos 1988; 4. Randall, Rogovin, and Shier 2000; 5. Ivins and Smith 1983.

tation period and males call more than females, whereas round-tailed, Belding's, and thirteen-lined males do not and females call more than males. Data on sciurids are reinforced by scattered data on primates, ungulates, rodents, and lagomorphs (table 6.4).

In addition, other strands of evidence support the importance of kin selection. First, in some species, alarm calling, like aspects of nest defense, is almost certainly related to offspring vulnerability and hence offspring survival (sec-

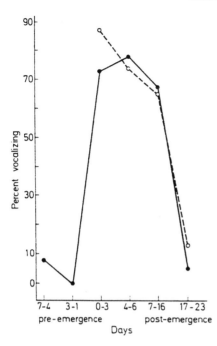

Figure 6.7 Percentage of parous thirteen-lined ground squirrel females (*solid line*) (12 < n < 22 females) and litters (*dashed line*) (8 < n < 17 litters) which vocalized at the approach of a human or dog as a function of emergence time (day 0 is the first day the litter was observed aboveground) (Schwagmeyer 1980).

tion 10.10). The percentage of parous thirteen-lined ground squirrels that call, for example, increases enormously as young emerge from their burrow despite those young being present underground for days beforehand (fig. 6.7). Second, nepotism appears to be limited by accessibility of kin that depends critically on mortality and dispersal (Sherman 1980a). Thus, male Belding's ground squirrels who disperse at the time that their mate's offspring are born seldom alarm call, and while mother-daughter and sister-sister pairs give alarm calls at predatory mammals, more distantly related females such as aunt-niece, first cousin-cousin, and grandmother-granddaughter pairs treat each other as they do unrelated females and do not call. For males and distantly related females co-occurrence in space and time is so infrequent that nepotistic behavior is rarely determined by selection (Sherman 1980a, 1981).

Third, not all age-sex classes give warning calls with equal probability. For instance, adult male and female vervet monkeys of high rank give alarm calls for longer periods to a stuffed leopard than low-ranking vervets (Cheney and Seyfarth 1981), and limited evidence suggests that males that gave the most alarm calls were more likely to have fathered youngsters in the group.

Misunderstandings over the meaning of nepotism have emerged in relation to the benefits that kin may obtain from hearing alarm calls. Shields (1980) and Blumstein and colleagues (1997) attempted to separate benefits obtained by descendant kin (that is, parental care) from those obtained by nondescen-

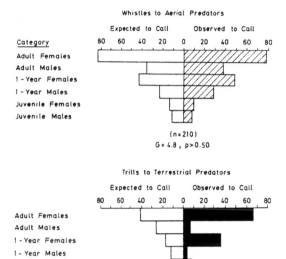

Figure 6.8 Expected and observed frequencies of alarm calling to aerial (upper) and terrestrial (lower) predators by various age-sex classes of Belding's ground squirrel. Expected values were determined from the frequencies with which age-sex classes were found aboveground before predators appeared. *n* refers to the number of individuals (Sherman 1985).

dant kin. Hamilton's rule, however, $rB > C$, does not distinguish between allelic copies that are transmitted through offspring or nondescendant relatives, since r is a probabilistic term in both cases (Dawkins 1979; Sherman 1980b; Hauber and Sherman 1998). Moreover, it is inappropriate to sum coefficients of relatedness between the caller and each recipient of the alarm call (Blumstein et al. 1997) because costs and benefits may differ among recipients, and inclusive fitness is maximized by concentrating benefits on closest relatives (Hauber and Sherman 1998). Sweeping these issues aside, it does seem that the pattern of alarm calling in mammals, at least as exemplified by sciurids, conforms broadly to expectations of kin-related benefits.

Explanations for alarm calling based on kin selection are compelling but alone cannot explain all alarm calling, even in the paradigm taxon, ground-dwelling sciurids (Taylor, Balph, and Balph 1990). Consider Belding's ground squirrels, which emit whistles to aerial predators but trills to terrestrial predators (Sherman 1985). Adult and yearling females trilled more than expected by chance, while adult and yearling males trilled less than expected (fig. 6.8), providing evidence for kin-related benefits. To an aerial predator, however, different ground squirrel age-sex classes whistled as expected under the null hypothesis (fig. 6.8). Thus, reproductive condition and presence of descendant and nondescendant kin had a marked effect on trilling to carnivores (table 6.3) but not on whistling to raptors. Taken together with data on predator-induced mortality indicating that predatory mammals were attracted to and killed more

Table 6.5 The relative vulnerability of individuals in each age-sex class to predation by different confirmed predators, compared with the relative frequency with which they gave first alarm calls to those predators

Age-sex class	Vulnerable to	Give alarm call to
Adult male	Leopard > Eagle > Baboon	Leopard** > Eagle* > Baboon
Adult	Leopard > Eagle > Baboon	Leopard* > Eagle** > Baboon
Juvenile	Baboon = Eagle > Leopard	Baboon > Leopard > Eagle

Source: Cheney and Seyfarth 1981.
* $P < 0.1$; ** $P < 0.05$.

callers that trilled than individuals that did not trill (table 6.1), whereas hawks rarely caught whistlers, results suggest that there are differential benefits and costs to trill and whistle alarm calls. That individuals farthest from cover whistled most and whistled while they ran, and that this resulted in pandemonium, suggests that whistles served phenotypically selfish ends for those individuals in the most danger, perhaps by confusing the predator. In contrast, the pattern of trills suggested kin-related benefits.

In another sciurid example, male Gunnison's prairie dogs call very frequently to a stuffed American badger model but in a fashion unrelated to the kinship of listeners nearby (Hoogland 1996). In a third example, again from a social mammal living with kin, vervet monkeys alarm called at leopards, martial eagles, and yellow baboons most often to those species to which they themselves were most vulnerable rather than to predators to whom kin were most vulnerable (table 6.5), again suggesting individual selection (Cheney and Seyfarth 1981). Thus, we cannot assume that alarm calls given in presence of kin have necessarily been shaped by kin selection (Shelley and Blumstein 2005). Moreover, many interesting, unanswered questions remain for mammals (Hoogland 1996; Stanford 1998): why do individuals with no kin call? Why do only some individuals with nearby kin call? And why do some species indulge in a cacophony of alarm calls, whereas in others only a single individual calls?

In most bird species, monogamous pairs raise offspring together without the help of conspecifics; this practice limits opportunities for kin-related benefits of alarm calls extending beyond offspring. There are many examples of parents increasing the rate of alarm calls after young have hatched (for example, long-toed lapwings and blacksmith plovers: Walters 1990; see chapter 10) and of young behaving adaptively on hearing the warning calls of their parents. For instance, in moustached warblers the probability that chicks crouch in the nest or move from the center to rim in preparation for jumping is correlated with the probability of the parent giving an alarm call (fig. 6.9). Parental alarm calls may also help young recognize predators (Kullberg and Lind 2002). Historically, however, most of the work on maternal and paternal

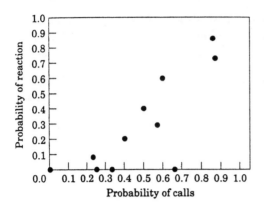

Figure 6.9 Probability of adult moustached warblers giving alarm calls and adaptive chick reactions (jumping and ducking) to alarm calls. Averages per day for an average of 6.2 nests are shown (Kleindorfer, Hoi, and Fessl 1996).

antipredator behavior in birds has focused on nest defense (chapter 10) rather than alarm calling per se.

Among cooperatively breeding birds, there is remarkably little information on alarm signals. The most well-worked species in relation to antipredator behavior is the Florida scrub jay. Scrub jays live on permanent territories in small family groups of three individuals on average, usually a breeding pair and offspring of previous breeding seasons (Woolfenden and FitzPatrick 1984). Individuals take turns as sentinels, that is, remaining vigilant and often perching at a prominent place while members of the group forage nearby (section 5.4). Sentinels give alarm calls at aerial predators and are usually the first bird to call. Although all birds participate in sentinel behavior, the male breeder contributes the most time, followed by the female breeder. Upon hearing the call, other birds may fly to a perch and look in the direction that the sentinel is looking; or if the sentinel detects a raptor at close range, it gives a different call and all birds dive for cover (McGowan and Woolfenden 1989; Hailman, McGowan, and Woolfenden 1994). Since group members are either breeders, dependent young, or offspring from previous years, alarm calls emitted by sentinels have the potential of benefiting kin, although this cannot be taken for granted (see also Griesser and Ekman 2004).

The majority of studies of alarm calling in birds have, however, stressed the structural characteristics of "seeet" or "high zee" and other alarm calls used by passerines assembling in feeding aggregations where nonrelatives collect. In these circumstances calls must be mutualistic, because kin are unlikely to be present and group composition is likely to vary, precluding reciprocal altruism, although this is less certain. It is not clear whether kin selection has been a more important force in shaping alarm calls in mammals than in birds or whether differences are simply an artifact of research effort.

Table 6.6 Examples of avian species responding to each others' alarm calls

Prey	Predator	Finding	Reference[a]
Western grebe and Forster's tern	Playback of tern alarm call	Grebes respond to tern alarm calls by leaving nest or showing alarm posture	1
Rolland's grebe, silvery grebe, and brown-hooded gull	Caracaras and grison	Grebes cover their eggs when they leave nests after hearing gull warning cries. Predation on eggs is lower with gulls present	2
Gulls, lapwings, and European golden plovers	Human predator	Lapwings and plovers leave sooner when gulls take off, and later when gulls do not, showing that gulls spot for them	3
Downy woodpecker, chickadee, and titmouse	Red-tailed hawk, sharp-shinned hawk models	Chickadees and titmice call more than woodpeckers; thus woodpeckers exploit others	4
Stonechat and 15 other species, including willow warbler	Assorted predators in the wild	Other species exploit stonechats that have greater flight distances	5
Chickadee, titmouse, and peripheral species, including black and white warbler	Accipters, including sharp-shinned hawk	Chickadees and titmice respond first	6
Willow tit, great tit, European robin, common redstart, and reed bunting	Not reported	Alarm calls of each species result in willow tits seeking cover or alarm calling	7

[a] 1. Neuchterlein 1981; 2. Burger 1984; 3. D. B. A. Thompson and Barnard 1983; 4. Sullivan 1985; 5. Greig-Smith 1981a; 6. Gaddis 1980; 7. Haftorn 2000.

6.5 Alarm calls between species

Whereas putative benefits of alarm calling among conspecifics have received much attention, selective pressures acting on prey responding to the alarm calls of other species have not been examined very thoroughly. Rather, the focus has simply been on the fact that sympatric species do respond to each other's alarm calls in mixed-species flocks, between species of mammals, and between birds and mammals (tables 6.6 and 6.7). Structurally, calls made by each of the species that associate together are usually different from each other; thus, they do not help to explain the convergence of alarm-call structure among sympatric species that do not necessarily associate (section 6.2).

Most of the bird examples stress that the early warning benefits of alarm calls are asymmetric for members of mixed-species flocks (table 6.6; see also Rasa 1983; P. N. Newton 1989), whereas mammal studies either have been descriptive (for example, Shriner 1998) or have examined whether species respond appropriately to different types of alarm calls given by heterospecifics (table 6.7). The most sophisticated of these latter studies is of wild Diana monkeys (Zuberbuhler 2000a). In the Tai Forest, Diana monkeys are preyed on principally by crowned hawk eagles and leopards but also by chimpanzees.

Table 6.7 Examples of mammalian species responding to other species' alarm calls

Prey	Predator	Finding	Reference[a]
Verreaux's sifaka and red-fronted lemur	Playbacks of each species' aerial and general alarm calls	Both lemur species respond to playbacks of both types of call with appropriate alarm calls and antipredator escape movements	1
Verreaux's sifaka and ring-tailed lemur	Playback calls of ring-tailed lemurs	Verreaux's sifakas respond to lemur alarm calls and vice versa, but only lemurs do so referentially	2
Vervet monkey and superb starling	Aerial and terrestrial predator playbacks	Vervets make appropriate responses, looking up or running into a tree in response to aerial and terrestrial playbacks, respectively	3
Diana monkey and Campbell's monkey	Playbacks of leopard and eagle alarm calls on males of other species	Diana and Campbell's monkeys respond to playback of other species' leopard or eagle calls as though original predators were present	4
Diana monkey and chimpanzee	Playbacks of two sorts of chimpanzee call	Some Diana monkeys give leopard alarm call when hear chimpanzees' leopard-specific alarm call	5
Putty-nosed monkey and Diana monkey	Playbacks of male putty-nosed monkey alarm calls made to eagles	Diana monkeys give eagle alarm calls in response to male putty-nosed monkey eagle alarm calls	6
3 *Cercopithecus* species	Aerial and terrestrial predators	Different species detect different sources of danger and respond to each other	7
Bonnet macaque, nilgiri langur, hanuman langur, and sambar	Playbacks of alarm calls	Bonnet macaques respond with flight and scanning to nilgiri and hanuman langur and sambar alarm calls	8
Langurs and chital	Observations	Chital alerted to langur alarm call more often than vice versa	9
Red colobus, red-tailed monkey, blue monkey, chimpanzee, and bushbuck	Observations	Red colobus alarm call in response to other species' alarm calls	10
Dwarf mongoose and 2 hornbill species	Raptors	Hornbills more likely to give warning call of flee before mongooses	11
Yellow-bellied marmot and golden-mantled ground squirrel	Playbacks of alarm calls	Marmots respond by running or raising heads and scanning when they hear ground squirrel calls	12
Thirteen-lined ground squirrel and red-winged blackbird	Observations	Ground squirrels respond by standing erect or fleeing to burrows	13

[a]1. Fichtel 2004; 2. Oda and Masataka 1996; Oda 1998; 3. Cheney and Seyfarth 1985; Hauser 1988b; Seyfarth and Cheney 1990; 4. Zuberbuhler 2000b, 2001; 5. Zuberbuhler 2000a; 6. Eckhardt and Zuberbuhler 2004; 7. Gautier-Hion, Quris, and Gautier 1983; 8. Ramakrishnan and Coss 2000a; 9. P. N. Newton 1989; 10. Stanford 1998; 11. Rasa 1983; 12. Shriner 1998; 13. Schwagmeyer 1980.

Table 6.8 Diana monkeys' vocal behavior in response to playbacks of predators

Playback stimulus	Number of groups giving alarm calls		
	Yes	No	Rate (%)
Males			
Chimpanzee social screams	0	15	0
Chimpanzee alarm calls	20	14	58.8
Leopard growls	13	0	100.0
Females			
Chimpanzee social screams	1	14	6.7
Chimpanzee alarm calls	15	19	44.1
Leopard growls	11	2	84.6

Source: Zuberbuhler 2000a.

They respond to leopards with conspicuous alarm calling but to chimpanzees with silent, cryptic behavior. An additional complication is that chimpanzees are also preyed on by leopards, to which they give loud and conspicuous alarm screams if they see them. When chimpanzee social screams were played to Diana monkeys, the latter remained silent, but when chimpanzee alarm screams were played about half the Diana groups switched from a chimpanzee-specific cryptic response to a leopard-specific conspicuous calling, suggesting some Diana monkeys assumed that a leopard was present (table 6.8). These cognizant groups tended to be those whose home range was in the resident chimpanzee core area. In a follow-up experiment, Diana monkeys were primed with chimpanzee alarm calls, and then were tested with leopard growls 5 minutes later to see if they anticipated a leopard. Some males responded by giving "leopard" alarm calls in response to the chimpanzee playback probe but gave few calls to the subsequent leopard growl (fig. 6.10b), suggesting that they were treating chimpanzee calls as leopard growls (fig. 6.10a). Other males remained silent when the chimpanzee call was played (fig. 6.10c) but responded loudly to leopard growls, suggesting that they were treating chimpanzee alarms as signifying only that chimpanzees were present (fig. 6.10d). Zuberbuhler (2000a) speculates that a single experience could be sufficient in acquiring knowledge about the causal link between the presence of a leopard and chimpanzee alarm calls.

Diana monkeys are also preyed on by crowned hawk eagles, as are sympatric Campbell's monkeys. Dianas respond to the alarm calls of Campbell's monkeys and vice versa (section 6.6.c), specifically producing a crowned hawk eagle alarm call in response to a playback of a crowned hawk eagle call from a male of the other species, or the call of the eagle itself; or responding with a leopard alarm call on hearing a leopard call from the other prey species or a leopard growl (Zuberbuhler 2000b, 2001). More subtly, Diana monkeys do not respond to either of the Campbell's monkey alarm calls if these calls are preceded by a boom that is uttered just before alarm calls in a number of circumstances,

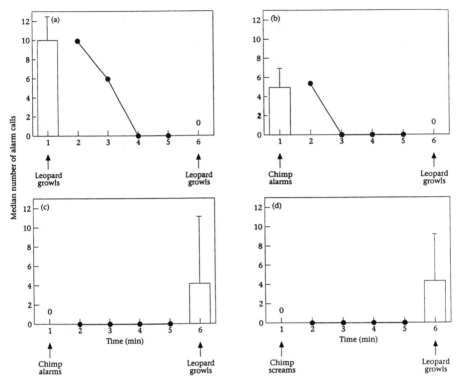

Figure 6.10 Median number (+ third interquartile) of male Diana monkey alarm calls in (*a*), baseline (*N* = 13 males); (*b*), conspicuous males (*N* = 14); (*c*), cryptic males (*N* = 10); and (*d*), control conditions (*N* = 15). Males who responded with alarm calls to chimpanzee alarm calls were termed "conspicuous," (*b*), while those who did not respond with alarm calls were termed "cryptic" (*c*) (Zuberbuhler 2000c).

including in response to falling trees or to far-off alarm calls of neighboring groups. Diana monkeys do not give booms themselves and show full responses (as measured by calling) to playbacks of Diana monkey alarms artificially preceded by a boom. In short, Diana monkeys are acutely sensitive to the extent of threat conveyed by heterospecific messages (Zuberbuhler 2002).

That mixed-species groups gain antipredator benefits from the alarm calls of heterospecifics does not necessarily indicate that species associate to avoid predation afforded by early warning from heterospecifics. Without being able to separate fitness benefits derived from enhanced foraging efficiency and from predator evasion, it is difficult to reach firm conclusions as to why species stay in each other's company, but there is some evidence that moustached monkey, white-nosed guenon, and crowned guenon associations in the wild bring considerable antipredator benefits through alarm signals. Moustached monkeys, in particular, benefit from alarm calls given by crowned guenon males to danger from above and from the early alarm reactions of white-nosed guenon males. In mixed associations, moustached monkeys are less likely to seek out dense, safe forest, and this allows them to feed in richer zones. Gautier-Hion, Quris, and Gautier (1983) argue that, for moustached monkeys, the most important benefit for polyspecific associations may be early warning by heterospecifics (Bshary and Noe 1997a).

6.6 Variation in alarm calls

There is ample confirmation that certain birds and mammals modulate their alarm calls according to the extent of danger and type of threat posed by a predator. Most of the research has been conducted on North American sciurids, assorted birds, and a handful of primates.

6.6.a Sciurids   Flying low and rapidly, raptors routinely approach ground squirrels undetected until they are close by, consequently constituting a high risk. In contrast, carnivores amble toward colonies slowly and represent a lower threat. As far back as 1966, it was known that Uinta ground squirrels "chirped" in response to airborne predators but "churred" at the approach of ground predators (Balph and Balph 1966), and this finding has been replicated in other ground-dwelling sciurids (table 6.9). In general, calls to aerial predators consist of one or two pure-note whistles, while multiple-note trills or chatters are given repeatedly to terrestrial predators (table 6.9), although there are possible exceptions (for example, Richardson's ground squirrel: Davis 1984). Recently, an ultrasonic alarm call has been reported in Richardson's ground squirrels (Wilson and Hare 2004). Sciurid alarm calls seem to indicate "response urgency" rather than the type of predator that is approaching (the latter of which is called "referential signaling": Macedonia and Evans 1993; Hauser 1996). For example, California ground squirrels that are chased at high speeds by carnivores sometimes produce whistles; raptors sighted at a distance, and hence of no immediate threat, sometimes elicit chatters (Leger, Owings, and Gelfand 1980; Owings and Leger 1980; Owings and Virginia 1978; Leger, Berney-Key, and Sherman 1984). Similar observations have been made on Belding's ground squirrels (table 6.10). Calls thereby convey information about the distance of the predator and hence the urgency with which escape is required.

Structure and temporal patterning of chirps or whistles and trills or chatters appear adapted to predatory context (table 6.11). Whistles are of short duration and may be difficult to locate, as might be expected where risks are high or where the predator is a raptor that finds such calls difficult to localize. Additionally, time constraints could limit number of calls (MacWhirter 1992). Chatters, however, are longer, thereby drawing attention to the sender, enabling conspecific recipients to note the caller's posture and orientation toward the predator (Owings and Virginia 1978; S. R. Robinson 1981).

Table 6.9 Instances of alarm calling in ground-dwelling sciurids

Common	Call	Threat	Reference[a]
Ground squirrels			
Uinta	Chirp	Aerial predator	1
	Churr	Terrestrial predator	
Californian	Whistle	Low-flying raptors	2
	Chatter/chat or long chat	Mammalian predators and snakes (distant aerial predators)	
Belding's	Chirp or Whistle	Aerial (and terrestrial) predator	3
	Trill or churr	Terrestrial (and aerial) predator	
Columbian	Shrill chirp	Aerial or ground predator	4
	Churr	High-intensity alarm	
	Soft chirp	Aftermath of encounter	
Arctic	Whistle	Aerial predator	5
	Chatter	Terrestrial predator	
Richardson's	Chirp	Aerial predator	6
	Whistle	Terrestrial predator	
	Ultrasonic	Terrestrial predator	
Round-tailed	Silence	Aerial predators	7
	Whistle or Peep	Ground predators	
Thirteen-lined	Silence	Aerial predators	8
	Trill	Aerial and terrestrial predators	
Prairie dogs			
Gunnison's	Barklike	Terrestrial predators	9
Black-tailed	Bark	Aerial and terrestrial predators	10
	Jump-yip	Snakes	
Marmots			
Alpine	Whistle	Aerial and terrestrial predators	11
Hoary	Chirp	Aerial predator	12
Golden	Call	Fewer notes with risk	13
Yellow-bellied	Whistle	Rate varies with risk	14
	Chuck	Rate varies with risk	
	Trill	High-risk situations	

Source: Based on Owings and Hennessy 1984.

Note: Number of calls listed are conservative.

[a] 1. Balph and Balph 1966; 2. Owings and Virginia 1978; Owings and Leger 1980; 3. L. W. Turner 1973; Sherman 1977, 1985; S. R. Robinson 1981; Leger, Berney-Key, and Sherman 1984; 4. Betts 1976; Harris, Murie, and Duncan 1983; Koeppl, Hoffmann, and Nadler 1978; MacWhirter 1992; 5. Melchior 1971; 6. Davis 1984; Wilson and Hare 2004; 7. Dunford 1977; 8. Matocha 1977; Schwagmeyer 1980; Schwagmeyer and Brown 1981; 9. Hoogland 1996; Slobodchikoff et al. 1991; Ackers and Slobodchikoff 1999; 10. King 1955; Owings and Owings 1979; W. J. Smith et al. 1976, 1977; Waring 1970; Halpin 1983; 11. Lenti-Boero 1992; Blumstein and Arnold 1995; 12. Taulman 1977; Noyes and Holmes 1979; 13. Blumstein 1995a, 1995b; 14. Blumstein and Armitage 1997a.

Listeners exhibit different escape behaviors on hearing these different calls. For instance, California ground squirrels are quicker to post following chatter-chat playback calls, usually given to terrestrial predators, than to playback whistles, usually given to raptors (Leger and Owings 1978; see also L. W. Turner 1973; Leger, Owings, and Boal 1979; Schwagmeyer and Brown 1981; M. A. Harris, Murie, and Duncan 1983).

Other species of sciurid have only one type of alarm call that they submit

Table 6.10 Number of chirps (89 bouts) and trills (320 bouts) given in different contexts by Belding's ground squirrels

	Chirps			Trills		
	One caller		Several	One caller		Several
	One chirp	Several chirps		One trill	Several trills	
Terrestrial predator	3	2	3	13	12	30
Aerial predator	30	7	15	25	5	9
Harmless animal	8	3	1	8	3	0
Social chase	3	3	0	16	4	1
Unknown chase	7	2	2	103	67	24

Source: S. R. Robinson 1981.
Note: Data show that chirps are usually given to aerial predators but not exclusively and that trills are given to both terrestrial and aerial predators.

Table 6.11 Comparison of alarm call design

Property	Aerial alarm calls	Ground alarm call
Structure	Pulse + second component (wide or narrow band)	Repeated broadband pulses
Amplitude	Low (after initial pulse) Very low after first call	Consistently high
Active space (detectability)	Small (especially second and subsequent calls)	Large
Localizability	Poor (especially second and subsequent calls)	Excellent
Audience effect	Yes	No
Duration of calling	Brief	Prolonged
Potential receivers	Nearby social companions	Conspecifics over large area + potential predator

Source: Evans 1997.

to both aerial and terrestrial predators (for example, round-tailed and thirteen-lined ground squirrels: Dunford 1977; Schwagmeyer 1980). Yet others have several alarm calls that are not tied to specific predator contexts (yellow-bellied marmots: Blumstein and Armitage 1997a). These species, as well as sciurids that have risk-specific types of calls, modulate the rate or duration at which they call to response urgency, generally calling at a higher rate when risk is high (Blumstein and Armitage 1997a; see also Ivins and Smith 1983). For example, the rate of repetitive calls of juvenile Richardson's ground squirrels is higher the closer that a threat is presented to them (fig. 6.11). Nonetheless, some remain silent under high risk (alpine marmot: Blumstein and Arnold 1995; golden marmot: Blumstein 1995a). Furthermore, listeners react to differences in call rate: Richardson's ground squirrels are more likely to post the greater number of trills uttered and the more callers that are involved (fig.

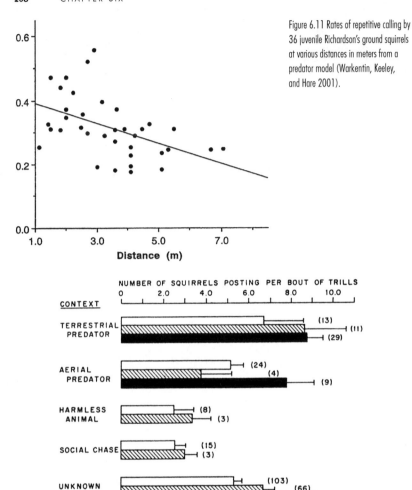

Figure 6.11 Rates of repetitive calling by 36 juvenile Richardson's ground squirrels at various distances in meters from a predator model (Warkentin, Keeley, and Hare 2001).

Figure 6.12 Mean number (+SE) of Belding's ground squirrels posting to different classes of trills given in different contexts. Open bars represent single trills; hatched bars, several trills by one caller; solid bars, trills by several callers. The number of trill bouts is in parentheses (S. R. Robinson 1981).

6.12; see also Weary and Kramer 1995; Warkentin, Keeley, and Hare 2001; but see Blumstein 1995b). Finally, among Gunnison's prairie dogs, the information encoded in alarm calls is very subtle. Discriminant function analyses show that individuals call differently at coyote and skunk models (Ackers and Slobodchikoff 1999) and even to humans wearing different-colored tee shirts and different clothing shapes (Slobodchikoff et al. 1991)! It is not clear whether Gunnison's calls convey more detailed information about threatening stimuli than calls of other species or whether the level of analyses is simply deeper than in other studies.

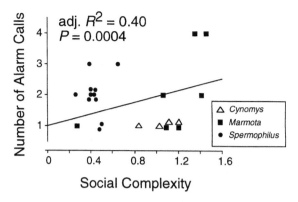

Figure 6.13 Relationship between social complexity and alarm-call repertoire size. *Cynomys* are prairie dogs, *Marmota* are marmots, and *Spermophilus* are ground squirrels (Blumstein and Armitage 1997b).

    The reasons that some sciurid species produce different types or rates of calls in response to varying risk while others do not are enigmatic. Rather than examining environmental and social constraints, or the auditory capabilities of different predators that might select for particular types, rates, or durations of alarm calls, studies have sought to identify the type of information (class of predator or degree of threat) carried in the message. Nonetheless, Sherman's (1985) study at Tioga Pass is instructive, because his data show that whistling tendencies, usually given to raptors, were not affected by age, sex, or kinship (fig. 6.8), and that whistlers were rarely captured (table 6.1), which suggests that whistling directly benefits the caller. In contrast, trills increased risk and benefited kin. Thus, Belding's ground squirrels appear to give two calls to predators for entirely different reasons (see also Klump and Curio 1983). Such arguments have not been brought to bear on other ground squirrel species as yet.
    In a wider ranging analysis, Blumstein and Armitage (1997b) tried to relate alarm-call repertoire size of 22 species of ground-dwelling sciurids to social complexity as measured by a formula that incorporated number of social group years that held a particular number of individuals, and time to natal dispersal. Applying phylogenetic controls to their analyses, they found that social complexity explained some variation in alarm-call repertoire size in marmots but not in prairie dogs or ground squirrels (fig. 6.13). They speculated that laryngeal morphology might constrain production of variable vocalizations in some species; that habitat structure might preclude discrimination of different calls at a distance; or competition for bandwidths with conspecifics might limit production of several calls. The most likely explanation that they arrived at, however, was that ground-dwelling sciurids have but one escape route, fleeing to a burrow, that reduces the importance of telling conspecifics about the type of predator approaching because escape strategies are the same regardless of threat (Macedonia and Evans 1993). This idea is intriguing, because closely related red squirrels living in a three-dimensional environment "seeet"

or "seeet-bark" at predatory birds but bark at terrestrial predators (Greene and Meagher 1998). Although it is unclear as to whether the two types of call are referential or convey response urgency, there is very little cross-taxon calling in red squirrels (only 1/21 "seeets" to a human and 2/42 barks to a bird model), giving the referential signaling idea an edge. If subsequent work bears this out, the contrast between arboreal and terrestrial sciurids would support an ecological explanation for referential signaling.

6.6.b Birds   For a long time, it had been known that galliformes produce different calls in response to aerial and terrestrial predators (for example, Goodwin 1953; Stokes 1961; H. W. Williams 1969; Maier 1982), but only careful experimental work on domestic chickens has shown that these calls are related to the spatial characteristics of threat and, further, that they produce different escape responses in listeners (Gyger, Marler, and Pickert 1987). Displaying raptor-shaped video images or a video of a raccoon to a cockerel, next to which a conspecific hen was confined, C. S. Evans, Evans, and Marler (1993) were able to elicit two sorts of alarm calls from the male. Hawk animations presented on an overhead screen elicited aerial alarm calls, whereas raccoon footage presented at ground level elicited only ground alarm calls. But when raccoon videos were played overhead, cockerels responded with aerial alarm calls (C. S. Evans and Marler 1995), indicating they were cueing in on spatial characteristics rather than type of predator. When calls were played to hens, they moved toward cover and scanned horizontally and upward after an aerial call but stayed where they were and scanned horizontally after a ground call (fig. 6.14).

Just as in sciurids, information about danger can be conveyed by the structure of avian alarm calls. Female Carolina wrens, for instance, give high-frequency "chirts" that are more frequently spaced when a hawk is moving (Morton and Shalter 1977), and Mexican chickadees give higher pitched calls in a high-risk situation, when hawks fly within 30 m of the flock (Ficken 1990). Under experimental conditions, cockerels utter more alarm calls to raptor-shaped silhouettes that subtend a greater angle on the retina (that is, that appear closer). Similarly, they call more when raptor stimuli are moved more rapidly across the overhead screen (Evans, Macedonia, and Marler 1993). Both protocols represent high-risk situations, information about which can be conveyed to hens.

Field studies have shown that birds in some orders other than galliformes alarm call differently either to different predators or to different threats normally associated with different predators (for example, Marler 1956; Jurisevic and Sanderson 1994). Superb starlings in Amboseli National Park, Kenya, give

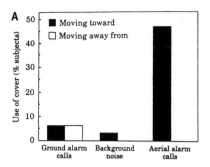

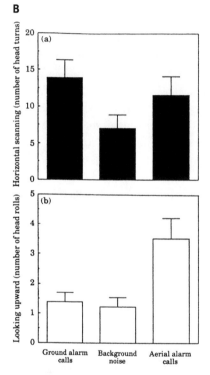

Figure 6.14 *A*, proportion of domestic hens moving toward, or away from, the covered area of the cage following playback of ground alarm calls, aerial calls, or a control sound composed of background noise present in alarm call recordings. *B*, visual monitoring of hens following playbacks. (*a*), Scanning in the horizontal plane (that is, moving the head in an arc from side to side): mean (+SE) number of head turns. (*b*), Looking upward (that is, rolling the head to fixate laterally with one eye): mean (+SE) number of head rolls (C. S. Evans, Evans, and Marler 1993).

one type of call to aerial and another to terrestrial predators (Cheney and Seyfarth 1985a; Seyfarth and Cheney 1990), although it must be noted that they give different calls to the same predator in different contexts, suggesting they might be signaling different escape strategies. Also, lapwings (*Vanellus* sp.) discriminate between similar predators, calling to harriers (*Circus* sp.) more often than to fish eagles, but it is not clear whether this simply reflects response urgency (Walters 1990). These species and galliformes live in a variety of habitats, so it is difficult to relate the evolution of predator-specific alarm calls to environmental conditions in birds. On the one hand, one could argue that rapid and appropriate responses to different predators would be selected for in dense brush, where sudden ambush is a problem; but one could equally argue that predator recognition would evolve in open-country species, for which the costs of false alarms are high, as prey are so visible much of the time (Walters 1990). Comparative analyses are presently hindered by lack of information on which bird species do and do not give acoustically variable responses to different classes of predator.

6.6.c Primates   From a cognitive psychologist's viewpoint, the most sophisticated kind of alarm calling is found in primates. In two classic papers, Seyfarth, Cheney, and Marler (1980a, 1980b) showed that free-living vervet monkeys in Amboseli National Park gave different sorts of alarm calls to different classes of predators (table 6.12). In a series of playback experiments in the field, vervets exhibited markedly different responses to different types of call: rapidly climbing a tree when they heard the playback of an alarm call given to a leopard; running out of a tree, heading into cover, and looking up when they heard an alarm call elicited by an eagle; and looking down when they heard a call given to a snake (table 6.13). These results could not be attributed to differences in alarm call amplitude, duration, or to age-sex class of callers or receivers. Vervets also responded to superb starlings' raptor alarm calls as if they were vervet "eagle" calls, as judged by habituating them to starling raptor calls

Table 6.12 Animals that elicited alarm calls for vervet monkeys

	Confirmed predators	Potential predators	Nonpredators
Mammalian carnivore alarms	Leopard	Lion Spotted hyena Cheetah Black-backed jackal	Warthog
Avian predator alarms	Martial eagle	African hawk eagle Black-chested snake eagle Tawny eagle Verreaux's eagle owl	African goshawk Bateleur Egyptian goose Gray heron Ground hornbill Lilac-breasted roller Marabou stork Pigeon Secretary bird Spoonbill Vulture Falling leaf
Snake alarms	Python	Cobra Black mamba Green mamba Puff adder	Tortoise Mouse
Baboon alarms	Baboon		

Source: Seyfarth and Cheney 1980.

Table 6.13 Responses of vervet monkeys to playbacks of leopard, eagle, and snake alarms

	Responses on ground					Responses in tree				
Alarm type	No. of trials	Run into tree	Run into cover	Look up	Look down	No. of into	Run higher tree	Run out of	Look up	Look down
Leopard	19	8(E)S*	2	4	1	10	4	0	3	4
Eagle	14	2	6LS	7(L)S	4	17	4	5(L)S	11(L)	12
Snake	19	2	2	2	14L*E*	9	2	0	5	9L*(E)

Source: Seyfarth, Cheney, and Marler 1980a.
Note: Entries indicate the number of trials in which at least one subject showed a given response for longer in the 10 sec after a playback than in the 10 sec before. Letters indicate where there was a statistically significant difference between responses to the two call types. For example, with monkeys on the ground, leopard (L) alarms caused one or more subjects to look down in 1 out of 19 trials. Thus snake alarms were significantly more likely to cause animals on the ground to look down than either leopard or eagle alarms. Letters in brackets: 0.05 $<$ p $<$ 0.1; letter alone: 0.01 $<$ p $<$ 0.05; starred letter: $p <$ 0.01.

and then witnessing an absence of responses to a vervet "eagle" alarm call. More subtly, starling terrestrial alarm calls were elicited by a wide range of mammals, reptiles, and birds, making them somewhat unreliable. Vervets treated them as such, failing to respond to vervet "eagle" calls after habituating to starling terrestrial predator calls (Seyfarth and Cheney 1990).

Parallel results have been found in certain lemurs and other cercopithecines. In a semicaptive colony, ring-tailed lemurs were more likely to climb lower in response to playbacks of calls given to raptors and to climb into trees if they heard a call normally given to carnivores. In contrast, cohabiting ruffed lemurs did not demonstrate informed responses to playbacks of their calls (Macedonia 1990; see also Sauther 1989). In free-living red-fronted lemurs and white sifakas, adults responded to playbacks of raptors and to aerial alarm calls to these birds by giving specific alarm calls, scanning the sky, and moving from exposed positions. In response to terrestrial predators, however, both species gave "woof" calls or roaring barks that they also utter in aggressive interactions with conspecifics or when they mob snakes (Fichtel and Kappeler 2002; Fichtel 2004). Thus, these two species apparently showed referential alarm calls to diurnal raptors but not to carnivores. Instead, red-fronted lemurs convey response urgency through "woof" frequency or amplitude (Fichtel and Hammerschmidt 2002; see also Fichtel and Hammerschmidt 2003). In free-living Diana monkeys, males give acoustically different calls to leopards and crowned hawk eagles; female and juvenile Dianas respond (give an alarm call of the appropriate type) to playbacks of these calls in the same way as they do to playbacks of leopard and crowned hawk eagle vocalizations, respectively (Zuberbuhler, Noe, and Seyfarth 1997). In free-living Campbell's monkeys, males approached a speaker regardless of whether it played vocalizations of crowned hawk eagle shrieks, alarm calls given by male Campbell's monkeys to crowned hawk eagles, or alarm calls to this eagle given by sympatric Diana monkeys. Also, the whole group approached the speaker if they heard leopard growls, or leopard alarms given by Campbell's or Diana monkey males (Zuberbuhler 2001). Finally, putty-nosed monkeys give "tocks" in response to playbacks of the shrieks of the same eagle species but "zecks" to leopard growls (Eckardt and Zuberbuhler 2004).

That at least seven species of primate voice acoustically different alarm calls to different classes of predators does not necessarily mean that calls denote different predators ("referential signaling"). As in birds, they could simply denote differential escape strategies normally associated with terrestrial and aerial predators. At least three lines of evidence speak against this. In vervets, subjects exposed to playback calls of an individual, made unreliable by repeated presentation of, say, a "leopard" alarm call in the absence of a real leopard, eventually failed to respond. But when an "eagle" alarm call given by the same individual was played, subjects' responses were reinstated. By failing to transfer habituation across call types, subjects must have been treating the referents of the two calls as different. If they had transferred habituation, they would have been paying attention to the caller's concern about a general predatory threat (Cheney and Seyfarth 1988). Second, semicaptive ring-tailed lemurs were exposed to aerial or terrestrial predators each in relatively benign or in dangerous situations. These were, respectively, a perched red-shouldered hawk or great horned owl on a traditional lemur path (benign), a raptor silhouette that glided close to them along an overhanging wire (perilous), a live dog on a leash (innocuous), and, finally, a hidden (but chained) dog that suddenly surprised the lemurs at their provisioning site (dangerous). In all four permutations, responses conformed to expectations generated by referential signaling rather than signaling response urgency and resembled patterns seen in natural settings: rasps and shrieks given to raptors irrespective of danger, and open-mouth clicks and yaps given to carnivores following flight whatever the degree of threat (Pereira and Macedonia 1991). Third, when Diana monkeys were exposed to "eagle" and "leopard" alarm calls played at close and far distances, and at elevations that were above and below them, they consistently responded (by alarm calling themselves) to predator category regardless of the proximity or direction of the playback (Zuberbuhler 2000c). Together, these three findings indicate that these primates give referential alarm calls rather than signaling response urgency.

Why should some primates signal referentially, some birds signal predator elevation, and the majority of sciurids signal response urgency (see Fichtel and Kappeler 2002 for a review)? It is easy to attribute these differences to differences in cognitive skills, but the ability of Gunnison's prairie dogs to change their calls in response to humans wearing different clothes suggests that this explanation is too simple. Failure to find referential signaling in barbary macaques, bonnet macaques, and yellow baboons raises the same concern (Fischer 1998; Hohmann 1989; Fischer et al. 2001). Instead, environmental factors, as yet poorly understood, may be involved. For example, vervets and ring-tailed lemurs are small-bodied compared with their avian and mammalian predators, whereas larger ruffed lemurs are relatively invulnerable to extant predators on Madagascar. In addition, vervets and ring-tails are far more terrestrial than ruffed lemurs and hence vulnerable to carnivore attack. Perhaps, then, there has been greater selection on signaling the nature of predatory threat in these species than in ruffed lemurs (Macedonia and Evans 1993)? Being small and terrestrial cannot be the only factor, however: ground-dwelling sciurids are terrestrial but do not signal referentially. Instead, it may be that sciurids have the same escape response to aerial and terrestrial predators, namely disappearing down a burrow, and therefore do not need to be informed of the type of predator approaching (see also Le Roux, Jackson, and Cherry 2001). In contrast, vervets and ring-tailed lemurs must respond either by fleeing into trees or descending to the ground and therefore require information about the class of predator (Macedonia and Evans 1993). Whether this argument extends to Diana and Campbell's monkeys is less clear, since they are more arboreal and respond in somewhat similar ways to aerial and terrestrial predators. Admittedly, these arguments are speculative, but they force us to consider ecological factors rather than cognitive abilities in explaining referential signaling and to look for further examples in species that have incompatible escape tactics.

6.7 Development of conspecific warning signals

6.7.a Ontogeny of response
Young mammals and birds respond to alarm calls differently from adults. The ontogeny of this behavior can be divided into four stages for convenience (fig. 6.15), and examples can be found of young animals attaining each of these. As an example of early detection and discrimination, infant Belding's ground squirrels show decreased heart rates in response to playbacks of whistle alarm calls but increased rates in response to playbacks of trills even before they emerge from their burrow, and experience of calls had little effect on this discrimination (Mateo 1996b). As illustrations of the response stage, mallard ducklings freeze upon initial exposure to maternal alarm calls (D. B. Miller, Hicinbothom, and Blaich 1990), and juvenile thirteen-lined ground squirrels run to a burrow or post there when they

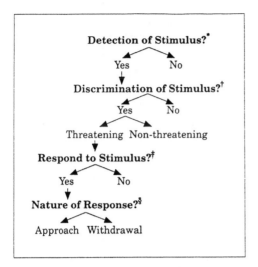

Figure 6.15 Four stages of the development of alarm-call response behavior. Stages also broadly apply to the expression of antipredator behavior by experienced individuals of any species (Mateo 1996a).

hear an alarm call, just like adults (Schwagmeyer and Brown 1981). Finally, regarding the fourth stage, infant squirrel monkeys reared in social isolation respond to alarm peeps, a warning call to bird predators, by fleeing to a surrogate mother; but on hearing yapping, the alarm call given to terrestrial predators, they avoid a cat model when it is presented simultaneously (Herzog and Hopf 1984).

The development of alarm-call response behavior has been documented particularly well in Belding's ground squirrels (Mateo 1996a; Mateo and Holmes 1999a, 1999b). In this species, juveniles become more selective in their responses as they grow older: newly emergent young freeze or enter a burrow in response to playbacks of all vocalizations, but within 5 days respond only to alarm calls. Responsivity to whistles, indicative of fast-moving predators, develops earlier than to trills that are associated with slow-moving predators, but responses to both are fully developed by the time juveniles disperse at 4 weeks. By dint of their own alarm response behavior, mothers but not unrelated females have a mild positive effect on the rate of development of appropriate responses, but mothers' behavior is not necessary for alarm calling to develop (Mateo and Holmes 1997). A similar result has been found in vervet monkeys (Seyfarth and Cheney 1986). There, infants show incorrect responses to referential alarm calls by looking down or rapidly climbing a tree in response to "eagle" alarm calls, for example, but are less likely to make a mistake if they look at an older animal before responding (table 6.14). Other studies have shown that juveniles overestimate risk compared with adults. As a case in point, juvenile bonnet macaques are more likely to flee from the sound of a

Table 6.14 The relation between an infant vervet monkey's response to alarm call playback and its behavior immediately prior to responding

	Looks at others before responding?	
	Yes	No
Adultlike response	17	8
"Wrong" response	2	7

Source: Seyfarth and Cheney 1986.
$\chi^2 = 5.51$, df $= 1$, $p < 0.02$.

juvenile alarm call and from a playback of a motorcycle engine than are adults (Ramakrishnan and Coss 2000b).

6.7.b Ontogeny of alarm calls   Compared with adults, juveniles call more to inappropriate objects in the environment (S. R. Robinson 1981). Detailed study of the ontogeny of alarm vocalizations in vervet monkeys has shown that infants are least likely to give calls out of all age classes but, more interestingly, gave "leopard" alarm calls to warthogs and other herbivores, and "eagle" alarms to gray herons, ground hornbills, and even falling leaves. Over time these became more specific (fig. 6.16). Even among infants, however, alarm calls were not random: "leopard" calls were given primarily to terrestrial mammals, "eagle" alarms to birds, and "snake" alarm calls to snakes (Seyfarth and Cheney 1980). Infants and juveniles were more likely to utter mistaken "eagle" alarm calls to bird species other than to dangerous martial and crowned hawk eagles if they saw the raptors hunting or if they encountered birds at close range (Seyfarth and Cheney 1986). Vervets became more discriminating in their calling behavior with age, perhaps guided in part by the vocal response of adult group members. On 12 out of 17 occasions when an immature was the first member of its group to call to a martial or crowned hawk eagle, adults followed with an alarm call of their own. By contrast, adults gave a second alarm in only 3 out of 60 cases when an immature called to a nonpredator. Such confirmation could act as a guide to young vervets (see also C. S. Evans 1997).

An alternative idea is that young animals do not make mistakes in alarm calling but call because they are subject to a greater range of predators owing to their small size and poor motor coordination compared with adults and thus require help. The argument here is that alarm calls recruit help and only secondarily warn conspecifics of danger. Thus, it has been argued that because adult black-tailed prairie dogs utter more-barklike yips to venomous rattlesnakes than to bull snakes, the more-barklike structure of juvenile calls may

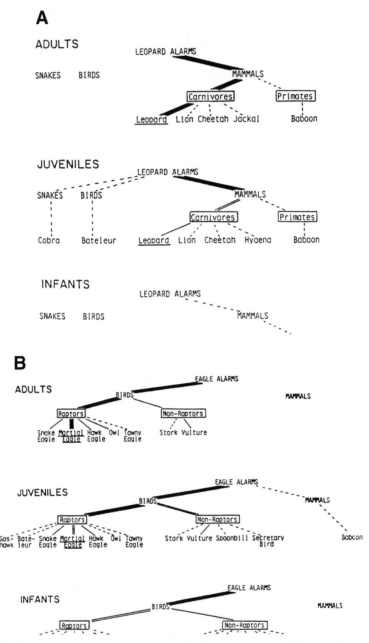

Figure 6.16 Hierarchical representation of the number of *A*, "leopard," and *B*, "eagle" alarms given by adult, juvenile, and infant vervet monkeys to different predator species. *Broken line*, 1–5 alarms; *single line*, 6–10; *double line*, 11–15; *solid line*, >15 (Seyfarth and Cheney 1980).

reflect their greater vulnerability to snakes compared with adults (Owings and Loughry 1985). Calling in this species may be an attempt to cajole others to monitor the snake.

That alarm calling behavior by juveniles reflects differential vulnerability to predators suffers from one important difficulty, however. That is, adults of some species disregard immature alarm calls but respond to those of adults. Thus, adult beavers rarely flee when juveniles slap their tail (Hodgdon and Larson 1973); adult Belding's ground squirrels are less likely to post than juveniles after hearing juvenile trills (S. R. Robinson 1981); adult California ground squirrels spend less time out of view after hearing playbacks of juvenile compared with adult whistles (Hanson and Coss 2001); and bonnet macaques are less likely to flee in response to juvenile than to adult alarm call playbacks—instead, they simply scan their surroundings (Ramakrishnan and Coss 2000b; see also Seyfarth and Cheney 1986; but see Blumstein and Daniel 2004). Adults appear to understand that juvenile warning calls are unreliable, because adults and juveniles have many of the same predators in each of these species. The proposal that juvenile calls serve to solicit aid rather than warn conspecifics in other species needs further investigation, however.

6.8 Use of warning signals in deception

Although adults ignore the warning cries of juveniles in some species, one might generally suppose that costs of failing to respond to a call signaling the presence of a predator are high. That said, it comes as less of a surprise to find that there are cases of birds and mammals using false alarm calls in competitive interactions over food and mates. One of the best examples is of two species of fly-catching birds in Amazonia, white-winged shrike tanagers and bluish-slate ant shrikes, that lead mixed-species flocks through the understory and forest canopy (Munn 1986a, 1986b). These two species are virtually always the first to give loud warning calls at the approach of bird-eating hawks. They obtain their food by diving after falling arthropods flushed by more active species or by chasing after arthropods at the same time as other birds. On these latter occasions, the two sentinel species sometimes give an alarm call that Munn interpreted as an attempt to distract the pursuing heterospecific. As sentinels remained motionless on partially concealed perches when giving calls to hawks that Munn could see, but when giving alarm calls in situations in which he saw no hawk they flew with other birds out in the open; as true and false alarms had similar acoustical structure; and as heterospecifics treated true and false calls as true when they were played back to them, there is strong support for the hypothesis that birds were giving false calls. Parallel findings have been reported for great tits that give alarm calls in the presence of spar-

Table 6.15 Number of different great tits giving false alarm calls during 30 minutes of observation in response to the presence of sparrows and food dispersion

| Food dispersion | Presence of sparrows | Number of great tits | | p-value |
		giving alarm calls	not giving alarm calls	
Concentrated	Present	10	2	<0.001
Concentrated	Absent	1	11	
Dispersed	Present	2	10	
Dispersed	Absent	0	12	0.24

Source: Moller 1988b.

rows feeding at a clumped food source (Moller 1988b, table 6.15); marsh and willow tits show similar behavior (Matsuoka 1980). In addition, colonial nesting barn swallow males fly around the colony alarm calling if they return to the nest and find the female absent; calling flushes members of the flock. Males only alarm call in this way during the egg laying period, when their certainty of paternity could be in doubt, and not during the incubation period (Moller 1989b).

These cases indicate that birds manipulate conspecifics and heterospecifics in a way envisaged by many followers of the model of Charnov and Krebs (1975), although the caller does not obtain antipredator benefits per se. These examples therefore constitute a separate category of cost-benefit analyses, with $B > C$ for the actor and $B < C$ for the recipient. The examples enumerated above make up a fairly comprehensive list of animals using alarm calls for deception, so why are there so few instances in which costs of disregarding danger are high? One possibility is that in social groups composed of the same subjects, individuals giving false information might soon be disbelieved. For instance, Richardson's ground squirrels soon discriminate between callers made reliable or unreliable by pairing alarm calls with or without a model predator (Hare and Atkins 2001). Yet in tests where vervet monkeys were repeatedly exposed to alarm calls with the same referent (for example, leopard), they did not transfer response habituation to new calls produced by the same individual with a different referent (for example, eagle). Vervets appear to make judgments according to the call's meaning rather than just to the signaler's identity (Cheney and Seyfarth 1988, 1991). In short, the possibility that deceivers would be ignored does not look strong in this species, where it has been investigated systematically, and suggests that it may be a poor explanation for why deception is rare. A second possibility is that it is difficult to make a false alarm call believable if it is easy to detect that no predator is present (Cheney and Seyfarth 1991). This might explain why the proportion of false alarms is so high among Amazonian flycatcher sentinels that live in dense forest habitat, where it is difficult to spot predators (56 out of 104 cases were false: Munn 1986a), and possibly why willow tits that live among alpine birch give a high proportion (38%–81%) of false calls (Haftorn 2000). In open-country species, where predators are more easily noted, individuals might not actively deceive but instead withhold information about predators, because it is far more difficult for others to distinguish whether a conspecific has failed to see a predator or is deceiving others. Low-ranking vervets do appear to withhold their alarm calls (Cheney and Seyfarth 1985b), but parallel evidence from other species is lacking. At present, the paucity of instances of animals using alarm calls to deceive others remains vexed.

As a result of so much high-quality research on conspecific warning calls, we are now in an enviable position of being able to pose several interesting questions: why do some species give calls at all, whereas others remain silent in the face of danger? We do not know. Sociality and the presence of kin seem to be factors promoting conspecific warning calls; reliance on crypticity and hiding may operate in the opposite direction. Even here, however, it is not clear why an alert posture or flight is insufficient for alerting conspecifics, or why calling is not always used by cryptic species after they have been discovered. Why do some species give quiet high-frequency conspecific alarm calls that are difficult to locate, whereas others shout loud calls that can be heard far beyond the range of group members? Why do some species signal response urgency, others signal class of predator, and others neither? To what extent do conspecific alarm calls perform double duty as pursuit-deterrent signals (section 7.6)? Given that predators are likely to take note and respond to such calls, it seems strange that prey give different calls to conspecifics and to predators.

The excitement of research on alarm calls is that broad, interesting questions can be tackled at both a comparative and an experimental level, and that careful experiments performed in the field are still uncovering sophisticated cognitive aspects of antipredator defenses closely tailored to the species' ecology.

6.9 Summary

Warning or alarm calls given in the presence of danger may inform conspecifics of danger or signal to the predator; the former are the focus of this chapter. Calls in small birds and mammals are high-frequency sounds with a gradual onset and offset that are difficult to discern. Laboratory evidence shows that aerial predators find it hard to locate the source of these sounds, supporting the idea that selection has acted to reduce localizability of alarm calls. Selection is predicted to structure alarm calls so that they can be heard by conspecifics but not by predators. Conforming with expectations, calls are often low in amplitude, difficult for predators to hear at a distance, easily attenuated by vegetation and turbulence, and not tailored to maximizing sound transmission, all of which would make it difficult to hear alarm calls clearly from far away.

Both anecdotal evidence and systematic observations indicate that survivorship costs of alarm calling are high against some predators but low against others, and depend on factors such as predator's hunting style and the extent to which prey are exposed. An early model suggesting that callers benefit more than listeners from alarm calls has been interpreted as alarm callers reducing the fitness of others; there is no evidence for this except perhaps in rare cases of deception. Instead, by causing listeners to increase vigilance or flee en masse, alarm calls may be mutually beneficial to sender and receivers but to differing extents. The evolution of warning calls through kin selection has received strong support from studies of ground squirrels. Across species, individuals with relatives present are more likely to call to ground predators than those with no kin, while a study of Belding's ground squirrels refuted competing hypotheses for alarm calling and confirmed predictions regarding kin-related benefits. Parallel evidence in birds is far weaker, but this may be an artifact of differential research effort on species living in permanent social groups.

Some pairs of sympatric species respond to each other's alarm calls, but antipredator benefits are often asymmetric, especially in birds. In some primates, species react differentially to different alarm calls given by heterospecifics to various types of threat.

Many sciurids utter different types of alarm call in response to aerial and terrestrial predators, but this is related to response urgency rather than class of predator. Call rate and duration are also related to risk. Birds similarly emit different calls to aerial and ground predators that signal different escape strategies. Studies of vervet monkeys show that they produce different alarm calls to leopards, eagles, and snakes, and that receivers respond differentially to these calls by, respectively, rapidly climbing trees, descending to the ground, or looking down. Parallel results have been found in some lemurs and other cercopithecines. Three independent tests demonstrate that such calls are referential rather than signaling response urgency. In contrast with sciurids, primates may need to signal about different classes of predators because they have mutually exclusive ways of avoiding predatory attack.

Responses to alarm calls develop rapidly in mammals. There appear to be innate predispositions to make appropriate responses, but their development is also influenced by maternal behavior. Young mammals at first call to inappropriate, harmless objects but later become more discriminating. Adults often disregard alarm calls of juveniles, suggesting they perceive them as unreliable.

There are cases of birds giving false warning calls in competitive interactions over food or mates, but the number of instances is small for reasons unknown. Deceptive alarm calls may constitute a case of the sender manipulating recipients' behavior.

7 Signals of Unprofitability

7.1 Introduction

Once a predator has seen prey and is weighing whether to approach or close in, its decision may be influenced by traits that make certain species or individuals unprofitable to pursue, attack, or consume. Prey may be unprofitable because they have detected an approaching predator, thereby eliminating the element of surprise; because they are in sufficiently good condition to outdistance a predator; because they have mouthparts, feet, or weapons that make them dangerous; because they have noxious defenses; or because they are unpalatable or poisonous. Individuals with such attributes will benefit by signaling their unprofitability to predators if it can reduce the probability of attack and hence time and energy costs of interaction and chance of death or injury (Leimar, Enquist, and Sillen-Tullberg 1986). Predators will benefit by responding to these signals and not attacking, thereby reducing their own time, energy, and injury costs (Guilford 1990; Hasson 1991).

Studies of morphological and behavioral signals of unprofitability have constituted different fields of biology. Generally, conspicuous morphological traits that signal prey are distasteful or noxious are called aposematic signals (Poulton 1890); these can be defined as increasing an individual's probability of discovery (Sillen-Tullberg and Bryant 1983). Classically, these are visual signals, usually warning colors that consist of a few bright colors set in a bold pattern that contrast with both the background and one another (Poulton 1898; Cott 1940), whereas distastefulness might be either a continuous or discrete trait (Leimar, Enquist, and Sillen-Tullberg 1986; Guilford 1994), be effective

Figure 7.0 (*Facing page*) Adult male impala stamps its front hoof. Foot stamping was initially thought to be an auditory signal given when a predator approaches, but it may principally be a visual signal, although it is still unknown whether it signals to conspecifics or the predator (reproduced by kind permission of Sheila Girling).

against some predators but not others (Endler and Mappes 2004), and is a relative concept dependent on the palatability of alternative prey (Sherratt 2003). Aposematic signals have received considerable theoretical and experimental attention, especially in insects, but have been reported only descriptively for homeothermic taxa. In contrast, behaviors that signal unprofitability, termed pursuit-deterrent signals, have received less theoretical and little experimental study and have been observed principally in mammals, birds, and lizards. In this chapter, I juxtapose these disparate areas of inquiry, because the distinction between behavioral and morphological signals is artificial, being based on history and convenience, and because both sorts of signals are intertwined: morphological signals are often exaggerated by conspicuous behavior and vice versa (Forsman and Appleqvist 1998). The key difference between aposematic signaling and pursuit deterrence is that aposematism normally relies on predators generalizing between different individuals of a given prey species (or prey morph) and concluding that if it has already experienced that prey, then other, similar-looking individuals will also be noxious or dangerous to handle. In contrast, pursuit-deterrent signals contain information about a particular individual animal that may be specific to that singular predator-prey interaction (or change little over time), but the information does not refer to other, similar-looking prey items (Ruxton, Sherratt, and Speed 2004).

The evolution of an unprofitable signal is problematic, because when it first appears in a population of cryptic conspecifics, it will be novel, so that predators will have to learn its meaning; and it will be rare, making reinforcement learning difficult (Mallet and Singer 1987; Endler 1988; Mallet and Joron 1999; Guilford 1988, 1990; Schuler and Roper 1992). Also, signals are, by definition, conspicuous and have been selected to attract a predator's notice (Darwin 1871; Maynard Smith and Harper 2003). Aposematic signals alert the predator that the prey is present and thereby pose a problem for how they originated, although this problem does not afflict pursuit-deterrent signals, because the predator already knows that prey is there. A large body of theoretical and empirical research has investigated how naïve and experienced predators react to aposematic signals; although some of the same problems bedevil pursuit-deterrent signals, they have never been studied. R. A. Fisher (1930) recognized the potential dangers of aposematism when it first arises in a population and proposed that it could spread if the predator became reluctant to sample similar-looking kin after an unpleasant experience. His reasoning is famous, because it predated W. D. Hamilton's (1964) work on kin selection. The relative importance of direct and indirect fitness benefits of aposematism is discussed in section 7.2. In recent years (for example, Speed 2000), it has become clear that predators exhibit aversions to aposematic prey during prey selection that may overcome costs of rarity and conspicuousness (section 7.3). Alleged examples of aposematism in birds are assessed in section 7.4, where there is some evidence that unpalatability may be widespread in avian taxa (Cott 1946/47) and that birds may signal other forms of unprofitability by means of bright coloration (Baker and Parker 1979). Section 7.5 discusses putative cases of aposematism in mammals.

Pursuit-deterrent signaling can be divided into situations in which individuals signal that they have seen an approaching predator—"perception advertisement"—and those in which they additionally signal their potential ability to outrun or escape from the predator—"quality advertisement." There have been only a few theoretical treatments of pursuit deterrence (Nur and Hasson 1984; Vega-Redondo and Hasson 1993; Bergstrom and Lachmann 2001). Most empirical studies consist of attempts to identify behaviors given in a predatory context as being signals to a predator rather than to conspecifics, and there is a now a small body of examples of homeotherms using pursuit-deterrent signals (section 7.6). Factors maintaining honest signaling in the context of prey-predator interactions, the circumstances under which predators respond to or ignore behavioral signals, and the way in which predators learn the meaning of pursuit-deterrent signals have hardly been studied.

7.2 The evolution of aposematism

7.2.a Individual selection   Three principal selective forces could be responsible for the evolution of signaling unprofitability (Guilford 1990; Servedio 1999). Some models show that aposematism can arise through individual selection when factors such as probability of detection and level of distastefulness are high (for example, Sillen-Tullberg and Bryant 1983; Engen, Jarvi, and Wiklund 1986). There is good experimental evidence demonstrating that insect prey, at least, can survive encounters with a predator. In an important paper (Wiklund and Jarvi 1982), aposematic adult insects and larvae were exposed to naïve birds of four different species; the overwhelming majority of prey items survived encounters, with many of the insects not even being touched (table 7.1; see also Jarvi, Sillen-Tullberg, and Wiklund 1981; Chai 1986; but see Pinheiro 1996). If novel aposematic prey are avoided by naïve predators, aposematism could spread in a population of cryptic prey despite being conspicuous and rare when it first appears. If innate avoidance is widespread, there is no need to impute additional selective routes for the maintenance of aposematism.

7.2.b Kin selection   Nonetheless, it is only relatively recently that aposematic coloration has been considered beneficial for individuals, even when predators first encounter them. For years, it was believed that aposematic prey would suffer death if attacked. R. A. Fisher (1930) proposed a solution to this problem by arguing that similarly colored kin would receive fewer attacks after a predator

Table 7.1 Responses of 4 species of hand-reared birds, Japanese quails, starlings, great tits, and blue tits, to 5 species of aposematic insects: swallowtail butterfly larvae (*Papillo machaon*), large white butterfly larvae (*Pieris brassicae*), burnet moths (*Zygaena filipendula*), ladybirds (*Coccinella septempunctata*), and firebugs (*Pyrrhocoris apterus*)

Bird species	No. birds tested	Insect species	A	B	C	D	E	F	G
		Swallowtail larva	5	12	0	0	1	17/18	12/13
Japanese quail	18	Firebug	8	8	2	0	0	16/18	8/10
		Ladybird	3	13	2	0	0	16/18	13/15
		Large white larva	6	10	0	0	0	16/16	10/10
Starling	16	Firebug	15	1	0	0	0	16/16	1/1
		Ladybird	13	3	0	0	0	16/16	3/3
		Swallowtail larva	5	0	0	0	0	5/5	0/0
Blue tit	5	Burnet moth	4	1	0	0	0	5/5	1/1
		Ladybird	5	0	0	0	0	5/5	0/0
		Burnet moth*	1	2	1	0	1	3/5	2/4
		Ladybird*	0	4	1	0	0	4/5	4/5
Great tit	8	Swallowtail larva*	2	0	0	1	0	2/3	0/1
		Large white larva^	0	1	1	0	1	1/3	1/3
		Firebug^	2	1	0	0	0	3/3	1/1

Source: Wiklund and Jarvi 1982.
Note: Column headings are as follows: A, Not touched; B, Seized and dropped; C, Seized and pecked; D, Partly eaten;
E, Entirely eaten; F, Surviving/number encountered by birds; G, Surviving/number seized by birds.
*refers to 5 birds, ^ to 3 birds, together they add up to 8 as shown.

had learned to associate a particular type of conspicuous coloration with un-palatability. To facilitate predator education, aposematic kin should be found in fairly large, clumped aggregations so that they would be detected and avoided by the same experienced predator (Sillen-Tullberg and Leimar 1988). Many warningly colored insect larvae are found in sib-groups, because eggs from which they hatch are laid in clusters, on the underside of leaves, for instance (J. R. G. Turner 1971; Edmunds 1974; Stamp 1980). Early models and discussion indicated that kin selection can drive the evolution of aposematism but only under restricted conditions, when aposematic morphs are not too conspicuous, when they hasten predator learning, and when the cryptic morphs are not too numerous so that predators will encounter aposematic families rapidly (Harvey and Greenwood 1978; Harvey et al. 1982; Harvey and Paxton 1981). Nonetheless, there are many aposematic species that live alone in nature (Tullrot and Sundberg 1991), such as the striped skunk. Either these species have lost their gregariousness after becoming unprofitable, or aposematism can arise in the absence of aggregation, as suggested by individual-selection models.

A formal attempt to disaggregate gregariousness from aposematism was made by Sillen-Tullberg (1988, 1993), who mapped solitary habits, gregariousness, warning, and cryptic coloration onto the larvae of different species of Papilionid, Pierid, and Nymphalid butterflies. Plotting changes in these traits onto phylogenetic trees, she found that gregariousness never arose before warning coloration but that gregariousness was more likely to arise in species with warningly colored rather than cryptic larvae, perhaps because they had less need to be solitary. She concluded that Fisher's argument for aposematism arising in aggregated prey cannot apply generally (see also Tullberg and Hunter 1996). Indeed, aggregation is unlikely to precede unpalatability, because in some circumstances aggregated prey are more likely to be discovered by predators than solitary prey.

A laboratory test of the association between aggregation and aposematism arrived at a different conclusion. Alatalo and Mappes (1996) presented great tits with straws filled with edible fat, onto which little flags with black and white printed symbols were attached. This novel prey allowed the researchers to circumvent the problem of birds avoiding classically colored black-, yellow-, and red-colored aposematic prey even on their first encounter as a consequence of not being naïve in an evolutionary sense. Straws could be made cryptic or conspicuous by altering the pattern of the background and could be rendered unpalatable by adding quinine to the fat; straws were then presented to tits alone or aggregated in groups of four straws. Gregariousness initially enhanced the effectiveness of unpalatability whether straws were made conspicuous or not, but after repeated trials, conspicuous unpalatable morphs "survived" better than cryptic unpalatable morphs whether they were aggregated or solitary. From these experiments, Alatalo and Mappes suggested an evolutionary route of unpalatability selecting for gregariousness, as predators would leave aggregations after tasting the first items, and then gregariousness selecting for warning coloration, to allow predators to learn to associate the signal with unpalatability. Unfortunately, unpalatable cryptic straws were presented with palatable ones that might have elevated tit attack rates on the former and so have exaggerated differences when comparisons were made with unpalatable conspicuous prey. When the experiment was repeated (Tullberg, Leimar, and Gamberale-Stille 2000), there was no support for conspicuous unpalatable straws showing greater "survival" than cryptic unpalatable straws in the aggregated condition. It is difficult to form a consensus from these conflicting results; at bare minimum, we can say that strong phylogenetic and some experimental evidence suggest that aposematism can evolve in the absence of gregariousness, which therefore adds weight to arguments based on individual selection.

7.2.c Synergistic selection   Guilford (1985, 1988, 1990) has argued that aggregation can promote the evolution of aposematism without individual prey being related. Where conspicuous individuals are recognized by a discriminating predator and not attacked (that is, favored), any other individual with this warning color will benefit. This is called synergistic selection, or the "green beard effect" (Dawkins 1976; see also Ridley and Grafen 1981). Green beard selection within members of the same species is somewhat implausible, because it demands a gene that codes for a phenotypic marker, recognizes the trait in others, and shows a tendency to be altruistic toward others with that trait. In Guilford's reconstruction, however, the predator takes responsibility for recognition and altruism (not attacking) rather than the prey's genome. Moreover, the aposematic trait in prey is likely to be honest because it will be costly to produce physiologically, and its bearers will incur some risk of attack (so-called probing). Across taxa, there is compelling but anecdotal support for such an idea, because many species use the same colors, red, orange, yellow and black, often juxtaposed in bands, to advertise unpalatability. Also, in Mullerian mimicry, different unpalatable species resemble each other (Turner, Kearney, and Exton 1984; Speed 1993; Mallet and Joron 1999). Synergistic selection at the within-species level may also operate. For example, adult wild-caught great tits kill fewer traditionally aposematically colored black and yellow mealworm larvae made distasteful with Tabasco sauce than they do pink larvae; pink is a novel warning color that they are unlikely to have encountered in nature (J. Mappes and Alatalo 1997). That tits find black and yellow more averse than pink suggests that predators recognize and avoid certain conspicuous coloration patterns and that unrelated conspecific (or heterospecific prey) could gain advantages through similar coloration whether they are aggregated or not.

7.3 Mechanisms by which predators select prey

7.3.a Single prey   When an aposematic form first appears in a cryptic population, it will be conspicuous and so incur a greater risk of discovery by predators, and it will be rare, preventing predators from learning its characteristics readily. Rare aposematic prey will therefore suffer very high per-capita mortality, because they are conspicuous but few in number and therefore a poor educational tool for predators. Some warningly colored prey may overcome this problem by being cryptic at a distance but conspicuous only close by (Windecker 1939; Endler 1978; Jarvi, Sillen-Tullberg, and Wiklund 1981; Gotmark and Unger 1994) or only in specific habitats (Gotmark and Hohlfalt 1995). Other species, such as *Schistocerca emarginata* grasshoppers that are facultatively unpalatable depending on diet, exhibit cryptic coloration at low densities, where they are less likely to be seen, but warning colors at high densities, where they might attract attention anyway (for example, Sword 1999; Sword et al. 2000; Despland and Simpson 2005). Environmentally determined aposematism may reduce initial costs of conspicuousness, because many individuals will exhibit warning colors at once (Sword 2002; Darst et al. 2005). Nonetheless, in perhaps the vast majority of instances of aposematism, novel warningly colored morphs will be faced with the dual problems of rarity and conspicuousness, and aposematism will only be evolutionarily stable if there is some sort of compensation (Leimar, Enquist, and Sillen-Tullberg 1986).

Compensation may be accrued if predators find novel prey items aversive (neophobia), because this might allow conspicuous morphs to survive long enough to enable aposematism to become established over evolutionary time. For example, naïve birds are hesitant about sampling aposematic prey (for example, Coppinger 1970; Schuler and Hesse 1985; Sillen-Tullberg 1985; Schuler and Roper 1992; table 7.1), and dietary conservatism may last for weeks or even months in some avian species (Marples, Roper, and Harper 1998; Marples and Kelly 1999). Experiments also show that novelty facilitates rapid avoidance learning (for example, T. J. Roper 1993; L. Lindstrom et al. 2001; Thomas et al. 2003). In addition, odors can foster aversions (C. Rowe 1999; but see Kauppinen and Mappes 2003). Pyrazine, found in many toxic insects, can produce avoidance of both novel and aposematically colored crumbs in naïve domestic chicks (Marples and Roper 1996; C. Rowe and Guilford 1996, 1999), and other odors, novel but unrelated to aposematic defenses, such as ethyl acetate, can do the same (Jetz, Rowe, and Guilford 2001).

More specifically, certain aspects of conspicuousness, such as contrast, bright colors, and patterns (Cott 1940), are also aversive when first encountered (for example, G. S. Caldwell and Rubinoff 1983). For example, domestic chicks hesitate longer before attacking and attack fewer aposematic insect larvae when larval coloration is conspicuous against the background than when it matches the background (Gamberale-Stille 2001; but see L. Lindstrom, Alatalo, and Mappes 1999). Possibly conspicuousness increases the time available for viewing the prey before attack, thus reducing error rate (Guilford 1986). Nonetheless, birds seem principally to attend to color rather than contrast (Gamberale-Stille and Guilford 2003). Indeed, birds have an innate avoidance of yellow-, red-, and black-colored prey as compared with brown prey (T. J. Roper 1990; L. Lindstrom, Alatalo, and Mappes 1999; Thomas et al. 2003), although this appears limited to certain sorts of prey, insects rather than fruit for instance (Gamberale-Stille and Tullberg 2001). Pattern is also important: black and yellow stripes are more aversive than either plain-colored, yellow, or bicolored but nonstriped black and yellow prey (T. J. Roper and Cook 1989); asymmetry in size, color, and shape appears to play an additional role (Forsman and Herrstrom 2004).

Conspicuous signals also confer a separate benefit on prey; namely, experienced predators learn to avoid aposematic prey faster than they do cryptic prey. In an early experiment, domestic chicks were offered blue or green distasteful crumbs on blue or green backgrounds (Gittleman and Harvey 1980; Gittleman, Harvey, and Greenwood 1980). Initially, chicks took more of the conspicuous crumbs (that mismatched the background), but by the end of the experiment fewer conspicuous than cryptic crumbs had been taken (fig. 7.1). Although birds may initially have pecked at higher rates at conspicuous items

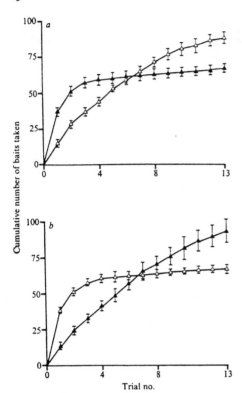

Figure 7.1 Cumulative number of baits taken by domestic chicks in successive trials. Experimental series: *a*, background green; *b*, background blue. Closed symbols represent experiments with blue food and open symbols green food. 95% confidence limits are shown (Gittleman and Harvey 1980).

(T. J. Roper and Wistow 1986), this cannot explain more-rapid avoidance learning entirely, as a single trial with a conspicuous noxious food item produces more-effective avoidance learning, too (T. J. Roper and Redston 1987). (This latter finding is interesting, because it obviates the need for kin selection: a single exposure to an aposematic item can reduce the probability of another similar-looking item being taken.)

Additional mechanisms are probably involved in more-rapid learning. For example, easier recognition of conspicuous items results in more time for observation, again reducing error rate (Guilford 1986; Gamberale-Stille 2000, 2001). Furthermore, predators may come to treat roughly similar-looking prey in the same fashion through the phenomenon of peak shift (Dawkins and Guilford 1995; Yachi and Higashi 1998; Jansson and Enquist 2003), in which animals generalize to more-positive aspects of a stimulus. For instance, even after one exposure to aposematic *Tropidothorax leucopterus* larvae, naïve domestic chicks exhibit a greater aversion toward instars that are larger and more aposematic than smaller instars—a mechanism that could be involved with evolution of increasing aposematism (Gamberale and Tullberg 1996a; Hagman and Forsman 2003; but see L. Lindstrom et al. 1999). (Incidentally, this may be one reason that small instars of some insects are cryptic, because they will gain little by being aposematic; but larger instars are aposematic because they will benefit from such a bias.)

Finally, conspicuousness may prevent predators from forgetting about the aversive properties of prey (Guilford 1990; Speed 2000, 2001). Simulations demonstrate, for example, that neophobia, learning rates, and reduction in forgetting may all act together to favor aposematism. Certainly, there is experimental evidence to show that the strong contrast between a prey item and its background makes an avoidance response more resistant to forgetting (T. J. Roper and Redston 1987) and retards the reversal of a learned avoidance response when palatable mimics are presented (T. J. Roper 1994).

All these mechanisms involve conspicuousness acting as an educational tool for predators. Experiments with students choosing cryptic and conspicuousness pixels on a computer screen show, however, that conspicuousness can evolve simply as a result of defended prey (pixels giving the student a negative score) becoming increasingly different from cryptic undefended prey (pixels with a positive score) (Sherratt and Beatty 2003). As defended prey evolved, any trait that enabled them to be distinguished from undefended prey (even including becoming cryptic) was favored; thus, conspicuousness could be viewed simply as a mechanism to set prey apart from undefended (and usually cryptic) prey (see also Jansson and Enquist 2003).

If, having become established in a population through any one of these mechanisms, aposematic morphs become more numerous than less conspicuous forms, they would enjoy lower per-capita mortality rates and higher fitness simply on the basis of being more numerous. This critical ratio of conspicuous to cryptic forms could be reduced if warning signals lower the costs of predator education in one of the ways outlined above. In addition, once an aposeme is at or near fixation in one area, it may replace cryptic morphs in neighboring populations and spread through the wider population as a whole (Mallet and Joron 1999).

7.3.b Aggregated prey   The problems of rarity and conspicuousness can also be reduced if aposematic prey are aggregated locally, because this may enable predators to learn more easily to avoid sampling them. Several mechanisms may be involved, just as they are for solitary conspicuous prey items. First, if prey is aggregated into a few large clumps, a given cluster is less likely to be found than if there are many small aggregations of prey within a predator's territory (Leimar, Enquist, and Sillen-Tullberg 1986; see chapter 8). Second, aggregation may enhance the effect of protective morphology (shields or horns) or chemical defense (for example, Cott 1940; Howard, Blum, and Fales 1983; Servedio 1999; chapter 9). Third, aggregation may result in a lower per-

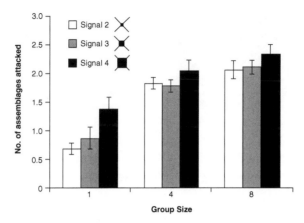

Figure 7.2 Detectability risk of each prey assemblage expressed as the number of prey assemblages attacked by great tits. Increasing signal strength of artificial prey caused an increase in the number of those assemblages that were attacked by birds, but signal strength caused little additional increase in detectability when in a group, so that costs of signal strength are lower for aggregated prey. Means and SEs are shown (Rippi et al. 2001).

capita predation risk if predators quit patches soon after sampling unpalatable prey (the dilution effect: Sillen-Tullberg and Leimar 1988; chapter 8), or if they are neophobic, because they will avoid a large number of prey at once. In addition, there is something inherently unpleasant about aggregations of live aposematic prey. Aggregations generate higher levels of avoidance in naïve birds than solitary prey (Sillen-Tullberg 1990; Gamberale and Tullberg 1996b, 1998; but see L. Lindstrom, Alatalo, and Mappes 1999), and this increases with size of aggregation (Gamberale and Tullberg 1998). For instance, using wild-caught great tits that were trained to eat almonds glued to pieces of paper that had a signal pasted on them, Rippi et al. (2001) gave birds the opportunity to choose between prey assemblages that differed in group size (1, 4, or 8 items) and in signal strength or detectability (crosses with a different-sized square in the center). Once again, this "novel world" circumvents confounding difficulties of ontogenetically naïve birds capitalizing on evolutionary experience of aposematic prey. They found that larger assemblages were attacked more but not as much as would be expected from their group size (the dilution effect), and that more-conspicuous prey suffered a greater number of attacks than less-conspicuous prey but not appreciably more when in groups (fig. 7.2). This means that the cost of strong signals might be lower for aggregations than for solitary prey. In a second experiment, with unpalatable prey, survival of unpalatable prey increased with group size, and there was a learned avoidance effect with very conspicuous signals. Thus, group size enhanced avoidance but required a strong signal to work (see also Gagliardo and Guilford 1993; Alatalo and Mappes 1996; Sherratt and Beatty 2003).

While these receiver-based theories of the mechanisms underlying the evolution of aposematism have found broad support, they do not explain why predators should suffer neophobia or learn about conspicuous prey so quickly; they just assume that predators are biased to respond to conspicuous signals in these ways. Sherratt (2002) modeled a situation in which conspicuousness provides a relative and reliable indication that the prey is defended simply because other predators have noticed it and avoided it. The greater conspicuousness of the model, the more likely it has been encountered by other predators but survived (Beddard 1892); and, as a consequence, conspicuous novel prey will de facto be defended, while inconspicuous novel prey are unlikely to be defended. Accordingly, naïve predators should treat conspicuous prey items with great respect. This leads to two interesting outcomes: a rapid runaway process between aposematic forms and increasing predator aversion, and establishment of aposematic forms even in realistic situations in which predators may attack, attack cautiously, or not attack at all (see Marples, Roper, and Harper 1998). This is the first model to suggest that other predators are the agents of selection that drive predatory responses to signals of unprofitability in prey.

7.4 Aposematism in birds

Very little attempt has yet been made to relate these conceptual advances in understanding the evolution of aposematism to birds, and no attempt has been made to connect them to mammals. In part, this disparity stems from a general idea that birds are not unprofitable prey items, despite substantial reviews that suggest that unpalatability (Cott 1946/47; Cott and Benson 1970; Dumbacher and Pruett-Jones 1996; Weldon and Rappole 1997) and unprofitability (Baker and Parker 1979) are widespread.

In an important yet neglected comparative study, Cott (1946/47) set up a large number of experiments in which he allowed hornets (*Vespa orientalis*) and domestic cats to choose between the flesh of female birds of different species that varied in their conspicuousness, and supplemented these data with people's reports about unpleasant-tasting birds. He found a negative association between visibility and edibility in 38 species of Palearctic birds (fig. 7.3). Similar results were found in a sample of 200 bird species from southern Africa (Cott and Benson 1970). He also extended conspicuousness to other aspects of invulnerability, including size, defensive weaponry, diet (because birds that have to handle large prey may be less subject to attack themselves), habitat (where arboreal species are deemed safest), boldness, and sociability (where gregarious species are safest), scoring each of these variables separately before summing them together. A scattergram of visibility and vulnerability yielded four classes, as determined by eye (fig. 7.4). Class A species were highly visible and vulnerable to attack but almost without exception were unpalat-

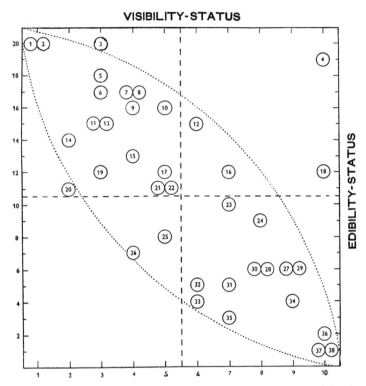

Figure 7.3 Visibility status relative to edibility status in 38 species of birds investigated in the Middle East. Dotted ellipse shows the general trend of the correlation, from species combining high edibility with low visibility to those combining distastefulness with conspicuousness. Numbers refer to identity of species (Cott 1946/47).

able (for example, Eurasian hoopoe); class B were also visible but were relatively immune from attack for reasons other than distastefulness, such as being large (for example, buff-backed heron); class C were least conspicuous and were palatable (such as the crested lark); and class D constituted a single species, the little owl, that was the only nonvulnerable bird that lacks conspicuous coloration, probably because of its nocturnal habits. Cott's study is outstanding, because it was one of the first to draw links between antipredator defenses and both habitat and sociality, and because it weaved many aspects of species' biology together to generate an index of vulnerability, still a rarity in the way we analyze antipredator behavior.

Almost 50 years later, Gotmark (1994b) reanalyzed Cott's data on unpalatability, but for each sex used several independent scores of plumage conspicuousness, employed statistical tests, and separated small and therefore vulnerable passerines from other taxa. He found that edibility was negatively correlated with four independently derived measures of plumage conspicu-

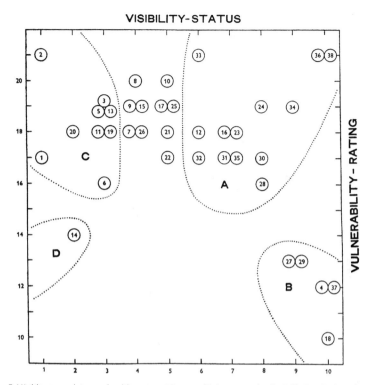

Figure 7.4 Visibility status relative to vulnerability rating in 38 species of birds investigated in the Middle East. Numbers refer to identity of species (Cott 1946/47).

ousness for 30 European passerine species and 87 African nonpasserine species, whether females or males were considered. For 105 African passerine species, palatability was negatively correlated with visibility and plumage conspicuousness for females, but not for males. Gotmark's analysis shows that across nonpasserines, comprising many taxonomic groups, conspicuousness signals distastefulness (although he did not control for phylogeny); but that in passerines, which are a single order (appendix), female but not male brightness signals distastefulness. Whether distastefulness as measured by Cott and Gotmark actually affects hunting decisions of predators remains to be seen.

Some birds are unquestionably toxic. Five endemic bird species of the genus *Pitohui* and the blue-capped ifrita from New Guinea have batrachotoxin alkaloids in their skin and feathers, especially the belly, breast, and leg regions; batrachotoxins are also found in phyllobatid poison-dart frogs. Pitohuis have sharply contrasting black and orange areas of their body, emit a strong odor, and are regarded as unfit to eat by local people (Dumbacher et al. 1992; Dumbacher, Spande, and Daly 2000). The association of warning colors and malodor parallels that in insect larvae (Rowe and Guilford 1996). The spur-winged goose (Eisner et al. 1990), Eurasian quail (Lewis, Metallinos-Katsaras, and Grivetti 1987), and red warbler (Escalante and Daly 1994) are also poisonous. Dumbacher and Pruett-Jones (1996) tabulate an additional six species that are reportedly toxic and as many as 96 species that are unpalatable and/or malodorous, although standards of evidence vary across species.

A second important attempt to link conspicuous coloration to unprofitability in birds was Baker and Parker's (1979) review of theories underlying bird coloration. After juxtaposing competing theories of sexual selection, threat displays, signaling to predators, and social signals other than threat, they combined nine regions of the body to generate a conspicuousness score for 516 species of birds. These were then matched with 17 variables related to morphology, latitude, general feeding ecology, gregariousness, extent of exposure during incubation, and measures of male parental care. Some of the associations hinted at conspicuous birds being unprofitable; for example, conspicuousness was found in gregarious species that show corporate vigilance (chapter 4) and are therefore difficult to approach undetected, in larger species that could inflict injury, and in birds defending an exposed nest site where they can see approaching predators, making attack unprofitable. These findings were reinforced by a negative association between conspicuousness and probability of predation on adults of 18 British passerines (Baker and Hounsome 1983; but see J. B. Reid 1984). Some of the findings fit the unprofitable-prey hypothesis only loosely, however, such as conspicuousness being associated with diurnality. While it is intuitively obvious that nocturnal species that are immobile during daylight hours should be cryptic, it is not obvious why diurnal birds should be conspicuous.

The broad scope of the unprofitable-prey hypothesis attracted withering criticism, primarily because it dismissed the importance of sexual selection in favor of unprofitability. For example, the hypothesis failed to explain the development of bright plumage prior to pair formation (M. Andersson 1983) or to discuss the association between conspicuousness and mating advantage (Lyon and Montgomerie 1985; Rytkonen et al. 1998), despite numerous studies to show that bright males attract mates more easily (reviewed in M. Andersson 1994). Given that several forces appear responsible for the evolution of coloration in homeotherms, including concealment, inter- and intraspecific communication (for example, during mate choice), and physiological factors, it is perhaps unreasonable for Baker and Parker, or their critics, to have expected that one hypothesis could satisfactorily account for coloration; nonetheless, it is premature to reject it entirely.

Experiments with live raptors in the field provide limited support for the unprofitable-prey hypothesis (table 7.2). When paired stuffed-bird mounts are set up in the field, raptors preferentially attack the less conspicuous of the

Table 7.2 Experimental studies of Baker and Parker's (1979) unprofitable-prey hypothesis in birds

Prey	Mount[a]	Predator	Finding	Reference[*]
Supports hypothesis				
Black-billed magpie and Eurasian jay	M	Goshawk	Fewer attacks on conspicuous magpie mounts than on cryptic jay mounts	1
F[b] European blackbird and F great spotted woodpecker	M	Sparrow hawk	Fewer attacks on conspicuous woodpecker mounts than cryptic blackbird mounts	2
Normal European blackbird and blackbird with wing patches added	M	Sparrow hawk	Fewer attacks on mounts with red wing patches than normal mounts	3
Normal European blackbird and blackbird with wing patches added	M	Goshawk	Fewer attacks on mounts with red wing patches than normal mounts	4
M and F pied flycatchers	M	Sparrow hawk	Fewer attacks on conspicuous M[b] mounts than dull F[b] mounts	5, 6, 7
M[b] and F chaffinches	L	Sparrow hawk	Fewer remains of conspicuous M than cryptic F at sparrow hawk nests	8
Refutes hypothesis				
Normal black-billed magpie and cryptic Eurasian jay	M	Goshawk	Attacks conspicuous magpie and cryptic jay at same rate	1
Normal black-billed magpie and magpie painted brown	M	Goshawk	More attacks on normal magpie mounts than brown mounts	1
Pied flycatcher	L	Sparrow hawk	More conspicuous M taken than dull M. Conspicuous M and dull F taken at same rate	9, 10
Normal fledgling great tits and fledglings painted red	L	Sparrow hawk	More remains of conspicuous red tits than normal tits at sparrow hawk nests	11
White wagtail and meadow pipit	M	Sparrow hawk	Attacks conspicuous wagtail and cryptic pipit at same rate	2
M and F chaffinches	M	Sparrow hawk	More attacks on conspicuous M than dull F	6
M and F chaffinches	M	Sparrow hawk	Attacks conspicuous M and cryptic F at same rate	8

[a] M: mount; L: live.
[b] F: female; M: male.
[*] 1. Gotmark 1997; 2. Gotmark and Unger 1994; 3. Gotmark 1996; 4. Gotmark 1994a; 5. Gotmark 1992; 6. Gotmark 1993; 7. Gotmark 1995; 8. Gotmark et al. 1997; 9. Slagsvold, Dale, and Kruszewicz 1995; 10. Slagsvold and Dale 1996; 11. Gotmark and Olsson 1997.

two mounts. In these experiments, conspicuous mounts comprise gaudy males, bright species, or dull birds given bright wing patches. In addition, remains of conspicuous male chaffinches taken in the wild are less likely to be found near the nests of sparrow hawks than feathers of cryptic females. On the other hand, an equal number of findings show the opposite effect (table 7.2), demonstrating that conspicuous birds suffer the same or greater number of attacks or mortality than dull birds. Several factors help to explain this contrasting set of results. First, birds that look conspicuous to human eyes when held in the hand may be cryptic at a distance or in different habitats, obfuscating comparisons. As illustrations, conspicuous white wagtails and cryptic meadow pipits were equally often detected first by volunteers asked to examine standardized photographs taken under different lighting conditions and against different backgrounds when photographs were first shown to them at a distance of 7 m and then brought closer (Gotmark and Unger 1994). Also, conspicuous male pied flycatcher mounts were detected by inexperienced people much more often than dull females when they were placed on the ground, but there was no difference when they were placed in trees; conversely, for chaffinches there was no difference in detectability on the ground, whereas conspicuous male mounts were seen first when perched in trees (Gotmark and Hohlfalt 1995).

Second, mounts may produce aberrant results because they fail to incorporate behavioral differences associated with conspicuousness (Slagsvold, Dale, and Kruszewicz 1995). As a case in point, male and female chaffinch mounts are attacked equally, but a greater percentage of female remains are found around sparrow hawk nests (Gotmark et al. 1997). Males sing more than females, perching at greater heights, and it is known that there is a negative correlation between mean height of sparrow hawk prey and relative predation risk (Gotmark and Post 1996). Moreover, females spend a greater proportion of time foraging on the ground. Foraging conflicts with vigilance (chapter 4), demands movement that may attract a sparrow hawk's attention, and occurs in exposed places. Furthermore, females have impaired flight ability during certain phases of reproduction (Kullberg, Houston, and Metcalfe 2002; Kullberg, Metcalfe, and Houston 2002). Most telling, perhaps, female pied flycatcher disappearance peaks during egg laying (0.53%/day) and is also high during nest building (0.43%/day) and nestling (0.36%/day) stages, but is very low during incubation (0.05%/day), probably because little time is spent outside the nest (Slagsvold and Dale 1996). Thus, conspicuous plumage may be a secondary factor influencing predation risk compared with behavior.

Finally, it may be difficult to extrapolate from experiments with mounts to the natural situation, because diverse ecological and demographic factors may override aposematically driven patterns of predation (Gotmark and Post

1996). These include prey body mass, because raptors take larger prey than expected on the basis of availability; habitat, where open-country species are more vulnerable than forest species (Rytkonen et al. 1998); and frequency-dependent predation, where species-specific predation rate increases as migrating prey of that species arrive in the area (Gotmark 1997) or, conversely, where attack rate on a given species declines as it becomes more common in the prey population (Gotmark and Post 1996). Generalizing from predation on mounts to predation on prey populations therefore demands caution, yet there are arguable grounds for thinking that raptors treat some brightly colored birds as unprofitable.

Returning to the Baker and Parker paper, it is still remarkably modern, considering it was published 25 years ago. Following a review of theories of bird coloration at the time and using an impressive number of species, they presented a great number of multiple regressions matching coloration of different parts of the body to independent ecological and social variables for western Palearctic birds, examining breeding and nonbreeding birds, males and females, and juveniles separately. They even tried separate analyses for different taxonomic groups in an early attempt to control for phylogeny. The text is filled with penetrating biological ideas and observations. For example, Baker and Parker suggested that dangerous prey have brightly colored bills and legs, and drew attention to the issue of naïve predators having to gain experience with unprofitable prey. Unfortunately, their refutation of sexual selection resulted in the paper being summarily dismissed, but there is a danger of throwing the baby out with the bathwater: there are many interesting ideas that are worthy of reanalysis with better data but that take confounding variables, such as operational sex ratio and phylogeny, into account.

7.4.a Mimicry in birds   In Mullerian mimicry, two or more species of equal or unequal unpalatability or unprofitability share a warning signal, and their ranges overlap (Turner, Kearney, and Exton 1984; Kapan 2001; Rowe, Lindstrom, and Lyytinen 2004; but see Speed 1993; MacDougall and Dawkins 1998). Examples of Mullerian mimicry are scarce in vertebrates (Wickler 1968), although snakes (for example, Greene and McDiarmid 1981), salamanders (for example, Brandon, Labanick, and Huheey 1979), and frogs (for example, Symula, Schulte, and Summers 2001) all show this form of mimicry. At present, there is one strong candidate example of Mullerian mimicry in birds. Five species of New Guinean pitohuis are conspicuous and toxic. Using a molecular phylogeny, Dumbacher and Fleischer (2001) showed that aposematism in three species could have arisen through shared ancestry but that in two sympatric species, the variable pitohui and hooded pitohui, the species evolved the same color pattern independently. Nonetheless, comparative evidence showing that different populations within a single species mimic different models, the most convincing evidence for Mullerian mimicry (Pough 1988), is lacking. Weaker avian candidate examples include the unpalatable southern black flycatcher and fork-tailed drongo in southern Africa, both of which are black and difficult to distinguish (Swynnerton 1916).

In Batesian mimicry, a palatable prey species gains protection through its resemblance to an unpalatable model, the mimic is less abundant than the model, and their ranges overlap (Turner 1977; Huheey 1988; but see Speed 1993). There are a number of possible cases, including the palatable rufous flycatcher that resembles the malodorous red-tailed ant thrush, and Finsch's rufous flycatcher-thrush mimicking the white-tailed ant thrush (Ziegler 1971); Kermadec's petrel and Herald's petrel resembling the pomarine jaeger and South Polar skua (Spear and Ainley 1993); and *Oriolus* orioles mimicking *Philemon* friarbirds in Australasia (Diamond 1982). Nonetheless, these examples hinge on similarity in plumage and behavior rather than on predators' similar responses to mimic and model as prey, and there is evidence from two of these examples, petrels and orioles, that mimics accrue benefits by respectively escaping kleptoparasitism and attacks by models rather than antipredator benefits. Another tantalizing example is of the burrowing owl, which nests and shelters in abandoned rodent burrows throughout much of the New World. Owls produce a vocal hiss that sounds remarkably like a rattlesnake's rattle, and the geographic ranges of each species show strong overlap (Rowe, Coss, and Owings 1986). In the laboratory, California ground squirrels from populations sympatric with burrowing owls but not with rattlesnakes are less discriminating when they hear rattle, hiss, and control sounds than ground squirrels from rattlesnake infested areas. The latter treated the owl hiss and rattle cautiously and responded with greater prudence to both than to control sounds. Rowe and colleagues argue that if ground squirrels are mistakenly treating owls as snakes, the sound may dupe American badgers, coyotes, and other predators to stay clear of owl burrows.

7.5 Aposematism in mammals

The venerable example of a mammal signaling its unprofitability is the striped skunk, which sports formidable patches of black and white fur running anterioproximally. This species sprays malodorous secretions from its anal gland that induce nausea and vomiting in mammalian predators (Johnson 1921; Lariviere and Messier 1996). Before discharging, skunks stand on their forelegs and lower their tail over their back to point their anus at the intruder. Skunks often assume a bipedal stance without discharging scent that acts as a deterrent to experienced predators while conserving scent (Walton and Lariviere 1994); skunks can spray only five or six times during an encounter (Cahalane 1961).

Seven species of mustelid have black and white coats, and all produce nox-

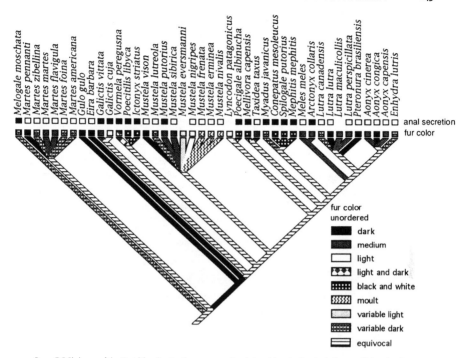

Figure 7.5 Phylogeny of the Mustelidae showing the reconstructed evolution of fur color (see key); "equivocal" branches denote ambiguities in character reconstruction. The row of boxes labeled "anal secretion" denotes whether the species possess a noxious anal-sac secretion (*black box*) or not (*white box*). Data for steppe polecat (*Mustela eversmanni*), Patagonian weasel (*Lyncodon patagonicus*), and spotted-necked otter (*Lutra maculicollis*) are missing for this character. (Ortolani and Caro 1996).

ious anal secretions (some of which produce a burning sensation), a significant association after controlling for phylogeny (fig. 7.5). Light tails are also associated with production of these secretions after controlling for shared ancestry (Ortolani and Caro 1996). Light colored tails are also related to the production of noxious anal secretions in herpestids, as are dark undersides, again after employing phylogenetic controls. Nonetheless, contrasting faces, white facial stripes, and black and white neck marks are not associated with production of noxious anal sac secretions in mustelids, viverrids, and herpestids, so there may be other explanations for these markings. One possibility is that they advertise other types of defenses (R. A. Johnstone 1996; Newman, Buesching, and Wolff 2005; see also Young 1971). For example, the ratel, a mustelid with black and white fur, a white cape, dark undersides, and contrasting face, is known to be particularly truculent, attacking all manner of heterospecifics (including moving vehicles) despite weighing only 8–14kg (Estes 1991). Another is that black markings around the eyes reduce reflected light entering the eyes in crepuscular carnivores (Ortolani 1999).

Independent evolution of black and white coloration in mustelids, viverrids, and herpestids hints at Mullerian mimicry. As a possible illustration, the ratel, zorilla, and striped weasel are all aposematic and sympatric across much of Sub-Saharan Africa (Estes 1991). This is still supposition, however, and Mullerian mimicry requires formal investigation in mammals, first examining different antipredator defenses separately and then in combination (see Cott 1946/47).

Similar to the sympatric ratel, cheetah cubs have dark undersides and a thick covering of long gray hair on their nape, shoulders, and back until they reach 3 months of age. Whether this is an example of Batesian mimicry of ratels (Eaton 1976) or of lion cubs (Percival 1924) or of camouflage (Caro 1994a) is controversial. Other possible examples of Batesian mimicry in mammals include aardwolves mimicking striped hyenas, each of which has dark vertical stripes on their sides; the former is strictly insectivorous with greatly reduced dentition, whereas the latter captures small vertebrates and scavenges using formidable teeth (Gingerich 1975; but see Goodhart 1975; Greene 1977).

Aposematism is also seen in rodents, where many species of porcupine have black, white, and gray quills that are dangerous to would-be predators (section 9.4). White markings on quills are particularly prominent at night, when porcupines are active. Some porcupines additionally rattle their quills while they are alarmed or even ambulating (for example, Cape porcupine), whereas others smell disgusting (for example, Brazilian tree porcupine), thereby advertising their presence (Cott 1940). Maned rats also bear a superficial resemblance to zorillas (Hinton 1947).

Some species of insectivores are aposematic, including hedgehogs, shrews, and gymnures (chapter 9). Hedgehogs are malodorous, their spines are often bicolored, and they make a lot of noise when moving about. Most species of shrew are unpalatable and are not eaten by terrestrial carnivores. They draw attention to themselves by frequent squeaking and emitting a strong odor when alarmed. Bluish-gray coats of some species are luminous at dusk and may serve to draw attention to their bearers (Cott 1940). Last, in Borneo, five species of tree shrew whose flesh is repugnant are mimicked by five species of palatable shrews; the resemblance is so great that examination of museum skins is insufficient to separate mimics (Cloudsley-Thompson 1980). These examples of aposematism in mammals draw out the point that there are syndromes of conspicuous morphological and behavioral signals.

7.6 Pursuit deterrence

Where species flee from predators, running away will be costly in terms of energy and lost opportunities for other activities. Thus, when a prey animal has a high probability of escaping from a predation attempt, it will pay to advise the predator that pursuit is likely to be unprofitable. Pursuing prey imposes similar costs on a predator, so it will pay a predator to take note of the signal and perhaps give up. These signals are called pursuit-deterrent signals (Woodland, Jaafar, and Knight 1980). The word *deterrence* has perhaps an overly forceful meaning, given that models indicate that signals do not always stop the predator from continuing its hunt. Also, deterrence in this context can be confused with deterrence when animals stand their ground and actively defend themselves or companions or perhaps intimidate a predator (chapter 1), but I do not use it in these ways here. Fortunately, examples of pursuit deterrence are increasingly cast into the two categories that obviate use of the word: perception advertisement and quality advertisement.

Further approach on the part of the predator may be unprofitable for two reasons then (ap Rhisiart 1989; Hasson, Hibbard, and Ceballos 1989; Hasson 1991; Caro 1995). (i) If the prey has seen the predator at a distance greater than the flight distance at which such prey are usually captured (section 12.6), or if prey are close to a refuge (section 3.4), it will pay the prey individual to signal that it has detected the predator and for the predator to give up. Prey will be selected to signal at this (or a greater) distance from the predator and no closer, perhaps because predators may occasionally probe the prey, that is attempt a capture if the predator is naïve or very hungry; because the predator can judge the distance as well as the prey; or because signaling might attract additional predators. The importance of these possibilities is unknown. Pursuit-deterrent signals are likely to be given only in the presence of predators; otherwise, prey might raise their chances of being detected if they signaled irrespective of circumstance, or perhaps signaling would interfere with other activities (Hasson, Hibbard, and Ceballos 1989). Evidence that lizards briefly tail wag when first emerging from a refuge in the absence of predators (W. E. Cooper 1998a) suggests that this idea may be too simplistic, however.

In contrast with many other areas of behavioral ecology, empirical evidence for pursuit deterrence has preceded theory (Caro 1995). Recently, however, a model for pursuit deterrence has been developed that synthesizes signal detection, foraging, signaling, and game theories (Bergstrom and Lachmann 2001). The model specifically addresses situations in which the prey signals when the predator is likely to be present but also occasionally makes mistakes, signaling when the predator is not there; and the model assigns a cost to signaling that results from the possibility of attracting secondary predators to the prey. The payoff matrix is shown in table 7.3. From this matrix, Bergstrom and Lachmann derived two necessary conditions for a signaling equilibrium, that is, that signals honestly reveal prey's awareness. First, the prey's sense of predation risk must reflect accurately the chance that the predator is present, and second, prey that sense that they are at risk of predation must be less likely to be captured if pursued than those that are unaware of the predation threat. A number of interesting observations emerge from the model, namely that more

Table 7.3 Payoffs to prey and predator

| | Predator present | | | | Predator absent | |
| | Signal | | No signal | | Signal | No signal |
	Chase	Do not chase	Chase	Do not chase	Chase	Do not chase
Prey	$(1 - c)(1 - t)$	$1 - c$	$1 - t$	1	$1 - c$	1
Predator	$t - d$	0	$t - d$	0	0	0

Source: Bergstrom and Lachmann 2001.
Note: c is the cost that the signaler incurs by attracting a secondary predator, t is the probability that the predator will capture the prey should it choose to give chase, and d is the cost of the chase to the predator.

prey will be willing to signal when predators are common than when predators are rare, and that greater pursuit costs to the predator will allow cheaper signals by the prey and vice versa. Getty (2002) expanded the model so that instead of assuming a simple opportunity cost of attacking prey (the payoffs of zero to "do not chase" in table 7.3), he showed that a predator should resume searching for nonsignaling prey whenever the opportunity costs of pursuing signalers are greater than the expected benefits of attacking a nonsignaler. By incorporating continuous time setting into the model rather than considering a series of rounds, Getty highlighted the likelihood of encountering additional prey (affected by prey density and predator search tactics) as being additional parameters, and his refinement therefore adds realism to the model.

(ii) If the prey has seen the predator at a distance at which prey are sometimes captured, or is at a distance from a refuge that it sometimes fails to reach in time, it will pay those particular prey individuals that can still escape under these circumstances to signal their escape ability to the predator. Such signals of individual quality demonstrate both that the individual has seen the predator and that it is likely to escape if pursuit occurs. In this situation, the predator understands that some prey are liable to capture at this distance, but does not know how a given individual stands in relation to the average member of the population; this kind of signaling is therefore open to cheating. If the prey gives a costly signal, however, only those individuals in sufficiently good condition will be able to sustain the cost and still escape (Zahavi 1975, 1977). Signals might only be given by individuals in good condition (the revealing handicap: Andersson 1986) or might be given for longer, more frequently, or at greater intensity by individuals in better condition (the strategic choice or condition-dependent handicap: Grafen 1990). The signal might be emitted during flight, temporarily reducing the individual's potential to escape by slowing it down or expending energy reserves. Alternatively, it might be given before the flight while the predator is still approaching; here the cost is a delayed start to the flight or consumption of energy. The few empirical examples of quality advertisement have uncovered signals given either before or during flight.

Vega-Redondo and Hasson (1993) modeled prey signaling their quality and found that signaling is evolutionarily stable if an attack on an individual in poor condition is more likely to succeed than on one in good condition; that signal intensity is related to condition and that the cost of increasing signal intensity is greater for the poor than the high-quality individual; that the predator is always successful when signal intensity is very low; and that escape is increasingly likely as signal intensity increases. Honesty is maintained, because if a low-quality individual finds itself among a group of high-quality animals, it will have a $1/n$ chance of being targeted; but it will pay the price of dishonesty because the predator has been allowed to approach too close before it "pretends" to be a high-quality individual and signals. By extension, large groups and few high-quality individuals in a population will foster cheats (Ruxton, Sherratt, and Speed 2004).

Perception advertisement signals principally differ from those of quality advertisement, because the former do not necessitate large costs and performance is not tied to individual condition. Also, predators are predicted to abandon their hunt in response to any perception advertisement but differentially abandon it in response to more rather than less intense or frequent quality advertisement signals. Unfortunately, studies have been inconsistent in measuring signal costs and condition of prey (table 7.4) and have taken to categorizing signals as perception advertisement if cost and condition were not measured and as quality advertisement if either of these variables were measured! The key to showing quality advertisement is that predators avoid pursuing individuals that signal most intently, that signal intensity is linked to differences in condition, and that the signal is costly.

In theory, an individual may signal perception in one situation but advertise its quality in another, depending on its distance from the predator and its condition. Also, the same behavior could be used to signal perception to a stalking predator that relies on surprise and to signal condition to a coursing predator that relies on stamina (ap Rhisiart 1989). Finally, some pursuit-deterrent signals may have multiple functions, including warning conspecifics or acquiring information about the predator (for example, Brown, Godin, and Pedersen 1999).

The evolution of pursuit deterrence is less problematic than the evolution of aposematism, because pursuit-deterrent signals do not alert the predator that prey is present; the predator has already taken note of the quarry. When pursuit-deterrent signals first arise, however, they will be rare; the circumstances and rate at which predators learn their meaning over the course of evolution and ontogeny are unknown and have received no attention as yet.

Although many prey animals give warning signals or show special alert postures when they first catch sight of predators, documentation of these signals

Table 7.4 Perception and quality advertisement signals in homeotherms

Behavior	Prey	Predator	Evidence[a]	Observations	Reference[b]
Signals with little or no assumed cost					
Bipedal stance	California ground squirrel	Pacific rattlesnake	D	Longer duration with large snakes	1
	Brown hare	Red fox	O	Fox gives up approach	2
Alert posture	Numerous artiodactyls	Numerous predators	O	Predator gives up approach	3
Tail flick	Eastern swamphen Common moorhen	Human and marsh harrier	D, E	More flicking close to danger, when harrier present and when vigilant	4, 5
Tail flag	White-tailed deer Eurasian dotterel	Human Common raven and mew gull	E D	Flag with no conspecifics Leave nest and flag to predators	6, 7, 8, 9, 10 11
Signals with low assumed cost					
Bark	Reeve's muntjac and many cervids	Various predator models	D	Given repeatedly and without kin	12, 13, 14
Snort	White-tailed deer Topi and other African ungulates	Human Human	E E, Ex	Snort with no conspecifics Snort less when move away, and snort with no conspecifics	6, 8, 10 15
Duet	Klipspringer	Black-backed jackal and other predators	D	Repeated calls from safe vantage	16
Long-distance call	Diana monkey and five other species	Playbacks of leopard, eagle, chimpanzee, and human	D, Ex	Call in response to playback of surprise predators. Leopard gives up when seen	17, 18
Signals with some assumed cost					
Inspection[c]	Thomson's gazelle	Cheetah	O	Cheetah moves farther after inspection	19
Tail flag	California ground squirrel	Pacific rattle-snake	O	Not measured	20
Foot drum	Banner-tailed kangaroo rat	Gopher snake	O	Snakes decrease stalking with increase in drumming	21, 22
Stot[c]	Topi and Grant's gazelle	Human	D	Stot at higher rate when in good condition	15
	Thomson's gazelle	Cheetah	O	Cheetah gives up hunt or fails to capture stotters	23
	Thomson's gazelle	Wild dog	D,O	Wild dogs select quarry with lower stotting rate; stot at higher rate when in good condition	24
Leap	Impala	Human	D	Impala in good condition more likely to leap	15
Song[c]	Skylark	Merlin	O	Merlin more likely to catch non-singing than poorly singing than full-song birds	25, 26

[a] D, argument based on design; E, elimination; Ex, experiment; O, observation.
[b] 1. Swaisgood, Owings, and Rowe 1999; 2. Holley 1993; 3. Hasson 1991; 4. Woodland, Jaafar, and Knight 1980; 5. Alvarez 1993; 6. Hirth and McCullough 1977; 7. Bildstein 1983; 8. LaGory 1987b; 9. W. P. Smith 1991; 10. Caro et al. 1995; 11. Byrkjedal 1987; 12. Schaller 1967; 13. Yahner 1980; 14. Reby, Cargnelutti, and Hewison 1999; 15. Caro 1994b; 16. Tilson and Norton 1981; 17. Zuberbuhler, Noe, and Seyfarth 1997; 18. Zuberbuhler, Jenny, and Bshary 1999; 19. FitzGibbon 1994; 20. Hersek and Owings 1993; 21. Randall and Stevens 1987; 22. Randall and Matocq 1997; 23. Caro 1986b; 24. FitzGibbon and Fanshawe 1988; 25. ap Rhisiart 1989; 26. Cresswell 1994c.
[c] Demonstrated cost.

is weak or insufficient for researchers to conclude that they signal to predators. Rather, the predator must be shown to give up further approach as a result of the prey's behavior (Caro 1995). Unfortunately, many putative examples of pursuit-deterrent signals are based on studies that use humans as surrogate predators; consequently, the predator's reaction to prey behavior cannot be observed. Instead, these studies rely on eliminating other possible functions, such as warning conspecifics, to conclude that the behavior informs the predator that it has been detected (for example, Caro et al. 1995; Randall and Matocq 1997; W. E. Cooper 2001).

7.6.a Low-cost perception advertisement signals   For stalking and ambushing predators that must get near prey before launching an attack, being observed may be sufficient to cause them to abandon their hunt. Thus, prolonged staring in their direction may act as a signal that they have been spotted. Many birds and mammals assume idiomatic postures that involve elevating their head, craning their neck, and becoming immobile. For example, brown hares stand bipedally with their ears erect and turn their white ventral surface toward a red fox if it approaches in the open (Holley 1993; see also Parker 1977). Foxes turn away from or pass by hares that assume this vigilant posture (perhaps exaggerated: Lotem, Wagner, and Balshine-Earn 1999). For prey, there is ambiguity in using vigilance as a perception advertisement, because prey may also stare at the origin of sudden noises, wind blowing through the trees, falling leaves, conspecifics, or sources of danger. Therefore, it has been argued that some prey have color patches that can only be seen by the predator when it is being stared at directly, such as the black and white face of the Beisa oryx (Hasson 1991) (although, in even-toed ungulates, conspicuous faces are not associated with being preyed on by stalking predators: Stoner, Caro, and Graham 2003). As orientation behavior, movement, or even running occurs so often throughout the day, there may have been selection for the production of specific signals that are given only in the presence of an approaching predator. One such example is eastern swamphens' rapidly raising and lowering their tail, called tail flicking, the first pursuit-deterrent signal to be described (Woodland, Jaafar, and Knight 1980). Like most perception advertisement signals, tail flicks are given more frequently when the predator (a human in this case) is neither too far off nor too close, between 90 m and 20 m in this field study, and is abandoned altogether when birds run off. It was argued that tail flicking signaled awareness of humans, because solitary swamphens as well as those in flocks persistently flicked their white underparts directly at people, and individuals closest to the source of danger tail flicked (but see Craig 1982). A number of other waterbirds tail flick (table 7.4), as well as deer and antelope, although ungulates may use it to signal absence of disturbance to conspecifics (LaGory 1981; Caro et al. 1995).

Some species of deer lift their tail vertically to expose a bright white rump patch, but in contrast with tail flicking, this is carried out even during rapid flight. The function of this tail flagging in white-tailed deer, the only species where it has been studied, is sometimes attributed to pursuit deterrence, but this conclusion has been reached only by elimination of other hypotheses and comparative evidence from artiodactyls still supports a conspecific warning-signal hypothesis (Caro et al. 2004); no study has yet examined whether natural predators abandon hunts specifically in response to tail flagging (Caro et al. 1995).

It is easy to speculate that vigilance and flight were the evolutionary precursors of more-exaggerated pursuit-deterrent signals.

7.6.b Auditory signals of perception advertisement   Many artiodactyls utter loud, harsh calls when they first detect danger (see Reby, Cargnelutti, and Hewison 1999; Caro et al. 2004 for reviews; see also Branch 1993). Typically, deer and antelope snort or bark once or twice before running off, and then turn to face the predator at the end of the flight and snort repeatedly, and occasionally foot stamp; snorts before and after the flight could have different adaptive significance (Yahner 1980). Solitary bovids and cervids frequently snort, implying that it is not a warning signal to conspecifics (table 7.5), and other group members are usually aware of danger when a snort is first heard (Caro 1994b; Caro et al. 1995). Nonetheless, auditory signals are also given by more than one individual: among small ungulates, individuals on neighboring territories will counterbark (Reby, Cargnelutti, and Hewison 1999); indeed, one of the first examples of pursuit deterrence was of klipspringer pairs barking in duet to black-backed jackals once they had fled to safety (Tilson

Table 7.5 Percentage of individual animals snorting at an approaching observer when individuals were alone or in groups

Species	Alone	In groups	p-value
Impala	11.1% (18)	1.8% (734)	$p < 0.1$
Hartebeest	40.0% (15)	6.7% (223)	$p < 0.001$
Wildebeest	50.0% (6)	4.4% (573)	$p < 0.001$
Topi	21.7% (46)	5.0% (339)	$p < 0.001$
Grant's gazelle	22.2% (9)	2.6% (422)	$p < 0.02$
Thomson's gazelle	10.5% (19)	0% (710)	$p < 0.001$

Source: Caro 1994b.
Note: Brackets denote number of individuals approached.

and Norton 1981). Anecdotal observations of predators (for example, Schaller 1967; Tilson and Norton 1981), studies that eliminate competing hypotheses (for example, LaGory 1987b; Caro 1994b), and an experiment in which a retreating person elicited fewer snorts from solitary male topi than during advances (Caro 1994b) suggest that snorting in ungulates informs the predator that it has been detected. There was no evidence that snorting in topis, hartebeests, wildebeests, Grant's gazelles, and impalas in better condition were more likely to snort than those in poor condition (Caro 1994b).

In addition, there is indirect quantitative evidence that predators abandon hunts in response to calls. Western red colobus, western black and white colobus, Diana monkeys, lesser white-nosed monkeys, Campbell's guenon, and sooty mangebeys in the Tai Forest also call at predators, but give a far greater number of calls on hearing the playback of a leopard growl than a chimpanzee pant-hoot, as for example by Diana monkeys in table 7.6. Leopards hunt using surprise, whereas chimpanzees pursue monkeys (Zuberbuhler, Noe, and Seyfarth 1997). By following a single radio-collared leopard, researchers found that the leopard spent less time in hiding after being detected by monkeys. As monkeys vocalize at such high rates on seeing a leopard, the data suggest that calling serves as a perception advertisement to leopards, and also to crowned hawk eagles that similarly hunt using surprise (Zuberbuhler, Jenny, and Bshary 1999).

That different species of ungulate even living in different habitats emit such similar snorts and barks, at least to the human ear, is intriguing. It suggests that there has been selection to make unprofitability easier to learn, echoing examples of Mullerian mimicry normally studied in visual signals.

7.6.c Inspection as perception advertisement   On first noticing a predator, a number of species of fish, birds, and mammals approach it, sometimes alone, sometimes in groups, provided that it is not too close (Hennessy and Owings 1978; Milinski 1987; Magurran and Seghers 1994; Randall and Boltas King 2001). Repeatedly they move toward and rush away from the predator. Inspection is generally thought to be costly in that it limits opportunities for other activities

Table 7.6 Male Diana monkey long-distance calls given and approaches by males in response to playbacks of the vocalizations of different predators

Stimulus	Predator type	At least one long-distance call given (*N* trials)			Approaches observed		
		Yes	No	Rate (%)	Yes	No	Rate (%)
Leopard	Surprise	18	0	100.0	15	13	27.8
Eagle	Surprise	16	0	100.0	10	6	62.5
Chimpanzee	Pursuit	1	12	7.7	0	13	0
Human	Pursuit	0	7	0	0	7	0

Source: Zuberbuhler, Noe, and Seyfarth 1997.

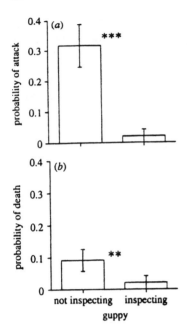

Figure 7.6 Mean and SE conditional probabilities of (a), attack; and (b), death incurred by noninspecting and inspecting guppies, given that they were approached by a predator (here, 24 cichlids). Guppies approached the predator when they were allowed to see it. **$p < 0.01$, ***$p < 0.001$ using a Wilcoxon matched-pairs test (Godin and Davis 1995a).

(Godin and Crossman 1994) and perhaps increases predation risk (Milinski et al. 1997). Across vertebrates, it appears there are multiple functions of inspection behavior that include maintaining vigilance (Magurran 1990b), learning the characteristics of a predator (Kruuk 1976; Fishman 1999), and even conveying information about the predator to noninspecting conspecifics (Pitcher, Green, and Magurran 1986). Perception advertisement may also be involved (Harvey and Greenwood 1978). For example, in the laboratory, pike attack European minnows at a lower rate when the minnows can see and inspect the predator, indicating that inspection limits but does not inhibit predatory attack (Magurran 1990b); and blue acara cichlids are less attentive to and are less likely to attack guppies that inspect them than those that do not (fig. 7.6), although it is not clear whether these are necessarily examples of perception or quality advertisement (Godin and Davis 1995b; Milinski and Boltshauser 1995).

The best-worked example of predator inspection in homeotherms is of free-living Thomson's gazelles inspecting cheetahs (FitzGibbon 1994; Fitz-Gibbon and Lazarus 1995). In gazelles, normally all the group approaches the predator but does not harass or mob it; inspection attracts attention of neighboring groups that run toward the predator and inspect it as well. Just as in fish, there are costs to inspection. Inspection bouts of cheetahs and lions averaged 0.8% and 3.4% of daylight hours, respectively; much of this time was spent running toward and away from predators and must have been energetically expensive. The probability for adults of being attacked and killed while

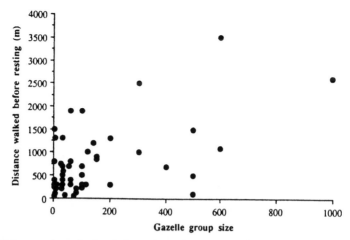

Figure 7.7 Scattergram showing the relationship between the size of the Thomson's gazelle group at the start of inspection behavior and the distance the cheetah subsequently moved (measured from the point at which the cheetah was first approached [$N = 50$, $r = 0.545$, $p < 0.001$]) (FitzGibbon 1994).

inspecting cheetahs was 1 in every 5000 approaches, but was 1 in 8926 if the cheetah was detected but not approached; corresponding figures for adolescent gazelles were 1 in 417 versus 1 in 949, respectively. Moreover, inspections were more likely in safer situations when groups were large and vegetation was low, reducing the chances of ambush. Why do gazelles incur these costs?

There are several benefits to inspection behavior, one of which is perception advertisement, although it is unlikely to be anything more than a supplementary reward. Remember, cheetahs selectively ignore more-vigilant gazelles that they can assess in the absence of inspection (section 4.6.c), so it is unclear as to why gazelles would need to devote up to 72 minutes to inspection to convince a cheetah that it had been detected. The clearest result to emerge from FitzGibbon's study was that cheetahs move farther between rests and successive hunts if they had been approached and followed by large groups (fig. 7.7). A cheetah probably quit the area because once some gazelles had started to follow it, all the gazelles in the area became alerted to its presence. Similarly, Eurasian kestrels move farther between hunting positions if they have been mobbed (Pettifor 1980) (section 11.5.b), and fish predators leave an area after being mobbed by damselfish (Ishihara 1987). Also, through inspection, gazelles were able to monitor the predator's movements, alert other gazelles to its presence, and perhaps benefit from dilution; moreover, given their propensity to inspect, young gazelles could learn about the predator.

Thomson's gazelles' behavior highlights multiple advantages of inspection (Magurran and Girling 1986; Dugatkin and Godin 1992), such as acquiring information about the threat and informing conspecifics about it, even advertising the prey's quality to mates, as well as deterring predators either through perception, quality advertisement, or perhaps confusion, and compelling the predator to move on. The study also shows that it is difficult to separate prey boldness, investigative behavior, predator inspection, predator harassment, and mobbing, as these are distinguished only by slight differences in behavior and perhaps risk, while the consequence for the predator is often the same. For example, inspection by a large group has the same upshot, namely moving the predator on, as some forms of mobbing in which individuals form a mobbing chorus but do not physically attack the predator. Multiple benefits and difficulties in operationally defining antipredator behavior patterns within and across species has impeded advances in understanding many behaviors dealt with in this chapter.

7.6.d Foot drumming as advertising predator monitoring

Kangaroo rats are specialized desert rodents that avoid bright illumination, have enlarged tympanic bullae to pick up sounds of approaching predators, and use bipedal locomotion for speedy and erratic escape from predators (Randall, Hatch, and Hekkala 1995). They are principally preyed on by owls and snakes. Some species, such as the banner-tailed kangaroo rat, foot drum in the presence of snakes or even the odor of snakes, as well as in territorial disputes (Randall, Hatch, and Hekkala 1995; Randall, Rogovin, and Shier 2000; Randall and Stevens 1987). Foot drumming consists of lifting the hind feet and repeatedly striking the proximal ends against the ground in a series called a foot roll. Foot rolls occur at up to 12 per minute and therefore appear to be energetically costly, although this has not been measured.

Banner-tailed kangaroo rats do not foot drum on first sighting a gopher snake, but they start to drum after they have approached the snake and continue to drum until the snake stops approaching or leaves the kangaroo rat mound that it has entered. By a process of eliminating other hypotheses, Randall and Matocq (1997) interpreted foot drumming as signaling that the rat is aware of the snake and is continuing to monitor it, and thus that the chances of successful capture are low. Certainly, the low-frequency vibrations of 200 Hz–2000 Hz must be very detectable and possibly aversive to snakes, yet it is difficult to be sure whether the initial approach (inspection) or the foot drumming informs the snake that it has been detected, or whether foot drumming principally causes the predator to move away to leave aversive noise (Randall and Boltas King 2001). While foot drumming occurs in the absence of conspecifics and does not attract them, mothers foot drum more than nonmothers, suggesting that it could also be a conspecific warning signal.

Tail flagging by ground squirrels in the presence of rattlesnakes may serve an analogous monitoring role, but here conspecifics are additionally involved

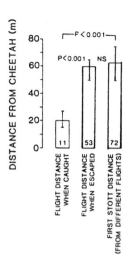

Figure 7.8 Estimated median flight distances (and interquartile ranges) of Thomson's gazelles from cheetahs in hunts where gazelles were or were not caught, and the distance from the cheetah that the first stot occurred (taken from a different data set). Flight distances taken from flights that ended in capture were significantly shorter than those from flights in which prey escaped. Numbers in bars refer to sample sizes (Caro 1986b).

(Owings and Owings 1979). After approaching a snake, ground squirrels repeatedly wave their tail, which may induce other members of the group to continue to monitor the snake visually (Hersek and Owings 1993). This may act as a tonic signal to maintain the snake's attention. While keeping track of snakes must be a major concern of all small rodents because snakes' behavior is so cryptic, in neither of these cases is it clear that hunting episodes fail as a consequence of the rodent signaling to the predator.

7.6.e Stotting as perception and quality advertisement   Stotting is a common but peculiar form of jumping observed in the flight of many species of Cervidae, Antilocapridae, and Bovidae, and even in Patagonian maras. During stotting the front and hind legs are stretched stiffly downward as the animal springs upward, and all four legs are brought forward slightly so that they land at the same time before the next stot (Walther 1969). Eleven hypotheses have been put forward for the function of stotting, but a review showed only three rising to the surface: perception advertisement, quality advertisement, and possibly distraction display by mothers (Caro 1986a). There are now reasonably good empirical data to support the first two.

Stotting has been investigated in relation to cheetahs, wild dogs, humans on foot, and vehicles. In Thomson's gazelles, stotting to cheetahs and humans is not particularly costly in terms of time, because it is a slow form of travel normally exhibited in slow flights rather than during chases, and it does not delay the onset of flights (Caro 1986b, 1994b). Stots normally occur at about the distance at which gazelles flee from cheetahs and manage to escape, that is, about 40 m beyond estimated flight distances of flights in which gazelles are subsequently caught (fig. 7.8). In response to adult Thomson's gazelles stot-

Table 7.7 Outcomes of 31 cheetah hunts in response to different Thomson's gazelles (adults, subadults, and fawns) that did or did not stot

| | Chase occurred | | | |
	Chase successful	Chase failed	Hunt abandoned	Totals
Quarry stots	0	2	5	7
Quarry does not stot	5	7	12	24
Totals	5	9	17	31

Source: Caro 1986b.

ting, cheetahs usually abandoned the hunt, though sometimes they chased the prey but failed to capture their quarry; chases were hardly ever successful. Stotting seems to have little time or survivorship cost, but the energy cost appears substantial. In topis, greater estimated heights of stots were negatively associated with flight speed (Caro 1994b).

When Thomson's gazelles stot in flights from a cheetah, they appear to advertise that they have seen it, because the cheetah usually gives up hunting (table 7.7). It is less likely, though not impossible, that gazelles are advertising their physical condition, because so few gazelles per group stot. This might be expected if they were informing the cheetah of their comparative health. Yet the idea cannot be discarded without taking measures of condition.

When Thomson's gazelles stot to wild dogs, a different picture emerges (FitzGibbon and Fanshawe 1988). To this predator, and to other coursers such as spotted hyenas, a much larger proportion of gazelles stot as dogs run toward the group but then cease stotting after one of their number has been selected. When hunting a herd, wild dogs selected and chased gazelles that stotted at lower rates, an average of 1.64 versus 1.86 stots/second; and when dogs switched the focus of their hunt from one gazelle to another, the second stotted at a lower rate in four out of five instances. Following selection, gazelles that escaped stotted for longer duration than those that were captured, despite chases of the latter type lasting longer. FitzGibbon and Fanshawe think that stotting to dogs is an honest signal of physical condition (quality advertisement), because mean stotting rates and percentage of fleers that stotted were higher in the wet season, when more food was available. In a separate study in which body condition was estimated and scored on a three-point scale (that was related to sward quality), topis and Grant's gazelles in good condition showed a greater number of stots than those in poorer condition, and the latter species stotted at higher rates as well (Caro 1994b), corroborating the idea that stotting is related to condition.

Cheetahs try to get close to Thomson's gazelles using a concealed approach and normally select the nearest to them as a consequence. They then use a short, very rapid chase to contact the prey but are rarely successful against adults that have already seen them. In contrast, wild dogs usually run toward Thomson's gazelles in full view and choose their quarry after the gazelles have started to run, and then chase them down over a longer distance but at a slower pace. Given these different hunting styles, it is not that surprising that Thomson's gazelles show different patterns of stotting to these two predators. A single or small number of animals stot to cheetahs and stop if a chase ensues. To dogs, a large proportion of the group exhibits high rates of stotting that continue during the chase until one gazelle has been selected.

A parallel and intriguing explanation is that cheetahs do not select their quarry on the basis of its condition, but wild dogs do. In a provocative paper, Temple (1987) suggested that predators capture substandard individuals from only those prey species that they find difficult to capture and kill. If wild dogs find it more difficult (have lower hunting success) to capture gazelles than do cheetahs and therefore select substandard prey, it may pay gazelles to advertise their good condition to dogs but less pressing to do so to cheetahs. Against this idea, empirical measures of hunting success show that dogs are very successful gazelle hunters (Schaller 1972), but there is some circularity here, because hunting success may be high because of condition-dependent prey selection.

Even if one accepts the argument that stotting is a perception advertisement to cheetahs and quality advertisement to dogs, nagging questions remain. If stotting is energetically costly, why is it not used as a quality advertisement signal to cheetahs? Also, what information is being sent to other predators, such as hyenas and humans on foot? Furthermore, stotting also occurs in other situations that do not fit well with these observations. For example, Thomson's gazelle mothers whose vulnerable neonates were being chased by cheetahs stotted at high rates between the cheetah and her offspring (Kruuk 1972). Mothers whose fawns escaped stotted at higher rates than mothers whose fawns were captured or than females without offspring. As cheetahs must have recognized that they had been seen (because the two gazelles were fleeing in front of them), mothers' stotting was unlikely to be perception advertisement; and as cheetahs were pursuing the neonates, it was unlikely that mothers would benefit from advertising their own quality, since they were never the focus of attack in these situations and could not defend themselves or their young against cheetahs. That high stotting rates by mothers is a way to distract the cheetah's attention from the youngster is one, as yet, unproven explanation (Caro 1986b).

High rates of stotting are also shown by neonate gazelles when they move out of hiding after being disturbed by a vehicle, and stotting rates are higher the farther they are from their mother. Perhaps they are signaling to their mother that they are changing position or need assistance, but again this requires further research (Caro 1986b). At present, multiple functions of stotting seem probable.

7.6.f Leaping as quality advertisement   Leaping is another gait shown by some artiodactyls during flight. Impala are famous for leaps that take them almost vertically over 1 m off the ground (but where legs are not held stiff and straight). Intuitively, leaping appears energetically costly; in impala, flight speeds are slower if leaping occurs, and flight speeds also decline as the estimated height of leaps increases. Data on leaping in African ungulates failed to support competing hypotheses concerning warning kin, manipulating conspecifics, facilitating escape, or startling or confusing the predator, by default leaving a pursuit-deterrent function intact. That impalas in good estimated body condition were more likely to leap in flights from a human than those in poorer condition lent weight to quality advertisement. Individuals in poor condition did not leap, suggesting a revealing handicap, but observations of natural predators' reactions to leaps are still required (Caro 1994b).

Leaping and stotting are interesting, because both have associated costs and both appear to signal physical condition, as estimated crudely. In contrast, snorting probably has low costs and does not appear to signal physical condition but just that the prey has been seen. Empirical data therefore conform to theoretical predictions. Leaping and stotting are witnessed in species living in open plains or plains-woodland habitats, where selection might favor signals of condition given that escape demands speed and maneuverability rather than crypsis (Caro 1994b). Why some species leap, others stot, and yet others do both is mysterious.

7.6.g Song as quality advertisement   Song is used as a quality advertisement signal in skylarks when they are pursued by merlins (ap Rhisiart 1989; Cresswell 1994c). If a skylark is attacked in flight and after it has been singled out, it sometimes begins to sing a loud, complex song as it starts its exhausting climb, often when the merlin is less than 10 m away. Song cannot therefore be perception advertisement, but it could be quality advertisement, provided that it is costly and merlins usually give up on hearing the song. The costs of song have not been measured directly in this system, but they may exist: song production involves oxygen consumption that is also used in flight muscles, and may demand exhalation that could reduce oxygen uptake during normal exercise respiration (ap Rhisiart 1989). Despite song lasting an average of only 13 seconds, singing only occurs in flying birds and not in birds that have alighted; this suggests that it is only given when it imposes considerable cost. Evidence of merlins abandoning their chase on the basis of song is compelling. Merlins were more likely to capture a skylark that did not sing during an attack; moreover, they chased nonsingers for a longer period of time than those that sang (fig. 7.9). Comparative mortality data showed that a skylark in good condition should always fly and sing, while a skylark in poor condition should always

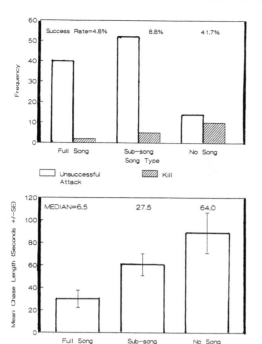

Figure 7.9 *A*, frequencies of capture of skylarks singing different song types when attacked by merlins; *B*, length of chases by merlins unsuccessfully attacking skylarks emitting different song types (Cresswell 1994c).

seek refuge on the ground and remain silent if at all possible. Currently there are no other compelling examples of song acting as a pursuit-deterrent signal, although it has been mentioned for red-breasted geese (Quinn and Kokorev 2002). Song is a different type of quality advertisement than stotting to wild dogs, because merlins had chosen their quarry before singing began (responses of natural predators to leaping have not been documented).

7.6.h Quality advertisement in poikilotherms   Several additional examples of quality advertisement signaling to predators are found in fish and lizards. As in many fish species, sticklebacks approach predators such as rainbow trout during predator inspection visits. Much of the research in this area has centered on controversial tests of how fish lead or defect during inspection and thus whether they conform to predictions of cooperation based on the prisoner's dilemma game. In one such study, Kulling and Milinski (1992) examined the physical condition of inspecting fish using a measure of ($100 \times$ weight/length)$^{3.66}$, which assumes that heavier fish of a given length are in better condition. Fish with a higher condition index approached the trout more closely. As fish in better condition are probably better able to escape, inspection could indicate quality advertisement.

Finally, pursuit-deterrent signals have been documented in at least three

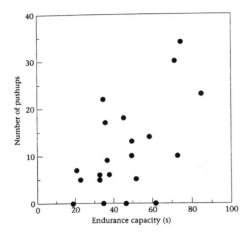

Figure 7.10 Relationship between signal intensity (measured as the number of pushups) in the presence of the snake model and endurance capacity (measured as time spent running) of Puerto Rican anoles (Leal 1999).

species of lizard; some of the earliest examples of pursuit deterrence involved lizards waving tails at sources of danger. As examples, the greater earless lizard bends its tail forward and undulates it slowly to display the black and white striped underside at approaching humans before it flees (Dial 1986), whereas the zebra-tailed lizard wags its black and white striped tail laterally when it is close to cover and a human is approaching (Hasson, Hibbard, and Ceballos 1989). Tail curling, lateral face-offs, extension and contraction of the throat (dewlapping), and up and down body movements in the vertical plane (pushups) are also believed to signal awareness of threat (for example, curly-tailed lizard: W. E. Cooper 2001; Puerto Rican giant anole: Leal and Rodriguez-Robles 1997a). Unfortunately, these studies rely on elimination to reach the conclusion that pursuit deterrence is involved, and none have shown that the behavior reduces the likelihood of predatory attack (W. E. Cooper 2000); male-male competition and displaying to females may also be involved.

In the Puerto Rican anole, however, lizards seem to use four behavior patterns to signal to predators: dewlapping, pushups, lateral face-offs, and predator inspection (Leal and Rodriguez-Robles 1995, 1997b), and pushups are correlated with endurance. Wild anoles were first observed during staged encounters with a stuffed colubrid snake skin, and the rate of pushups was recorded. Then the same individual anoles were subsequently brought into the laboratory, and their endurance was measured as the time and distance spent running around a circular racetrack. There was a strong correlation between pushup frequency and the number of seconds spent running (fig. 7.10), as well as with distance. Results therefore suggest that the Puerto Rican anole is communicating that it has detected the snake and its physical ability to escape; long periods of signaling may reduce the glycolysis necessary to power short, fast flights (Leal 1999). It is not known if the other pursuit-deterrent signals are also used in quality advertisement, and it is unclear why lizards (and ungulates) should employ several pursuit-deterrent signals simultaneously. One hypothesis proposes that they are backup signals allowing the predator to assess more accurately the prey's condition; another is that each signal conveys a different aspect of the signaler's condition (R. A. Johnstone 1996). The former seems improbable, because it is unlikely that all displays would similarly reduce energy available for muscles used in flight; displays that involve leg muscles are a-priori likely to be most honest. As yet, however, there are no experimental data to separate these possibilities.

In conclusion, aposematism and pursuit deterrence are different ways of signaling unprofitability, the first through a more permanent, morphological mechanism, the latter through a facultative behavioral method. The conventional view is that these are not alternatives, because aposematism signals unprofitability across situations (individuals such as adult monarch butterflies [*Danaus plexippus*] are thought to be toxic wherever they are), whereas pursuit-deterrent signals are context specific (prey only signal to predators where they are at a sufficient distance to escape). As a result, the two processes have been examined as separate phenomena. Nonetheless, this may be too simplistic, because aposematic coloration is facultative in some species such as *Schistocerca gregaria* locusts (Sword 1999), and some unpalatable species possess only weak warning signals (Endler and Mappes 2004). Moreover, quality advertisement signals demonstrating condition may be expressed over and over again in different prey-predator interactions if the prey's condition changes little over time. Erosion of the permanent versus facultative dichotomy and the way in which morphology (coloration) is linked to physiology (toxicity) and behavior in aposematism, and behavior is linked to morphology and physiology (muscle strength and vigor) in quality advertisement, suggest that these topics could be more closely related than previously believed. If considered together, additional questions, such as how pursuit-deterrent signals first spread through a population, or whether aposematic signals are targeted at some predator species but not others, would be raised.

7.7 Summary

For predators, prey may be unprofitable if they are dangerous, noxious, or difficult to catch. Prey may signal that they are unpalatable using warning coloration, called aposematism, or that they are unlikely to be captured by the predator using pursuit-deterrent behavioral signals. The first is an obligate species- or sex-specific trait, the second a facultative individual behavior that varies according to circumstance.

Aposematism might evolve through individual selection, given that naïve bird predators avoid warningly colored prey; by means of kin selection, because predators may learn about unpalatability more effectively if aposematic prey are aggregated and if warningly colored kin are found in groups; and through synergistic selection, in which conspicuous unpalatable individuals are avoided because the predator recognizes and avoids the same phenotype. For aposematism to evolve, it must overcome initial problems of being rare and conspicuous in a population of cryptic prey. A number of mechanisms show that it may be advantageous to be aposematic. These include predator avoidance of novelty; novelty facilitating rapid learning; avoidance of contrast, bright colors, and certain color patterns; more-rapid avoidance learning with conspicuous prey; and slower rates of forgetting. Aposematic prey may also benefit by being aggregated. These mechanisms of predatory aversion may have evolved because conspicuousness provides a reliable indication that prey is defended, since it has been avoided by other predators.

Certain birds are poisonous, and many are unpalatable as judged by distastefulness to hornets, cats, and humans. Unpalatability is correlated with plumage brightness in both sexes in European and African birds, and conspicuousness may additionally signal aspects of invulnerability such as size, weaponry, and gregariousness (the unprofitable-prey hypothesis). Experiments measuring rates of attack on conspicuous and dull stuffed-bird mounts placed in the field generate mixed support for the unprofitable-prey hypothesis. Some show that conspicuous mounts are avoided by sparrow hawks and goshawks, while others show the converse. Conflicting results may be due, in part, to behavioral, demographic, and ecological factors that override conspicuousness.

Aposematism has been described in carnivores, notably skunks, rodents, and insectivores. Many of these species have black and white markings, emit noxious odors especially when alarmed, and are noisy, thereby attracting attention.

When prey has a high probability of escaping, it will pay prey to signal unprofitability to predators. Pursuit-deterrent signals can be divided into those that indicate that the predator has been seen and has lost the element of surprise (pursuit advertisement) and those in which the prey has not only seen the predator but is in good enough physical condition to outrun it (quality advertisement). While pursuit-deterrent signal costs are rarely measured, alert postures and tail flicking apparently serve as low-cost visual perception advertisement signals. Snorting and barking in ungulates and calling in some primates probably act as auditory perception advertisement signals. Inspection behavior in ungulates and foot drumming in some rodents may show predators that they have been detected and are being monitored. Stotting is probably an energetically costly behavior that may signal perception advertisement to cheetahs, which use a concealed approach to get close to prey, and quality advertisement to wild dogs, which run down their prey over long distances; mothers and neonate gazelles may stot for other reasons. Leaping in ungulates potentially signals physical condition. Song is probably a quality advertisement signal that skylarks use when closely pursued by merlins in flight. Perception advertisement and quality advertisement signals have also been identified in fish inspection behavior and in head-bobbing and dewlapping behavior of lizards. Research into pursuit deterrence is hampered by difficulties in measuring both prey's condition in natural settings and the cost of signaling.

8 Antipredator Benefits of Grouping

8.1 Introduction

While a prey individual can avoid being the object of predatory attack by signaling its distastefulness or unprofitability to predators, it can also avoid being the target by assuming membership of a group. Two key influential papers in 1971 alerted biologists to the antipredator benefits of group living (W. D. Hamilton 1971; Vine 1971) in situations in which a predator kills only one prey individual per attack. Under these circumstances, an individual that seeks refuge in a group size n will only experience a probability of $1/n$ of being captured. Broadly defined, this is called the dilution effect (section 8.2). Subsequent work has identified other antipredator benefits of group membership (see Morgan and Godin 1985) that include transmission of information through the group about the presence of danger (section 8.3), erratic movements of individuals during flight that may confuse the predator (section 8.4), capitalizing on the time the predator wastes on capturing and consuming other group members (section 8.5), and yet other, miscellaneous factors (section 8.6). The purpose of this chapter is to explore the ways in which grouping can lower risks of being chosen as the target of a predatory attack, pursuit, and capture. The ways in which grouped prey can deter predators or injure them is left to chapter 11.

The principal contribution of W. D. Hamilton's (1971) paper, however, was the idea that not all individuals in a group are equally vulnerable to predation. Those on the periphery are in more dangerous positions in regard to most

Figure 8.0 (*Facing page*) A small group of closely related squirrel monkeys watch from a palm frond. Groups of individuals obtain antipredator benefits in numerous ways, a chief one of which is dilution of predation risk (reproduced by kind permission of Sheila Girling).

kinds of predatory attack, and there will hence be constant attempts by prey individuals to move from the periphery to center of the group. The title of his paper, "Geometry for the Selfish Herd," reflected the striving of individuals to place another group member between themselves and the predator. Much empirical work, especially with fish, has demonstrated the disadvantages of occupying peripheral positions in a group or of being a straggler (section 8.7). In homeotherms, the most concerted empirical observations on the antipredator costs and benefits of position in groups come from nesting colonies of seabirds.

Two other issues that pertain to the evolution of group living are also dealt with here. First, there is a long-standing debate over the relative importance of antipredator benefits and foraging advantages in the evolution of sociality in primates (section 8.8). Second, if grouping has specific antipredator benefits that include both vigilance (chapter 4) and group defense (chapter 11), we would expect it to be associated with certain species, peculiar lifestyles that facilitate group activities, and particular environments where recourse to other antipredator defenses is more difficult. Behavioral and ecological correlates of grouping as an antipredator behavior are addressed in section 8.9.

8.1.a Definition of groups   Groups are difficult to define, because individuals are found at varying distances from one another that may be more or less clustered across the landscape. In studies that compare predation rates of individuals living inside and outside groups, group members are often contrasted with stragglers (Morgan and Godin 1985), strays (Milinski 1977a), solitaries (FitzGibbon 1990c), outliers (I. J. Patterson 1965), and isolates (R. H. Taylor 1962). These terms carry connotations: a stray implies that an individual has wandered away from its normal group, whereas an outlier suggests an individual that is loosely attached to a group, and it is not always clear as to whether these individuals should be treated as a group members. Cluster analysis can help resolve these problems analytically after data on individual locations have been collected (Strauss 2001), but researchers prefer unambiguous, operational definitions that can be employed in the field or laboratory on a minute-to-minute basis as group sizes change. As a consequence, definitions are often ad hoc and vary greatly between studies according to body size of the species, whether it lives in two- or three-dimensional space, the size of the landscape, and rate of individual movements. In general, such definitions are based on absolute distances between individuals, such as less than 50 m from the nearest individual, or number of body lengths from the nearest neighbor (Stankowich 2003).

8.2 The dilution effect

W. D. Hamilton (1971) imagined a group of frogs sitting around the edge of a pond and a snake that once a day appears from the bottom and snatches the nearest frog. Starting from a random distribution, each frog will attempt to move so as to place itself in a narrow gap between two others in order that another individual has a higher probability of being nearest to the snake, which may emerge from anywhere in the pond. As a result, a tight group of frogs will quickly form. Two concepts are embedded in this idea: (i) that selfish avoidance of a predator will lead to aggregation and (ii) that the probability of being captured is greater at the margins of a group. The first concept assumes that the predator takes only one prey individual (or a number less than the size of a group: Brock and Riffenburgh 1960) and is no more likely to encounter a cluster of individuals than a solitary animal. If this is the case, the probability of an individual being captured is $1/n$, where n is group size. Individuals are assumed to have an equal probability of being taken, perhaps because they are continually moving to the center of the cluster (Bertram 1978a; Viscido, Miller, and Wethey 2001; but see Hogstad 1995b for a counterexample). Computer simulations have confirmed that simple movement rules, such as moving toward the nearest neighbor, do lead to aggregation (T. L. Morton et al. 1994). According to theory, then, animals should show a greater tendency to form groups under heightened risk of predation, and this observation has been made numerous times (table 8.1; fig. 8.1).

Hamilton's model referred to risk after a predator had encountered prey, and it cannot explain why less than three individuals would associate, even though this is common in nature. The conundrum can be solved if one imag-

Table 8.1 Variation in group size within species of mammals according to predation risk

Species	Observation	Reference[a]
Eastern gray kangaroo	Form larger groups when red foxes are present	1
Moose	Group size positively correlated with distance from cover	2
White-tailed deer	Form larger groups in open habitat where easily seen than in forest and wooded pasture	3
Mule deer	Group together in presence of coyotes	4
Musk ox	Typical group sizes are positively correlated with wolf densities	5
Mountain sheep	Form larger groups as terrain becomes less rugged and more accessible	6

[a] 1. Banks 2001; 2. Molvar and Bowyer 1994; 3. LaGory 1987a; 4. Lingle 2001; 5. Heard 1992; 6. Warrick and Krausman 1987.

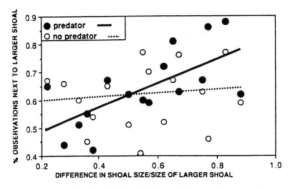

Figure 8.1 Effects of predator pressure on shoal size discrimination in fathead minnows under experimental conditions. Percentage of observations when fathead minnows chose the larger shoal regressed against the difference between sizes of shoals/size of larger shoal. Minnows showed a stronger preference for large shoals when the difference between shoal sizes was large, but only under predation risk. Largemouth bass predator present: $y = 0.430 + 0.389x$, $r^2 = 0.366$, $P < 0.01$; no predator: $y = 0.066 + 0.572x$, $r^2 = 0.014$, $P = 0.65$ (Hager and Helfman 1991).

ines each prey individual having a limited domain of danger, essentially a circular area around it corresponding to the distance that it could be detected or that a predator must approach undiscovered to have any chance of success. Now, if a prey individual approaches another to a distance less than this radius, it will reduce its domain of danger (each domain will be squashed). By this means, two individuals will reduce the chance of being detected or a predator encountering them unexpectedly (James, Bennett, and Krause 2004).

Hamilton's verbal model can be sharpened by separating the probability of the prey being encountered and the probability of being eaten once the group is encountered (G. F. Turner and Pitcher 1986; Inman and Krebs 1987). If in table 8.2 a group of 200 individuals is 200 times more likely to be encountered as a solitary individual, there is a strong per-capita cost to grouping (b). If a group is as likely to be encountered as a solitary but the probability of an individual being eaten is the same as if the solitary was encountered, there is no benefit to grouping (c). Similarly, there will be no net benefit if the probability of being eaten postencounter is lowered but groups are more likely to be found than solitaries (d). Only if groups and solitaries are encountered as often as each other and the probability of being eaten is lowered in groups (e) will it pay to form groups. As payoffs are only manifest when groups are no more likely to be located than solitaries and when the per-capita probability of capture is lowered in groups, it is inappropriate to refer to this as a dilution effect, since encounter rates and dilution are both involved (Inman and Krebs 1987). The product of avoidance and dilution used in this narrow sense has been called attack abatement by Turner and Pitcher (1986; see also Wrona and

Table 8.2 Analyses of the components determining individual predation risk in solitary and grouped individuals, with and without the encounter and dilution effects

	Probability of encounter	Probability of any one individual being eaten once group encountered	Overall individual predation risk
(a) Solitary individual	x	y	$1xy$
(b) Group of 200, no encounter effect and no dilution effect	$200x$	y	$200xy$
(c) Group of 200, encounter effect only	x	y	$1xy$
(d) Group of 200, dilution effect only	$200x$	$1/200y$	$1xy$
(e) Group of 200, encounter effect and dilution effect	x	$1/200y$	$1/200xy$

Source: Inman and Krebs 1987.
Note: The encounter effect assumes that the probability of encountering a group is independent of group size. The dilution effect assumes that only one individual is taken from a group once it is encountered.

Dixon 1991) as well as encounter-dilution, but unfortunately this combination of the two phenomena is often referred to simply as the dilution effect. In most contexts, then, it is helpful to be explicit about whether the broad- or narrow-sense use of the term is being employed.

8.2.a Rates of encounter   Before examining evidence of whether groups of prey are more or less likely to be encountered than solitaries, it is necessary to address the idea that a group is less likely to be noticed by a passing predator than any one of an equivalent number of scattered solitary individuals, because the argument is based on group selection (Pitcher and Parrish 1993). While a group may indeed pass unnoticed, the probability of an individual being seen when it is alone or in a group is the same (other things being equal and assuming that it does not hide behind conspecifics). Nonetheless, it is not fair to say that a group is as likely to be encountered as a solitary, because groups occupy a larger area (or volume, in aquatic species) and are therefore more likely to be encountered by a predator searching at random (Vine 1973). Moreover, where predators locate prey visually, groups of prey are likely to be somewhat easier to detect than single individuals, because they occupy a greater proportion of the visual field (Cullen 1960). In open habitats, this will be particularly evident, but it should also pertain in dense habitat, where it is more likely that some member of the group will be in the predator's line of sight and groups may make more noise than solitaries. Furthermore, once predators encounter a prey item, they may focus subsequent searching to that vicinity (area-concentrated search: J. N. M. Smith 1974), thereby increas-

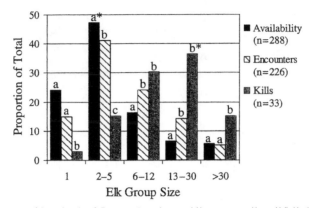

Figure 8.2 Proportion of the total number of elk groups in 5 size classes available to, encountered by, and killed by 2 wolf packs in winters of 1977–99 in Banff National Park, Canada. Significant differences between elk groups in a group size class that were available to, encountered by, and killed by wolves are indicated by different letters, where $P < 0.05$ (*$P < 0.1$); α levels were adjusted for experimentwise error rates ($0.05/n$ categories) (Hebblewhite and Pletscher 2002).

ing the likelihood of detection once one group member is found. Positive correlations between predation attempts and nest densities have been found for many species, for example (Tinbergen, Impekoven, and Franck 1967; Croze 1970; Krebs 1971; Fretwell 1972; Goransson et al. 1975; Dunn 1977; Emslie, Karnovsky, and Trivelpiece 1995). Finally, some groups of prey are continually followed by predators for days (Pitcher 1980). On the other hand, one can visualize situations in which grouping might result in predator avoidance if a predator left an area as a result of encountering few prey because they were clustered in groups (R. J. Taylor 1984). It is generally accepted, however (but on the basis of very little evidence), that encounter rates will not increase as fast as group size in almost any system after group size grows beyond two.

Some data bear indirectly on this issue. In an experimental study of blue acara cichlids choosing between shoals of different size of guppy prey, the predator attacked the larger shoal first, attacked it more often, and spent longer nearer it. This was almost certainly due to visual conspicuousness, because the cichlid preferred the smaller shoal to the larger if the latter was kept at a lower temperature that reduced individual fish movements (Krause and Godin 1995). Among mammals, evidence of predators encountering groups of differing size is very limited. Elk herds of < 5 animals were less likely to be encountered by wolves than herds of > 6 based on their availability in Banff National Park, Canada, possibly because they were less easy to smell or were less predictably located (fig. 8.2). Kill rates followed a similar pattern, resulting in per-capita predation risk peaking for individuals in herds of between 13 and

30 animals (Hebblewhite and Pletscher 2002). More commonly, field studies document predators' preferences for prey groups of different size after they have been encountered rather than measuring the probability of actually encountering groups of differing size (for example, Cresswell and Quinn 2004; Quinn and Cresswell 2004). As an illustration, cheetahs preferentially elect to hunt smaller groups of Thomson's gazelles that come within 1 km of the hunting cheetah (FitzGibbon 1990c); such preferences are likely to be influenced by the probability of past successful attacks (over evolutionary time and within the predator's lifetime) that are themselves influenced by group size (see sections 8.3 and 8.4). In general, differential encounter probabilities of solitary and grouped prey will be influenced by the extent to which predators employ visual, auditory, or olfactory cues to locate prey; whether they hunt by day or night; and environmental conditions: forest or shoreline, clear or muddy water. These factors have yet to be investigated in any detail.

8.2.b Reduced risk of capture   Empirical evidence concerning the dilution effect in the narrow sense, that is, a lower per-capita risk of being killed in larger groups, is more extensive than data pertaining to the encounter effect. Research in aquatic systems (for example, P. F. Major 1978) shows that per-capita risk of capture declines as a reciprocal of group size, as expected by Hamilton's model (fig. 8.3). Support for reduced risk of capture following an encounter has also been found in terrestrial systems. Lions prefer to hunt smaller groups of prey (Scheel 1993b), and in colonies of free-tailed bats emerging from roosts, per-capita risk of predation by raptors is lower when $>$ 100 bats rather than $<$ 100 bats emerge at once (Fenton et al. 1994). More broadly, field studies that compare mortality of solitary animals with those living in groups frequently show that it is more dangerous to be alone (table 8.3), but these studies often infer predation risk rather than witness it, conflate dispersal into unfamiliar and therefore dangerous areas with living alone (but see Isbell, Cheney, and Seyfarth 1993), and perforce combine encounter rate with risk of capture. Even in studies in which group-size-related predation rates are compelling, the reasons that per-capita risk declines with group size are often opaque. To start to understand them, one has to measure the probability of attack, hunting success per attack, and number of prey individuals within the group. Thus, for example, individual dunlin and least sandpiper in larger flocks suffer lower per-capita risk from merlins than those in smaller flocks, but for different reasons than do redshanks in larger flocks predated by sparrow hawks and peregrines (see table 8.4). While these sorts of analyses indicate the stage at which grouping reduces (or increases) risk, even they do not address the mechanisms by which this may occur.

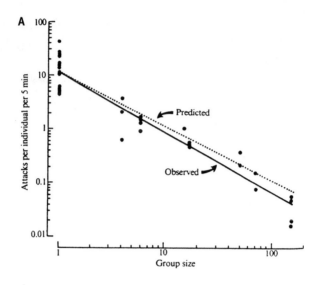

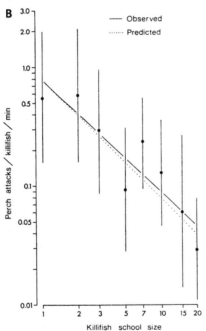

Figure 8.3 *A*, relationship between rate of attack by juvenile Pacific sardines per individual ocean skater (*Halobates robustus*) and the size of the *Halobates* flotilla on Santa Cruz Island, Galapagos. Rate of attack is plotted as the number of attacks in a 5-minute observation period, divided by the size of the group. The observed slope (*solid line*) is $-1.118 + 0.123$ (95% confidence limits, $n = 33$) (Foster and Treherne 1981). *B*, relationship between the rate of attack by white perch in the laboratory per individual banded killifish and killifish school size. The observed slope (*solid line*) is $-0.945 + 0.117$ (95% confidence limits). Means and standard deviations are each based on 10 replicates (Morgan and Godin 1985). In both graphs, the dotted lines represent the expected slope of -1.000, assuming a simple dilution effect.

Table 8.3 Studies of mammals suggesting greater predation-induced mortality in solitary animals or those living in small groups, compared with those in larger groups

Prey species	Principal predators	Findings	Reference[a]
Gray-cheeked mangabey	Chimpanzees, leopards, golden cats, crowned hawk eagles	Isolated males were twice as likely to die as males in groups	1
Yellow baboon	Leopard	Monthly mortality rates were 3–10 times higher for males when alone than in groups	2
Meerkat	Black-backed jackals	Adult mortality declined with group size, juvenile mortality declined with group size in area where predators were present	3
White-nosed coati	Jaguars, pumas	Predation rates higher on solo coatis than coatis in groups. Predation rates negatively correlated with group size at one site	4
Black-tailed prairie dog	Not given	Lower survival of solitary dispersers compared with residents in a group, with 73% of deaths due to predation	5
Yellow-bellied marmot	Coyotes, American badgers	Solitary dispersers suffer higher mortality than philopatric marmots of comparable age	6

[a] 1. Olupot and Waser 2001; 2. Alberts and Altmann 1995; 3. Clutton-Brock, Gaynor, et al. 1999; 4. Haas and Valenzuela 2002; 5. Garrett and Franklin 1988; 6. Van Vuren and Armitage 1994.

Table 8.4 Predator-prey interactions where attack frequency (AF), hunting success (HS), and individual risk of being killed (Risk) are positively (+), negatively (−), or not (0) correlated with flock size in birds

Prey	Flock size	Predator	AF	HS	Risk	Reference[b]
Eider duck	1–20	Herring and great black-backed gulls			−	1
Common redshank	1–100	European sparrow hawk and peregrine	+	−	−	2
Dunlin	1–50+	Merlin	−	−+[a]	−	3
Dunlin	50–10000+	Merlin		0		4
Least sandpiper	1–3000	Merlin	−+[a]	−	−	5
Wood pigeon	1–50+	Goshawk		−		6
Guianian cock-of-the-rock	1–61	Raptors	−			7
Bank swallow	3032–5074	Hobby	+	+	−	8
Chaffinch and brambling	10–10000+	European sparrow hawk	+	+	0	9

Note: Gaps denote that the variable was not measured. All studies refer to solitary hunting predators.
[a] Lowest against intermediate flock sizes.
[b] 1. Munro and Bedard 1977; 2. Cresswell 1994a; 3. Page and Whitacre 1975; 4. Buchanan et al. 1988; 5. Kus 1986; 6. Kenward 1978; 7. Trail 1987; 8. Szep and Barta 1992; 9. A. Lindstrom 1989.

Currently, we know of at least four separate mechanisms that enable grouped prey to avoid being targeted by a predator (sections 8.3 to 8.6), but it is important to note that the dilution effect in the broad sense (that is, attack abatement) is simply a numerical reduction in risk. Not only is it oblivious to mechanism, but it would still apply if there were no additional antipredator benefits accruing to group members, provided that a predator encounters groups disproportionately less than the size of the group and strikes only once. Thus, dilution has the potential to be a significant antipredator force in those situations in which prey become unavailable after the first attack, either because the predator relies heavily on surprise to get close to prey, or the prey can reach physical cover and become unavailable; or where one prey item is sufficiently large to satisfy the predator's food requirements.

8.3 The Trafalgar effect

Where a predator approaches a group of prey from the outside, those nearest the predator may become aware of it first and alarm call or change their behavior, thereby alerting other members of the group that would otherwise be unaware of danger. If the rate of transmission of information across the group is faster than the predator's speed of approach, individuals on the far side of the group will be alerted earlier than if they were on their own and may gain more time to escape. Treherne and Foster (1981) coined the term *Trafalgar effect* because of the signals that were sent between ships to Admiral Nelson before the Battle of Trafalgar, informing him that the French and Spanish combined fleet was leaving Cadiz, even though it was beyond the horizon of his flagship, HMS *Victory*. Treherne and Foster showed that marine isopods, *Halobates robustus*, increased their movement velocity in response to a model predator, and that this change in movement spread across the group more rapidly than the speed at which the predator approached (fig. 8.4). The phenomenon is analogous to Pitcher and Parrish's (1993) idea that other group members will gain information about a predator as soon as it attacks one group member. Other examples of the Trafalgar effect include warning calls given by group members and movements such as flight initiation, which are used as a cue to threat of danger (for example, yellow-eyed juncos and tree sparrows flying from a rubber ball: Lima 1995a [section 4.6.a]); common redshanks flying from sparrow hawks: Hilton, Cresswell, and Ruxton 1999). Strangely, alarm calls or departures are never referred to in the context of the Trafalgar effect despite benefiting group members through transmission of information.

Evidence that information passes through groups more rapidly as group size increases is lacking. In banded killifish, there was no relation between the difference in the time that the first (nearest) and last fish responded to an ap-

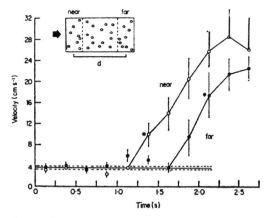

Figure 8.4 The Trafalgar effect. Rates of increase in movement velocity of *Halobates robustus* in two zones, near (*open circles*) and far (*filled circles*), of the flotilla. The initial mean velocity of individuals in the undisturbed flotilla is indicated by the continuous horizontal line, and 95% confidence intervals by the two broken lines. The points at which the mean velocities of individuals in near and far zones differed significantly from that of the undisturbed flotilla are shown with a star and were used to estimate the time taken for avoidance behavior (increased movement) to be conducted across distance *d* in the inset. In this example, *d* = 29.6 cm and *t* = 0.75 seconds, giving a conduction velocity of 39.5 cm/sec (Treherne and Foster 1981).

proaching fish model across shoals of different size (Godin and Morgan 1985). Furthermore, when flying from raptors, redshanks in larger flocks actually took longer to take flight, as measured from the time that the first bird flew off, than those in smaller flocks (Hilton, Cresswell, and Ruxton 1999). It is plausible but not demonstrated that individuals in larger herds, flocks, or shoals detect an approaching predator sooner but delay their flight response (Lazarus 1979) in order to reduce the costs of responding to false alarms (Treisman 1975). In sum, current evidence shows that benefits of the Trafalgar effect are manifest in comparisons between solitaries and groups but not between groups of differing size.

8.4 The confusion effect

The second and better substantiated mechanism by which individuals might gain antipredator benefits from grouping is by confusing the predator during flight (R. C. Miller 1922; Welty 1934). Noting that the flight path of insects is often much more direct and simple when they leave a perch of their own volition than when they are disturbed by a predator, Humphries and Driver (1967) drew attention to erratic and unpredictable zigzagging, looping, or bouncing flight paths of individuals in many taxonomic groups. They suggested that these "protean displays" serve to confuse and disorient the predator that is trying to follow the prey visually or is in pursuit, and that this either increases the predators' reaction time or reduces speed of pursuit. Under this hypothesis, solitary individuals can disorient the predator, although there are methodological hurdles to demonstrating that predators become disoriented following erratic displays.

In groups, an individual should also be able to benefit from its own erratic behavior (although Humphries and Driver did not discuss this). Even in the absence of erratic movements, however, a predator might find it difficult to follow a single animal through a group if the quarry's minute-to-minute movements or flight path crossed over those of others. Certainly, humans experience great difficulties in changing their attention quickly between two objects (Treisman 1969; Broadbent 1971). Compounding this, erratic behavior by many individuals simultaneously might additionally hamper the predator from following a single prey individual through the flock or herd. The first experimental evidence of the confusion effect, as it is called, is usually attributed to two sets of experiments: Neill and Cullen (1974) showed that the ratio of contacts to captures of squid (*Loligo vulgaris*), cuttlefish (*Sepia officinalis*), pike, and perch declined with group size of their fish prey and described attackers as being distracted to different targets again and again. In a subsequent experiment, in which different numbers of *Daphnia magna* were confined to test tubes set next to each other in an aquarium so as to simulate a swarm, Milinski (1977a, 1977b, 1984) demonstrated that three-spined sticklebacks preferred to attack stragglers. He interpreted this preference as the predator attempting to avoid being confused by the swarm. Furthermore, hungry sticklebacks were poorer at detecting a predator of their own (a model common kingfisher) when feeding upon a high- as opposed to a low-density swarm; the researchers saw this as the stickleback having to sacrifice predator vigilance in order to overcome the confusion effect (Heller and Milinski 1979).

There are many plausible mechanisms involved in the confusion effect, including difficulties in distinguishing individuals when selecting a target, in aiming at the quarry, in pursuing it, in concentrating attention on it, or in filtering complex visual information if individuals alternate different-colored surfaces or different color patches in and out of synchrony (Milinski 1977b; Ohguchi 1981). Ohguchi (1981) systematically examined the factors responsible for the sticklebacks' reluctance to attack a swarm of *Daphnia* using the same experimental protocol as Milinski; this isolated potential mechanisms to either fixating or keeping attention on individual prey items (because the fish could not chase the water fleas from the test tubes). Factors that reduced predatory attempts were increased movement velocity of prey, movement of prey in the same direction, reduced distance between group members, increased group size, and uniformity of group members. In a key experiment, *Daphnia* confined to test tubes were placed so as to move at right angles and across one another. Compared with both a single control individual that moved vertically and another control in which two water fleas moved without cross movement, sticklebacks attacked the experimental *Daphnia* for a shorter period of time, at a lower rate during the first 20 seconds of the experiment, and less frequently. This demonstrated that simultaneous cross movements hampered the fish in fixating on individual water fleas and thus decreased rates of attack. Similarly, leopard geckos and common marmosets took longer to catch one out of several mealworms than one presented alone (Schradin 2000).

There are numerous anecdotes of homeotherms fleeing in an erratic fashion. For example, Kitchen (1974) wrote,

> Pronghorn flight behavior seems geared to confusing a predator, making detection of vulnerable individuals difficult. Erect rump patches and white side markings enhance this confusion, as fleeing herds present a series of flashing white rosettes that obscure the outlines of individuals. Individuals make sudden, erratic movements within the herd, and the rump patch accentuates them, making it more difficult to follow a single pronghorn. Entire herds make unpredictable turns especially when hard pressed. When a turn is made, the alternating flashes of brown from the side and leg, and white from the side markings and rump thoroughly confuse the eye of an observer. On 3 occasions, I chased pronghorns on horseback and was never able to keep track of an individual after a sudden turn. (p. 84)

Some shorebirds show peculiar patterns of flocking behavior during predator evasion, including flashing, where dark-colored dorsal and light-colored ventral surfaces are alternately and synchronously exposed by flock members through changes in flight direction or tilts in body position; rippling, where flock members use delayed timing to shift their body position, which produces a rippling effect; and columnar flight, where birds coalesce into a towering vertical column that undulates throughout its length. We do not know whether these evasive tactics confuse predators (they might deter them, for instance), nor is it clear whether they reduce hunting success. There are some data, but it is difficult to know what to make of them. In a field study of merlin predation on dunlins, hunts ended in failure on 77.5% of all occasions ($N = 111$ flights); flashing resulted in escape in 85% of 53 flights; flashing and rippling in a 91% escape rate ($N = 23$ flights); and flashing, rippling, and columnar flight in a 71% escape rate ($N = 7$ flights) (Buchanan et al. 1988). Moreover, across Charadriiformes, species that flock have more extensive patches of white on their back and covert than nonflocking species (Brooke 1998). While these examples suggest visual confusion, they are not yet compelling, and they do not address the mechanisms underlying confusion. Early records of bush tit flocks uttering a shrill, quavering cacophony as a hawk flew over intimate that the sound prevents the raptor locating individual prey (Grinnell 1903; R. C. Miller 1922; see also C. H. Brown 1982), but again the inference is circumstantial.

8.4.a Oddity and confusion   As the confusion effect relies on group members behaving synchronously in a relatively large, tightly knit group, there will be selection on individuals to behave similarly and to resemble each other morphologically; otherwise, the predator may be able to choose and follow one of them through the group. Selection against oddity is an inevitable consequence of confusion (Ohguchi 1978, 1981), although oddity can additionally be selected against without the confusion effect (Mueller 1975; section 2.5). Clear experimental results demonstrating the costs of being an odd individual when the confusion effect is operating derive from a study of largemouth bass preying on Mississippi silvery minnows (Landeau and Terborgh 1986). First, minnows in large groups are far less vulnerable to predation than those that are alone (table 8.5a). As a bass often chases one minnow for a short distance, then switches to another and still another as fleeing minnows cross its path, and is most successful when it separates single minnows from the group, this is reasonably clear support of the confusion effect. Second, bass preferentially attack (table 8.5b) and kill (table 8.5c) odd prey whether they are naturally brown in color or dyed blue.

These observations set the stage for conflict between odd individuals and those with a common phenotype. Odd fish gain an advantage by joining a group despite being singled out for attack, because lone fish are captured in every trial (table 8.5a) versus 50% of cases when they are the odd individual in a group (table 8.5c). Nevertheless, modeling shows that odd individuals suffer proportionately higher predation in large groups than they do in small groups (Krakauer 1995). For the common phenotype in small shoals, the presence of one or two odd fish in a school of eight increased attack rate by threefold and a capture by five times. When shoal size reached 15, attack and kill rates were similar, whether odd individuals were present or not. Thus, odd fish would prefer to join common fish in small shoals, but common fish would prefer to accept them if they were in a large shoal.

When offered choices, odd fish behave as though they are taking account of their oddity. For instance, (i) under predation risk, lone brook sticklebacks or fathead minnows prefer to associate with their own species than with heterospecifics, where they might stand out (Mathis and Chivers 2003). (ii) Sticklebacks spend more time close to shoals containing fish that best match their own body size (Ranta, Lindstrom, and Peuhkuri 1992), and within a shoal, individual minnows are found closer to individuals of their own size than to individuals that differ in size (Theodorakis 1989). The effect may be stronger for

Table 8.5 Aspects of predation of silvery minnows by largemouth bass

(a) Effect of prey school on attack rate and number of attacks per kill

No. prey in school	No. trials	Total time(s)	Total attacks	No. attacks per min.	% Trials ending in capture	Attacks per kill
1	12	128	18	8.6	100	1.5
2	10	339	33	5.8	100	3.3
4	8	643	54	5.0	88	7.7
8	12	3297	66	1.2	17	33
15	9	2593	58	1.3	11	58

(b) Attacks directed toward dyed-blue minnows when present as 1 to 7 individuals in shoals of 8

No. blue minnows in shoal of 8	No. trials	Total no. attacks	No. (%) of attacks on blue minnows		Probability χ^2
			Observed	Expected	
1	14	260	c93 (36)	33 (12.5)	<0.001
2	9	193	76 (39)	48 (25)	<0.001
3	7	83	29 (35)	31 (37.5)	NS
4	11	113	54 (48)	57 (50)	NS
5	8	55	34 (62)	34 (62.5)	NS
6	12	131	73 (56)	98 (75)	<0.001
7	9	116	85 (73)	101 (87.5)	<0.001

(c) Captures of blue versus undyed prey in shoals of 8

No. blue minnows in shoal of 8	Total no. captures	Prey phenotype		Probability (binomial)
		Odd	Common	
1	12	6	6	<0.05
2	8	5	3	<0.05
6	11	6	5	<0.05
7	8	3	5	<0.05

Source: Landeau and Terborgh 1986.

large fish that may stand out more in a shoal of small fish than vice versa (Svensson, Barber, and Forsgren 2000). (iii) Odd fish will leave shoals under threat of predation. When confronted with a model predator in the field, stoplight parrotfish leave mixed-species shoals of striped and stoplight parrotfish and hide by themselves in coral (Wolf 1985; see also Allan and Pitcher 1986). (iv) When body size, species composition, and shoal size are pitted against each other, banded killifish prefer to shoal with individuals of their own size under threat of predation, overriding considerations about species or shoal

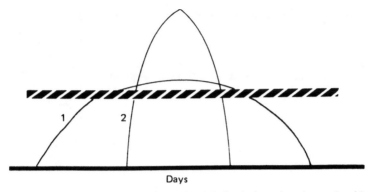

Figure 8.5 1. Hatching distribution of a relatively asynchronous colony. 2. Hatching distribution of a synchronous colony of the same size. Slashed line is the predation threshold for nonrecruitable predator(s). Predator(s) can consume all individuals below the line. Only when the number of individuals hatching on a given day exceeds this threshold will certain young survive. Thus, some individuals in the synchronous colony produce more surviving young than individuals in the asynchronous colony (Gochfeld 1980, after Darling 1938).

size (Krause and Godin 1994). Data from homeotherms are far less sophisticated. Currently, they consist simply of observations that marked animals disappear rapidly from wild populations (for example, wildebeest whose horns were painted white: Kruuk 1972) and that odd-looking individuals are preferentially targeted in laboratory settings (for example, house mice by hawks: Mueller 1975). Evidence that odd individuals choose to associate with individuals of similar phenotype has not been collected.

8.5 Predator "swamping"

Individuals in groups may also gain advantages if the predator takes a long time to kill, consume, or digest prey. This is because once a kill has been made, the rest of the group can escape while the predator handles the prey (Darling 1938; R. J. Taylor 1976, 1979; Barnard 1984). Many terms are used synonymously for this mechanism. *Predator swamping or saturating* refers chiefly to the benefits gained by being in a local population that is greater than the number of individuals that all predators can consume at any given time (fig. 8.5). Depending on the time frame, predator swamping may allow some individuals to move away from the predator, quit the area altogether, pass through a vulnerable life-history stage, disperse, or grow to a size beyond which they cannot be killed. This was the mechanism originally envisaged by Darling (1938). *Predator satiation or glutting* refers to the same mechanisms as they affect the prey but emphasizes the cumulative time that one or more predators take to digest prey (a proximate concept) rather than the predators' inability to consume all the prey (the outcome of this delay). *Time wasting* refers to a broader proximate factor: the length of time it takes a predator to handle prey (searching, chasing, subduing, and consuming). In an extreme case, prey do not necessarily have to be located at all. For example, where yellow-rumped and red-rumped caciques construct and lay eggs in nests that are intermingled with empty nests (Feekes 1981; S. K. Robinson 1985), predators such as snakes might take so long searching for eggs and chicks that they consume less chicks over the course of a cacique breeding season than if they encountered young every time. Also, slow searching might give adults time to escape. For convenience, I will lump all these terms under the heading of Predator Swamping.

At first sight, these three forms of predator swamping seem highly relevant to flocking birds, colonial nesters, birds and mammals that crèche their young, and herd-living ungulates. For example, as soon as a cheetah breaks cover and starts to pursue a Thomson's gazelle, other gazelles become aware of its presence whether or not it makes a kill. After watching intently, the group slowly drifts away; being aware of the predator and quitting the area reduce opportunities for a second attack. Additionally, if a kill is made, the cheetah may not hunt again (in that area) for over a day (Caro 1994a). Indeed, predator swamping has been proposed to explain synchronous behavior in diverse taxa, including emergence of periodical cicadas (M. Lloyd and Dybas 1966), marching Mormon crickets (*Anabrus simplex*) (Sword, Lorch, and Gwynne 2005), metamorphosing anurans (S. J. Arnold and Wassersug 1978), parturition in ungulates (Kruuk 1972), and guillemots leaving their breeding cliffs (Daan and Tinbergen 1979), each of which constitutes a highly vulnerable period in the life cycle.

For predator swamping to be an important mechanism in the evolution of permanent groups or temporary aggregations, predators must not be much larger than their prey; otherwise, they may have rapid handling times that will demote the importance of this mechanism. Furthermore, there must be forces preventing a buildup of predator populations in the vicinity of the prey group (Wittenberger and Hunt 1985) such as territorial behavior between predators or shortage of predator nest sites. Alternatively, there could be periodic surges in prey numbers (a short breeding season, for example) restricting predator numerical increase. Without this condition, predator numbers could wash out benefits of swamping, because additional individual predators will continue to kill regardless of how long a given predator takes to handle prey. In one of the few studies to monitor predator offtake, Nisbet (1975) showed that great horned owls took a similar biomass of common tern chicks from a colony throughout the breeding season despite a hundredfold increase in biomass of available tern prey. At the start of the season, owls took 100% of chicks, but this had declined to 2%–3% by the end. Similarly, densities of American kestrels and loggerhead shrikes within a 1 km radius of cliff swallow colonies increased only fivefold for a twentyfold increase in swallow colony size; moreover, attack

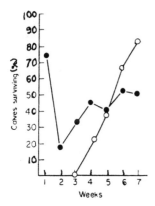

Figure 8.6 Survival of wildebeest calves in Ngorongoro Crater, Tanzania, in small herds (*filled circles*) and aggregations (*open circles*) in 1973 (Estes 1976).

rates on different-sized colonies were similar (Wilkinson and English-Loeb 1982). Nonetheless, accounts of predators collecting at breeding colonies (Kruuk 1964) or in areas of high prey density (Kruuk 1972) indicate that generalizations may be foolhardy. Demonstrating unequivocally that predator swamping is occurring therefore faces a number of challenges; consequently, there are few studies to show that per-capita predation rates are lower on large groups than on small groups solely as a result of this effect.

8.5.a Reproductive synchrony   When individuals give birth synchronously in time and in space (for example, Gross and MacMillan 1981), as, for instance, in colonial nesting birds (for example, red-winged blackbirds: R. J. Robertson 1973; black-headed gulls: I. J. Patterson 1965; see Gochfeld 1980 for a review) or ungulate herds (for example, wildebeest: Estes 1976), vulnerable offspring in some of these species are thought to benefit from the presence of many other offspring, because predators will become satiated after feasting on other youngsters (Darling 1938; Lack 1968; but see C. S. Haas 1997). Temporal synchrony is important because it allows numbers to build up far more rapidly in groups than would occur if breeding occurred throughout the year. Thus, for example, calf survival of wildebeest in Ngorongoro Crater, Tanzania, increased in both small and large aggregations over the 3-week calving period (fig. 8.6). Nonetheless, there is persistent debate over the role of predation or food being responsible for tight birth seasonality in ungulates, and no consensus on the proximate cues (photoperiod, temperature, nutrition) that might trigger parturition under the food hypothesis. Rutberg (1987) tested predictions arising from the two hypotheses in 27 species of ruminants. He found that species exhibited shorter birth seasons with increasing latitude and with increasingly variable precipitation, but this was almost entirely due to hider species; follower species exhibited no clear association between seasonality and length of birth season (fig. 8.7). The correlations confirm the overall importance of cli-

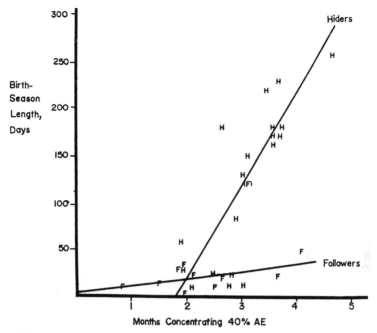

Figure 8.7 The relationship between birth-season length and seasonality (40% of actual evapotranspiration, or AE), with separate regressions calculated for species with concealed young (hiders, H) and precocial young (followers, F). For hiders, $r^2 = 0.63$, $p < 0.00005$; for followers, $r^2 = 0.30$, $p = 0.10$. The F in brackets denotes African buffalo (Rutberg 1987).

mate in determining birth seasonality, and they support the antipredation hypothesis because follower species would be expected to benefit from synchrony of births through predator saturation, and perhaps confusion and group defense; whereas hider species, which rely on crypsis (section 2.2), would not and might even suffer from it if high densities of hidden calves led to concentrated search effort by predators. Rutberg views synchronous breeding as evolving in response to seasonal factors but then tightening in response to predation in those species that can benefit from predator swamping.

Among colonial nesting birds, those nests where eggs are laid early or late in the season have reduced success compared with those laid at the season's peak, and in some species this stems from predation (Ashmole 1963; I. J. Patterson 1965; R. J. Robertson 1973; Nisbet 1975; Veen 1977; see Burger 1981; Wittenberger and Hunt 1985 for reviews; fig. 8.8), although the mechanism by which this occurs, dilution or group mobbing (chapter 11), is difficult to disentangle (S. K. Robinson 1985; Westneat 1992). In other groups, such as troops of primates, ungulate herds, and flocks of birds, where corporate vigilance rises with group size (chapter 4), it is similarly difficult to isolate the role of predator swamping as being a key advantage to grouping (Crook 1965; Jarman

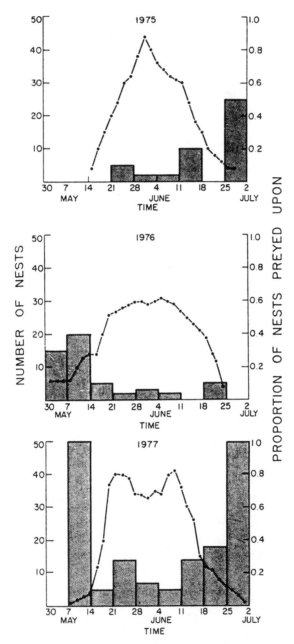

Figure 8.8 Number of nests of yellow warblers and 16 other species available to predators (*dotted line*) and the proportion of nests preyed on (*bars*) over time for 3 breeding seasons 1975–77 (Clark and Robertson 1979).

1974; Rolland, Danchin, and de Fraipont 1998). While swamping may play a greater role in the formation of crèches, given that immatures are poor at deterring or detecting predators themselves, the small body size of crèche members (juveniles by definition) perhaps predisposes them to multiple attack that would reduce benefits of dilution.

8.6 Miscellaneous mechanisms

Additional mechanisms by which grouping can reduce the chance of an individual being the victim of an attack are likely to emerge in other taxa and other environments. One interesting illustration is yarding behavior, shown by wintering white-tailed deer at the northern limit of their range (Messier and Barrette 1985). In December, over 600 deer move 40 km to winter in a relatively small area (36 km^2). Close proximity of conspecifics in these areas creates a network of trails not found in snowbound outlying areas. A greater proportion of deer kills occurs in areas of low deer density than in areas of high or medium density. Tracks in snow show that coyotes depend on the opportunity to corner quarry in deep snow, where the deer is harassed until it becomes exhausted. To escape, the deer must use a runway through the snow, along which it can quickly flee. Messier and Barrette think that the network of runways in areas where deer are grouped is the key to reducing coyote predation.

Grouping can also provide members with an antipredator benefit after an attack. For example, sparrow hawks preferentially target those common redshanks that leave the mudflats later than other individuals; as a result, there is no greater delay in taking off in larger groups than in smaller flocks. In essence, Hamilton's assumption of individuals being at equal risk in a group does not apply here (see section 8.7). Nonetheless, larger flocks return to feed more rapidly than smaller flocks, suggesting that the dilution effect ameliorates opportunity costs of real and false alarms but does not reduce predation risk during the course of real alarms (Cresswell, Hilton, and Ruxton 2000).

8.7 Position in the group

Up to this point, all proposed benefits of grouping can in theory be shared equally among members of a group, but this is unlikely to be the case in nature (Eshel 1978). Hamilton's key contribution in his 1971 paper was to highlight the peculiar difficulties faced by prey individuals (frogs) closest to a hypothetical (snake) predator, because the frog's domain of danger (the part of the perimeter of the pond on which an emerging snake would find that frog closest to it) is half that of the gap between its nearest neighbors. This domain can be reduced by jumping to a narrower gap between two conspecifics (fig. 8.9). In two dimensions, the domain will be a polygon centered on an individual, the sides of which are half the distance to each nearest neighbor (Voronoi tes-

Figure 8.9 *Above:* a frog reduces its domain of danger from the length of the solid line to that of the dotted line by jumping as denoted by the arrow. *Below:* movements of frogs in the positions shown above that will reduce their domains of danger (W. D. Hamilton 1971).

sellation), so that it encompasses all points closest to that individual than any other. De facto, peripheral individuals will have a greater domain of danger than others. Vine (1971) extended Hamilton's model to cases in which a predator attacks prey from the outside rather than center of the group. He argued that if a predator approached a line of prey from a distance, it might only see one animal at the end of the line, but a predator approaching nearer the center of the group would see increasing numbers of prey individuals, thereby reducing the probability of anyone being taken. Therefore, it would pay prey to move to the center of the line. When extended to two dimensions, individuals on the circumference of a group will be at risk of being detected by a distant predator, but those in the center of the group will enjoy a reduced risk of predation simply because they will be farther away and out of sight (exclusive of whether they are hidden by conspecifics). As the direction of the predator's approach is unknown, any prey individual on the periphery will be at greater risk than those in the center of the group (T. L. Morton et al. 1994; see also Viscido, Miller, and Wethey 2001; James, Bennett, and Krause 2004).

A separate but related argument is that individuals in central positions actually remain hidden behind conspecifics on the periphery (G. C. Williams 1964). Thus, although the predator is close enough to see all the individuals in the group, its view of central group members (and those on the far side) is blocked. While no systematic data exist on this mechanism, Estes and Estes (1979) noted that wildebeest usually interposed themselves between their calves and a human observer and that calves were difficult to see behind their mother's beard.

Given asymmetrical potential benefits of group position, we might expect more-competitive individuals to occupy central positions. Thus, dominant individuals (Barta, Flynn, and Giraldeau 1997), larger individuals (Rayor and

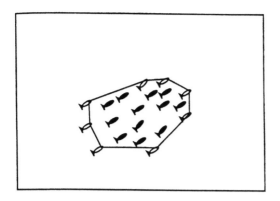

Figure 8.10 A school of fish. Peripheral individuals are unfilled, and central individuals are filled (Krause 1994).

Uetz 1993; P. J. Jakobsen and Johnsen 1988), or individuals likely to suffer greater costs, such as reproductive females (Rayor and Uetz 1990), are likely to be found nearer the center. Nonetheless, position in the group is influenced by foraging decisions, too. Hungry fish go to the front of shoals, where they have higher feeding rates (Krause, Bumann, and Todt 1992; Krause 1993a), for example; thus, any comprehensive explanation of group position must incorporate foraging decisions, antipredator considerations, and competition among group members.

What is the empirical evidence to support Hamilton's proposal that peripheral individuals suffer greater risk of predation than central individuals, also called marginal predation? A preliminary hurdle of defining regions within a group has to be jumped first. Individuals on the periphery have been classified as those at the vertices of the smallest closed polygon that encloses all members of the flock or herd (fig. 8.10), but other definitions have been employed, including individuals being more than the mean distance from the geographic center; individuals in the outer of two concentric rings of equal area, the outermost of which circumscribes all group members; individuals that are not entirely enclosed by n layers of other group members (Stankowich 2003); and even absolute measurements, such as individuals being 30–40 cm into the school (K. Parrish, Strand, and Lott 1989). In some studies, individuals on the edge and those in small groups are placed in one category, thereby conflating group position with group membership. Despite these differences in classification, most studies show remarkable agreement that peripheral individuals are at greater risk than central individuals.

Good data on the advantages of being in the core of the group come from invertebrates (Rayor and Uetz 1990; Romey 1995) and fish. When Milinski (1977b) manipulated densities of *Daphnia* within a swarm using test tubes filled with different numbers of water fleas, sticklebacks first attacked the densest part of the swarm (that is usually at the center) but subsequently delivered more bites to the periphery. In addition, stray *Daphnia* separated from the swarm received higher rates of lunges than the swarm itself. Milinski interpreted the temporal change in preference as hungry fish first attempting to maximize food intake and then switching to easier targets at the periphery as they started to become satiated. If one attributes these results to the confusion effect, as did Milinski, then sticklebacks are exerting marginal predation to avoid confusion and not for the reasons outlined by Hamilton and Vine that identified predators as taking the nearest prey (see also Jakobsen and Johnsen 1988). Therefore, there are at least two underlying mechanisms for marginal predation.

In fish, some experimental studies have shown that peripheral individuals suffer greater risk than central ones (for example, European minnows: Krause 1993b), whereas other studies have shown that central individuals are at greater risk (for example, Atlantic silversides: J. K. Parrish 1989). The difference is attributable, in part, to predator hunting style. Comparing slower moving, solitary, stalking predators to fast moving, pelagic, group hunters in the wild, J. K. Parrish (1993) found that stalkers targeted individual flat-iron herring, whereas pelagic predators either picked off stragglers or accelerated into the shoal and rammed central fish. As being in the center of the shoal carries predation costs from specific predators, species that shoal must either be those that are not hunted by predators that go after central fish, must be the outcome of a compromise to avoid predation from a number of fish predators, or must realize advantages to shoaling besides antipredator benefits (such as food: Krause 1993c; Hoare et al. 2004).

Predation on central individuals questions a core assumption of Hamilton and Vine: that a predator simply takes the nearest prey individual. If predators do make considered decisions about which individuals to attack, then there is far less reason to assume that peripheral individuals are subject to greatest risk as theory predicts. Given that some mammalian and avian predators choose prey on the basis of age (FitzGibbon and Fanshawe 1989), sex (FitzGibbon 1990c), and condition (Temple 1987), the question of whether central members of a group of homeotherms are safer than edge individuals is open. For convenience, it is easier to examine groups of stationary and mobile homeotherms separately when weighing the evidence on position effects (Krause 1994).

8.7.a Colonially nesting birds   Patterns of predation on colonially nesting birds provide indirect empirical support for individuals being at greater risk on the periphery of groups. Without question, peripheral nests suffer greater risk than central nests in numerous taxa, and in some, solitary nesters are at greater risk than colonial nesters (table 8.6); but attributing these differences

Table 8.6 Relative risk of predation according to position in colonies of nesting birds

| | Predator | Relative predation risk | | Measure | Reference[a] |
		Within group	Lone vs. group		
Adelie penguin	Antarctic skua, brown skua, kelp gull, giant petrel	P > C	L > G	BS	1, 2, 3
Double-crested and pelagic cormorant	Northwestern crow and glaucous-winged gull	P > C		*	4
White pelican	Not given	P > C		ND	5
Common shelduck	Herring gull		G > L	BS	6
Gray heron	Corvids	P = C		BS	7
Great blue heron	Raven, bald eagle, raccoon	P > C		NS	8
Black-headed gull	Fox, hedgehog, crow, gulls	P > C	L > G	BS	9, 10
Kittiwake	Not given	P > C		BS	11
Least tern	Black-crowned night heron, crow	C > P		NS	12
Common tern	Herring gull	P > C		BS	13
Royal and cayenne terns	Kelp gull	P > C		ER	14
Royal tern	Laughing gull	P > C		ER	15
Sandwich tern	Herring gull and black-headed gull	P > C		ER	16, 17
Brewer's blackbird	Mainly mammals, snakes	P = C		ND	18
Fieldfare	Crow, woodpecker, jay, squirrel, mustelid	P > C	L > G	ER	19, 20, 21
Cliff swallow	Bull snake, rattlesnake	P > C		NS	22

Note: P denotes periphery; C, center; L, lone; G, Group; BS, breeding success; ER, egg predation rate; NS, nesting success; ND, no experimental data.

[a] 1. Taylor 1962; 2. Teneza 1971; 3. Emslie, Karnovsky, and Trivelpiece 1995; 4. Siegel-Causey and Hunt 1981; 5. Schaller 1964; 6. Pienkowski and Evans 1982; 7. van Vessem and Draulans 1986; 8. Simpson, Smith, and Kelsall 1987; 9. Kruuk 1964; 10. I. J. Patterson 1965; 11. Coulson 1968; 12. Brunton 1997; 13. Becker 1995; 14. Yorio and Quintana 1997; 15. Buckley and Buckley 1977; 16. Fuchs 1977; 17. Veen 1977; 18. Horn 1968; 19. M. Andersson and Wiklund 1978; 20. Wiklund 1982; 21. Wiklund and Andersson 1994; 22. C. R. Brown and Brown 1987.

* Eggs, chicks, or food that adults had vomited on the ground.

simply to predators taking the first prey individual that they encounter is problematic (Burger 1981). First, nest density declines from the center outward, and predators may choose nests on this basis rather than on which nest they first encounter (Wittenberger and Hunt 1985). Second, in some species, older and hence more experienced birds breed earlier and thereby occupy central nests (Coulson 1968); they may be better at driving predators away.

Table 8.7 (a) Number (mean ± SE) of bank swallows counted in photographs of central and edge mobs at various colonies. (b) Deterrence of wild blue jays by bank swallow mobs at a 172-burrow colony

(a)

	Number of birds in		Significance of
Colony size	Central mobs	Edge mobs	center-edge differences
114	23.15 ± 1.17 ($N = 3$)	16.42 ± 1.50 ($N = 2$)	$P < 0.01$
172	61.50 ± 4.63 ($N = 2$)	45.10 ± 2.07 ($N = 2$)	$P < 0.01$
198	41.20 ± 3.49 ($N = 2$)	26.83 ± 3.75 ($N = 1$)	$P < 0.05$
451	29.13 ± 2.12 ($N = 2$)	22.57 ± 2.17 ($N = 3$)	$P < 0.05$

Note: Mobs were elicited by a stuffed weasel.

(b)

	Distance from burrows to tethered young swallow		
	0–1 m	9–11 m	18–20 m
Times jays attack and kill the young bird (N)	0	2	7
Times jays attack but are unsuccessful in killing the young bird (N)	0	2	1
Times jays are deterred from attacking the young bird (N)	5	0	0
Successful attacks at each distance (%)	0	50	88

Source: Hoogland and Sherman 1976 (for [a] and [b]).

Third, colonial nesters mob predators, which deters attack (Wiklund and Andersson 1994; section 11.5). As mobs form when neighboring birds alight and approach a predator, there is a tendency for mobs to be larger in the center than at the periphery of colonies (see Gilchrist 1999 for a review and table 8.7a), and predators are more effectively deterred by mobs adjacent to clusters of nests than by those at some distance (table 8.7b). Fourth, different types of predators may concentrate on different areas of the colony. For example, black-crowned night herons take chicks and eggs from the center of least tern colonies at night and can thereby avoid the mobbing attentions of terns. On the other hand, American crows are diurnal and are forced to take eggs from the periphery of the colony to mollify mob attacks. As a result of these factors, predation risk is greater at the center of the colony (Brunton 1997; table 8.6). Looking more widely, mammalian predators that approach colonies from the outside are more likely to pick off peripheral individuals than avian predators approaching from overhead (Kruuk 1964). In short, while patterns of nest

Table 8.8 Studies comparing mortality risk on the periphery and center of groups in mammals and birds

Species	Relative predation risk		Attack rate	Capture rate	Reference[a]
	Within group	Stray vs. group			
Mammals					
Mule deer	P > C	S > G	X	X	1
Thomson's gazelle	P = C	S > G	X	X	2
Uganda kob	P > C	S = G		X	3
Birds					
Sage grouse	P = C			X	4
Guianan cock-off-the rock	P = C			X	5
Chaffinch and brambling	P = C		X	X	6

Note: P, periphery; C, center; S, solitary; G, group.
[a] 1. Lingle 2001; 2. FitzGibbon 1990c; 3. Balmford and Turyaho 1992; 4. Wiley 1973; 5. Trail 1987; 6. Lindstrom 1989.

predation on bird colonies superficially support the selfish-herd hypothesis, the mechanism is likely to be due to mobbing rather than predators taking the nearest prey item that they encounter, as in Hamilton's original formulation.

8.7.b Flocks and herds   Positional studies of mobile birds are few and currently show no consistent differences in risk of capture between the center and periphery of flocks (table 8.8). For example, on cock-off-the-rock leks, five central-court males were taken by predators, but none were taken from peripheral courts. Unfortunately, this study underscores the small sample sizes in these studies. In mammals, peripheral individuals appear more at risk, but the number of studies is still embarrassingly few. The best mammalian study of marginal predation, to date, is FitzGibbon's (1990c) study of cheetahs hunting Thomson's gazelles that live in mobile but mostly sedentary foraging groups. Of the 43 hunts where she noted the position of the target gazelle, cheetahs chased an animal on the periphery of the herd (defined as having no group members within a semicircle of one side of them) on 83.7% of occasions. This was considerably more than the mean percentage of peripheral individuals available in hunted groups (59.6%). Cheetahs also preferred gazelles that were farther from their nearest neighbors ($X = 19.3$ m), as compared with an average 5.0 m nearest neighbor distance drawn from randomly sampled individuals; peripheral individuals were significantly farther from their nearest neighbors than central individuals. Interestingly, while solitary male Thomson's gazelles were better at escaping cheetahs than solitary females, males' predilection to be on the periphery of groups and their large nearest neighbor distances predisposed them to being hunted. While this study dem-

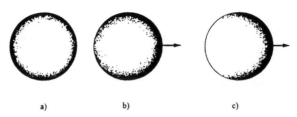

a)   b)   c)

Figure 8.11 Simulations of per-capita predation risk (10,000 predation events per model) of a prey group of 100 individuals which are randomly distributed within a circular area defining the group shape. (*a*) Group and predator are stationary (sensu W. D. Hamilton 1971). (*b*) Both the group and predator continuously move with equal velocity. (*c*) Group moves and predator is stationary. The predator's starting position is random relative to the group position, and the predator generally attacks the nearest prey. Black dots show the position of attacked group members for each case, indicating which group positions were most vulnerable to predation (Bumann, Krause, and Rubenstein 1997).

onstrates the import of marginal predation, other predators of Thomson's gazelles in the same ecosystem, such as jackals and olive baboons, do not select the nearest individuals to them but instead target fawns (Walther 1969). Thus, the relative benefits of position in the group (considering simply antipredator advantages of grouping and not foraging benefits) may accrue differentially to each age-sex class (D. A. Collins 1984). As to where an individual chooses to disport in the group will therefore depend in part on population sizes of predator species in the local community and on their prey preferences.

Hamilton's model dealt with a stationary group of prey; Bumann, Krause, and Rubenstein (1997) considered individual predation risk in groups of prey that were moving in one direction and that were attacked either by a mobile predator or a sit-and-wait stationary predator that was outside the group in both cases. Simulations showed that whereas peripheral individuals are always more vulnerable than central individuals, in Hamilton's stationary group they are equally vulnerable wherever they are (fig. 8.11a), but in the moving group they are more vulnerable at the front than at the rear (fig. 8.11b,c) and particularly so if the predator is stationary (fig. 8.11c). There is limited empirical support for individuals being more vulnerable at the front than the rear (for example, roach and chub: Krause 1993a; Krause, Ruxton, and Rubenstein 1998; Bumann, Krause, and Ruxton 2002). Unfortunately, there is rather little that individuals at the front can do about their vulnerability. If they changed their position or direction of movement, the group would stop moving forward; this occurs in fish shoals after being frightened. Alternatively, leading individuals could reduce inter-individual distances, and this is also observed in moving fish shoals. Individuals in the middle and rear could attempt to hide behind individuals at the front, thereby producing a compressed, elongate group structure, again as observed in fish schools. That high-ranking male baboons lead troops into new areas and to water holes where predators may lurk and that they often defend the troop suggests that they are able or willing to bear the burden of heightened predation risk of being at the front of group movements (Rhine and Westlund 1981).

8.8 Primate groups

While there are clear theoretical reasons for believing that increasing group size confers antipredator benefits on group members, patterns of grouping need not be an evolutionary outcome of predation pressure (Alexander 1974). For example, among seabirds, where pelagic or offshore feeders nest in dense colonies on inaccessible rock stacks (for example, auks, petrels), inshore feeders (gulls and terns) nest in small colonies on headlands or on islands, and marine waders nest on the ground at the intertidal zone (Lack 1968), grouping patterns are better explained as information centers than responses to predation pressure. Put simply, birds feeding closest to their nest are least gregarious, while those feeding farthest offshore nest in largest colonies. A predation hypothesis based on grouping would have predicted the opposite: solitary nesters would be found on inaccessible sea cliffs that are free from mammalian predators, but inshore areas would contain huge colonies of ground-nesting species (Clode 1993). It is in primates, however, where the debate over the importance of predation in shaping group size has been most active and long running.

Interest in the ecological causes of grouping patterns in primates began in 1966, when Crook and Gartlan (1966) classified primate species into five categories based principally on differences in mean group size (see Terborgh and Janson 1986; Janson 1992, 2000 for historical reviews). These grades were more or less associated with ecological, behavioral, and morphological variables, including activity cycle, habitat, territoriality, and sexual dimorphism. These analyses were improved upon using statistical comparisons across a wider array of species by Clutton-Brock and Harvey (1977a, 1977b), who reiterated that nocturnal species are smaller than diurnal species and that they live in smaller social groups, and that terrestrial species are larger than arboreal species, on average, and generally live in larger social units. Subsequently, Wrangham (1980) put forward a deductive model that feeding benefits were the key determinant of group size. Specifically, where food is patchy, individuals in larger groups will be able to displace smaller groups from feeding patches, and thus group size will increase up to a point at which competition over food with other group members becomes too great. The alternative proposition is that group size increases in response to threat of predation (Terborgh 1983) due, in primates, to three factors: increased vigilance (chapter 4), dilution (this chapter), and deterrence (chapter 11). Antipredator evolutionary arguments for

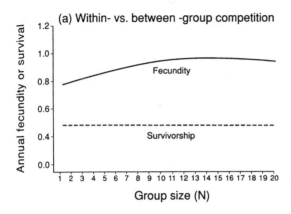

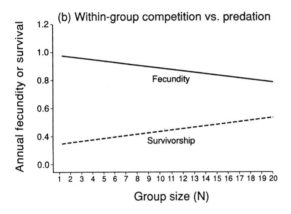

Figure 8.12 Expected relationships between group size and fecundity or survival when cost of increasing group size is greater within-group competition for food, and the benefit is (*a*), increased access to food patches (benefits of gaining access to food patches); or (*b*), reduced risk of predation (Janson 1992, redrawn from van Schaik 1983).

group size in primates normally run these three advantages together, although they are sometimes acknowledged as distinct (for example, Dunbar 1988) and are occasionally examined separately (for example, Stanford 1998).

In a key paper, van Schaik (1983) put forward a way of pitting the foraging and antipredation arguments against each other. He argued that if foraging benefits predominated, annual fecundity should increase up to a certain group size as individuals gained increasing access to food patches, but it should then plateau or decline as within-group competition outweighed the benefits of between-group competition; adult survival would be unaffected, however. If antipredator benefits predominated, adult survival should show a steady increase with group size, but per-capita food intake should decline as a result of within-group feeding competition. While very few primate studies had measured per-capita food intake across group sizes, per-capita fecundity should parallel food intake (fig. 8.12). Van Schaik found that out of 27 regression analy-

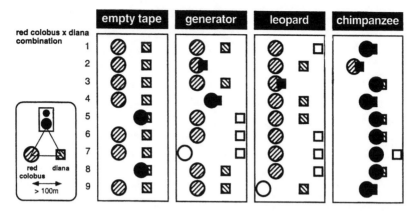

Figure 8.13 Position of 2 groups of red colobus and Diana monkeys 30 minutes after a playback experiment. Distance between cores of groups at the start of the experiment was > 100 m. *Hatched symbol:* group did not move; *open symbol:* group moved > 50 m away from the partner group; *solid symbol:* group approached the partner group and intermingled (Noe and Bshary 1997).

ses plotting average number of infants per female against group size, there were five positive correlations, only one of which was significantly different from zero. Twenty-two of the 27 showed declining fecundity with group size, and although only two of these were significant, the chance of finding so many negative slopes is small (see also Dunbar 1988). This suggested that explanations for primate grouping patterns were unlikely to lie in the benefits incurred through between-group competition over food patches, but it did not satisfactorily show that predation was responsible.

Subsequent indirect evidence lent support to the antipredation idea, however. As illustrations, baboon populations living in high-predation-risk habitats (grassland and savannahs, where few tree refuges are available) have higher mean group sizes than populations living in woodland or forest habitats, where more trees are present (Dunbar 1988; see also Stacey 1986). Group sizes of long-tailed macaques are higher on mainland Sumatra, where felids occur, than on the island of Simeulue, where there are no cats (van Schaik and van Noordwijk 1985). In Tai Forest National Park, Diana monkeys have contact calls that keep group members together in the trees; calls increase when predation threat is elevated (in poor visibility or following alarm calls: Uster and Zuberbuhler 2001). In the same forest, where red colobus but not Diana monkeys are frequently killed by chimpanzees, playbacks of chimpanzee sounds, but not of other sounds, lead red colobus to approach and intermingle with Diana monkeys (fig. 8.13), and they spend more time in such mixed-species associations as chimpanzee density rises (C. A. Chapman and Chapman 2000).

Nonetheless, lack of direct measures of predation is an underlying problem in trying to link the evolution of primate grouping patterns to antipredator defenses. In an effort to tackle this, Cheney and Wrangham (1987) sent a questionnaire to primate fieldworkers, asking them to estimate predation rates (percentage of population killed per year). Given that most instances of predation were inferred but not observed, predation rates vary across habitats, predators may leave an area where primate researchers work (Isbell and Young 1993a), and human predation was not considered, they regarded the data as preliminary. That said, there was a strong negative correlation between group size and estimated predation rates in monkeys and prosimians (apes were excluded from the questionnaire) (Isbell 1994). A separate analysis of categorical rankings of predation scores demonstrated that all 14 species living in group sizes of > 20 suffered low or no predation, whereas 5 out of 13 species living in groups of ≤ 20 suffered high predation (C. M. Anderson 1986). (Nonetheless, correlations between six grades of predation danger and observed instances combined showed positive correlations with group size within species, although many were nonsignificant.) In short, the association between large groups and low predation rates could be used to corral support for large group size being an effective defense mechanism to reduce predation.

Perhaps because the topic has received so much attention, many problems have been raised over this relationship. First, there are methodological issues: the validity of the original predation rate data; absence of phylogenetic controls, resulting in over-representation of species in certain genera; and lack of attention to confounding variables. For example, estimated predation rate is negatively correlated with both group size and body size, which are themselves correlated (Isbell 1994). Some of these issues have now been tackled: for instance, independent contrasts confirm the two negative correlations above, although they disappear when nocturnal species are dropped from the analysis (R. A. Hill and Dunbar 1998). Second, there are alternative explanations for the data: under high rates of predation, predators may decimate groups, resulting in groups of small size (for example, leopards and vervet monkeys: Isbell 1990; chimpanzees and red colobus: Stanford 1995). Also, groups living in thick vegetation, where the benefits of group vigilance are limited, may forage in small patches under heightened risk of predation to avoid being heard by the predator (Boesch 1991).

Third, and importantly, predation rate may not be a good proxy for predation risk, the selective pressure to which group size is meant to have responded (Vermeij 1982). Predation rate is the level of successful predatory attack that prey are unable to modify after they have implemented antipredator defenses. On the other hand, risk is a composite of the prey's vulnerability and probability of an attack, whatever its outcome (R. A. Hill and Dunbar 1998; Janson 1998). Recently, researchers have recognized the difference between rate and

Table 8.9 Calculating predation risk

	Swakop Region		Namib Region	
	Bed	Woodland	Plains	Hills
Visibility (d_{vis}) (m)*	80 (40->100)	30 (5–75)	65 (30->100)	60 (20–70)
Distance to refuge (R) (m)	45 (20–60)	10 (5–20)	70 (30->100)	75 (0->100)
Ground slope	0°	0° (0–10°)	0 (0–30°)	45 (30–70°)
Faidherbia albida density+	0 (0–1)	1 (0–13)	0	0
Prosopis glandulosa density+	1 (0–6)	13 (3–29)	0 (0–1)	0

Attack risk (r_A)[a]

	Ambush distance d_{amb} (m)	Attack risk: proportion of cases of visibility $< d_{amb}$			
Leopard	10	0	0.5	0	0
Lion	25	0	0.75	0.13	0.25

Capture risk (r_C)[b]

	Attack speed (m/s)	Capture risk: proportion of cases where refuge distance $> R_{max}$			
Leopard	8.1	0 (80 m)	0 (30 m)	0.5 (65 m)	0.63 (60 m)
Lion	13.7	0.5 (35 m)	0.88 (10 m)	1.0 (25 m)	0.75 (25 m)

Source: Cowlishaw 1997a.
* Medians (and range); $N = 8$ in each case.
+ Two most important trees in baboon diet; density/0.25 ha.
[a] r_A is calculated as the proportion of sampling points where visibility fell below d_{amb} in one or more directions at that point. $N = 8$ in each habitat. d_{amb} is the midrange of common ambush distances to the nearest 5 m.
[b] r_C is calculated as the proportion of sampling points where the nearest refuge was equal to or greater than R_{max} (the distance within which prey must be able to reach a refuge in order to be safe from predators) for that predator in that habitat (value in brackets). $N = 8$ in each habitat. All distances to nearest 5 m.

risk and have tried to estimate risk and then relate it to group size. At a basic level, risk can be estimated very crudely by noting whether a primate gives an alarm call or is attacked by a given predator. Such variables can then be partitioned by habitat, time of day, and age-sex class of primate to give measures of differential risk (Cowlishaw 1994). Cowlishaw (1997a, 1998) also devised a more sophisticated measure of risk using predator (lion and leopard) hunting strategies against yellow baboons in Namibia. He calculated risk of attack (r_A) by measuring visibility distances and then calculating the frequency with which they fell below a predator's ambush distance (d_{amb}) gleaned from the literature (table 8.9). He calculated capture risk (r_C) through the critical distance (d_{crit}) from the predator that prey must exceed in order to escape, where

$$d_{crit} = R (k - 1)$$

R is distance to a refuge as measured in the field, and $k = v_{pred}/(v_{pred} - v_{prey})$, where v is running velocity taken from the literature (4 m/sec for baboons;

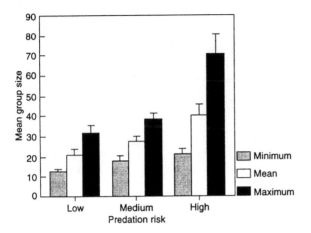

Figure 8.14 Mean and standard error for the minimum, mean, and maximum group sizes of populations under low, medium, and high predation risk (Hill and Lee 1998).

13.7 m/sec for lions, 8.1 m/sec calculated for leopards). Although d_{crit} varies with R, visibility (d_{vis}) is crucial, because it determines maximum detection distance. Given this, prey must remain within a maximum distance to a refuge R_{max}, where

$$R_{max} < d_{vis}(k - 1)$$

Values of R_{max} can then be determined for each habitat, as d_{vis} is different in each, and the frequencies with which they are equaled or exceeded can then be calculated. After making these calculations, Cowlishaw went on to determine whether baboons distributed themselves across habitats according to foraging opportunities or predation risk. He found that they spent more time feeding, resting, and grooming in low-risk poor-food habitats. In theory, risk of detection by predators could also be calculated in a similar fashion, but this may be less important in a species that moves in conspicuous groups.

Hill and Lee (1998) constructed another measure of predation risk and this time related it to group size in cercopithecoid primates. *Low risk* referred to predators being present with no predation observed or suspected; *medium* denoted interactions between predators and primate prey but infrequent predation; *high* referred to high to regular contact between predators and prey coupled with observed or suspected predation. Measures of group size clearly increased with predation risk (fig. 8.14). As predator threat is likely to be highly site specific, analyses across populations seem well justified. While these measures of risk are approximate, they identify the selective force to which antipredator defenses can be expected to respond. As increasing group size can potentially confer several antipredator benefits in primates, associations between risk and group size suggest that group size has indeed been shaped by predation risk in this primate family.

These types of analyses do not attempt to test which benefit of grouping has been responsible for the association between group size and predation, but several studies have found that the number of males in primate groups increases under threat of predation (for example, van Schaik and Hostermann 1994; Hill and Lee 1998; Treves 1999b; review in Stanford 1998). As males in several primate species have been seen to attack predators and drive them away, predator deterrence is clearly an antipredator benefit of increased group size in some primate species, although the factors associated with its taxonomic distribution remain unexplored. More generally, while most primate studies nowadays recognize that group sizes are likely to be influenced by feeding considerations, predation, and threats from conspecifics (Treves and Chapman 1996), research usually focuses on the importance of one over others, and there have been remarkably few attempts to assess their relative importance in shaping grouping patterns within or between populations.

8.9 Ecocorrelates of antipredator grouping in homeotherms

Group living is thought to be associated with living in particular habitats and with certain types of antipredator behavior in birds and mammals (Crook 1965; Crook and Gartlan 1966; Kaufmann 1974). For example, Jarman (1974) divided African antelope species into five classes based upon their feeding style: (a) browsers, (b) species that select new leaves and shrubs, (c) species that select a range of grasses and browse, (d) species feeding unselectively on grasses, and (e) species that feed unselectively on a wide range of grasses and browse. He suggested that most small-bodied antelopes fell into classes (a) and (b), because their small mouth and muzzle facilitated specialization on nutritious plant parts, whereas large-bodied species fell mostly into classes (d) and (e) because they lacked the oral morphology to feed selectively on new shoots and fruits. Next he reasoned that competition over clumped high-quality plants would result in territoriality in selective feeders, but that a widespread supply of grasses would reduce competition and allow coarse feeders to form groups. Further, grouping would be determined by positive selection in open habitats as a result of predation risk because of enhanced vigilance and benefits outlined in this chapter.

Jarman further dichotomized antipredator behavior into (A) flight or counterattacking the predator and (B) hiding or freezing on detecting a predator, and hypothesized that antipredator behavior would be modified by group structure and the environment such that species living in large groups in open habitats would employ defense A, whereas those in small groups in closed environments would use defense B. Generally, the data supported his claims, and most of his qualitative observations have now been verified in phylogenetically controlled comparisons within this (Brashares, Garland, and Arcese 2000) and related artiodactyls clades (Caro et al. 2004).

Parallel observations have been made for primates (Crook and Gartlan 1966; Terborgh 1983; van Schaik and van Hooff 1983; Terborgh and Janson 1986; Cheney and Wrangham 1987), where it has been noted repeatedly that small ($<$ 1 kg) species suffer higher rates of predation than large ones because they are subject to a greater number of predator species. Small species tend to be nocturnal and to live alone, and rely on crypsis to escape detection. Predators must locate nocturnal primates by sound and approach without being heard by prey, so it is disadvantageous for nocturnal species to live in (noisy) groups. Also, small species are usually arboreal, which may be a means to reduce risk of predation (but see Isbell 1994). Larger species tend to be diurnal and to live in groups, and rely on corporate vigilance, deterrence, and perhaps dilution to escape predation. Terrestrial species are thought to be subject to high rates of predation (Dunbar 1988). While many of these associations have been verified statistically, taking phylogeny into account (Clutton-Brock and Harvey 1977a, 1977b), there are a number of difficulties. First, food may confound these associations. Thus, nocturnal primates may be small because they need to be able to travel to and reach food on thin terminal twigs; such constraints are removed from terrestrial species that can afford to be large. Second, nocturnal species may live alone because of difficulties in maintaining contact with conspecifics rather than predation pressure (Clutton-Brock and Harvey 1977a). Third, the situation may be more complicated than at first glance. Once relative ranging costs are taken into account, group size in arboreal primates rises with body weight and then declines at 5 kg, and a similar pattern is found in terrestrial species, with group size declining above 15 kg–20 kg. Thus, large species may, in both habitats, also rely on crypticity, just as small species do (Janson and Goldsmith 1995). Put another way, very large arboreal primates that are relatively free from predation, such as orangutans, can afford to be solitary.

Antipredator syndromes have also been documented in birds. For example, in the Monteverde forest, Costa Rica, species that are difficult to discover and approach unobserved are solitary. These are birds that forage on the ground where shade is plentiful, hummingbirds feeding on the wing, and sentinel foragers that sit motionless. In contrast, active, arboreal birds that are more vulnerable are gregarious (Buskirk 1976). Similarly, in Guianan forest, nectivorous and frugivorous birds are usually solitary, insectivorous species often forage in pairs, and frugivores and insectivores foraging in the understory or upper canopy live in monospecific or multispecies flocks (Thiollay and Jullien 1998).

In all these mammalian and avian taxa, food requirements, predation pressure, body size, or some combination are thought to drive patterns of grouping, and then either grouping or habitat or both force species to adopt different antipredator defenses, such as crypticity among solitary species or reliance on corporate vigilance in group-living species. Thus, there are syndromes of solitary, selective, small species living in forests that avoid detection by remaining inconspicuous, and social, unselective, large species living in open habitats that rely on vigilance and group defense to take early evasive action and deter predators. Although we can roughly describe the pattern of antipredator syndromes across a few taxa, that is, the species that rely on avoiding detection, on early warning, and on attacking the predator, we are much less confident about which factors (body size, habitat, or food) are immediately responsible for the evolution of antipredator defenses. In addition, we do not understand the interaction of these factors. For instance, in Guianan forest, the important determinant of flocking may not be the food source but the searching technique used to locate food that constrains vigilance. For a sit-and-wait carnivore, vigilance is not limited, which reduces the importance of relying on corporate vigilance. As another example, very small body size or fast flight seen in hummingbirds may make them unprofitable prey, thereby reducing the necessity of resorting to group-related antipredator defenses.

In conclusion, there has been little effort to identify or tease apart dependent and independent variables that characterize antipredator syndromes, and systematically examine interactions between variables, let alone identify whether particular variables, such as small body size, differentially influence the probability of predation in different taxa.

Returning to the benefits of grouping, outstanding questions are concerned with the relative importance of different mechanisms, including vigilance (chapter 4); mobbing and group defense (chapter 11), by which grouping reduces individual predation risk; and related to this, the import of dilution risk per se. Prominence of various mechanisms likely vary enormously across prey taxa and between the same prey threatened by different predator species, making generalizations difficult for some time to come.

8.10 Summary

Group membership can reduce the probability that an individual will be chosen as the target of a predatory attack. Depending on the species, groups may be difficult to describe, but in practice, researchers rely on nearest neighbor or absolute distances to define groups. If a predator only captures one member of a group and encounters groups less than in direct proportion to group size, then individuals will suffer a lower probability of being captured as group size increases (the dilution effect). It seems unlikely that predators encounter groups a good deal more often than solitary animals, although there is no strong evidence on this point as yet. Regardless, data from birds and mammals show that there is a lower per-capita risk of being killed in large groups.

Several mechanisms may reduce the likelihood of an individual being a target of a predatory attack when it is in a group. For example, individuals may become alerted to the presence of a predator early, because other group members nearer to the predator become agitated or give warning signals (the Trafalgar effect). Predators may also find it difficult to follow a single individual through a group (the confusion effect). Experiments with sticklebacks attempting to catch water fleas show that prey's simultaneous movements hamper fish from fixating on individual targets; data from homeotherms are more circumstantial. The confusion effect selects against oddity, because unique individuals will be easier for a predator to follow through a group. Where predators take some time to handle prey, other prey group members may have time to move away in the short term or, in the longer term, pass through a vulnerable life history stage before the predator attacks again. This is called predator "swamping," and is thought to be an important mechanism in the evolution of colonial nesting in seabirds and reproductive synchrony in colonial birds and group-living ungulates.

All these mechanisms assume that risk is spread equally among group members, but individuals at the edge of groups will be more vulnerable to predation than those in the center, because they are nearest to a predator approaching from the outside. Individuals are therefore expected to compete for central positions. Data from aquatic invertebrates and fish show that animals are usually more vulnerable on the periphery of groups. Data from colonially nesting seabirds point the same way, but this may derive, in part, from reduced mobbing in areas where nests are scarce. In ungulate herds, peripheral individuals suffer higher rates of predation from stalking predators than do central individuals.

Explanations for the evolution of patterns of grouping in primates have contrasted benefits gained as a result of between-group competition over food with antipredator benefits realized through enhanced vigilance, deterrence, and the dilution effect. Indirect evidence supports the importance of antipredator benefits, but there is a dearth of good data on predation rates across species. Group size does increase with increasing predation risk, however, and this may be a factor to which antipredator defenses have responded over the course of evolution. Group living is associated with particular habitats and types of antipredator behavior in birds and mammals. Typically, open-country species that rely on vigilance and flight live in large groups, while forest-living species that rely on crypticity to escape predation live alone. This pattern is repeated in passerines, ungulates, and primates, but the evolutionary drivers of these antipredator syndromes have not been identified satisfactorily.

9 Morphological and Physiological Defenses

9.1 Introduction

If a predatory sequence advances to the point at which a predator either sees or has approached prey, its decision over whether to give chase, seize, or try to subdue prey will be influenced by prey morphology, because this will affect both the costs of predatory attack and the likelihood of pursuit and subjugation being successful. Perhaps the morphological trait of most overarching significance at this juncture is body size, because larger prey animals are more difficult to subdue and are more likely to cause a predator injury. Thus, with some provisos, larger prey species will suffer predation from fewer predator species, possibly reducing the number of predatory attempts that they experience (section 9.2). Although it is a truism that larger quadrupeds run faster than smaller ones, we might also expect a trade-off between increased size and locomotor performance. The relationship between these variables is surprisingly complex in inter- and intraspecific comparisons in mammals, although it is more clear-cut in intraspecific comparisons in birds. Thus, there is no obvious general evolutionary trade-off between the ability to deter or injure a predator and to flee from it effectively. Other aspects of morphology include wing shape in birds and bipedalism in mammals which have huge effects on locomotion and hence escape mechanisms open to prey (section 9.3).

Aside from such considerations, there are some specific morphological traits concerned with self-defense that are found in restricted mammalian taxa. These include spines and quills in porcupines, hedgehogs, and echidnas, and dermal plates in armadillos and pangolins; these traits often characterize

Figure 9.0 (*Facing page*) European hedgehog drinking. The hedgehog epidermis is covered in dense spines that can stand considerable force. Before or during contact by a predator, the animal can roll itself into a ball, making it difficult to gain access to the vulnerable belly and face (reproduced by kind permission of Sheila Girling).

all the species within a particular family. Such forms of armor appear so obviously designed to thwart predatory attack that they are assumed to be antipredator defenses on the basis of rather little evidence. There are quite clear trade-offs between carrying armature and flight speed, so it is instructive to understand the circumstances that favor the evolution of armor over flight (sections 9.4 and 9.5). Other homeotherms have morphological traits that are principally used in obtaining food, such as talons in raptors or sharp teeth in carnivores, but can be used to great effect in stopping a predatory attack from being successful (section 9.6). These feeding adaptations are thereby co-opted for defense. Yet other dangerous weapons are principally used in the context of intrasexual contests between males but may also be employed to great effect in deterring predators. Thus, antlers, horns, tusks, and spurs can dissuade predators from pressing home an attack or, in a close encounter, prevent injury or death (section 9.7).

Turning to physiological characteristics that may lower the probability of being consumed, certain homeotherms smell unpleasant and are alternatively or additionally disgusting to eat. Some of these species are aposematic in that they use bad odor to advertise their unpalatability or their ability to defend themselves, whereas in others the odor itself may act as a predator deterrent. A few prey species anoint themselves with strong-smelling substances that may make them distasteful (section 9.8). A diverse group of mammals living in close association with poisonous snakes have evolved resistance to snake venom. The extent of resistance is closely tied to prey-predator sympatry and hence constitutes an unusual putative example of predator-prey coevolution in homeotherms (section 9.9). Finally, life history traits that may serve to ameliorate the effects of predation are examined in section 9.10.

9.2 Body size

Across numerous taxa, larger predator species take larger prey (for example, mammals: Rosenzweig 1966; raptors: Schoener 1968; snakes: H. W. Greene 1983; vertebrates: Vezina 1985; carnivores: Dayan and Simberloff 1996; ungulates: Sinclair, Mduma, and Brashares 2003). The relationship between predator size and prey size can be derived from the rate at which predators capture prey, biomechanics of performance limitation, or foraging theory (Emerson, Greene, and Charnov 1994). So, for example, in regards to the last, if handling time increases rapidly with prey body size (fig. 9.1a), a predator will choose a smaller prey item (W_{m1}) than it can physically take (W_{m2}), because the handling time of the latter is so great. Instead, it will focus on prey at the point where the cost curve crosses the ingestion rate line. If the handling time curve and predator ingestion rate each change with predator size, then, using this

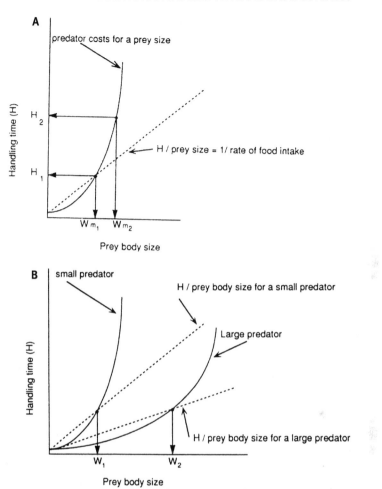

Figure 9.1 *A*, the theoretical relationship between handling time (*H*) and prey body size (*W_m*) for a given prey type (*H*/prey body size). Predator costs for a prey body size are shown as a sharply rising power function. Dotted line represents 1/rate of food intake (or handling time/prey body size = constant). *B*, the theoretical relationship between *H* and *W_m* for a small and large predator with similarly shaped cost functions. Dotted lines represent 1/rate of food intake (Emerson, Greene, and Charnov 1994).

argument, we can expect to see an increase in prey size with size of predator (fig. 9.1b) and can derive a relationship between the two. Owing to differences in handling costs, however, predators with different feeding methods show different size-specific increases in mean prey size with increasing predator body weight that range from 0.7 to 1.2 ordinary least-squares slopes when plotting prey size against predator size.

In some organisms, very large body size reduces susceptibility to predation. Prey size in fish is limited by predator gape size, so that very large fish cannot be taken by other species (Hall and Werner 1977; Wainwright 1987), thereby driving rapid juvenile growth rates in prey (but see Lankford, Billerbeck, and Conover 2001). For mammals and birds pursued by snakes, the same principle may apply, but advantages of body size may be reduced in relation to avian but especially mammalian predators that suffocate, disembowel, or rip their prey apart (but see Dayan et al. 1990). This proviso might be particularly germane when predators are carnivores or group-hunting species. More generally, however, one can speculate that an individual of a large prey species will be subject either to predation from fewer predator species or to fewer predation attempts than an individual of a smaller prey species; the latter idea seems quite likely, given that larger predators live at lower densities than smaller predators.

For very large mammals indeed, there is an ill-defined argument that they are somehow immune from predation, a position that stems in part from the danger that these species pose to humans on foot and the paucity of extant large predators that might be able to tackle very large prey (Peters 1983; Owen-Smith 1988). Nevertheless, massive herbivores such as hippopotami, elephants, and giraffes show sophisticated antipredator defenses, including pattern blending and vigilance; adults are documented as succumbing to predation, and offspring are subject to frequent predatory attack (for example, Schaller 1967, 1972; Berger and Cunningham 1994), indicating that predation is by no means annulled, although it may be diminished. In sum, the case for assuming that large body size is an antipredator benefit in mammals is likely but poorly established at present.

Within species, a larger individual may be subject to fewer predation attempts (particularly by predators roughly equivalent in size to prey) than a smaller prey individual, because it should be able to fight off an attack with more alacrity. Thus, there may be selection to appear as large as possible. For example, many carnivores have broad dorsal capes and narrow middorsal crests that, when erect, increase apparent size. A study of 55 species of African carnivores showed that species with long dorsal hairs were poor climbers, unable to seek refuge in trees from predators. In addition, long-haired viverrids occupied open habitats, whereas short-haired species lived in woodlands and forests, where there are more opportunities for escape through crypsis. Finally, and most interesting, relative dorsal hair length was greater in species that had a larger number of sympatric carnivores that were 20% smaller to 50% larger than the focal species and might thereby be a danger to them (fig. 9.2). All these findings give one to understand that manes and crests are an antipredator defense to make animals appear bigger to predators (Wemmer and Wilson 1983). In a parallel example, standing birds may lift one or both wings vertically to increase apparent size.

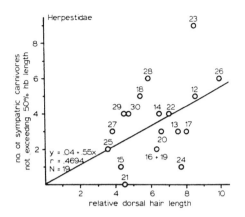

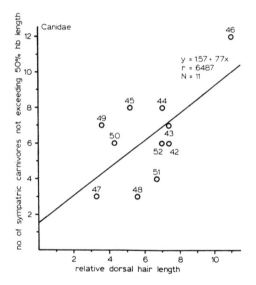

Figure 9.2 Correlations between relative dorsal hair length of the torso and number of sympatric carnivores not exceeding 50% of head and body length in the Herpestidae (*above*) and Canidae (*below*) (Wemmer and Wilson 1983).

9.2.a Body size and locomotor performance

Large body size has enormous ramifications for behavior, life history, physiology, and morphology (S. J. Gould 1966; Schmidt-Nielsen 1984; Peters 1983; J. H. Brown and West 2000), including many aspects of locomotor performance (Pedley 1977); but locomotor relationships have been studied chiefly in amphibians and reptiles, where larger species generally swim, jump, and run faster than smaller animals (Jayne and Bennett 1990; Garland and Losos 1994). In birds, flight dynamic theory predicts that smaller species should have greater power output necessary for takeoff than larger birds (Tobalske and Dial 2000). Once airborne, however, wing loading (mass divided by wing area) comes into play and is negatively associated with escape flight speed (Burns and Ydenberg 2002).

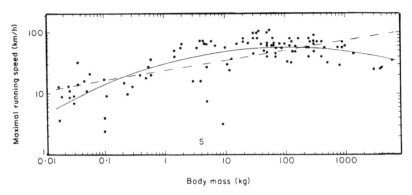

Figure 9.3 Maximal running speeds of 107 species of mammals. Dashed line represents allometric equation for all mammals except sloth (represented by S), with a slope of 0.17. Solid curve represents a polynomial regression equation that fits the data significantly better. An optimal body mass for running ability is suggested: approximately 119 kg (Garland 1983a).

In mammals, the relationship between body size and speed is complex (Garland 1983a; Garland, Geiser, and Baudinette 1988). Across all mammals, maximal running speeds scale at Mass$^{0.17}$, increasing up to 56 km/hr at a body size of 119 kg but then declining (fig. 9.3). An alternative measure, maximum relative running speed (body lengths/second), shows a negative but nearly independent relationship with body size for small species but a strong negative association for large species similar to absolute running speed (Iriarte-Diaz 2002). Within specific taxa such as artiodactyls, carnivores, and rodents, maximum running speed is independent of body weight, so that the relationship between body mass and maximum running speed drops out in analyses that control for phylogeny (Garland and Janis 1993). Moreover, at any given body size there is great variability in running speeds in certain species that adopt means of defense other than flight, such as skunks and porcupines being slow for their body weight. Similarly, while energetic costs of transport increase with body size, energetic costs vary enormously between species, because costs are a function of the product of the distance moved per day and cost of transport divided by daily energy expenditure, and distance moved per day is extraordinarily variable between species (C. R. Taylor 1977; Garland 1983b). These complexities suggest that there is no simple trade-off between possible antipredator benefits of being a large mammal and either flight speed or energetic costs of flight from predators.

Within a species of mammal, evidence of a similar trade-off is also mixed. Running speed declines with body mass in Belding's ground squirrels (Trombulak 1989), whereas in male yellow-pine chipmunks and adult golden marmots, running speed is unrelated to body weight, although longer individual chipmunks run faster than shorter ones (Blumstein 1992; Schulte-Hostedde and Millar 2002).

Table 9.1 Some flight variables important for prey capture and their dependence on total predator mass, *M*

Flight variable	Dependence on mass *M*
Maximum linear acceleration in flapping flight	Decreases faster than $M^{-1/3}$
Maximum speed in horizontal flapping flight	Decreases as *M* increases
Terminal speed in dive	$\alpha M^{1/6}$
Maximum rate of climb	Decreases faster than $M^{-1/3}$
Maximum angular roll acceleration	$\alpha M^{-2/5}$
Turning ability; 1/(minimum turning radius)	$\alpha M^{-1/5}$

Source: M. Andersson and Norberg 1981.

Turning to intraspecific comparisons in birds, there is stronger evidence of a trade-off between body weight and flight performance (as described in chapter 12). Generally, heavier individuals show reduced flight takeoff speed and acceleration in routine but not escape flights. For example, among predatory birds, a number of flight performance variables scale with body mass (table 9.1). M. Andersson and Norberg (1981) used these relationships between size and aspects of flight to help explain reversed sexual dimorphism in predatory groups such as raptors, owls, and skuas, where females are routinely larger than males. They noted that these species need to be extremely maneuverable to capture avian prey, so it may pay these predators to be small. Against this, large body size may be important in defending the nest against egg and chick predators. Given the trade-off between maneuverability and size, it may pay one sex to be large and defend the nest and the other to be small and catch prey for its partner. Andersson and Norberg argued that the female should be large to reduce the cost of carrying eggs prior to laying, and from a proximate standpoint, the female will be near the nest when she begins to lay. The female may subsequently supplement offsprings' diet with larger prey items near fledging, when their demands are highest and she is free to leave the nest because offspring can defend themselves.

Young animals face particular antipredator challenges, because juveniles are smaller than adults and hence easier to subdue. Furthermore, they are less able to accelerate as a result of both scaling effects and poor functional capacities of bones and muscles. Young animals also have less stamina than adults and consume more energy during running. This is because they must take strides more frequently and so their muscles have to develop force more rapidly, and also because they have a lower capacity to store energy while running. Last, juveniles are clumsy perhaps because of immaturity of the nervous system or lack of experience, but these factors are not well understood. Although most juvenile mammals take time to reach adult performance abilities, juveniles in a few species are equivalent to adults very soon after birth (for example, Marsh 1988). Young black-tailed jackrabbits of only 30% adult body mass have equivalent takeoff velocity from a standing jump and can accelerate more rapidly than adults because they have relatively stronger gastrocnemius muscles and relatively thicker bones than adults (Carrier 1996).

Larger prey size confers antipredator advantages because it lengthens handling time and, in general, allows animals to run faster up to a certain size limit, but there may be predation costs to increasing size simply because of greater conspicuousness. Across species, there is roughly linear scaling between acuity and body size, because the relationship between eye size and body size is hypoallometric but is hyperallometric for acuity and eye size. The distance at which a prey animal is detected probably scales with body size by a factor > 1, however, because both prey size and visual acuity of predators both scale by factors > 0.5 with predator body size. Thus, larger prey species are disproportionately more likely to be detected farther away than small prey species (Kiltie 2000). In a study in which people were required to "catch" (mouse-click) prey of different sizes moving across a computer screen at different speeds, catchability depended on body size, so that smaller "prey" were more likely to escape at a given "speed." If this is a general phenomenon in predators, it might help explain why there is a preponderance of species smaller than the mass-flight speed optimum (van Damme and van Dooren 1999).

Other factors that influence maximum running speed include incline (Reichman and Aitchison 1981), sex (Blumstein 1992), and age (Garland 1985).

9.3 Forms of locomotion

Most birds are capable of flight, and their ability to take off using flapping flight, take off at a high speed, and take off at a steep angle all reduce the probability of being captured by a terrestrial predator (Lima 1993), although a few well-known extant species are flightless and escape by running fast (for example, ostriches) or hiding (for example, brown kiwis). A number of observations have been made on aspects of wing morphology. Wings with a high aspect ratio (large span compared with area) minimize flight costs when cruising at relatively high speeds and are expected to be found in birds that undertake long-distance migratory flights (Marchetti, Price, and Richman 1995; Voelke 2001). Broad wings with low wing loading are expected in slow-flying, maneuverable species (Saville 1957). Similarly, a rounded wing tip can maximize thrust from flapping and produce lift toward the wing tip, whereas a pointed wing minimizes distal wing weight and inertia, promoting speed of wing motion (Rayner 1993). As an illustration, least sandpipers and western sandpipers are roughly the same size and use the same aerial escape tactic, but least sandpipers have less pointed wing tips. Accordingly, least sandpipers are 40% faster at takeoff (Burns and Ydenberg 2002). Species with rounded wing tips usually manage a steeper takeoff angle (Swaddle and Lockwood 2003) and, perhaps as a consequence, enjoy low predation risk (Swaddle and Lockwood 1998). Birds that have recently molted take off at a slower speed than before molting (Swaddle, Williams, and Rayner 1999). Birds with longer tarsi tend to hop rather than walk, and birds with shorter femora take off more quickly; species with these types of legs have low predation rates, probably reflecting their prowess at escaping (Swaddle and Lockwood 1998).

Most of the work on the energetics of bird flight has been carried out in relation to migration rather than escaping from predators (for example, Hedenstrom and Alerstam 1994, 1996; Welham 1994; Pennycuick 1997). Such studies either test optimization hypotheses, including minimizing the ratio of power to speed or maximizing distance traveled per unit of energy expended; examine the effects of environmental factors such as wind speed, time of day, and body weight (section 12.7); or simply investigate species that exhibit high levels of flight performance (Irschick and Garland 2001). Nonetheless, in relation to predation, if a linear horizontal chase has been initiated, it is best to accelerate as fast as possible to a maximum speed that is constrained by the ability of flight muscles to deliver power anaerobically, or simply to fly at a maximum speed attained by aerobic muscles if this speed is faster than that of the predator (Hedenstrom and Alerstam 1995). Alternatively, if a bird is chased by a larger avian predator, it may be best to outclimb it (Hedenstrom and Alerstam 1992). Another option may be for prey to use a last-minute tight turn to evade the predator (Howland 1974). Ornithologists have documented that different types of escape predominate in different bird species (Lima 1993), and these are just starting to be mapped onto different wing morphologies systematically across species.

Most terrestrial mammals are quadrupeds, but some are occasionally (for example, mountain gorillas or giant pangolins) or always bipedal (for example, spring hares) or even pentapedal (for example, kangaroos at slow speeds: M. B. Bennett 1987). The majority of mammals rely on running fast to escape predators, although other species of mammals climb, burrow, glide, or fly (chapter 12). Among quadruped mammals, differences in limb morphology enable some species to specialize in different types of locomotion (for example, massive femoral muscles for vertical jumping in caracals) and employ different flight speeds. Longer limb length predicts faster running speeds in cursorial mammals, for example (Garland and Janis 1993). Other factors such as lung capacity may be involved in flight performance, too (for example, large lung capacity for sustaining high speeds over a long distance in the pronghorn antelope: Lindstedt et al. 1991). For quadruped mammals of similar body weight, energetic costs of flight remain surprisingly similar across species with different limb configurations (C. R. Taylor et al. 1974; Garland, Geiser, and Baudinette 1988).

Among mammals with bipedal limb morphology, such as kangaroo rats, individuals switch from quadrupedal hopping when moving slowly, to bipedal hopping when moving rapidly, to powerful hopping with frequent and unpredictable changes in direction when escaping that allows them to dodge predators or jump clear of them (Bartholomew and Caswell 1951; Taraborelli, Corbalan, and Giannoni 2003). Although small bipedal species may run faster than small quadrupeds, large bipeds (for example, kangaroos) are not particularly fast for their size (Garland 1983a). Furthermore, while one might think that bipedal hopping was more efficient than quadrupedal locomotion because of the kinetic energy stored in elastic elements used to propel the animal forward and upward, oxygen consumption of bipeds is no less than that of quadrupeds (S. D. Thompson et al. 1980); indeed, in red kangaroos, oxygen consumption exceeds that of quadrupeds at slow speeds, although dropping below that expected at high speeds of more than 18 km/hr (Dawson and Taylor 1973). More likely, the key evolutionary innovation for small bipeds is longer running endurance and more-erratic escape behavior than quadrupeds (Djawdan and Garland 1988; Djawdan 1993), enabling them to escape raptors and canids; this in turn may allow them to exploit more-open habitats than quadrupeds (chapter 3).

In some instances, different types of escape are incompatible with each other. As an illustration, sprinting speed and stamina are negatively correlated with each other because of biomechanical constraints; massive muscles produce high power for sprinting but increase costs of locomotion, and fast-twitch muscles fibers for speed compete with slow-twitch muscle fibers for endurance in the limited space available (for example, Vanhooydonck, Van Damme, and Aerts 2001). In contrast, some locomotor patterns coevolve together: sprinting and jumping in lizards are both associated with longer hindlimb lengths, for example (Losos 1990).

9.4 Spines and quills

Quills and spines are found in three groups of mammals: monotremes (echidnas), insectivores (hedgehogs and tenrecs), and rodents (New World rats and mice [Cricetidae], Old World rats and mice [Muridae], spiny dormice [Platacanthomyidae], Old World porcupines [Hystricidae], New World porcupines [Erethizontidae], spiny rats [Echimyidae], and cane rats [Thryonomyidae]). They represent a classic example of convergent evolution. Quills and spines of porcupines, hedgehogs, and echidnas have been examined in some detail:

porcupine quills are long and slender and may break off at their ends; hedge-hog and tenrec spines are short with internal septa that prevent buckling; and echidna quills are massive, short, blunt, and thick in cross section and filled with foamlike material (Vincent and Owers 1986). Differences in mechanical design suggest differences in function.

The four genera of New World porcupines are arboreal and respond to predators by climbing trees, freezing, or sometimes retreating into narrow crevices or caves, although they may use their spiny prehensile tail to lash out at predators (W. P. Taylor 1935). Old World porcupines consist of three genera that are nocturnal and mostly terrestrial; some carry few spines, others sport numerous spines and long quills. When alarmed they run to burrows but will run sideways or backward toward an intruder. Porcupines from both families erect their quills and rattle them if a predator approaches closely (Chapman and Roze 1997) and will stamp their feet and bite their teeth together rapidly to produce a "clacking" sound (Jori, Lopez-Bejar, and Houben 1998); they also may emit odor (G. Li, Roze, and Locke 1997). Black, white, and brown bands adorn quills, and the effect of this display likely serves to warn a predator to stay away (aposematism). Contact with a porcupine can result in tens to hundreds of quills becoming lodged in a predator (Shadle 1947), because the tips of short porcupine quills have retrose overlapping scales that face backward (Shadle and Po-Chedley 1949), and erect quills are easily released from the porcupine dermis (Roze 2002). Once embedded, quills move quickly inward through muscle or vital organ but sometimes pass out through the other side of the body later in time. There are many records of lions, tigers, and mountain lions shot by hunters who discovered quills still embedded in their face or paws (for example, Quick 1953).

While one could be forgiven for thinking that porcupine quills are a superb antipredator morphological adaptation, species such as wolverines and bob-cats prey on North American porcupines, and fishers specialize on killing and eating them (R. A. Powell and Brander 1977). Like porcupines, fishers are ar-boreal, but they have a weasel-shaped face and move close to the substrate, al-lowing them to attack the porcupine's face, which carries few quills, and then to flip the porcupine onto its back and attack its ventrum. Taken together with observations that North American porcupines exhibit risk-sensitive foraging by avoiding moonlight, and that juveniles inhabit low-risk areas, these find-ings imply that porcupines are still subject to considerable predation pressure (Alkon and Saltz 1988; Sweitzer and Berger 1992).

On encountering a predator, hedgehogs (five genera) will make for a bur-row. If a refuge is unavailable, they erect their spines and, with further provo-cation, roll into a ball. Special muscles, including the penniculus carnosus and orbicularis, completely cover the dorsal surface of the animal and when contracted keep the body compressed, resulting in an impenetrable exterior of sharp spines. European hedgehogs can remain rolled up for hours and are difficult to tackle in this position. Hedgehogs use their spines offensively against predators by jumping backward or butting the face of the opponent, and also employ hissing, snorting, or even screaming (Reeve 1994). Additionally, spines may be used to run at and harass European adders, causing them to strike again and again until the time when the hedgehog can bite the snake, then kill and eat it (Herter 1965). While hedgehog spines plainly reduce predation, their design suggests that they also act to protect the owner from impact if it falls out of a tree or comes into hard contact with a predator, because the spines buckle elastically under high force. Both observations show that spines are not solely used in defense.

Tenrecs, eight genera of insectivores found exclusively in Madagascar, also sport a variable number of spines and show a very diverse array of antipredator defenses (Macdonald 1984). For example, the streaked tentrec is a small (140 mm long) animal covered with black and yellow quills; vigorous movement causes the quill tips to rub together and produce high-frequency sounds, suggesting aposematism, especially when it is conducted with vocalizing and foot stamping. In the common tenrec, the animal advances toward a predator while gaping, hissing, and head-bucking, which can drive detachable erectile spines that are concentrated around its neck into a predator (Marshall and Eisenberg 1996). Greater and lesser hedgehog tenrecs curl up into a spiny ball when threatened. Yet the ring-tailed mongoose, a Madagascan viverrid, is able to dispatch a streaked tenrec by gingerly pinning the back down or rolling it over, again showing that spines are not entirely effective (Eisenberg and Gould 1970).

The two species of echidna have spines that may be short and stout or long and thin (M. Griffiths 1978). Antipredator behavior consists of rolling into a ball, using powerful claws to dig into soil until only the spines are protruding, clasping root structures if available, or wedging the body into a crevice from which an echidna is almost impossible to dislodge. Indirect evidence indicates that spines are effective against predators. In Western Australia, introduced red foxes and domestic cats have decimated native mammals, but the short-beaked echidna remains relatively unaffected, possibly as a consequence of its spiny defense (Abensperg-Traun 1991).

Across vertebrates, spines are found in a great number of species, including lizards, pufferfish, and sticklebacks. Fish with and without spines show differential risk taking under threat of predation, as armored species occupy more-open habitats and show shorter reactive distances to a predator (Abrahams 1995). Nevertheless, spines and pelvic girdles can reduce the speed at which a fish conducts a rapid escape response, such that fish populations do not possess armor if subject to very high predation pressure (Andraso and Barron 1995; see also T. C. Grand 2000). Unfortunately, there is no parallel work on the balance between antipredator morphology and behavior for any group of mammals that carry spines.

9.5 Dermal plates and thickened skin

Other species of mammals, principally pangolins and armadillos, have thickened dermal plates that appear to provide defense against predatory attack. The pangolins, or scaly anteaters, consist of a single family (Manidae) from the Old World with seven species, some of which live in burrows whereas others are arboreal. Most of the dorsal surface of the body and thick tail is covered in sharp, moveable scales. Defense consists of tail lashing or rolling into a tight ball, the armored limbs and tail protecting the soft underparts. If a predator approaches too close, scales may be stirred and the pangolin may discharge anal secretions, providing further discouragement (Nowak 1999).

There are eight genera and 20 species of armadillo restricted to the New World. Modified skin provides a double-layered covering of bone and horn over the upper surface and sides. The covering consists of bands or plates connected by flexible skin. In addition, the head, limbs, and often tail are shielded by bony rings or plates. Defense includes running into dense vegetation, rapid digging, or diving into a burrow and anchoring firmly there. If overtaken, however, some species draw in their feet so that the edges of the armor plates contact the ground, whereas other species roll into a ball. If provoked, the Southern three-banded armadillo rolls up but then snaps its shell together tightly, discouraging further investigation (Nowak 1999). Despite these defense mechanisms, armadillos constitute a significant proportion of the diet of jaguars (between 8% and 54% occurrence across studies: Nunez, Miller, and Lindzey 2000), showing that dermal plates are not entirely effective in halting predation from large predators.

Other miscellaneous mammals have thickened skin that may be used for deterring attack or preventing sharp teeth and claws from penetrating. Rhinoceroses have thick skin that may be furrowed or pleated, especially evident in the Indian rhinoceros; but other very large herbivores, including elephants and hippopotami, have very thick skins, too. Some sheep have thickened skin around the neck area that may protect them from both lunging strikes of opponents' horns and attacks by predators. Ratels and European badgers have a tough and loose epidermis that is difficult to penetrate (Neal and Cheeseman 1996; Nowak 1999).

Body armature, defined more generally as dermal shields and spines, is found in a subset of mammals of middling body weight (fig. 9.4) ranging from

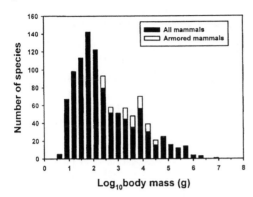

Figure 9.4 The frequency distribution of body sizes of armored (*open bars*) and nonarmored mammals (*filled bars*). The nonarmored mammals ($N = 718$) represent the complete assemblage of mammals from the Neararctic, Australian, and Afrotropical subregion zoogeographic zones. The armored mammals ($N = 70$) represent all known armored mammals from all zoogeographic regions of the world (Lovegrove 2001).

the pink fairy armadillo to the giant armadillo. Very generally, one may surmise that small mammals can rely on crypsis and large mammals on active defense, but intermediate-sized mammals cannot deploy either and must therefore employ novel defense mechanisms. Lovegrove (2001) sees armor as being one of several methods by which middle-sized plantigrade mammals that cannot run as fast as digitigrade or unguligrade mammals reduce predation. Examining surface-dwelling plantigrade species of intermediate body mass or larger (> 358 g), he found that 23.8% of 80 species were armored and that a large proportion of the rest were either arboreal (31.3%), aquatic (5.0%), or rock dwelling (5.0%), possibly reflecting a response to terrestrial predation pressure. Of the terrestrial species (26.3%), over two thirds were neotropical, where forest habitat might allow crypsis to be effective. These speculative arguments are exciting, because they cast light on how predation pressure shapes mammalian lifestyles.

One can view armor as a necessary but heavy form of defense that reduces flight speed, or one can think of slow-moving species affording to carry armor; the direction of causation over evolutionary time is unclear. Whatever the case, armored species have lower basal metabolic rates than nonarmored species due to an absence of selection for fast movement (Lovegrove 2001). From an ecological perspective, inability to run fast may constrain risky behavior and habitat use. For example, among cordylid lizards that have varying degrees of armature, species with longer spines run more slowly and for shorter distances and are therefore forced to remain close to refuges (Losos et al. 2002). In effect, the repeated observation that armor reduces but does not eliminate predation means that armored species still have to employ risk-sensitive behavior. Thus, on the one hand, armor may be viewed as a liberating defense allowing species to forage in habitat that would otherwise be too dangerous, but on the other it can be seen as a constraint limiting other escape possibilities and keeping animals close to a refuge; these alternatives are not mutually exclusive (E. B. McLean and Godin 1989; Abrahams 1995; but see T. C. Grand 2000).

9.6 Weapons used for feeding

Feeding morphologies in carnivorous and frugivorous birds include ample claws, sharp talons, hooked beaks of varying size, and sharp bills, all of which can be used against a predator. For example, raptors defend their nest by diving toward an intruder and veering away at the end or actually striking with the flat of the foot or raking with hind claws (I. Newton 1979). When held in the hand, raptors and parrots can be extremely dangerous, especially larger species. In species of predatory birds where females remain on the nest and defend it, antipredator benefits of body size coupled with weaponry may be one factor underlying reversed sexual dimorphism (M. Andersson and Norberg 1981).

The dentition and sharp claws of many mammals, including rodents, primates, and carnivores, may similarly deter predators, prolong handling time, or increase risk of injury. Rapid body twisting, combined with attempts to bite using teeth and claws, can make handling almost any mammal problematic. Indeed, most mammals bare their teeth under threat as a signal to the attacker, be it heterospecific or conspecific. Selected examples of this sort of mammalian defense include the sharp teeth of shrews and solenodons that, combined with toxic saliva, may act as effective antipredator defense (Edstrom 1992); attempts by Burchell's zebras to chase and bite spotted hyenas (Kruuk 1972); and the large, hooked, sharp claws that sloths and anteaters use to strike out at predators but that are normally used for moving upside down along branches or breaking open termite nests respectively (Nowak 1999). Although it would be absurd to argue that these morphological traits principally function in self-defense, the extent to which sharp teeth and claws are evolutionarily maintained as a consequence of reducing mortality through predator defense is an open question.

Animals may also use other parts of their anatomy unrelated to feeding to defend themselves. Ungulates routinely strike out with their hooves at predators when cornered or standing their ground, mule deer fending off coyotes with their forelegs, for example (Lingle and Pellis 2002), and zebras delivering sharp kicks at spotted hyenas that pursue them too closely (Kruuk 1972).

9.7 Sexually selected weaponry

Considerable effort has been given to understanding the evolution of sexually selected characteristics that are advantageous in intrasexual competition or in mate choice by members of the opposite sex (M. Andersson 1994), but the relationship between these and predation has usually been couched in terms of sexually selected behavior increasing predation risk (for example, Endler 1987; Magnhagen 1991; Jakobsson, Brick, and Kullberg 1995; Sih 1995; Jennions and Petrie 1997; Zuk and Kolluru 1998). Thus, male-male combat or mate guarding may expose males to enhanced predation risk, exposing gaudy ornaments may reduce crypticity, and the act of comparing or staying close to males may increase the time that females are subject to predation. Almost unrecognized are the antipredator benefits of sexually selected traits, which is somewhat surprising given that early work considered some of these traits, such as antlers in deer (Darwin 1871) and manes in lions (Schaller 1972), as having an antipredator function. The principal morphological structures in mammals that are selected through male-male competition and could serve a secondary role in driving predators away or deterring them from attack are deer antlers, antelope and rhinoceros horns, suid and elephant tusks, and enlarged canines in many taxa but most obviously primates and musk deer. In birds, these are spurs (M. Andersson 1994).

Antlers are morphological structures in deer that are used in intraspecific combat (Clutton-Brock 1982; Clutton-Brock, Albon, and Harvey 1980) and, aside from caribou, are found only in males. Antlers are shed and regrown each year; thus, their ability to defend cervids against predatory attack is necessarily limited. Nonetheless, there are records of predators being impaled and dying on the antlers of deer (Nelson and Mech 1985). Use of minerals by the developing embryo may preclude the development of antlers in female cervids, because they have to be grown repeatedly. Female caribou are unusual in having offspring that follow them from birth, and this may tip the balance toward developing antlers as a mechanism of keeping predators at bay (Clutton-Brock, Guinness, and Albon 1982). The relationship between following and hiding and female weaponry in artiodactyls needs closer attention.

Horns in antelopes and pronghorn are permanent bony structures, allowing greater opportunity for use as antipredator devices. While horns in males are regarded as being adapted for intrasexual fighting (M. Andersson 1994), there are numerous, often lurid accounts of large male ungulates, such as buffalo, using their massive horns and frontal shields to toss and gore predators such as lions, spotted hyenas, and human hunters (for example, Selous 1908); but other bovids, including eland and wildebeest, employ these weapons against predators to good effect as well (for example, Kruuk 1972; Estes 1991; Caro 1994a). In some species, horns are found only in males, but in other species both sexes carry them; the adaptive significance of female horns poses a problem, because female-female fighting is rare. While it is possible that female horns result from a correlated response to selection in males, there is more to it than that, as female horns are found only in species weighing more than 20 kg and appear to have evolved independently several times (Packer

Table 9.2 Distribution of female horns in bovid subfamilies and tribes

Subfamily	Tribe	No. of genera	Component genera have female horns?
Bovinae		10	Mixture
	Tragelaphini	3	Mixture
	Boselaphini	2	All hornless
	Bovini	5	All horned
Cephalophinae	Cephalophini	2	Mixture
Hippotraginae		5	Mixture
	Reduncini	2	All hornless
	Hippotragini	3	All horned
Alcelaphinae	Alcelaphini	5	Mixture[a]
Antilopinae		16	Mixture
	Neotragini	8	Mixture[b]
	Antilopini	8	Mixture[c]
Caprinae		11	Mixture
	Rupicaprini	4	All horned
	Ovibovini	2	All horned
	Caprini	5	All horned[d]

Source: Kittie 1985.
[a] Includes impala.
[b] Includes rhebok; one population of klipspringer is reported to be female-horned.
[c] Includes saiga antelope.
[d] Some populations of mouflon are female-hornless.

1983; table 9.2). In addition, female horns are often thinner and straighter than those of males but are of the same length, making them effective in stabbing (Packer 1983; Kiltie 1985; Caro et al. 2003). Noting that there are many anecdotal observations of females defending their offspring against attack by aerial predators using spiked horns, Packer (1983) suggested that female horns may be used as antipredator devices, especially in light of their absence in females of smaller species, such as duikers, that often use crypsis to avoid predators. An alternative hypothesis, that female horns are thinner because of the need to shunt minerals to the fetus or for lactation, seems untenable, because female horns do not break more often than those of males.

In all five species of rhinoceros, females have horns as well as males. Whereas males are known to use their horns in displaying to other males and in sparring, the function of female horns is more enigmatic. Fortuitous data collected during the course of a dehorning operation in Namibia suggest that female black rhinoceros apply their horns to defend offspring against spotted hyenas. Of three females that had been dehorned in an area where spotted hyenas lived, all lost their calves, whereas none of the three dehorned females that lived in an area free of hyenas and none of the four intact females that lived sympatrically with hyenas and occasional lions did (Berger and Cunningham 1994).

The nine wild pig species have well-developed, upturned canines that males use in skirmishing with each other, employing fighting styles ranging from shoving to slashing each other with tusks (Macdonald 1984). Tusks would similarly be effective in protecting animals against predators, because they are very sharp and can be used to great effect when an animal is cornered. Warthogs, for instance, bolt for a burrow and tuck themselves in with their armored head and tusks facing outward, making it difficult for a lion to extract them without injury (Ratner 1975). Elephants have tusks that are used in fights between males (Poole 1989). Males usually spar with their tusks, but in escalated fights a tusk can penetrate a vital organ, causing death. Elephants are known to kill lions and human hunters using a combination of weight and tusks.

In many mammals, males have elongated canine teeth compared with females which they use in displays and fights against male opponents (Packer 1979). Canines are administered to inflict deep wounds in savage fights between males, as in species such as musk deer and hyraxes. In primates, relative canine length is greater in polygynous species than in monogamous species, and additionally in polygynous species that have multimale male as opposed to single-male groupings, paralleling the increased competition that might be expected between males over access to females (Harvey, Kavanagh, and Clutton-Brock 1978). Though canine length is a sexually selected trait in males, in certain circumstances canines are used to deliver deep bites against predators, as, for example, when baboons attempt to drive leopards away (Altmann and Altmann 1970; Cowlishaw 1994).

Several of the suids have toughened head shields, facial warts, and, on their shoulders, thickened skin and matted hair to protect them against opponents. Manes protect males in intraspecific fights in other species, too, including hamadryas baboons, equids, and lions. Mane length is a signal of fighting success and short-term health in lions; other males are more likely to approach short-maned than long-maned models (West and Packer 2002). While lion mane length (and mane color) is a sexually selected trait, it may additionally confer advantages by limiting wounding in the neck region (Bertram 1978b). If so, it is conceivable that a single male attacked by a pack of hyenas could reduce injury with the help of its mane. The evolutionary forces maintaining manes in other mammals are unknown.

In birds, bony spurs are found on the tarsometatarsus in three subfamilies of Phasianidae among Galliformes (table 9.3), whereas alar spurs, extrusions from the carpal joint, are found in diverse waterbirds: Anhimidae, Anatidae, Jacanidae, Charadridae, and Chionidae (Davidson 1985). Across species, spurs may be single or multiple and vary in shape from a blunt knob to a sharp-tipped sword. Of 113 species measured by Davidson, 72 had one spur on each leg in males but none in females, and 25 had multiple spurs in males and none

Table 9.3 Number of spurred species in various families and subfamilies of Galliformes

	Total species	Spurred species	% spurred
Megapodiidae	12	0	0
Cracidae	44	0	0
Tetraonidae	18	0	0
Phasianidae			
Odontophorinae	33	0	0
Phasianinae	152	108	71.1
Numidinae	7	3	42.9
Meleagridinae	2	2	100.0
Total	268	113	42.2

Source: Davidson 1985.

in females, whereas 16 species had single or multiple spurs per leg in both sexes. Spur length increases with avian body size. Observations (at cockfights, for example!) show that males try to wound opponents by jumping into the air and thrusting their pointed spurs forward to penetrate the skin or to deliver a stunning blow. Spurs' predominance in males, their preponderance in polygamous as opposed to monogamous species (Davidson 1985; but see M. S. Sullivan and Hillgarth 1993), and observations of fighting behavior indicate that they are sexually selected. They can, however, be used to similar devastating effect in response to predatory attack by means of flying leaps and sharp downward leg thrusts toward a predator's face and eyes. Antipredator defense could potentially be involved in maintaining spurs in females of some species as well as secondarily promoting their appearance in males.

Duck-billed platypuses have elongated keratinous hollow spurs on their forefeet that are highly developed in adult males but not in females. Ducts transport venom from the crural gland to the tip of the spur that can then be injected into other animals. Spurs are presumed to be used in fights between males over females because of platypuses' sexual dimorphism and seasonal variation in glandular activity associated with breeding, but they may additionally be used in defense against predators. Venom injected intravenously into mice causes convulsions, paralysis, and death and has similar effects to that of shrew salivary venom. Envenomation causes agonizing pain to humans and can kill a dog, so it could presumably be used in defense against marsupial and placental carnivores. In other monotremes, the echidna species, males have vestigial spurs that do not appear to be functional (Burrell 1927; M. Griffiths 1978).

Finally, secondary sexual ornaments involved in female choice could double as defensive devices against predators. For example, there has been controversy over whether spurs are also an intersexually selected characteristic upon which females choose their mate (von Schantz et al. 1989; Badyaev et al. 1998). Spurs, ruffs, and manes or long tail feathers might be used to dull a full-blown predatory attack (as when a fox is left with a mouthful of chicken feathers), but as yet there is no firm evidence on this point.

Despite these arguments for sexually selected traits conferring antipredator benefits on their bearers, no data have been collected on whether predation rates are lower on males than on females in these species as a result of heightened physical defense.

9.8 Malodor and unpalatability

Mammals frequently defecate and urinate when attacked, and the novel or repugnant odor may play a role in deterring assault. For example, the pentail shrew rolls onto its back, gapes to bare its sharp teeth, hisses, and defecates and urinates when disturbed (E. Gould 1978). Some mammals produce odiferous secretions continuously from skin or from anal glands that can be detected at a distance (R. E. Brown and Macdonald 1985). In general, these secretions are thought to be involved in intraspecific communication; but they could additionally advertise that prey is distasteful or dangerous and thus constitute an aposematic signal, or they could directly repel a predator by inducing vomiting or pain or both. While to the human nose there is considerable variation in the type, strength, and noxiousness of mammalian scent, there are very few data available on predators' responses to the odors of potential prey, making it difficult to assign the production of mammalian scents to different functional categories.

Some of the strongest-smelling mammals include species of shrews that produce a repulsive odor from their flank and ventral glands that may advertise their sharp bite and venomous saliva or additionally make them unpalatable (Churchfield 1990). Shrews are unpalatable to most mammalian predators, including foxes and cats (Macdonald 1977), although they form a small part of the diet of weasels and stoats (Eadie 1938). Shrews constitute a larger proportion of the diet of avian predators that have a poorer sense of smell; common shrews make up 5% of the diet of tawny owls, for instance (Southern 1954).

Carnivores, particularly mustelids and viverrids, produce strong-smelling secretions from a variety of glands, including anal pouches, auxillary anal glands, anal sacs, and subcaudal glands (Macdonald 1985). There is a close association between noxious secretions and prominent black and white markings (Ortolani and Caro 1996) in mustelids, but it is not clear whether the coloration signals distastefulness as a consequence of the whole animal smelling unpleasant, or whether coloration and smell are signals of difficulty in subduing the carnivore. Two anecdotes speak to the latter idea, as European badgers were a historical delicacy in Britain and, along with many other mustelids, have an extremely powerful bite; their jaws are notoriously difficult to prize apart when

Table 9.4 Composition of major volatile components of anal sac secretion from 4 species of North American skunk

Compound	Amount (%)			
	Hooded skunk	Striped skunk	Spotted skunk	Hog-nosed skunk
(E)-2-Butene-1-thiol	32	38–40	30–36	71
3-Methyl-1-butanethiol	39	18–26	48–66	
S-(E)-2-Butenyl thioacetate	16	12–18		17
S-3-Methylbutanyl thioacetate	7	2–3		
Phenylmethanethiol	0.3		0.2–0.8	1
2-Phenylethanethiol	1.4		2–5	
2-Methylquinoline	2.3	4–11	0.3–0.9	2
Bis[(E)-2-butenyl] disulphide	trace		trace	3
S-Phenylmethyl thioacetate	trace			
(E)-2-Butenyl 3-methlybutyl disulphide	0.5	0.2–1.6	0.2–0.6	
Bis(3-methylbutyl) disulphide	trace		0.1–0.2	
S-2-Phenylethyl thioacetate	0.2			
2-Quinolinemethanethiol	1.3	4–12	0.2–0.3	0.5
S-2-Quinolinemethyl thioacetate	trace	1–4		

Source: Wood et al. 2002.
Note: Compounds less than 1% were not identified.

clamped to an aggressor. A badger lodged in the opening of its sett and showing its black and white face is a formidable prey item (Neal and Cheeseman 1996). These suggest that malodor is a signal of pugnacity rather than distastefulness. The greater moonrat may present a parallel but even less understood illustration of the same phenomenon. It gives off a pungent odor of rotten garlic, is strikingly colored black and white or off-yellow, and has very sharp teeth. When cornered, it delivers a severe bite characterized by a sudden forceful closure followed by circular twisting movements of the head (Gould 1978).

The most famous example of malodor in mammals, the anal sac secretion of skunks, consists of volatile components (table 9.4) that are sprayed accurately at predators that approach to within 5 m but can hit a target at 30 m (Lariviere and Messier 1996). The spray induces vomiting and groaning in some domestic dogs that attempt to remove it by rubbing their head and shoulders on the ground. When sprayed at a person's eyes, it produces a burning sensation and tears (Cuyler 1924). Like the European badger, skunk flesh is palatable if not contaminated by musk, but skunks do not have a reputation as being ferocious, so it seems that the spray itself acts to deter predators as opposed to advertising that an unpalatable meal or dangerous prey lies ahead. In short, and on the basis of very little data, malodor appears to serve several functions in mammals, including signaling unpalatability and/or danger and

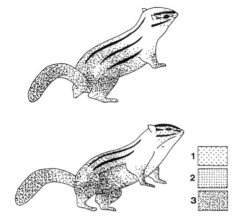

Figure 9.5 The body areas to which the application movements were directed during 113 bouts of snake-scent application behavior in Siberian chipmunks. 1 = single application; 2 = double application; 3 = quadruple application (Kobayashi and Watanabe 1986).

acting as a deterrent in its own right (reliable indices of quality: Maynard Smith and Harper 2003).

Eighty genera of birds from 17 orders are reported as malodorous to humans (Weldon and Rappole 1997). For example, certain bird taxa, including fulmars and petrels, smell and taste unpleasant because they smell of fish oil or whale excrement at least to some (but not all) people (for example, J. Fisher 1952). It is not clear if these species are repulsive to mammalian and avian predators as well, but if they are, naïve individuals would probably have to consume some individuals before learning to avoid them. Thus, it is surprising that fulmars and petrels are gray and white rather than aposematic; perhaps they cannot afford bright coloration when fishing. Dull-colored waterbuck are reputed to be distasteful, too (Estes 1991).

A rather intriguing case of malodor in birds is found in eider ducks, northern shovelers, and a few other Anatidae. When surprised from their nest, these birds defecate on their eggs. In choice tests, both ferrets and Norway rats refused food soiled with very small quantities of such excrement, whereas they would eat food soiled with eider duck feces outside the breeding season. The repellant appears therefore to be specifically adapted for nest defense (Swennen 1968). Common waxbills add carnivore scat to their nest throughout incubation and nestling periods, and experiments show greater artificial nest survival with addition of scat (Schuetz 2005).

Some species of mammals that are palatable anoint themselves with strong-smelling substances that they chew first and then daub on their body with saliva. Self-anointing may make an animal smell different and elicit a fear reaction in potential predators (Weldon 2004). Siberian chipmunks will chew snake skin and then lick their body or apply snake feces or snake urine–soaked sawdust to parts of their body (fig. 9.5). This may deter snakes from killing them. Alternatively, or additionally, self-anointment may make the animal distasteful.

Some evidence suggests that rat snakes avoid eating house mice that have had snake-derived substances applied to them (Kobayashi and Watanabe 1986; see also Xu et al. 1995). Mammals that are already armored, such as tenrec and hedgehog species, chew toxic substances, including toad skin, and then wash their spines with saliva (E. D. Brodie Jr. 1977; Holst 1985). While the function of this behavior is not well established among different species and may not necessarily be related to antipredator behavior, the relative importance of choosing volatile substances that falsely advertise the presence of another species, selecting a malodorous substance, or choosing a toxic substance with which to cover the body probably depends on alternative prey defenses, palatability, and type of predator involved.

A handful of mammals secrete white fluid from glands surrounding their eyelids when disturbed. These include tenrecs (Poduschka 1974), pacaranas (L. R. Collins and Eisenberg 1980), and mountain beavers (Hackmann, Zamora, and Stauber 1990), but its function is unknown.

9.9 Venom resistance

Several species of small mammals are resistant to the venom of sympatric snakes in an analogous way to which venomous and nonvenomous snakes are resistant to snake venoms (Perez et al. 1978). These mammals include Virginia possums (Kilmon 1976; R. M. Werner and Vick 1977), European hedgehogs (Herter 1965; de Wit and Westrom 1987), prairie voles (de Wit 1982), gray wood rats (de Wit 1982), hispid cotton rats (Perez, Pichyangkul, and Garcia 1979), and Egyptian and Indian gray mongooses (H. E. Hinton 1967); indeed, 16 out of 40 species of homeothermic mammals tested had an antihemorrhagic factor in their sera (Perez et al. 1978). Venom resistance is quite specific to different snake species. For example, Virginia possums survive venom of eastern diamondback rattlesnakes and copperheads but not Indian cobras or puff adders (Werner and Vick 1977). Some of these mammals that have antihemorrhagic components in their sera, including the striped skunk, European hedgehog, or common opossum, kill and eat venomous snakes, but most are simply prey of snakes, indicating that resistance is usually a mechanism for surviving predatory strikes rather than coping with a counterattack during a predation attempt on the snake. Some of these species can survive multiple strikes from a venomous snake (Coss and Owings 1985).

Degree of venom resistance varies across geographic ranges of some of these mammals (Perez et al. 1978), and the pattern is tied closely to the presence of snakes in the prey's habitat (see also E. D. Brodie Jr., Ridenhour, and Brodie 2002). In a series of studies on ground squirrels, venomous Pacific rattlesnakes, and nonvenomous gopher snakes, Coss and his colleagues explored the geographic mosaic of venom resistance and its correlates, the develop-

Table 9.5 California ground squirrel (*Spermophilus beecheyi* [*S.b.*]) populations and corresponding rattlesnake densities shown with respect to concomitant RIA serum titer magnitudes

Population	County or region	Subspecies	Rattlesnake density scale	Serum titer magnitude*
Folsom	El Dorado	*S. b. fisheri*	3 (high)	2.03
Sunol	Alameda	*S. b. beecheyi*	3 (high)	1.47
Walnut Creek	Contra Costa	*S. b. beecheyi*	3 (high)	1.58
Chico	Butte	*S. b. douglasii*	2 (moderate)	0.73
Woodacre	Marin	*S. b. douglasii*	2 (moderate)	1.28
Richmond	Contra Costa	*S. b. beecheyi*	2 (moderate)	1.21
Davis	Yolo	*S. b. douglasii*	1 (rare)	1.51
Tracy	San Joaquin	*S. b. fisheri*	1 (rare)	1.37
Petaluma	Sonoma	*S. b. douglasii*	1 (rare)	1.13
Vallejo	Solano	*S. b. douglasii*	1 (rare)	0.71
Holland Tract	Solano delta	*S. b. fisheri*	1 (absent)	0.92
Ryer Island	Solano delta	*S. b. fisheri*	1 (absent)	0.78
Mt. Shasta	Siskiyou	*S. b. douglasii*	1 (absent)	0.94
Lake Tahoe	El Dorado	*S. b. sierrae*	1 (absent)	1.15
Arctic	Mt. McKinley	*S. p. ablusus*	1 (absent)	1.00

Source: Poran, Coss, and Benjamini 1987.
*Titer magnitude = ratio of the average of 1:250 and 1:1000 serum dilutions of each California ground squirrel population relative to the arctic population serum dilutions.

ment of resistance in ground squirrel pups, and antipredator behavior patterns associated with different degrees of venom resistance (Coss and Owings 1985; Coss 1999 for reviews).

In the Central Valley of California, USA, rattlesnakes are found in the surrounding foothills of the Sierra Nevada and Coast Range but not on the floor of the valley, which was subject to repeated flooding until recently. Comparisons of 14 populations of California ground squirrels using in vitro squirrel serum–venom binding as quantified by radio immunoassay showed that populations living among high rattlesnake densities had higher serum titers than those from areas where rattlesnakes were rare or absent (table 9.5), and that sera from arctic ground squirrels from central Alaska, which is free of snakes, were 3.5–5.3 times less effective in neutralizing venom than sera from California ground squirrels. These relationships indicate that intraspecific variation is probably a consequence of differential natural selection due to predation pressure from Pacific rattlesnakes (Poran, Coss, and Benjamini 1987).

Young ground squirrels are born without venom resistance but develop it rapidly between 14 and 30 days of age, 2 weeks before they emerge from the burrow. Nonetheless, they are still very susceptible to snake predation aboveground, because they have a small body mass, lack muscular strength to harass and thereby halt a snake's advance, and so are more likely to be envenomated than an adult. As a consequence, the average size of ground squirrels usually taken by rattlesnakes is 20%–50% that of adult weight (Fitch 1949; Poran and Coss 1990). In areas with large snake populations, 9-week-old ground squirrels can differentiate rattlesnakes from gopher snakes and show adultlike bold behavior in the face of snakes. In a Central Valley population that has not experienced rattlesnakes for 900 years (Davis, California) and shows low venom resistance, young laboratory-born ground squirrels throw more substrate and tail flag more at a gopher snake than at a rattlesnake than do young from a Coast Range population (Winters, California) where rattlesnakes are prevalent and resistance is high. Absence of rattlesnake pressure for only a relatively short time seemed therefore to have lowered inhibition in harassing gopher snakes (Coss et al. 1993).

Population differences in resistance to rattlesnake venom are reflected in population differences in behavior. Arctic ground squirrels that have been isolated from snakes for 3 million years and have no venom resistance do not treat rattlesnakes as being dangerous. Thus, belowground, arctic ground squirrels are less likely to throw substrate at, approach, or harass a gopher snake than California ground squirrels; aboveground, they are more willing to keep a close distance to the snake (Goldthwaite, Coss, and Owings 1990). When two populations of California ground squirrels that diverged 228,000 years ago — one that experiences very high snake densities and has high resistance (Folsom Lake, California) and the other that encounters few snakes and has lower resistance (Sierra Valley, California) — were compared, however, the Sierra Valley animals built more plugs in their burrow and were less likely to severely harass (bite) rattlesnakes in keeping with their low venom-neutralizing capabilities (Towers and Coss 1990). This cautious approach following relatively recent loss of resistance contrasts with the foolhardy approach of arctic ground squirrels that lost resistance much longer ago (section 1.6). More generally, these findings caution against assuming that changes in resistance are paralleled by changes in behavioral response.

9.10 Life history characteristics

Life history evolution is tied closely to mortality risk of both adults and young (Cole 1954; Law 1979; Reznick, Bryga, and Endler 1990; Charnov 1993). To date, most comparative analyses that attempt to relate life history variables to mortality schedules do not distinguish between sources of intrinsic mortality, such as costs of reproduction, and extrinsic sources, such as climate or predation; nor do they differentiate sources of extrinsic mortality from one another. That

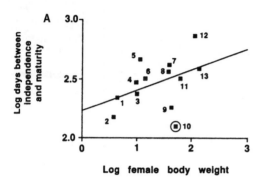

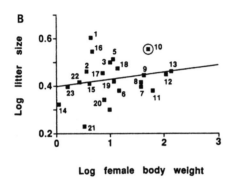

Figure 9.6 A, $\log_{10}$ number of days between independence and maturity, and B, $\log_{10}$ average litter size plotted against $\log_{10}$ female body weight in kilograms for felids. Species number 10 (*circled*) is the cheetah (Caro 1994a).

said, mammalian species with high rates of juvenile mortality for their body weight are weaned early, reach maturity early, and have relatively short gestation lengths, large litters, and small young. Species with high adult mortality simply reach maturity early (Promislow and Harvey 1990). Rapid onset of reproductive activity is necessary for species with short life expectancy, while offspring variables are probably geared to coping with stochastic juvenile mortality, such that a large litter of very small neonates will result in high fitness in good years, but will minimize maternal losses in bad years.

Within selected mammalian taxa, the same pattern holds. Thus, among felids, cheetahs have extremely high juvenile mortality, at least in the Serengeti Plains (95% mortality before independence. of which 73% stems from predation), greater than any other nonhunted felid. Cheetahs reach maturity earlier, as judged by the length of time between independence and first breeding, and they have larger litters than other felids for their body weight (fig. 9.6). Across primates, where phylogenetic controls have been used, high rates of mortality are associated with early age of first reproduction and high birthrates; specifically, there is a negative correlation between mortality prior to reproduction and length of the juvenile (prereproductive) period, and a positive correlation between adult mortality and birthrate (Ross and Jones 1999).

In birds, clutch size has classically been related to food limitation: the hypothesis formulated by Lack (1968) states that the number of eggs per clutch represents an optimum between offspring number and offspring survival. Life history theory, however, predicts that clutch and litter size are related to mortality risk (Harvey, Promislow, and Read 1989; Promislow and Harvey 1990). Nowadays, clutch size is seen to be molded by both food availability and predation on eggs and nestlings, with clutch size being lower when predation rate is higher (Lima 1987c), not only because smaller clutches of nestlings might attract less attention from predators but because it may be possible to complete them faster (Ricklefs 1984). Indeed, high predation rate may select for an increasing number of nesting attempts per season irrespective of whether previous attempts were successful (Martin 1992b).

Variation in developmental rates can be also partially explained by mortality schedules in both birds and mammals (Lack 1968; Ricklefs 1969a; Case 1978; Promislow and Harvey 1990). In birds that occupy predator-free islands of Hawaii, USA, and New Zealand, for example, nesting periods are longer than species on adjacent predator-rich continents of North America and Australia. Furthermore, incubation and nestling periods of island species that nest in exposed sites, but not those that nest in holes, are longer than related species from continental areas. In contrast, bird species introduced less than 170 years ago have similar incubation and nestling periods to their country of origin. Since developmental rates are associated with developmental periods, it seems that there has been selection to speed up development where risk of nest predation is high (Bosque and Bosque 1995; see also T. E. Martin 1995). Directly relating life history changes to age-specific patterns of predation is complex, because interactions can be expected between variables (Martin 1992b), and there is still some work to do in relation to both classes of homeotherm.

Although this chapter could be construed as a ragbag of antipredator defenses linked only by the fact that they are not behavioral, they represent a class of defenses that are brought to bear toward the end of the predatory sequence. They are also defenses that some predators may perceive immediately as being unprofitable. A predator noting a particular species of bird may have already learned that it will be too fast or too dangerous to subdue. This is easy to visualize when thinking of an experienced lion watching a Cape porcupine amble by, but the principle may pervade decisions that predators make about many other prey species. Unfortunately, to date, most of the defenses discussed in this chapter have been treated as isolated, unusual phenomena outside the mainstream, but they deserve to be integrated into the antipredator defense literature.

9.11 Summary

Larger prey species are attacked by larger predator species, so increased body size might reduce the number of potential predator species and the number of predatory attempts on larger species. Particularly large mammals are unlikely to be immune from predatory attack, however. Animals may attempt to make themselves appear larger by erecting crests and dorsal capes. In mammals, flight speed increases with body size up to 119 kg but then declines; the relationship between intraspecific variation in body size and flight speed is similarly equivocal. Young animals that are necessarily smaller than adults face a number of antipredator challenges, including weak acceleration and being easier to handle. Larger animals may be easier to detect than smaller ones, reducing the advantages of body size in some instances.

Wing and leg morphologies are associated with different sorts of locomotion and hence forms of predator escape in birds and mammals. Quadrupedal mammals generally rely on fast flight to escape predators, whereas bipedal mammals employ sharp turns and jump clear of their attackers.

Echidnas and certain species of insectivore and rodent have quills or spines that are used as protection against predators. Porcupine quills are aposematic in coloration and become embedded easily in an attacker; hedgehogs roll into a defensive ball; tenrecs use a variety of defenses, including release of quills and curling up; and echidnas wedge themselves into crevices or dig into soil. Pangolins and armadillos carry body armor, namely thickened dermal plates, whereas some species, such as rhinoceroses, have thickened skin. Dermal shields and spines are found in middle-sized mammals that cannot rely on crypsis or large body size to reduce predation. Body armature may allow species to inhabit dangerous habitats but may also force them to stay close to refuges because they are slow at flight.

Weapons used in prey capture, teeth and talons, are employed to fight off predators. Sexually selected traits, such as antlers in deer, tusks in suids and elephants, canines in primates, and spurs in birds, may also be employed in antipredator defense. Horns in female antelopes and rhinoceroses can be used to drive predators away.

Mammals sometimes defecate and urinate when attacked, whereas some are malodorous, such as skunks. Malodor may signal unpalatability or pugnacity, or may act as a predator deterrent in its own right. Some mammals, such as European hedgehogs, chew strong-smelling substances and then daub their fur with saliva, a possible defense mechanism.

Certain mammals are resistant to the venom of particular snake species. Extent of venom resistance is tied to the length of time that venomous snakes have been present, as demonstrated in California ground squirrels and Pacific rattlesnakes. Relatively recent loss of resistance over evolutionary time is associated with cautious behavior to snakes, but loss of resistance far back in time is associated with reckless behavior.

Mammals with high juvenile mortality show early age of maturity but larger litters and smaller litter weights in order, respectively, to escape the stage of high predation risk, to maximize fitness if their litter escapes predation, and to minimize losses if their litter is predated. Clutch size is lower and developmental rates are higher under high predation risk in birds.

10 Nest Defense

10.1 Introduction

If a predator achieves close proximity to its prey, and prey cannot or will not flee, the prey is forced to confront the danger by mounting direct defense. For convenience, these can be divided into cases in which an individual animal, often a parent, defends itself or its offspring, and cases in which groups of individuals defend themselves. These are treated separately in this and the next chapter, respectively.

Nest predation is the principal cause of nesting mortality for many species of birds (Ricklefs 1969b; Nilsson 1984; T. E. Martin 1993a), particularly passerines (T. E. Martin and Roper 1988; Holway 1991; Filliater, Breitwisch, and Nealen 1994), and can reduce annual (T. E. Martin 1993b) and lifetime reproductive success (Wiklund 1995) substantially. For example, nest predation accounts for 25.4% of the variance in lifetime reproductive success in merlins, surpassed only by breeding lifespan. An average male merlin, itself a predator, may experience nest predation with a probability of 30%; indeed, in one study, all males that bred more than four times lost a nest to predators at least once (Wiklund 1996). It is hardly surprising, therefore, that birds have evolved many types of behavior to defend their nest against predators.

Nest defense is defined as

> behavior that decreases the probability that a predator will harm the contents of the nest (eggs or chicks) while simultaneously increasing the probability of injury or death to the parent. (Montgomerie and Weatherhead 1988, p. 170)

Figure 10.0 (*Facing page*) Andean flamingoes nest on open salt pans that provide no cover for an approaching predator. These birds nest in huge colonies, reaping a variety of group-related benefits (reproduced by kind permission of Sheila Girling).

Nest defense has been studied principally in birds (and has been defined in relation to this group), but the evolutionary forces shaping nest defense apply equally well to other taxa where parents defend offspring, including fish and mammals (Clutton-Brock 1991).

The driving force behind interest in nest defense is that it is seen as a central component of parental investment which weighs long-term survival of the parent, and hence its ability to reproduce again, against survival of the current brood, exactly as Trivers envisaged parental investment in his classic 1972 paper. Nest defense, therefore, has been used, along with chick provisioning and nestling begging, as a tool to study predictions that arise out of parental investment theory, although the two activities are not necessarily congruent (Rytkonen et al. 1995), and discussion of nest defense has sometimes been divorced from direct fitness payoff considerations. The first part of this chapter focuses directly on costs and benefits of nest defense, whereas the second part explores aspects of nest defense decision making based on parental investment.

Many different types of behavior are considered under the rubric of nest defense, and several methods are used to elicit and record them quantitatively. Behavior patterns range from those that are relatively specific, shown only when birds are nesting and a predator is approaching, such as the broken wing display, to behavior seen in a wider variety of contexts, such as calling, approaching, scolding, dive bombing, and physically attacking the predator (section 10.2). Distraction displays are particularly interesting, because they dupe predators into following an apparently injured adult bird away from the nest site (section 10.3).

Surprisingly, the costs of nest defense for parents are sometimes based on arguments from design or anecdotes, because few studies have quantified predation attempts on defending adults in the wild. Defense may also be costly for offspring, as it can attract a predator's attention to the nest site (section 10.4). That offspring survival is enhanced as a result of nest defense is well established for some species of birds. Two mechanisms seem to be involved, driving predators away and silencing offspring (section 10.5), but more generally, the mechanisms by which nest defense improves offspring survival through their effects on predators' behavior have received scant attention.

In a thoughtful and influential review, Montgomerie and Weatherhead (1988) enumerated parental and offspring characteristics that might affect the intensity of nest defense based on evolutionary considerations, and I have used the structure of their review as a basis for organizing the second part of the chapter. Most obviously, nest defense may be influenced by the type of predator involved, notably by the risk that it imposes on eggs or fledglings, the risk that it imposes on parents, and the ability of parents to drive it away (section 10.6). As nest defense constitutes an important component of parental investment, the benefits of protecting current offspring must be weighed against a parent's future reproduction. Consequently, the probability of parents being able to renest within a breeding season or during their lifetime will affect number of offspring produced in the future and should influence defense of the current brood (section 10.7). Parental sex should also influence nest defense, because risk of counterattack by the predator, availability of future partners in a renesting attempt, life history characteristics, confidence of parenthood, and ability to raise offspring alone may all vary by sex (section 10.8). Joint defense by parents could influence costs and benefits of individual parental defense as well (section 10.9).

Offspring characteristics include age of offspring, because costs of replacing offspring, offspring reproductive value, and offspring vulnerability all change with age (section 10.10). In addition, brood size (section 10.11) and offspring condition (section 10.12) should influence parental defense, because they are related to the reproductive value of the brood. A supplementary hypothesis to that of offspring condition states that offspring will be harmed through lack of warmth necessary for incubation or thermoregulation; or through lack of provisioning with food if parents are actively defending their nest (section 10.12). Unfortunately, many of these parental and offspring factors can equally well explain the consistent rise in nest defense between incubation and hatching and between hatching and fledging seen in so many species, making it difficult to distinguish between them. Even now, there are relatively few studies that have attempted to account for confounding variables in explaining changes in nest defense over the breeding cycle.

In contrast with studies of nest defense in birds, studies of maternal defense of offspring in mammals against predators (as opposed to conspecifics) are far more limited and focus principally on the benefits, namely driving predators away (section 10.13).

10.2 Scope of nest defense activities

Birds react to the approach of a predator in seemingly disparate ways. These range from sitting "tight" until the predator is very close; flying away from the nest ("flushing"), sometimes explosively; remaining near the nest but performing specific behaviors, such as holding the wings vertically, or appearing to have a broken wing; giving calls from a perch nearby; repeatedly flying away and resettling as the predator approaches; hovering over the predator; and diving and striking the predator about the face (Simmons 1955). An individual bird may perform only one or some of these actions depending on numerous factors, including its species (for example, Kruuk 1964; Weidinger 2002), its sex (Wiklund and Stigh 1983), how well the nest is concealed (for example, Cresswell 1997a; Murphy, Cummings, and Palmer 1997), and type of predator involved (for example, Weidinger 2002; Kleindorfer, Fessl, and Hoi 2005). At a restricted set of distances, when the predator is not too close and not too far from the nest, the decision as to whether a parent attempts to remain cryptic or performs conspicuous actions must depend in part on the way in which the predator locates a nest (Gottfried and Thompson 1978). If it uses visual cues, including observing the parent leaving the nest, sitting tight may be the best option, but if it uses olfactory cues or the cries of begging young, loud calling and hopping between nearby perches may be in order. If the predator discovers the nest, the parent may start to dive and strike the predator. In addition, each category of behavior shows great variation (Walters 1990). As illustrations, birds may leave their nest rapidly, or slowly and close to the ground (for example, Westmoreland 1989), or they may utter different calls that apparently serve different functions (for example, East 1981; Wiklund and Stigh 1983). Usually, intensities of different facets of predator defense are highly correlated within individuals (Curio and Regelmann 1985; Redondo and Carranza 1989; Pavel et al. 2000; but see Gunness and Weatherhead 2002).

While Montgomerie and Weatherhead's definition of avian nest defense is the most accepted and cited, it is formulated strictly in cost-benefit terms, and it is difficult to demonstrate a cost for some behaviors that seem to fall under the rubric of nest defense. For example, the definition excludes some parental activities that are elicited when the nest is in danger, such as flushing, as these likely carry no survivorship cost for the parent, although the negative correlation between flushing distance and the extent to which nests are concealed suggests that flushing is designed to minimize the likelihood of predators locating the nest (Burhans and Thompson 2001). Similarly, birds may be slower to reenter their nest and arrive using an indirect approach in the presence of a predator (Wheelwright, Lawler, and Weinstein 1997).

Among precocial species, whose young leave the nest soon after hatching, parents also show a diverse suite of specific behaviors if predators approach. For instance, grouse and wader hens give warning calls and then freeze if avian predators fly overhead; the mother's call induces chicks to stay immobile until they receive a signal to join her, or it prompts offspring to scatter. In response to mammalian predators, mothers may scold or circle them or show distraction displays (Simmons 1955). Parents of precocial species may actively defend the brood, just like altricial species; they may also move their brood repeatedly (for example, Robel 1969; Hilden and Vuolanto 1972). Sonerud (1985a) argues that it pays parents of precocial species to move their brood under threat from aerial predators, because predatory birds return to a site where they have been successful earlier, but movement may be less important when responding to mammalian predators that rely more on olfactory cues.

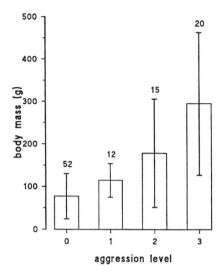

Figure 10.1 Mean body mass for waders grouped by level of aggressive nest defense. 0 = no reports on aggressive nest defense; 1 = usually does not attack predators, but exceptions have been reported; 2 = attacks only reported from the chick period, or unspecified; 3 = attacks reported also from the incubation period. Standard deviation and number of species given for each category (Larsen, Sordahl, and Byrkjedal 1996).

Nonetheless, moving young may be beneficial for escaping mammalian predators as well. For example, cheetah mothers carry cubs to new lairs before cubs are mobile, perhaps because their odor attracts larger carnivores (Laurenson 1993).

The distribution of nest defense across species has been investigated systematically in one group, Holarctic waders (Larsen, Sordahl, and Byrkjedal 1996). Waders experience very heavy nest predation, but against predators only some species defend their broods aggressively, defined in their review as an aerial attack whose function is to alter the direction of an advancing predator by posing a threat to it. Recognizing four levels of aggressive nest defense, Larsen and colleagues categorized 111 wader species and matched them to 10 morphological, behavioral, and life history variables. They found that aggressive nest defense was performed mainly by larger species, genera, and families (fig. 10.1; see also Gochfeld 1984) and by species and genera with biparental rather than uniparental care. Large waders are undoubtedly able to drive predators away better than small waders, because they pose a greater risk to the predator, or because they are in less danger from the predator; the critical threshold for active nest defense appears to be 100 g–150 g (Larsen 1991). The reason biparental care, or more specifically the presence of the off-duty parent on the territory, is linked to aggressive nest defense is because the commotion caused by two parents makes it less likely that the predator's attention will be drawn to the nest when the incubating bird leaves to defend it. Indeed, aggressive nest defense may select for monogamy (where two parents are present) and may explain the low frequency of polygyny (where only the female may be present) in large but not in small wader species (Larsen 1991). Three additional factors had a minor effect on promoting aggressive nest defense: nests being more detectable, birds being colonial, and lack of alternative (rodent) prey for predators, all of which might put nests more at risk.

10.2.a The study of nest defense   Nest defense is usually studied experimentally by recording birds' behavior while a human observer approaches the nest; after placing a mounted specimen of a predator near the nest; or, more indirectly, by placing eggs in artificial or used nests and documenting their loss, and then comparing these with natural nests, thereby uncovering the effects of parental activity. Each method has various problems associated with it that have led to debate about the validity of some of the conclusions arising from studies of nest defense (see chapter 3).

Use of human observers is potentially problematic, because birds may react in different ways to humans than to natural predators (Simmons 1952). In addition, nest defense is more intense if the observer looks at the nest rather than stares at the bird, and when the observer is familiar to the parents (table 10.1). When mounted predator specimens are used other potential problems arise because they elicit more vigorous defense than caged live specimens (table 10.1). All these issues potentially confound comparisons between stud-

Table 10.1 Male red-winged blackbird responses (means) to (a), an observer staring in different directions; (b), familiar and novel humans; and (c), a mounted and live crow

Response variables			p-value
(a)	Individual staring at bird	Individual staring at nest	
Hovers in 3 minutes	6.9	13.6	<0.005
Dives in 3 minutes	6.8	19.3	<0.001
Strikes in 3 minutes	0	6.9	<0.001
Nearest distance (m)	3.2	0.9	<0.001
(b)	Novel individual	Familiar individual	
Hovers in 3 minutes	4.8	10.9	<0.005
Dives in 3 minutes	4.1	12.2	<0.005
Strikes in 3 minutes	0.9	5.9	<0.005
Nearest distance (m)	6.3	1.6	<0.005
(c)	Mounted crow	Live crow	
Hovers in 3 minutes	14.6	8.1	<0.05
Dives in 3 minutes	46.0	33.8	<0.01
Strikes in 3 minutes	39.6	10.3	<0.001
Nearest distance (m)	0	0	—

Source: R. L. Knight and Temple 1986a.

ies. Moreover, none of these experimental techniques illuminate the means by which defense under natural conditions dissuades predators from taking nest contents, because real predators are not involved.

Artificial nest experiments are used in studies of nest defense in order to tease out effects of parental activity on egg survival. Quail, zebra finch, or artificial clay or plasticine eggs are placed either in artificial nests that are laid out by the researcher in habitats of interest, or in artificial nests concealed in the same vegetation as natural nests. Alternatively, artificial eggs are placed in used nests after offspring have fledged and parents have departed. Subsequent comparison with the fate of eggs from natural nests can, it is argued, shed light on the effects of parental defense during the incubation stage. Depending on the realism of both nests and eggs, however, such experiments suffer from many drawbacks (for example, R. E. Major and Kendal 1996; Butler and Rotella 1998; G. R. Wilson, Brittingham, and Goodrich 1998; Buler and Hamilton 2000; Davison and Bollinger 2000; Rangen, Clark, and Hobson 2000; Weidinger 2001; Zanette 2002; Burke et al. 2004). For example, visually orienting predators may locate wicker nests more easily than real nests, and be more likely to see eggs without an adult sitting. Large eggs may preclude predation by small rodents. "Laying date," clutch size, egg color, and egg and nest odor may additionally influence predation risk on artificial devices. The principal problem is that artificial nests suffer higher rates of predation than natural nests in most studies, and the latter are more susceptible to snakes, whereas artificial nests are more susceptible to avian predators in particular. In a compelling study in which natural and artificial nests were videotaped, three species of snake were principally responsible for predation on the former, whereas raccoons and a single American crow predated most of the artificial nests despite daily overall predation rates being almost identical (Thompson and Burhans 2004). Furthermore, experiments with artificial nests cannot distinguish beneficial aspects of nest defense from negative aspects of parental activity whereby predators, snakes in particular, are drawn to the nest by watching parents arrive and feed their young. While artificial nests generate large sample sizes and are easier to locate than natural nests, there is now a strong consensus that making inferences about nest predation or nest defense between habitats or between years within the same habitat is unwise using artificial nests because of the unknown confounding effects of habitat and predator communities.

It is rare for researchers to check the validity of their methods. In an exception, Komdeur and Kats (1999) examined factors affecting nest defense in Seychelles warblers. They tested the effects of visitation rate by observers; the effects of using artificial nests made of sisal by comparing them with old war-

Table 10.2 The effect of visitation rate, nest type, and egg type on egg loss of Seychelles warblers in high-quality habitat on Cousin Island (January 1997)

Variable	% egg loss	n	p-value
Visitation rate of nests with incubated warbler egg			
Morning and evening, days 1–3	10.0	10	NS
Morning day 1, evening day 3	15.4	13	
Nest type			
Artificial nest with artificial egg	70.0	10	NS
Old warbler nest with artificial egg	80.0	10	
Egg type			
Artificial egg incubated by warbler	11.1	9	NS
Warbler egg incubated by warbler	10.0	10	

Source: Komdeur and Kats 1999.
Note: Egg loss was measured between 1800 and 1900 hrs on day 3. p-value refers to Chi-square tests; NS denotes not significant.

bler nests from which young had hatched the week before the experiment; and the effects of artificial eggs made of candle wax by comparing the fate of artificial and natural warbler eggs placed in natural warbler nests 2–3 hours after natural egg loss, a period when adults will continue to incubate eggs. They found no differences in the fates of eggs after 3 days (table 10.2), indicating experimental techniques did not influence extent of egg loss unduly.

The most common variables measured in studies of nest defense are (i) distance to the predator, including minimum and average distance over a period of time (for example, Regelmann and Curio 1983); (ii) direct aggression, often lumping hovering, dives, and strikes (for example, R. L. Knight and Temple 1988); (iii) latency in displaying a response (for example, Curio, Regelmann, and Zimmerman 1984); and (iv) calling rate, where different calls are frequently pooled together (for example, McLean, Smith, and Stewart 1986; see Redondo 1989 for a review). Attempts to measure these aspects of nest defense quantitatively fall into three categories: assigning scores to different behavior patterns, ranking different behavior patterns, and focusing on one or a few aspects of nest defense. Assigning high scores to behaviors that appear most dangerous (for example, Biermann and Robertson 1981) is subjective and is used infrequently. Ranking behavioral responses and then using them separately as an index of defense or summing them on a single scale (for example, M. Andersson, Wiklund, and Rundgren 1980; Biermann and Robertson 1981; Hakkarainen and Korpimaki 1994; table 10.3) still involves some subjectivity, assumes that different components have the same function, and often leads to a behavior ranked as 2 carrying twice as much cost as that ranked as 1. As in most studies of behavior, the most cautious approach treats the different variables separately (for example, Curio 1980; Breitwisch 1988).

Table 10.3 Ranked call and behavioral responses of the meadow pipit to a stuffed stoat

Ranked response	Description
Call responses	
0	No calling by parent
1	Chirping "chutt" or "chitt" calls; used to maintain contact rather than express anxiety
2	Alarm calls near nest; a persistent "stitt-itt"
3	Distress and alarm calls near nest
Behavior	
1	Parent watches the nest and stuffed stoat from a distance
2	Parent walks or lands close to the nest and stuffed stoat and watches them
3	Parent displays slight fluttering flight around the nest and stuffed stoat
4	Parent hovers over the stuffed stoat
5	Parent intensively dives or strikes at the stuffed stoat

Source: Pavel and Bures 2001.

10.3 Distraction displays

Distraction displays are an ill-defined category of behavior patterns that appear to reduce the likelihood of the predator finding a nest or young. Semantic debate over what types of behavior constitute a distraction or diversionary display has proceeded for some time (Jourdain 1936, 1937; Nice 1943; Armstrong 1947, 1949; Williamson 1950; Simmons 1952, 1955), but the most recent and thorough survey lumps of all of these behaviors and others under a general heading of nest protection behavior (table 10.4). This list includes both cryptic behavior and aggressive attacks against predators that are close to or have found the nest, as well as classic distraction and displacement activities that appear to deceive a predator, such as "rodent runs," in which the parent runs off in a crouched position resembling a small rodent; erratic fluttering, in which the bird makes convulsive attempts to run, fly, and jump; and false brooding, in which the bird squats on the ground, apparently incubating (Gochfeld 1984). These behaviors are seen in charadriiformes, galliformes, anseriformes, columbiformes, and passerines, and have even been reported in a mammal (W. H. Hudson 1892), although they are extremely rare in this class.

One of the major impediments to studying distraction displays is their variability within individuals, within species, and between species. For instance, Gochfeld (1984) summarizes an archetype of injury feigning, the broken wing act, thus:

> The surfbird gives an injury feign that is characteristic of many calidridine sandpipers (e.g., the Knots). When flushed from the nest, it runs with wings half spread, and with the fanned tail dragging on the ground. Killdeer extend one wing over the back, beating the other in the dust, with the tail spread and the breast and tail pressed to the ground. Thick-billed plovers display with spread tail, creeping forward, beating one or both wings on the ground, characteristic of the Charadrius

Table 10.4 Categories of nest protection behavior in shorebirds

Habitat and nest site selection
    Protected site
        Avoidance of ungulate herds or human activity
    Cryptic site
    Spacing out

Cryptic nest building and parental care
    Incubation and feeding of young
    Cryptic young hide or remain immobile

Predator-induced nest protection behavior
    Nest-departure behavior
        Early surreptitious departure
          Fly-away trick
        Sitting tight
        Special behaviors
          Egg and chick covering
          Chick carrying
    Distraction displays
        Directly related to departure
          Explosive departure
          Impeded flight and incapacitation displays
          Running behaviors
            Crouched run
            Upright or rapid run
            Wing-out run
            Rodent run
        Other distraction displays
          Injury feigning or broken wing
          Tail flagging
          Agitated circling flight
          Erratic fluttering
        Displacement activities
          False brooding
          False feeding
          False swimming
          Pseudosleeping
          False maintenance behavior
          False reproductive behavior
    Aggressive behavior
        Approach
        Aggressive circling and scolding
        Mobbing and attacking
        Ungulate display
    Parent(s) leads young away

Source: Gochfeld 1984.

plovers. The display of the three-banded plover is described for a bird that approached the intruder to within a meter, calling excitedly, then shuffled away in a crouch, jerking its head, with the tail raised (reminiscent of woodcocks), and its wings extended to varying degree. The spotted sandpiper shows an apparently helpless fluttering over the ground with its wings partially extended and its tail fanned and dragging the ground. Occasionally, like a killdeer, it spreads its wings fully, waves them, and rolls from side to side. Sanderlings creep away from the nest with the tail down, beating fully spread wings on the ground.

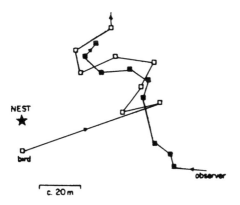

Without illustration or rigorous description, it is difficult to determine whether these are different ways of describing the same display, or whether species actually differ in displays. Written descriptions clearly capture the peak of display activity, and seldom do writers emphasize the rapid qualitative and quantitative transitions in behavior. (p. 313)

Many types of distraction display appear to render the performer easier to catch: for example, pretending that a wing is broken, or repeatedly opening and closing the bill and flopping on the ground with outspread wings, mimicking the death throes of a bird (Armstrong 1956). An important component of some aspects of distraction display is entrapment, whereby the parent monitors closely the responses of the intruder to its own behavior, periodically returning to it if the predator ceases to follow the displaying bird or loses attention. For example, stonechats leave the nest and repeatedly perch at some distance from it, thereby, with luck, drawing the intruder's attention and movements toward the parent and away from the nest (fig. 10.2; see also R. L. Knight and Temple 1986b). In this way, distraction displays resemble aspects of teaching, in which parents closely adjust their behavior to that of another individual, usually their offspring (Caro and Hauser 1992).

By design, therefore, the principal function of distraction displays appears to be to draw the predator away from the nest by means of deception, and while there are accounts of dogs and people being lured away (for example, Armstrong 1947; Duffey, Creasey, and Williamson 1950), reports of natural predators being deceived by these behaviors are less well documented (see Gochfeld 1984); thus, we have little idea of the magnitude of the benefits of distraction displays (Brunton 1990). Moreover, we have no more than an anecdotal appreciation of the costs for either parents or offspring. For example, parents are occasionally taken by predators while performing a distraction display (Brunton

1986; Amat and Masero 2004), and some predators initiate searches for offspring immediately after the parent starts to display (Sonerud 1988).

Little systematic work has been carried out on the factors predisposing species and individuals to perform distraction displays other than informed but descriptive accounts of where we might expect to see this sort of defense. In an early but important attempt, Armstrong (1954) outlined six factors that predispose birds to show injury feigning: (i) when the nest or hiding place of young is on open or exposed terrain, because prey can see predators from a distance and have time to fly from the nest to an open area to display; (ii) when the nest is accessible to nonavian predators, so injury feigning should be found mainly in ground-nesting birds; (iii) when the nest is insubstantial or inconspicuous, as these nests characterize exposed nests on the ground; (iv) among birds that nest alone, because their nest is more likely to be cryptic; (v) when predation principally occurs during daylight, as displays will be ineffective at night, and when predation is exercised by mammalian or reptilian predators, as avian predators may see cryptic nests easily and could follow parents in the air between bouts of distraction; and (vi) in northern latitudes, because extended daylight during the breeding season allows predators to witness displays for a longer period. It would be most helpful to conduct a comparative analysis of distraction displays similar to that of Larsen, Sordahl, and Byrkjedal's (1996) analysis to tease out the social and ecological correlates of aggressive nest defense (see section 10.2). Moreover, the observation that some predators disregard distraction displays and search for the nest suggests that only some predator species or individuals are duped, which parallels evolutionary arms races seen between nest parasites and their hosts (Davies 2000). Perhaps the diversity of distraction displays reflects evolutionary responses to a predator disbelieving one display and the prey trying something novel. Alternatively, different behaviors may be aimed at different predator species that are differentially gullible.

Among mammals, somewhat similar forms of distraction behavior have been noted, as when a Thomson's gazelle mother runs between her fleeing fawn and a pursuing predator, often stotting at high intensity (section 7.6.e).

10.4 Costs of nest defense
Most forms of nest defense, aside from sitting tight on the nest, necessarily entail time and energy costs, although these have never been measured. Nest defense, by Montgomerie and Weatherhead's (1988) definition, entails a heightened probability of injury or death to the parent, but there is little direct evidence to support this supposition. A handful of anecdotes suggests that nest defense and mobbing, taken together, result in occasional death (Sordahl 1990; section 11.4). For instance, Brunton (1986) saw a red fox dispatch a killdeer while it feigned injury near its nest. Indirect evidence of a survivorship

Table 10.5 Differences in responses of stonechats to the presence of birds of prey and nest predators in their territories

	Birds of prey	Nest predators
Stonechat perched exposed		
Many loud calls	5	85
0–5 quiet calls	7	2
Stonechat hidden		
Many loud calls	0	0
0–5 quiet calls	36	13

Source: Greig-Smith 1980.

cost comes from observations of birds being more cautious in keeping their distance in situations in which risk of predation is high, as when escape is restricted (chapter 11). On the other hand, Greig-Smith (1980) suggested that because small birds maintain such high levels of vigilance anyway, they would be unlikely to incur much greater risk of predatory attack when scolding predators from exposed perches. The disagreement highlights the idea that only certain aspects of nest defense may be risky. Studies that argue that reproduction entails predation risk (Magnhagen 1991), as, for example, when rhinoceros auklets curtail provisioning their chicks early in areas of high predation risk (Harfenist and Ydenberg 1995), do not show that nest defense per se is risky, because lowering parental activity at the nest site may simply be a means to enhance crypticity (T. E. Martin, Scott, and Menge 2000; Ghalambor and Martin 2002; section 3.3). In short, nest defense is assumed rather than demonstrated to carry a survivorship cost, and the issue needs more work.

As an aside, the definition of nest defense, sensu stricto, may be overly restrictive. For example, many parent birds alarm call and dive at aerial predators that are a threat to their nestlings but not to themselves. For example, pied flycatcher parents attack model great spotted woodpeckers vociferously, because woodpeckers peck into nest holes and eat the young even though they represent a minimal threat to parents (Curio 1975; Dale, Gustavsen, and Slagsvold 1996). Similarly, adult stonechats give relatively few loud calls to birds of prey that can kill them but loud calls to nest predators, especially when they call from an exposed site (table 10.5). An additional point is that a parent dying in defense of its nest toward the end of the breeding season may incur less cost than might appear if its chances of surviving the winter and breeding again are very low.

10.5 Benefits of nest defense

10.5.a Driving predators away   Parents that are more aggressive toward potential predators at their nest have greater nesting success, implying that nest defense prevents predators from stealing nest contents (table 10.6). Though most of these studies are experimental, the evidence is somewhat indirect, be-

Table 10.6 Studies showing that intensity of nest defense is greater at successful nests than unsuccessful nests

Species	Predator	Outcome	Reference[a]
Merlin	Stuffed raven	Breeding failure in nests where male defense low	1
Eastern kingbird	Human	Aggression scores of adults greater at successful nests	2
Northern mockingbird	Human	Pairs in which males launched close attacks showed greater nesting success	3
Stonechat	Human	Higher rates of calling in successful pairs than unsuccessful pairs	4
Seychelles warbler	Seychelles fody	Unguarded artificial nests experienced 7.1 times higher egg loss rate than guarded nests	5
		Increase in time spent guarding eggs by males results in increased hatching success	5
Red-winged blackbird	Mounts of raccoon and red-tailed hawk and model of American crow	Nest success higher where males had high call rates and females had high rates of calls, dives, and strikes	6
	Human	Nests that were defended more vigorously were successful	7

[a] 1. Wiklund 1990b; 2. Blancher and Robertson 1982; 3. Breitwisch 1988; 4. Greig-Smith 1980; 5. Komdeur and Kats 1999; 6. R. L. Knight and Temple 1988; 7. Weatherhead 1990.

cause natural predators were not seen being driven from the nest. In theory, heightened aggression shown toward a model predator or person could be associated with other aspects of parental quality that enhance reproductive success. In one study of Seychelles warblers, however, comparative data, observations, and experiments provide a solid case for excluding most confounding variables. On two Seychelles islands, Cousin and Cousine, Seychelles fodies are the main nest predator of the Seychelles warbler, but on a third, Aride, fodies are absent. As expected, rates of egg loss differ between the islands, being 20.4% on Cousin and 29.4% on Cousine but 0% on Aride. The proportion of time that male warblers spend guarding nests reflects the differential risk of predation: 43.2%, 41.1%, and 4.9%, respectively. When fodies were close to natural nests (within 2.5 m) and at least one warbler was present, antagonistic interactions were observed in 62.0% of cases, and in every case the fody was chased away. In addition, on one island, Cousin, rate of loss of eggs from artificial nests not tended by warbler pairs was far higher than eggs incubated and guarded by breeding pairs (75.0% versus 10.5% after 3 days), and, overall, unguarded nests experienced a 7.1-times higher egg loss rate than guarded nests. Finally, the more time that males spent guarding the nest, the lower the fraction of eggs that were lost (Komdeur and Kats 1999).

Other evidence can be marshaled for supposing that nest defense drives predators off. In a study of European blackbird nest defense, Cresswell (1997b) found that when blackbirds were present, the probability of failure was independent of the nests' detectability; but when quail eggs were placed in the same nest in July after the adults had departed, probability of failure was now determined by nest detectability and height. Apparently, blackbird behavior must have been compensating for a nest's relative vulnerability caused by its location. Given that another behavior seen at the nest site, incubation, increases detectability, if anything, the best explanation for compensatory behavior lies in nest defense.

10.5.b Silencing offspring   A second, less recognized benefit of nest defense is to silence offspring, an advantage only realized after the young hatch. In both precocial and altricial species, parents sometimes give a specific warning call that causes offspring to cease making contact or begging calls, respectively. Begging calls do, without doubt, attract mammalian predators. As illustrations, artificial nests baited with quail eggs from which playbacks of tree swallow begging calls were played were paired against silent artificial nest controls set 50 m away. In trials on both the ground and in nest boxes, experimental nests were usually lost to predators first (Leech and Leonard 1997). In a similar experiment, where Haskell (1994) played western bluebird calls from artificial nests, he could only find such an effect for ground nests; but, in addition, ground nests were even more likely to be predated if begging calls were played at high rather than low rates.

As a consequence of this cost of offspring begging, there appears to have been selection to reduce localizability of nestling calls made on the ground, where predation risk is high. Compared with begging calls of four tree-nesting warbler species, the frequencies of which center around 6 kHz, young of five other ground-nesting warbler species call at around 10 kHz and modulate their vocalizations less rapidly, both of which promote greater attenuation in the environment and so reduce localizability (Haskell 1999; see also Redondo and Arias de Reyna 1988). Similarly, when tree- and ground-nesting species' begging calls were played from artificial nests placed on the ground or in trees, ground species' calls incurred lower predation costs from eastern chipmunks when they emanated from the ground than did tree species' calls; the converse effect was not observed when each type of call was played from nests in trees (Haskell 1999). Parental calls that silence offspring would therefore be of obvious benefit in ground-nesting birds, where predation risk from mammals is high.

Parental alarm calls that silence offspring have been noted in a number of species. Red-winged blackbird screams make nestlings stop begging and

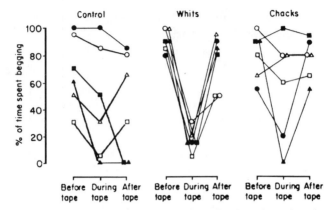

Figure 10.3 Results of a laboratory experiment to determine the effects of adult calls on begging by stonechat nestlings. Control consists of background noise. Chicks from the same nest have symbols of the same shape. "Whits" prevent nestlings from begging (Greig-Smith 1980).

crouch low in the nest (R. L. Knight and Temple 1988), European robin "seeep" calls induce young to remain still and silent (East 1981), American goldfinch "BB" calls cause a decrease in nestling conspicuousness (R. L. Knight and Temple 1986b), and stonechat "whits" prevent nestlings from begging (Greig-Smith 1980; see also L. Halupka 1999). These calls apparently have a specific function, because in each of these species parents additionally emit other types of call that do not influence offspring behavior (fig. 10.3). Warning calls of precocial species cause mobile young to remain silent and freeze in the same fashion.

10.6 Effects of predation risk on nest defense

The intensity with which a parent bird defends its nest should be a compromise between the survival of the current brood and the survival of the parent and its future broods (Trivers 1972, 1974; Carlisle 1982; Winkler 1987). Therefore, the level of nest defense should be responsive to (i) the threat to current nest contents, (ii) the ability of the parent to drive the predator away, and (iii) the threat to the parent itself (Greig-Smith 1980; R. L. Knight and Temple 1988; Montgomerie and Weatherhead 1988). Nest defense does vary according to the type of predator that is threatening the nest (for example, Kruuk 1964; Curio 1975; Veen 1977; Greig-Smith 1980; Buitron 1983; but see R. L. Knight and Temple 1986b), and there is evidence that nest defense is responsive to each of these three factors.

(i) One way in which it can be shown that parents respond to threat to offspring is to match responsiveness to different predators to the risk they pose to offspring of different ages. Black-billed magpies, for example, tailor their maximum response toward different predators to periods when their off-

Table 10.7 Summary of the responses of black-billed magpies to different predators according to the reproductive stage of the magpies

| Type of predator | Stage in greatest danger | | | | | | Stage of maximum response by parents | Effectiveness of response |
| | Eggs | Nestlings | | Fledglings | | Adult | | |
		Young	Old	Young	Old			
Squirrels	+	+	?				Eggs	Always
Crows	+	+	?				Eggs	Usually
Coyotes				+			All stages but especially young and fledglings	Usually
Flying raptors			?	++	+	+	Nestlings	Often
Perched raptors	?	+	+	++	+	+	All stages?	Usually

Source: Buitron 1983.
Note: + denotes vigorous attack; ++ very vigorous attack; ? unknown.

spring are most susceptible to those threats (table 10.7; see also Verbeek 1972; T. L. Patterson, Petrinovich, and James 1980).

Nest site exposure also affects nest defense, as would be expected if parents are responding to predation risk on young (Redondo 1989). Within species, inaccessible nests built high in trees elicit lower levels of nest defense than nests built lower down (Carrillo and Aparicio 2001), and exposed nests engender more distraction displays than cryptic nests (Hobson, Bouchart, and Sealy 1988). Across three *Acrocephalus* warbler species, those that nest lower show stronger reactions to snakes, whereas those that nest higher show greater responses to a Western marsh harrier mount (Kleindorfer, Fessl, and Hoi 2005); and across altricial birds, species with more exposed nest sites exhibit reduced parental activity around the nest that mollifies increased predation risk (Martin, Scott, and Menge 2000; see also Burhans and Thompson 2001).

(ii) That nest defense is related to the ability to see off a predator is demonstrated most obviously in the extreme case of parents giving up defending their nest. For example, parent white-crowned sparrows can do little to halt the advance of determined garter snakes, which are very effective predators on nestlings, and they barely attempt to defend their nest from these predators (T. L. Patterson, Petrinovich, and James 1980). Birds also tailor responses to aerial and ground predators that are deterred by different types of nest defense. For example, killdeer feign injury principally to mammalian predators but not to aerial predators, which can likely locate the nest contents easily from the air. Indeed, in the injury-feigning incidents, mammals followed killdeer on 76/146 occasions, but avian predators followed them on only 11/132 occasions, suggesting birds are less easily duped (Brunton 1990). Across species, response similarly varies by predation threat. For instance, the conspicuous nests of song thrushes attract the attention of corvids, but thrushes repel them through active defense. In contrast, blackcap nests are susceptible to both corvid and rodent predation, but parents are too small to drive corvids away; instead they remain on their nest, increasing crypticity and preventing mice from stealing eggs or chicks. Finally, yellowhammer nests are mainly preyed on by rodents, and, like blackcaps, they sit tight, keeping rodents away from nest contents (Weidinger 2002).

(iii) That nest defense is sensitive to threat imposed on the parent itself is supported by species in which adults remain cautious in the face of predators on adult birds but bold when confronting nestling predators. For instance, pied flycatchers never attack a stuffed sparrow hawk mount, a species that will kill adults, but remain in adjacent vegetation alarm calling, whereas they harass and strike a model great spotted woodpecker that predates nestlings but is no threat to parents (Dale, Gustavsen, and Slagsvold 1996; see also Curio, Klump, and Regelmann 1983; Hudson and Newborn 1990; Wheelwright, Lawler, and Weinstein 1997).

It might also be expected that the risk that predators impose will be related to the size of the predator versus the size of its adult prey, and the extent to which each party is armed. While heavier female tawny owls show higher levels of nest defense toward a person than do lighter owls (Wallin 1987), supporting this view, other studies contradict it (for example, merlins against a stuffed raven: Wiklund 1990a; Eurasian kestrels against a stuffed pine marten: Tolonen and Korpimaki 1995; great tits against a model great spotted woodpecker: Radford and Blakey 2000). It is possible that these nest predators are simply too well armed for relative body size to be a consideration.

There have been attempts to assess the relative importance of risk to offspring, ability to drive predators away from the nest, and risk to parents in relation to patterns of nest defense, but at present they are still unsophisticated. For example, because R. L. Knight and Temple (1988) found that red-winged blackbirds were most aggressive toward a model American crow followed by a red-tailed hawk mount and then a raccoon mount, they dismissed the "threat to adult" explanation as being important, because only hawks are a threat to adults; and the "threat to nest contents" explanation, because only crows and raccoons are dangerous for eggs. As birds in general are more effective at driving off bird than mammal predators (for example, Kruuk 1964; Roell and Bossema 1982; Buitron 1983), Knight and Temple's findings tentatively support "the ability of parents to drive predators off" explanation, at least for red-winged blackbirds.

Nonetheless, we might expect that the extent to which parents of different species weigh the three explanations will depend on life histories. As a crude illustration, species with a long life expectancy are likely to hold adult survival

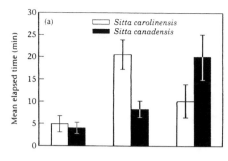

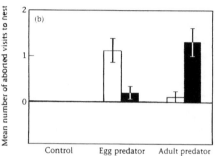

Control   Egg predator   Adult predator

Figure 10.4 (*a*), mean + SE elapsed time (min.) until the nest was visited by male *Sitta carolinensis* (white-breasted nuthatch) or *Sitta canadensis* (red-breasted nuthatch) in response to 3 different models: control junco, egg predator wren, and adult predator sharp-shinned hawk. Interaction between model type and species, $p < 0.05$. (*b*), mean + SE number of aborted visits to the nest by males in response to the 3 different models, as in (*a*). Interaction between model type and species, $p < 0.001$ (Ghalambor and Martin 2000).

in high esteem. Take, for example, two species of nuthatch that differ in reproductive effort: the white-breasted nuthatch is more fecund and has lower survival prospects than the red-breasted nuthatch. Ghalambor and Martin (2000) manipulated stage-specific predation risk by presenting models of a predator on adults, a sharp-shinned hawk, and an egg predator, a house wren, as well as a control model, a dark-eyed junco, along with vocalizations of each species to both species of nuthatch when males were feeding incubating females on the nest. Risk taking was measured as the male's willingness to feed the female (elapsed time to next feeding episode) and aborted visits to the female. In response to an egg predator, both measures showed that males of the white-breasted nuthatch were more cautious and hence placed greater value on their current offspring than did males of the red-breasted species. In response to an adult predator, males of the red-breasted species were more cautious than white-breasted nuthatches, placing greater value on their own survival and future reproduction (fig. 10.4), reflecting the red-breasted species' enhanced survival prospects (that is, greater reproductive value: Curio 1988a). Parallel results are obtained when comparing bird species from the Northern Hemisphere that have lower adult survival and larger clutches than those from the Southern Hemisphere. Again, parents of shorter-lived northern species decrease their feeding rates more when predation is directed at offspring (by a jay), but southern species decrease feeding rates more when the threat of predation is against parents (from a hawk) (Ghalambor and Martin 2001). These findings indicate that species' differences in nest defense toward a particular predator should be expected on the basis of life history considerations alone.

10.7 Parent's renesting potential

The optimal intensity of nest defense at any given point in time will be the maximum difference between the parent's fitness costs and benefits as defined in reference to the parent's reproductive value (V_x) at age x. Reproductive value is determined by fecundity of present and future broods (b) and the probability that the parent will survive to each future breeding period (l). Thus,

$$V_x = b_x + \Sigma \, (l_t/l_x) \, b_t,$$

where b_x is current reproduction at age x and $\Sigma \, (l_t/l_x) \, b_t$ is the residual reproductive value accrued from all future broods. Benefits of defending a current brood increase the fitness of b_x, while costs of defense decrease the probability that the parent will survive (l_t/l_x) to age t (Montgomerie and Weatherhead 1988). Numerous parental and offspring factors influence the balance between expected benefits for the current brood and future costs, the latter mediated principally through parents' probability of survival (Biermann and Robertson 1983; Montgomerie and Weatherhead 1988; Redondo 1989).

Turning first to future costs for the parent imposed by defense of the current brood, a parent's likelihood of nesting again should affect the willingness with which a parent engages in brood defense, because a bird with high renesting potential should risk less for its current brood than a bird with low renesting potential (Curio, Regelmann, and Zimmerman 1984; Ghalambor and Martin 2000, 2001). Renesting may be possible within the same breeding season or later on in the bird's lifetime. Thus, within a breeding season, birds should increase their level of nest defense as the opportunity to renest declines (Barash 1975b), and over their lifetime, nest defense should change (usually increase) as life expectancy alters with age (usually declines) (G. C. Williams 1966; Clutton-Brock 1984).

10.7.a Renesting potential within breeding seasons   Evidence that parents defend broods more as the breeding season progresses (and the potential to relay declines) is mixed. One problem is that the increase in nest defense over the breeding season, as reported in so many studies (table 10.8), can be explained in many ways other than by a decline in renesting potential—most obviously by increasing reproductive value of the current brood as it ages, but also by later clutches being smaller than earlier ones, older (larger) offspring being more profitable and more conspicuous to predators, or even a buildup of predators in the area over time (Redondo 1989). Second, in some species, nest defense is clearly unaffected by renesting potential (for example, stonechats:

Table 10.8 Changes in nest defense during the breeding cycle

Species	A or P[a]	Predator	Effect with offspring age	Reference*
Increases in nest defense with offspring age				
Chinstrap penguin	?P	Human	Increase rate of pecks between incubation and nestling stages	1
Northern shoveler	P	Human	Reduced flushing distance during incubation	2
Mallard	P	Human	Reduced flushing distance during incubation	2
Gadwall	P	Human	Reduced flushing distance during incubation	2
Red-tailed hawk	A	Human	Increased call rate with age of nestlings	3
Spanish imperial eagle	A	Human	Reduced distance of approach over breeding cycle	4
Eurasian kestrel	A	Human	Increase in nest defense over breeding cycle	5
Willow ptarmigan	P	Human	F, but not M, defense increases with age during incubation	6
Baird's sandpiper	P	Human	Increase in flutter attacks over incubation	7
Killdeer	P	Human	Risk-taking index increases during incubation then declines after hatching	8
Herring gull	A	Human	Return more rapidly to colony during nestling than incubation stage	9
Mourning dove	A	Human	Increased intensity of flight displays over nestling period	10
Eastern screech owl	A	Human	M show more flights, dives, and calls from incubation to nestling stages	11
Tawny owl	A	Human	Defense levels increase over breeding cycle	12
Eastern kingbird	A	Crow[b]	Become more aggressive from incubation to fledgling stages	13
Black-billed magpie	A	Human	Increase in more vigorous nest defense over breeding cycle	14
	A	Human	Increasing calls and proximity of approach over nestling stage	15
American robin	A	Blue jay[b] and rubber snake	Only attacks snake after chicks hatch	16
Fieldfare	A	Human	Increase nest defense during nestling stage	17
Gray catbird	A	Blue jay[b] and rubber snake	More aggressive to jay after hatching	16
Brown thrasher	A	Blue jay[b] and rubber snake	More aggressive to jay after hatching	16
Pied flycatcher	A	Models of birds	Increase calling throughout incubation and nestling stages	18
European robin	A	Human	Increase in "tic" and "seeep" calls from hatching onward	19

Table 10.8 *(continued)*

Species	A or Pᵃ	Predator	Effect with offspring age	Reference*
Stonechat	A	Human	"Whits" and "chats" low during incubation but increase during nestling stage	20
Barn swallow	A	Human and screech owl[b]	Increase defense intensity between incubation and nestling stages	21
		Human	Alarm calls of F[c], but not M[c], increase during incubation; alarm calls of M and F increase over nestling stage	22
		Great horned owl[b]	Increase in defense between incubation and nestling stages but only for first, not second, broods	23
Moustached warbler	A	Human	Probability of alarm calling increases with age of nestlings	24
Aquatic warbler	A	Hen harrier[b], polecat[b]	Increase in calling rate to both and flight rate to harrier over breeding cycle	25
Great tit	A	Raptor, cat, and mobbing calls	Reduced latency to approach and shorter approach distance with nestlings than eggs	26
Willow tit	A	Stoat[b]	Increased calling rate and reduced approach distance over breeding cycle	27
Meadow pipit	A	Stoat[b]	Increase in calling and intensity of display and decline in approach distance in M over nestling stage	28
Indigo bunting	A	Human	Increase in number of alarm calls between incubation and nestling stages	29
Swamp sparrow	A	Distress calls of adults	Parents react more quickly, get closer, give more calls and distraction displays during fledgling period	30
Song sparrow	A	Distress calls of adults	Parents react more quickly, get closer, give more calls and distraction displays during fledgling period	30
White-throated sparrow	A	Distress calls of adults	Parents react more quickly, get closer, give more calls and distraction displays during fledgling period	30
White-crowned sparrow	A	Kestrel[b], jay[b], live snake	Increase in defense against kestrel and jay over nest cycle but increase during incubation and decline during nestling stage to snake	31
Savannah sparrow	A	Human	Increase approach distance and number of calls over nestling stage	32
American goldfinch	A	American kestrel[b] and blue jay[b]	Parents give more "BB calls" when experimentally given older nestlings	33
Yellow warbler	A	Common grackle	More calls during nestling than incubation stage	34
Red-winged blackbird	A	Human	M increase aggression over breeding cycle	35

Table 10.8 *(continued)*

Species	A or P[a]	Predator	Effect with offspring age	Reference*
No changes in nest defense with offspring age				
Wigeon	P	Human	No effect	2
Blue-winged teal	P	Human	No effect	2
Lesser scaup	P	Human	No effect	2
Merlin	A	Raven[b]	Little difference in nest defense between incubation and nestling stages	36
Eurasian kestrel	A	Pine marten[b]	No difference between incubation and nestling stages	37
Red grouse	P	Dog	No effect of brood age	38
	P	Dog	No effect on distraction display	39
Black-billed magpie	A	Coyote, crow	No strong effects over breeding cycle	40
American robin	A	Human	No changes in calls, approaches, dives, or strikes during incubation stage	41
Northern mockingbird	A	Human	No effect on approach distance during incubation; M, but not F, show decline in approach distance during nestling stage	42
Meadow pipit	A	Hen harrier[b]	Constant defense during breeding cycle	43
		Human	No difference in defense intensity, flushing, or settling distances between incubation and nestling stages	44
Yellow warbler	A	Gray squirrel[b]	Calls and approach distance do not change over breeding cycle	45
Northern cardinal	A	Blue jay and rubber snake	Equally aggressive to jay before and after chicks hatch	16
Red-winged blackbird	A	Human, raccoon	Generally no changes in calls, approaches, dives, or strikes over incubation	41
	A	Human	Increase in alarm calls and decline in approach distance with nestling age	46
Decreases in nest defense with age				
Alpine accentor	A	Human	Decline in distraction displays	47

Note: Changes refer to one sex or both.

[a] A altricial; P precocial.

[b] Mount or model.

[c] F female, M male.

* 1. Vinuela, Amat, and Ferrer 1995; 2. Forbes et al. 1994; 3. Andersen 1990; 4. Ferrer, Garcia, and Cardenas 1990; 5. Carrillo and Aparico 2001; 6. K. Martin and Horn 1993; 7. Reid and Montgomerie 1985; 8. Brunton 1990; 9. Kilpi 1987; 10. Westmoreland 1989; 11. Sproat and Ritchison 1993; 12. Wallin 1987; 13. Blancher and Robertson 1982; 14. Erpino 1968; 15. Redondo and Carranza 1989; 16. Gottfried 1979; 17. Andersson, Wiklund, and Rundgren 1980; 18. Curio 1975; 19. East 1981; 20. Greig-Smith 1980; 21. Shields 1984; 22. Moller 1984; 23. Smith and Graves 1978; 24. Kleindorfer, Hoi, and Fessl 1996; 25. L. Halupka 1999; 26. Regelmann and Curio 1983; Curio, Regelmann, and Zimmermann 1984, 1985; 27. Rytkonen, Orell, and Koivula 1993, 1995; 28. Pavel and Bures 2001; 29. Westneat 1989; 30. Stefanski and Falls 1972a; 31. Patterson, Petrinovich, and James 1980; 32. Weatherhead 1979; 33. Knight and Temple 1986b; 34. Gill and Sealy 1996; 35. Weatherhead 1990; 36. Wiklund 1990b; 37. Tolonen and Korpimaki 1995; 38. P. J. Hudson and Newborn 1990; 39. Pedersen and Steen 1985; 40. Buitron 1983; 41. R. L. Knight and Temple 1986c; 42. Breitwisch 1988; 43. K. Halupka and Halupka 1997; 44. Pavel et al. 2000; 45. Hobson, Bouchart, and Sealy 1988; 46. Weatherhead 1989; 47. Barash 1975b.

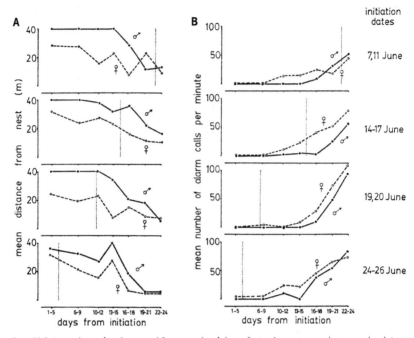

Figure 10.5 *A*, mean distance from the nest, and *B*, mean number of alarm calls given by parent savannah sparrows when their nest was threatened, plotted against days since initiation of egg laying. Vertical line shows last date for renesting (Weatherhead 1979).

Greig-Smith 1980; song sparrows: Weatherhead 1989; Eurasian kestrels: Tolonen and Korpimaki 1995). In savannah sparrows, for instance, the form of nest defense as measured by either mean distance to which birds approached or the rate at which they alarm called to a human was the same whatever the laying date (fig. 10.5), and the number of days before renesting potential reached zero explained less than 3% of the variation for both male and female parents (Weatherhead 1979).

Third, in yet other species, nest defense declines over the course of the breeding season, contesting the "reduced ability to renest" concept. For example, nest defense levels drop progressively for tawny owls as the date that they start to breed is delayed in spring (although this may have been caused by a progressive decline in brood size: Wallin 1987). Declines are principally seen in female parents: compared with defending first broods, nest defense declines in second broods in female willow ptarmigan (K. Martin and Horn 1993), merlin (Wiklund 1990a), and willow tits (Rytkonen, Orell, and Koivula 1993), while male defense remains unchanged (but see Curio, Regelmann, and Zimmerman 1984, and Onnebrink and Curio 1991 for an opposite example in great tits). Declining nest defense by mothers is opposite of that expected by the renesting-potential hypothesis, because females likely have a lower renesting potential than males due to costs of egg laying. That said, it is unclear why a pattern of increasing maternal defense is not more widespread within seasons, given that females always lay eggs. At present, the relationship between nest defense and renesting potential is opaque, and it is not even clear whether species that raise only a single clutch per season defend nests more assiduously than species raising the first of several broods. In passing, renesting potential may also lie at the root of greater willingness to defend artificially male-biased broods, as found in great tits, because sons are more expensive to rear than daughters and may therefore cost more to replace than daughters (Radford and Blakey 2000; but see Lessells, Oddie, and Mateman 1998).

10.7.b Renesting potential over lifetimes   Birds can be broadly classed as short-lived, where mortality is not age dependent, and long-lived, where mortality is age dependent. The first group should demonstrate no changes in nest defense with age, but the second should show an increase as their residual reproductive value declines. As expected, pied flycatchers, yellow warblers, song sparrows, and great tits, all small short-lived passerines, demonstrate no effect of parental age on the intensity of nest defense (Curio 1975; Hobson, Bouchart, and Sealy 1988; Weatherhead 1989; Radford and Blakey 2000; see also Breitwisch 1988), whereas California gulls, which live up to 18 years, do show increasing nest defense with parental age (Pugusek 1983). In other systematic interspecific tests, high adult survival is associated with lowered defense against nest predators in waterfowl (Forbes et al. 1994) and passerines (Ghalambor and Martin 2000, 2001; see section 10.6). Unfortunately, age effects will be confounded with greater experience of predators, which is likely to lower costs of nest defense and increase offspring survivorship, and with more efficient provisioning, both of which may promote increased investment, one aspect of which is nest defense (Montgomerie and Weatherhead 1988). Experience can therefore be viewed as altering cost-benefit ratios for parents. At a proximate level, experience may make parents bolder, because they have survived past bouts of defense, or even make them less interested in defending nests as a result of habituation.

Parents in good condition may have enhanced survivorship prospects within or across seasons compared with those in poor condition, thereby reducing their willingness to defend current broods. Alternatively, such parents may be more willing to attack predators in defense of their nest because they are better able to escape counterattack. Again, evidence is equivocal. At an extreme, individuals in very poor condition are unlikely to take on predators; but more generally, nest defense is positively correlated with body condition in female willow ptarmigan (K. Martin and Horn 1993) and in tawny owls (Wallin 1987), although there is no effect of body condition on nest defense in great tits (Radford and Blakey 2000).

10.8 Parental sex

There are marked sex differences in forms of nest defense, but no clear pattern or general understanding of these differences (table 10.9; see Gochfeld 1984 for shorebirds). In some species, females show more of one type of nest defense but males more of another (Wiklund and Stigh 1983), whereas in other species, patterns of nest defense in each sex change differentially as the breeding period advances, making them difficult to categorize (for example, merlins: Wiklund 1990b). Several different explanations could account for sex differences in defense, but there has been depressingly little experimental exploration of these possibilities 15 years after they were summarized by Montgomerie and Weatherhead (1988).

Given the extent of extra-pair copulations in birds (Griffith, Owens, and Thuman 2002; Westneat and Stewart 2003), confidence in parenthood is likely to be less for males than for females in many species, resulting in lower levels of offspring defense by males than by females; yet much of the support for this explanation lies simply in post-hoc arguments stating that male nest defense is less than that of females (Hobson, Bouchart, and Sealy 1988; Weatherhead 1989). In one experimental test in which males' confidence of paternity was reduced by separating male tree swallows from their mate for 1 or 3 days during their mate's fertile period and allowing males to observe and hear extra-pair copulations, no significant differences in subsequent male nest defense were observed (fig. 10.6), providing little support for this explanation.

In situations in which one sex has greater opportunities to renest than the other, that sex would be expected to defend its current brood less. To take an obvious illustration, if the adult sex ratio is biased toward males, opportunities for remating will be greater for females, and so females should defend nests less than males (Breitwisch 1988). Differences in longevity between the sexes and hence renesting potential could also drive differences in nest defense (Biermann and Robertson 1983). Alternatively, the challenge of nesting could be higher for one sex than the other, such that one sex (usually females) would suffer a greater decline in their residual reproductive value by renesting than would males. This is the mechanism that Rytkonen, Orell, and Koivula (1993) believe is occurring in willow tits, where the female defends the eggs more in the early stages of nesting, when renesting is still possible, but drops below that of the male when it is too late in the year to renest (fig. 10.7); other explanations, including differences in confidence of parenthood or in risk, were eliminated in this study (see also Weatherhead 1979).

A related idea is that one sex apparently defends the nest more than the other, but only because it is protecting or advertising to its mate in order to facilitate reproduction again together at a later date. One sex may value its mate more than vice versa if the sex ratio is biased, making ease of re-pairing asym-

Table 10.9 Sex differences in nest defense

Species	Predator	Behavior	Reference[a]
Males defend more than females			
Rough-legged buzzard	Mount of snowy owl	M defense responses always greater than F	1
Killdeer	Human and natural predators	M distraction displays closer to humans; more intense defense to natural predators	2
Eastern screech owl	Human	M make more calls, flights, and dives	3
Snowy owl	Human	Attacks by M more prevalent	4
Northern mockingbird	Human	M approach closer, follow to edge of territory, and strike more during nestling stage	5
Seychelles warbler	Seychelles fody	M chase fodies more than F	6
Great tit	Live pigmy owl	M approaches closer in presence of mate	7
	Mount of great spotted woodpecker	M calls more in defense of first broods	8
Meadow pipit	Stuffed stoat	M show direct attacks on predator when nestlings present	9
Swamp sparrow Song sparrow White-crowned sparrow	Playbacks of distress calls of adults	M react more quickly, approach closer, call and move more than F	10
Red-winged blackbird	Models of red- tailed hawk and American crow	M approach, hover over, and strike hawk and crow crow models more frequently	11
No differences between sexes			
Black-billed magpie	Raptors, crows, coyotes, squirrels	Defense approximately equal but slightly more by M	12
Fieldfare	Human	M and F respond similarly	13
Stonechat	Human	No differences in any aspect of nest defense	14
Females defend more than males			
Red-tailed hawk	Human	F approach closer, called and dived more	15
Red grouse	Domestic dog	F perform high- and low-risk distraction displays more	16
Snowy owl	Human	F calling and distraction displays more prevalent	4
Tawny owl	Human	F take more risks than M	17
Willow tit	Stuffed stoat	F approach closer and call more	18
Song sparrow	Human	F give more alarm calls and approach closer	19
Dark-eyed junco	Mount of eastern chipmunk	F arrive earlier and show more physical contact	20
Savannah sparrow	Human	F alarm call more and approach closer	21
American goldfinch	Mounted American kestrel and blue jay	F five more "BB calls" to silence offspring and distract predator	22
Yellow warbler	Mount of gray squirrel	F arrive first, stay longer, and are more persistent	23
	Common grackle	F spend more time close to predator and call more	24

Note: M, males; F, females.

[a] 1. S. Andersson and Wiklund 1987; 2. Brunton 1990; 3. Sproat and Ritchison 1993; 4. Wiklund and Stigh 1983; 5. Breitwisch 1988; 6. Komdeur and Kats 1999; 7. Curio 1980, Regelmann and Curio 1983, 1986a; 8. Onnebrink and Curio 1991, Curio and Onnebrink 1995; 9. Pavel and Bures 2001; 10. Stefanski and Falls 1972a; 11. R. L. Knight and Temple 1988; 12. Buitron 1983; 13. Andersson, Wiklund, and Rundgren 1980; 14. Greig-Smith 1980; 15. Andersen 1990; 16. P. J. Hudson and Newborn 1990; 17. Wallin 1987; 18. Rytkonen, Orell, and Koivula 1993; 19. Weatherhead 1989; 20. Cawthorn et al. 1998; 21. Weatherhead 1979; 22. R. L. Knight and Temple 1986b; 23. Hobson, Bouchart, and Sealy 1988; 24. Gill and Sealy 1996.

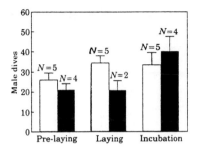

Figure 10.6 Mean (+ SE) number of dives made by male tree swallows toward a human observer during two 3-minute visits to the nest box in relation to experimental treatment. Pre-laying and laying periods were when the female was fertile; incubation was when she was no longer fertile and no differences would be expected. *Open histograms,* males held in captivity for 1 day; *filled histograms,* for 3 days (Whittingham, Dunn, and Robertson 1993).

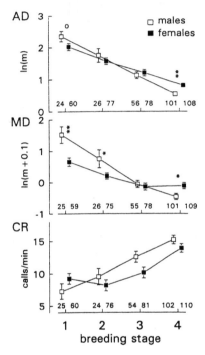

Figure 10.7 Sexual differences in willow tit defense measures (*AD* = average distance birds approach a mounted stoat; *MD* = minimum distance; *CR* = calling rate) at 4 breeding stages: 1 = start, 2 = end of incubation; 3 = start, 4 = end of nestling period. Means + SE and sample sizes are presented. Significant within-stage differences by *t*-test are indicated by asterisks: $^{o}p < 0.1$, $^{*}P < 0.05$, $^{**}p < 0.01$, $^{***}p < 0.001$. *Open squares,* males; *filled squares,* females (Rytkonen, Orell, and Koivula 1993).

metrical. Certainly, males in some species are more willing to defend their mate than their brood when forced to choose (K. Martin 1984). That male great tits indulge in more nest defense when they are in the presence of their mate, but females show no such predisposition, raises the possibility that nest defense is in some circumstances a display of quality and could be a factor in mate choice (Regelmann and Curio 1983) or mate retention (Curio, Regelmann, and Zimmerman 1984). In short, nest defense may have functions in addition to defense of offspring.

Sex differences in nest defense may also arise if one sex is unable to raise the brood alone. In this situation, the sex that could raise young by itself would suffer disproportionately higher costs when taking risks and would therefore reduce its intensity of nest defense. While this idea might explain relatively low female nest defense in the early part of the nesting cycle in those species in which the female incubates the eggs, it has never been examined seriously.

Finally, sex differences might arise if defense incurs greater risks for one sex than the other. Reduced female defense during incubation might occur because females are drained as a result of egg laying, for example (Reid and Montgomerie 1985). The most obvious sex differences in risk stem from sexual dimorphism, so, naively, one might expect the smaller sex to take fewer risks than the larger, because it would be less able to defend itself from counterattack (Andersson and Norberg 1981). Among raptors and predatory strigiformes, where females are normally larger than males, females do show greater nest defense in some species (for example, red-tailed hawks: Andersen 1990; tawny owls: Wallin 1987), but in others, males take greater risks than females (for example, rough-legged buzzards: M. Andersson and Wiklund 1987; snowy owls: Wiklund and Stigh 1983). The explanation can be rescued if one subscribes to the idea that males' smaller size enables then to outmaneuver avian predators better (making defense less risky), but this explanation is awkward, because it cannot explain greater female defense in other species of raptors and owls, and absence of greater female defense in other avian taxa in which females are smaller than males. Perhaps smaller defenders suffer less risks when attacking large predators than larger defenders because they can dodge them, whereas, against small predators, larger defenders accrue greater benefits than smaller because they can see them off more easily and suffer less risk of injury (Montgomerie and Weatherhead 1988); at present this remains an interesting supposition only.

10.9 Parental interactions

In species in which biparental care or cooperative breeding occurs, antipredator benefits that offspring receive from one parent may affect that which they could receive from an additional adult defending the nest (Chase 1980; Houston and Davies 1985; Winkler 1987; Clutton-Brock 1991). Benefits will depend on the shape of the relationship between the numbers of defenders and the probability of the predator leaving. The latter is likely to depend on the relative size and armaments of each party and hence will vary according to specific predator-prey species pairs. While we know that parents often chastise a predator together—for example, great tits confronted a live pigmy owl jointly on 77.2% ($N = 158$ pairs) of occasions, the male did so alone in 15.2%, the female in 3.2%, and neither in 4.4% of cases (Regelmann and Curio 1986b)—we do not know precisely how the number of defending adults influences the probability of the nest not being destroyed (benefit) or the probability of predator counterattack (cost) (P. J. Hudson and Newborn 1990) for any defender-predator species pair, although in general it is known that more defenders are better able to drive predators away (chapter 11). Thus, even though birds seem predisposed to join their defending mate, we cannot tell whether a bird's personal level of nest defense will be influenced by the presence of a defending mate or drop below or rise above what it would be if it was acting alone, or if individuals will be selected to match each other's effort, as some models predict (Houston and Davies 1985). Empirical observations show a variety of patterns. Some species show significant correlations in defense effort between the male and female of a pair (for example, northern mockingbirds: Breitwisch 1988; song sparrows: Weatherhead 1989), whereas others exhibit no correlated response (for example, willow ptarmigan: K. Martin and Horn 1993; yellow warblers: Gill and Sealy 1996). We do not know whether these early discrepant results simply reflect differential risk from various predators or different underlying rules.

10.10 Offspring age

Turning to the benefits accrued from defense of the current brood, a number of factors affect the reproductive value of the brood, namely its age, number of offspring, and the quality of those offspring, and hence the extent to which parents should defend (Carlisle 1985; Montgomerie and Weatherhead 1988). The age of the brood is important because as offspring grow older, the amount of parental investment necessary to replace the clutch and bring it to the same age as the bereaved clutch increases (Trivers 1972; R. Dawkins and Carlisle 1976; Boucher 1977). Second, the probability that older offspring will survive to breeding age is greater for older than younger offspring, because the instantaneous probability of mortality declines with age (the brood-value hypothesis: T. L. Patterson, Petrinovich, and James 1980). Since the relative difference between an offspring's probability of survival and that of the parent diminishes with offspring age, parents will be prepared to invest more in nest defense (M. Andersson, Wiklund, and Rundgren 1980; Winkler 1987). Third, offspring vulnerability is likely to change with their age, although the relationship could be highly variable (Harvey and Greenwood 1978). Offspring may become more noticeable to predators with age, because vegetation around the nest site becomes trampled, their increasingly boisterous behavior attracts predators, or rising parental feeding rates attract attention (Perrins 1965; Greig-Smith 1980). In addition, offspring's larger size could make them a more valuable resource for predators. Conversely, if found, offspring may be more difficult to kill as they grow older, because their increased size and strength make them more formidable prey (Onnebrink and Curio 1991). As a minimum of two and sometimes three underlying factors point to an increase in nest defense with offspring age (irrespective of any regard to renesting potential: Reid and Montgomerie 1985), it is hardly surprising that the majority of studies of nest defense

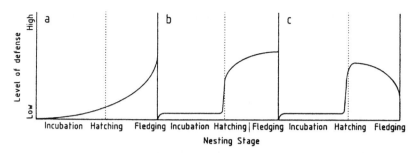

Figure 10.8 Graphical representation of predictions of models of parental defense. One model predicts that levels of parental defense correlate with offspring age, *a*. Another model predicts that levels of parental defense correlate with offspring vulnerability and thus predicts different relationships for altricial species, *b*, where vulnerability continues to increase after hatching until fledging, and for precocial species, *c*, where vulnerability peaks at hatching (Brunton 1990).

show a consistent increase in nest defense with offspring age (table 10.8, see also Gochfeld 1984). Unfortunately, most of these studies cannot distinguish the possible evolutionary mechanisms underlying these patterns (Clutton-Brock 1991), and many of them cannot dissociate these increases from that expected under the renesting-within-breeding-seasons hypothesis (section 10.7.a., but see Forbes et al. 1994).

Given that there are several evolutionary causes potentially underlying increases in parental defense with offspring age, it is difficult to predict when and how defense should change with offspring age. If reproductive value is principally involved, defense should increase steadily over the breeding cycle (fig. 10.8a). If defense is tied to offspring vulnerability, it should be low during incubation but increase sharply at hatching and remain high in altricial birds, because nestlings attract attention (fig. 10.8b). In precocial birds, however, defense should start to decline following hatching as offspring mobility enables them to scatter and hide, so that the whole brood is unlikely to be taken at once (fig. 10.8c). While there is qualitative support for all three relationships (table 10.8), studies that have attempted to distinguish between possibilities have usually found that the steady increase in defense with offspring age is attributable to increasing offspring value. For instance, in a study of Baird's sandpipers, which breed only once a season (thereby eliminating confounding effects of renesting potential in that season), and for which nest conspicuousness and incubation patterns are invariant before young leave the nest at hatching (thereby eliminating confounding effects of changing offspring vulnerability), parents showed an increase in distraction display to humans throughout incubation (Reid and Montgomerie 1985).

At a proximate level, parental nest defense may increase over the nesting cycle as a result of increasing parental interactions with young, particularly after hatching (the feedback hypothesis of McLean and Rhodes 1991). In a number of birds, females defend younger nestlings more intensely than do male parents, but males defend older nestlings more intensely (Pavel and Bures 2001). This is consistent with the feedback hypothesis, because females in many species incubate the eggs and will therefore be present as the young hatch.

10.10.a Past and future parental investment   Two avenues of experimental studies have investigated why defense so often increases with offspring age: those that have examined the influence of parental investment and those that have pitted offspring vulnerability against offspring age. The first stems from a historical debate as to whether parents use cumulative past investment in offspring (the so-called Concorde fallacy) or expected future benefits in making parental investment decisions (Trivers 1972; Dawkins and Carlisle 1976; Boucher 1977). The issue is largely forgotten now, because although decisions should logically be founded on future benefits, the extent of past investment will predict future investment in organisms with a fixed resource budget and de facto will be strongly correlated with future benefits (Maynard Smith 1977; Coleman and Gross 1991; Winkler 1991). Animals may therefore use past investment as a rule of thumb for decisions about future investment. Experimental tests of decisions being based on past versus future investment using nest defense as a measure of parental investment support both positions. Some studies clearly show that manipulation of expected benefits (clutch size) alters patterns of nest defense (R. J. Robertson and Biermann 1979; see also Ackerman and Eadie 2003). Other studies that manipulate past investment (length of incubation or clutch size) show that birds tailor nest defense to past investment (Weatherhead 1979, 1982; Rytkonen, Orell, and Koivula 1995). For example, when red-winged blackbird clutches of equal size that had been initiated 4 – 8 days apart from each other were exchanged between parents (thereby manipulating past investment), parents gave more alarm calls in the late nestling period to an approaching human if they had spent more time incubating eggs, although this effect was only partially replicated using approach distances. Thus, length of investment of one kind or another certainly influences intensity of nest defense irrespective of offspring vulnerability.

10.10.b Changes in offspring vulnerability   Offspring vulnerability is composed of two rudiments: vulnerability to detection and vulnerability to attack. The first is synonymous with conspicuousness that is, itself, composed of invariant elements, such as nest seclusion, and changing elements, such as parental visits to the nest and offspring's begging calls. The second depends on changing factors, such as offspring's ability to sense danger, its strength, and speed. At present, there is scant evidence that changes in nest defense with offspring age can be explained by changes in offspring vulnerability. First, it is not necessarily true that nestling vulnerability in altricial species consistently rises following hatching, as outlined in figure 10.8.b. For example, begging calls sometimes have their widest and most localizable acoustic range when nestlings are young; begging call rates may reach their zenith some time before fledging; and parental feeding visits may peak long before fledging. Moreover, begging calls may not be audible more than 20 m away and avian predators may not use auditory cues to locate nests (Cresswell 1997b). Given these nuances, consistently steady rises in nest defense documented across so many studies (table 10.8) are unlikely to correspond to increasing vulnerability (Redondo and Carranza 1989). Second, experimental studies challenge the importance of offspring vulnerability. When great tit nest holes were widened to allow predators easier access, it did not result in pairs defending their brood more vigorously against a dummy great spotted woodpecker (Onnebrink and Curio 1991). Third, two studies show that nest defense actually rises when offspring vulnerability declines! In both black-billed magpies and moustached warblers, young gain the ability to jump from the nest about two-thirds of the way through the nestling period, either abandoning it for good or hopping back in later. In both species, parents show a marked increase in calling rate just at the time that offspring are capable of jumping, which suggests that they are informing young of a predator being present (Redondo and Carranza 1989; Kleindorfer, Hoi, and Fessl 1996). Viewed another way, offspring's sudden ability to escape can be seen as an increase in their reproductive value, to which parental nest defense responds. In contrast with these studies and in support of the vulnerability hypothesis, patterns of injury feigning in male and female killdeer increased from laying to hatching but then showed a decline, mimicking the expected shape of the curve for precocial birds quite closely (fig. 10.8c) (Brunton 1990).

10.10.c Revisitation hypothesis   A methodological and proximate explanation of increases in nest defense with offspring age is that parents progressively lose their fear of experimental predators over time, because humans usually retreat or the predator model is removed after parents have defended their nest (R. L. Knight and Temple 1986a, 1986c). This positive reinforcement could result in an apparent increase in nest defense with offspring age that, in reality, is driven by an increased number of experimental visits. As examples, nest defense behavior of different female American robins and different male and female red-winged blackbirds visited only once did not increase over the course of the nesting cycle, but those visited repeatedly did (R. L. Knight and Temple 1986a). If this were a general concern, it would mean that empirical work on age-related aspects of nest defense using artificial predators could not be used as evidence to bolster theory.

Since 1986, however, many studies have been unable to support the revisitation hypothesis. For example, number of previous visits was not associated with nest defense in song sparrows (Weatherhead 1989) or willow tits (Rytkonen, Koivula, and Orell 1990); flushing distances in female ducks (Forbes et al. 1994) or mourning doves (Westmoreland 1989); and alarm calls in indigo buntings (Westneat 1989). Moreover, that nest defense behavior declines between the first and second breeding attempts in stonechats (Greig-Smith 1980), barn swallows (Moller 1984), great tits (Curio, Regelmann, and Zimmerman 1984), and song sparrows (Weatherhead 1989) speaks against the revisitation hypothesis.

Nonetheless, some species, in some circumstances, do become bolder with increasing exposure (Breitwisch 1988; Ferrer, Garcia, and Cadenas 1990; Lord et al. 2001; Gunness and Weatherhead 2002), and repeated sampling of the same parents raises problems of statistical independence, so it has become common practice to present results from only a single encounter with a given set of parents or at least check whether nest defense taken from repeated encounters matches those taken from first encounters with a threatening stimulus.

10.11 Offspring number

In most circumstances, the benefits of parental defense are not shared or depreciated among different offspring in a brood or litter; rather, an act of defense benefits either (i) the whole brood at once or (ii) simply a single offspring or fixed portion of the brood (Lazarus and Inglis 1986; Clutton-Brock 1991). (i) In the first case (termed the brood loss case of unshared investment), the benefits of parental defense will increase with number of offspring, so that we can expect intensity of nest defense to increase with brood size. In extremely risky situations in which the parent's behavior puts it at high risk, however, nest defense may decline with larger broods, as more young will be lost if the parent dies (Curio, Regelmann, and Zimmerman 1984). (ii) In the second case (termed fixed loss), intensity of parental defense will be unrelated to brood size. For both altricial and precocial species defending a clutch of eggs, or for altricial parents defending nestlings, especially when they are defending them against mammalian or snake predators, the brood loss case will apply. Only for precocial species defending mobile broods or for predation attempts by some avian predators that take just single offspring will the fixed loss case apply. Thus, in general, parental defense is expected to increase with offspring number as benefits to parents increase.

Several studies have attempted to match levels of offspring defense to offspring number but with mixed results (table 10.10a). One explanation is that some species view experimental predators as being very dangerous and are therefore unwilling to increase effort in defending larger clutches, whereas other species see them as less risky and alter defense in response to brood size. For example, Lazarus and Inglis (1986) argued that blue jays may represent a greater threat to adults of small birds than do snakes, and that this could account for lack of increase in nest defense with clutch size in the four altricial species studied by Gottfried (1979) (table 10.10a). Such reasoning could conceivably explain the increase in nest defense with clutch size in three species of duck in response to a human but its absence in three other species (Forbes et al. 1994) (table 10.10a). Regrettably, actual and perceived risks to parents are unknown in virtually every study, particularly in regard to humans.

Unfortunately, it is inappropriate to try to match levels of nest defense to expected benefits as measured by clutch or brood sizes in natural populations, because parents in good condition or living in resource-rich habitats may lay more eggs and additionally have more time or energy to devote to nest defense (Lessels 1991). A small clutch might also be associated with reduced nest defense for other reasons. Cresswell (1997a) suggested that diminished clutch size might be an adaptive response to local predation risk, because it will be associated with reduced nest size and will shorten nest building, egg laying, and chick rearing periods, thereby minimizing exposure time. Similarly, a smaller nest and fewer feeding visits will reduce conspicuousness. It is therefore imperative to manipulate clutch or brood size experimentally.

In studies in which egg or nestling number in altricial species is manipulated, some support the notion that parents are willing to take greater risks when offspring number and hence current reproductive value increases (table 10.10b). For example, Robertson and Biermann (1979) demonstrated higher intensities of nest defense toward a rubber snake among female red-winged blackbirds that had eggs added to the nests than those from which eggs had been subtracted (fig. 10.9). These results similarly hold for nestlings in some other species (table 10.10b), and findings cannot be attributed to any differences in chick weights caused by manipulation (Lambrechts et al. 2000).

Nonetheless, the failure to find consistent increases in nest defense following brood enlargement across species (table 10.10b) demands some explanation. Curio and Onnebrink (1995) argue that among great tits, the minimum distance that parents will approach a model woodpecker, a bird that threatens only chicks, is longer in artificially reduced broods compared with controls; but when facing a tawny owl, an adult predator, it is similar in experimental and control broods because of the high risk it imposes on parents. Another explanation for the failure to find an effect of clutch manipulation on nest defense is that parents may not be able to assess the survival prospects of their young, for example, if the adults are migratory and spend little time in the breeding site (Tolonen and Korpimaki 1995).

Table 10.10 Effects of brood size on nest defense

Species	Predator	A or P[a]	E or F[b]	Effect[c]	Reference[d]
(a) Natural broods					
Wigeon	Human	P	E	No effect	1
Northern shoveler	Human	P	E	Increase	1
Blue-winged teal	Human	P	E	Increase	1
Mallard	Human	P	E	Increase	1, 2
Gadwall	Human	P	E	No effect	1
Lesser scaup	Human	P	E	No effect	1
Red grouse	Dog	P	E F	No effect Increase in low risk defense	3
Snowy owl	Human	A	F	No effect	4
Tawny owl	Human	A	F	Increase	5
American robin	Blue jay mount and rubber snake	A	E F	No effect Increase	6
Gray catbird	Blue jay mount and rubber snake	A	E F	No effect No effect	6
Northern mockingbird	Human	A	E F	No effect No effect	7
Brown thrasher	Blue jay mount and rubber snake	A	E F	No effect Increase	6
Stonechat	Human	A	F	Whits: Increase Chats: No effect	8
Great tit	Pigmy owl	A	F	1st brood: No effect 2nd brood: Increase	9
Great tit	Great spotted woodpecker	A	F	Male: Increase	10
Willow tit	Stoat mount	A	E F	No effect No effect	11
Meadow pipit	Human	A	?	No effect	12
Northern cardinal	Blue jay mount and rubber snake	A	E F	No effect Increase	6
(b) Manipulated broods					
Merlin	Raven mount	A	E[e] F[e] F	No effect Increase Increase	13
Eurasian kestrel	Pine marten mount	A	E[e] F[e] E F	Males: No effect Females: Increase Males: No effect Females: No effect No effect No effect	14
Great tit	Tawny owl	A	F	No effect	15
Great tit	Pigmy owl	A	E	Decline with reduced brood size	16
Great tit	Woodpecker	A	F	Decline with reduced brood size	15
Blue tit	Red squirrel mount	A	E	Decline with reduced brood size	17

Table 10.10 *(continued)*

Species	Predator	A or P[a]	E or F[b]	Effect[c]	Reference[d]
American goldfinch	American kestrel and blue jay mounts	A	E and F	Increase with more offspring, decline with less offspring	18
Red-winged blackbird	Rubber snake	A	E	Increase	19
			F	No effect	

[a] A altricial; P precocial.
[b] E eggs; F fledglings.
[c] Changes in nest defense as clutch or brood size increases.
[d] 1. Forbes et al. 1994; 2. Albrecht and Klvana 2004; 3. P. J. Hudson and Newborn 1990; 4. Wiklund and Stigh 1983; 5. Wallin 1987; 6. Gottfried 1979; 7. Breitwisch 1988; 8. Greig-Smith 1980; 9. Curio, Regelmann, and Zimmerman 1984, 1985; Regelmann and Curio 1983; 10. Radford and Blakey 2000; 11. Rytkonen et al. 1995; 12. Pavel et al. 2000; 13. Wiklund 1990b; 14. Tolonen and Korpimaki 1995; 15. Curio and Onnebrink 1995; 16. Windt and Curio 1986; 17. Lambrechts et al. 2000; 18. R. L. Knight and Temple 1986b; 19. Robertson and Biermann 1979.
[e] Natural brood sizes.

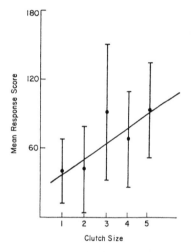

Figure 10.9 Mean (and SD) response scores of female red-winged blackbirds to a predator model placed on nests with 1–5 eggs. The mean was calculated from all tests of each clutch size over the breeding season (R. J. Robertson and Biermann 1979).

10.12 Offspring condition

The reproductive value of the current brood will be influenced by the probability that it will survive once it has fledged. Consequently, broods in better condition or of better quality should be defended more strongly by their parents than those in poor condition or poor quality (Carlisle 1982; Montgomerie and Weatherhead 1988). Unfortunately, it is not clear whether parents should compare offspring quality to their own brood or to that of the population. Theoretically, parents might reduce investment in superior offspring if returns on parental expenditure are lower than for inferior offspring (Clutton-Brock 1991), but this is unlikely to apply to nest defense, where a given parental act will probably alter the chances of a superior or inferior brood surviving in the same way. In many temperate birds, clutches produced late in the breeding season have a lower survival probability than earlier clutches, and there is some evidence that parents defend later broods less assiduously than

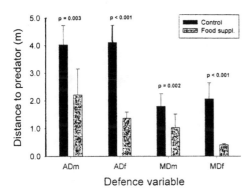

Figure 10.10 Nest defense (mean + SE) of female (*f*) and male (*m*) great tit parents in food-supplemented (*stippled bars*) and control broods (*solid bars*). Defense variables were average (*AD*) and minimum (*MD*) distance to predator model, proportion of time bird moved (*MOD*), and alarm called (*CALL*). The difference between supplemented and control groups was tested with nonparametric resampling tests. One-tailed *P*-values are presented; significant differences at $\alpha = 0.05$ after sequential Bonferroni correction are indicated (Rytkonen 2002).

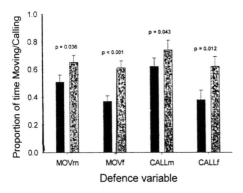

early ones. Snowy owls reduce nest defense during the breeding season irrespective of brood size which matches reduced survivorship of owlets (Wallin 1987).

Nestling weight at fledgling is a good predictor of offspring survival to next season in temperate latitudes. Studies that have matched nest defense to offspring mass have shown mixed results, however. Against the hypothesis, nest defense did not increase with nestling weight in great tits (Curio and Regelmann 1987) or willow tits (Rytkonen et al. 1995). In favor of the hypothesis, blue tits showed more intense nest defense of heavier than lighter chicks (Lambrechts et al. 2000), and parent great tits took greater risks for food-supplemented broods than for controls (fig. 10.10). In Tengmalm's owls, males, although not females, showed greater intensities of nest defense in years when vole (*Microtus* and *Clethrionomys*) populations were increasing and hence owlet survival prospects were strong (Hakkarainen and Korpimaki 1994). In short, no clear pattern has yet emerged between patterns of nest defense and the offspring quality aspect of reproductive value of the brood.

10.12.a Harm-to-offspring hypothesis   Parent birds giving a distraction display, perching, hovering over, or diving at predators, leave their offspring temporarily and must forgo the chance of feeding them. Eggs or nestlings that are entirely dependent on their parents might be harmed if not brooded or fed for a period of time. If parental defense of nests is geared to minimizing harm to offspring, parents should take greater risks and return to the nest sooner when their nestlings are in poor condition despite the predator still being present, because the marginal benefit from curtailing nest defense and returning to feed vulnerable offspring or keep them warm is greater than for less vulnerable offspring (Dale, Gustavsen, and Slagsvold 1996). For the same reason, they should take greater risks for young as opposed to old offspring (ibid.). In contrast with these predictions, the offspring quality component of the brood value hypotheses predicts parents should take increased risks for offspring in good condition, while the offspring age component stresses greater risks for older offspring. Actually, the harm-to-offspring hypothesis and the brood-value or reproductive-value hypothesis are not mutually exclusive conceptually, because the former focuses on risks or costs to offspring while the latter focuses on benefits, but they lead to opposing empirical predictions.

There is some support for the harm-to-offspring hypothesis. Brood defense increases in bad weather (Regelmann and Curio 1983) and under cold conditions (Bures and Pavel 1997). For example, female meadow pipits return to their nest sooner after responding to a stuffed European weasel when their chicks are cold (table 10.11). In an experimental test of the hypothesis, the condition of offspring of pied flycatchers was manipulated by taking chicks from nests and holding them without food for three hours (Listoen, Karlsen, and Slagsvold 2000). Under threat of nest predation from a stuffed sparrow hawk, parents took less time to enter the nest box if offspring had been starved that day than if they had been fed, and even took somewhat greater risks if off-

Table 10.11 Paired comparison of elapsed time (s) until the first entry of female parent meadow pipits to the nest

Nest	Nestling condition	
	Poor	Good
1	625	902
2	200	1030
3	125	280
4	217	1105
5	104	335
6	325	564
7	75	552

Source: Bures and Pavel 1997.
Note: Trials were carried out at 7 nests when nestling condition was good and when it was poor, as measured by chick cloacal temperature.

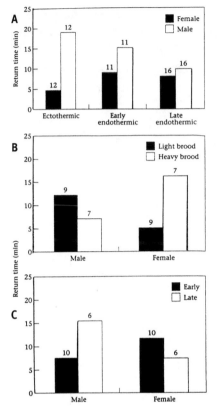

Figure 10.11 Median time collared flycatcher females and males took to return to the nest after presentation of a stuffed sparrow hawk in relation to A, 3 physiological phases of the nestling period: ectothermic 3- to 4-day-old nestlings, early endothermic 6- to 7-day-old nestlings, and late endothermic 10- to 11-day-old nestlings; to B, body mass of nestlings in the late endothermic phase, and C, whether the brood was early (first egg laid on or before median laying date) or late (started after median laying date). Sample sizes are given above the bars (Michl et al. 2000).

spring had been starved the previous day. While other explanations are possible, such as attempting to quiet begging chicks, the study supports the harm-to-offspring hypothesis.

In the collared flycatcher, male parents take greater risks, as measured by return time under predation risk, for older and heavier nestlings and broods laid earlier, while female parents show the reverse pattern (fig. 10.11). Older nestlings represent greater reproductive value to parents, because their probability of survival improves with age, as does the cost of replacing them; heavier offspring are more likely to survive and so have greater reproductive value; and broods laid earlier are more likely to survive. Thus, patterns of defense in male collared flycatchers follow the brood-value hypothesis, while those of females follow the harm-to-offspring hypothesis. Sex differences in defense may result from death of nestlings being more costly to the female because she incubates and broods, forcing her to take greater risks for offspring that are less likely to survive. Males, on the other hand, with opportunities for extra-pair copulations and polygyny, may be willing only to take risks in defense of high-quality offspring (Michl et al. 2000).

The study of nest defense in birds shows that it is the most theoretically well-grounded aspect of antipredator defense in homeotherms, because it takes its predictions directly from parental investment theory using parental defense as a testing ground. While empirical results are still at an early stage of development and often fail to provide clear support for or against predictions, it is probably just a matter of time before sufficient data are amassed to address these issues. How disheartening that comparable progress has yet to be made in setting up so many and such clear predictions in other realms of antipredator defenses.

10.13 Parental defense in mammals

Attempts to understand the extent, costs, and benefits of behavior shown by mammalian parents when defending their offspring against predators lag far behind those for birds. In some species, patterns of defense rise during pregnancy and even further during lactation (for example, Garner and Morrison 1980), and maternal defense of offspring has been documented for many species but usually for defense against conspecifics, often infanticidal males (Maestripieri 1992). Accounts of mothers defending offspring against predators are primarily confined to large ungulates (table 10.12). Most of these ex-

Table 10.12 Examples of maternal defense of offspring against predators in larger herbivores

Prey	Predator	Reference[a]	
Burchell's zebra	Spotted hyena	1	
Black rhinoceros	Spotted hyena	2	
Moose	Wolf	3	
White-tailed deer	Coyote	4	
Mule deer	Coyote	5	
Pronghorn	Coyote	6	
Wildebeest	Cheetah, spotted hyena, wild dog	7 1 8[b]	
Thomson's gazelle	Golden jackal	9	
African buffalo	Lion	10	
Mountain goat	Wolf	11	
Musk ox	Wolf	12	
Bighorn sheep	Coyote, bobcat	13 14	

[a]1. Kruuk 1972; 2. Berger and Cunningham 1994; 3. Stephenson and Ballenberghe 1995; 4. Garner and Morrison 1980; 5. Hamlin and Schweitzer 1979; 6. Lipetz and Bekoff 1980; 7. Caro 1994a; 8. Creel and Creel 2002; 9. Lamprecht 1978; 10. Schaller 1972; 11. Cote, Peracino, and Simard 1997; 12. Gray 1987; 13. Berger 1978b; 14. Hornocker 1969.
[b]Mothers had no effect in reducing predation success on calves.

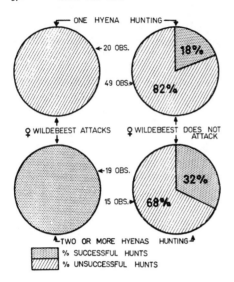

Figure 10.12 The effect of attacks by wildebeest cows on the success of different numbers of spotted hyenas chasing their calves. Pie charts above refer to 1 hyena hunting; those below, to 2 or more. Pie charts on the left refer to wildebeest mothers attacking the predator(s); those on the right, to not attacking (Kruuk 1972).

amples are of mothers defending fawns or calves against terrestrial carnivores, but mothers will also attempt to drive raptors from young (Packer 1983). Defense in these species consists of approaching, chasing, and butting the predator and striking out with the forelegs (L. W. Richardson et al. 1983); Burchell's zebras have even been seen biting spotted hyenas and killing them (A. Cullen 1969). In some species, males take an active role in defense of putative offspring (Marion and Sexton 1979; Kruuk 1972; Hauser 1986).

Defense of offspring can be highly effective. For example, 100% of attempts to capture wildebeest calves by single spotted hyenas failed because the mother attacked the hyena (fig. 10.12). Similarly, defending moose, Thomson's gazelle, mule deer, and wildebeest mothers are able to prevent wolves, golden jackals, coyotes, and cheetahs, respectively, from capturing their offspring (D. Mech 1970; Lamprecht 1978; Hamlin and Schweitzer 1979; Caro 1994a). Indeed, carnivores hunting in pairs or larger groups gain advantages from one predator facing the defending mother while the other tackles the calf (fig. 10.12; Caro 1994a). There are many reports of mothers being killed in defense of calves (for example, Creel and Creel 2002).

Few studies have made any effort to examine the evolutionary forces underlying parental defense in mammals as they have in birds. White-tailed deer vary their defense of fawns according to expected costs (the average physical condition of reproductive females in the population) and benefits (probability of fawn recruitment) (W. P. Smith 1987). Vervet monkeys respond to leopard alarm calls at different intensities as offspring become older, but the pattern is the reverse of what is expected from the trajectory of infant mortality, making interpretation awkward (Hauser 1988a). California ground squirrel mothers are less inclined to engage in risky behavior in response to playbacks of rattlesnake rattles than nonmaternal females and males, and risk-taking was negatively correlated with offspring age, suggesting that offspring vulnerability may be more important than offspring value in structuring antipredator behavior (Swaisgood, Owings, and Rowe 1999; Swaisgood, Rowe, and Owings 1999, 2003).

In cheetahs, mothers show more-offensive behavior—that is, stalking, chasing, or slapping other predators—when their cubs are young (2–5 months old) than when they are older, suggesting mothers are tailoring risky defense to offspring vulnerability because young cubs are extremely susceptible to predation (Caro 1994a; Laurenson 1994). Cheetah mothers' behavior is also sensitive to personal risk, however. Against lions and male cheetahs, which are dangerous for adult cheetahs, mothers do not go on the offensive more to protect young than older cubs; but against spotted hyenas and jackals, which pose less of a risk to adults, mothers are more offensive when protecting young than older cubs (Caro 1994a). Finally, mothers are more offensive against spotted hyenas if they had litters of 3 or 4 young cubs as opposed to litters of 1 or 2. The relative immobility of young cubs means that more offspring are likely to be taken in an attack on larger litters. Later, however, when cubs are older and more mobile, only one might be taken during an attack, and at this stage mothers' antipredator behavior becomes independent of litter size (Caro 1987). These defensive patterns correspond to those uncovered for birds, potentially indicating close parallels in the mechanisms driving offspring defense in the two classes.

10.14 Summary

Nest predation is a key component of offspring mortality in birds, and parents defend their nest against predators in many ways that range from scolding predators, to direct attacks, to attempting to draw predators from the nest. Nest defense is studied experimentally by recording birds' behavior when a human approaches the nest or when a model predator is placed close to the nest, or else by comparing artificial nests with natural nests, thereby teasing out the effect of parental activities. Each method has potential problems, but they are frequently disregarded.

Distraction displays are a type of nest defense that, by design, appear to dupe the predator, drawing it away from the nest. They include the broken wing display and false brooding and are commonly seen in shorebirds.

Nest defense is assumed to be dangerous, although there are surprisingly few reports of birds being killed while defending their nest. The benefits of nest defense are to drive predators away and to silence offspring; begging calls of nestlings attract mammalian predators.

The intensity with which parents defend nests is a trade-off between survival of the current brood and the survival of the parent and its probability of raising future clutches, and is thus influenced by the threat to the current brood, the ability of parents to drive the predator away, and the risk of death or injury to the parent. Consequently, parents respond differentially according to offspring vulnerability and to the threat posed by different predators. These, in turn, are influenced by life history considerations. Short-lived species take greater risks against predators of young than long-lived species, which are more likely to be able to breed again. Optimal intensity of nest defense is influenced by the future costs of defending the current brood, which in turn are affected by the likelihood that parents can renest in the same season or be able to breed in subsequent seasons. Thus, nest defense is expected to increase over the course of the breeding season as opportunities to renest diminish, although empirical evidence for this is mixed. While long-lived birds show more defense with age as expected because of diminishing opportunities to breed again, the effect is confounded by increasing experience.

There are many hypotheses as to why male and female parents should show different patterns of nest defense. These include differential opportunities to renest, differential costs of reproduction, asymmetries in confidence of parenthood, and differential predation risk. Sex-specific patterns of nest defense in nature have yet to be convincingly related to theory. While parents often defend their nest together, the costs and benefits of joint defense are poorly understood and cannot be referred to current models of parental investment at present.

Nest defense is expected to increase with offspring age, irrespective of renesting potential, as the amount of investment necessary to replace the clutch rises, as the value of the brood increases, and because vulnerability may increase with age. A swath of empirical studies show that nest defense does increase with age of the young. While studies indicate that increasing value of the brood with age drives patterns of increasing defense, there is little support for parents matching defense to offspring vulnerability. A third idea suggests that increasing defense is simply a methodological consequence of positive reinforcement that the parents obtain when threats are removed from their nest, but it is not well supported.

Defense is also expected to increase with number of offspring in situations in which predators take the whole brood. Experimental manipulations of brood size, as opposed to natural brood sizes, confirm that defense increases with offspring number. Defense is expected to increase with offspring quality or condition because this also reflects brood value, but there is no strong evidence for or against the hypothesis. Conversely, the harm-to-offspring hypothesis suggests that parents will take greater risks in defense of offspring in poor condition because they will suffer greater harm if parents stay away from the nest for long periods. This hypothesis has some empirical support.

Studies of parental defense in mammals, as they relate to predators, are principally descriptive and demonstrate that ungulate mothers can often drive single but not groups of carnivores away from their calves. The relatively few studies that have viewed offspring defense in mammals as a form of parental investment have found similar results as those in birds.

11 Mobbing and Group Defense

11.1 Introduction

Mobbing refers to a constellation of behavior patterns that involve prey animals approaching, observing, and usually harassing a predator before a predatory attack ensues. In descriptive terms, mobbing includes specific vocalizations that can often grade into warning signals (chapter 6), and advances toward the predator that can vary from inspection behavior (section 7.6.c), to close following, to haranguing the predator, to physical attack (M. J. Smith and Graves 1978; Berger 1979). Mobbing is principally a characteristic of groups and can hence be regarded as a type of group defense or attack, although it is sometimes initiated by a solitary individual and thus elides into forms of individual or even offspring defense (chapter 10). Frequently, several different prey species will mob a predator at the same time (for example, Pettifor 1990; Pavey and Smyth 1998), and mobbing vocalizations of one species carry meaning for others (Hurd 1996; M. S. Ficken 2000). Functionally, the significance of mobbing is, like so many antipredator behaviors, under discussion and likely has several consequences for members of the mob. Finally, and linked to this, mobbing has diverse outcomes for predators that vary from being struck, pecked, bitten, and even killed by prey, to abandoning a hunt, leaving the area, or sometimes even capturing a mobber. For all these reasons, mobbing is difficult to define (section 11.2). Furthermore, there is great variability in mobbing responses between and within species (section 11.3)

Mobbing is an intriguing form of defense, first because it necessitates pred-

---

Figure 11.0 (*Facing page*) Musk oxen form a defensive circle, with horns and bosses facing outward. There are few alternative options available in the tundra, because musk oxen cannot hide when a predator is in the vicinity due to little cover, and flight is impeded by snow; instead, they use body weight and weaponry to deter attack (reproduced by kind permission of Sheila Girling).

ator recognition (Altmann 1956) and has therefore been used as a tool for exploring the salient features of predator recognition (chapter 1). Second, certain species have specialized behavior patterns used only while mobbing, such as fieldfares bombarding a predator with feces, thereby destroying water-repellant insulating properties of plumage (Haland 1989; Wiklund and Andersson 1994). Third, mobbing brings prey closer to the predator and is therefore assumed to raise the probability of capture, although evidence for increased risk is mostly indirect (section 11.4). Assuming costs are involved, there must be compensatory benefits, and it is the nature of these benefits that forms the central focus of inquiry into mobbing. In a seminal paper, Curio (1978) formulated nine nonmutually exclusive hypotheses for the adaptive significance of mobbing behavior. When data from different species are set side by side, there is evidence to support a large number of them (section 11.5). On the basis of this evidence, we are forced into accepting that many selective pressures act to maintain mobbing in homeotherms.

Mobbing, broadly characterized, occurs principally in birds, although aspects are seen in insects (Seeley, Seeley, and Akratanakul 1982), fish (Ishihara 1987; R. J. F. Smith 1997), and mammals (Loughry 1988). Instances of mobbing may be more widespread in avian prey than in mammalian prey because birds are often attacked by diurnal predators, whereas mammals are chiefly preyed on by nocturnal or crepuscular mammals, which are more difficult for prey to see. Thus, virtually all the empirical data used to test hypotheses about mobbing derive from birds, with documentation being still at a descriptive stage for mammals. Given that there is considerable indirect and some direct evidence that mobbing deters predators from approaching adults or young, and because nesting birds, notably ground-nesting waterfowl and seabirds, mob predators in groups and have higher reproductive success when nesting in colonies than alone, mobbing is thought to be one of the driving forces behind the evolution of colonial nesting (section 11.6). Moreover, certain relatively defenseless species gain protection by nesting in close proximity to raptors or pugnacious species that drive predators from their own nests (section 11.7); thus, mobbing is seen as an important benefit of mixed-species nesting associations.

In mammals, group defense consists of animals jointly defending themselves against a predator or attacking it (depending on the prey species, type of predator, and nature of the threat), but it can easily turn into predator inspection, harassment, or monitoring and is therefore closely analogous to mobbing (section 11.8). Studies of group defense in mammals are somewhat restricted taxonomically, but they suggest that mammals use group defense when crypsis is difficult and flight is ill advised.

11.2 Definition of mobbing

Hartley (1950) was the first to offer a definition of mobbing:

> A mobbing is a demonstration made by a bird against a potential or supposed en-
> emy belonging to another and more powerful species; it is initiated by the member
> of the weaker species, and is not a reaction to an attack upon the person, mate, nest,
> eggs or young of the bird which begins it. (p. 315)

The proactive nature of mobbing prior to a predatory attack is a key feature of
this definition; it draws attention to the fact that prey will spontaneously respond to predators that are not hunting. The advantage of this definition is
that it is descriptive rather than functional; any restricted functional definition will be problematic, because there seem to be so many consequences of
mobbing. Its disadvantage is that a demonstration could also refer to warning
signals that alert others to the presence of a predator and, as a consequence,
cause them to flee, become vigilant, or stay quiet rather than mob. While
warning calls are sometimes structurally different from calls used to recruit
others to harass a predator (Greig-Smith 1980; Klump and Shalter 1984; R. L.
Knight and Temple 1988; Stone and Trost 1991), distinguishing them from
mobbing calls can be difficult in practice (Kruuk 1976; Walters 1990).

Shields (1984) provided a more comprehensive definition of avian mobbing:

> [A]n approach towards a potentially dangerous predator (whether it is actively
> hunting or not), followed by frequent position changes with most movements cen-
> tred on the predator. Relatively stereotyped visual displays and loud and localizable
> vocal displays usually accompany the locomotion. Often mobbing includes swoops
> or runs at a potential predator and it may involve direct attack, with physical con-
> tact by the mobber. (p. 132)

Again, the definition is structural and includes physical contact, reflecting
those instances in which prey strike the predator (incidentally linking it to aspects of group defense in mammals). Localizability was included, because
Marler (1959) had proposed that calls given while mobbing covered a wide frequency range and had abrupt onsets and ends, making them easy for conspecifics to localize. Subsequent work (Ficken and Popp 1996) has questioned
this premise. Across 52 passerines, approximately half do not emit brief, wide-
frequency-band calls while mobbing; some of the other species utter more
than one mobbing call, only one of which has a wide-frequency range (but see
K. J. Jones and Hill 2001). It is worth noting that neither definition refers to
what appears to be mobbing induced by dead or injured conspecifics or by
brood parasites (I. G. McLean 1987).

It is difficult to distinguish mobbing from nest defense because (i) a nesting bird may react to a predator that enters its territory but is far from the nest site, (ii) both parents or even helpers at the nest may respond together, or, alternatively, (iii) many of the behaviors used to characterize mobbing may be performed by a single individual on its own. Thus, criteria of nesting or not nesting, individual or group behavior, predators hunting or not hunting, or even a combination of these are inadequate to separate strictly these two phenomena. Instead, it may be more useful to abandon dichotomizing behavior and instead consider offspring protection as an additional benefit to defensive behavior, and conspecifics as reducing the costs and increasing the effectiveness of harassment. In this chapter, however, I generally use being in a group of two or more as the principal criterion for distinguishing mobbing from nest defense.

11.3 Variation in mobbing behavior

A further impediment to a satisfactory definition of mobbing is its variability (for example, Slagsvold 1985). First, closely related species show considerable differences in their propensity to mob. For example, carrion crows, rooks, Eurasian jackdaws, and black-billed magpies react quite differently to a caged carrion crow, a species that is a threat to all of their nests in the wild. Free-ranging magpies attack the caged bird most often, followed by rooks, but jackdaws never do; rooks scold and circle the crow more than other species; a crow hardly ever returns to its nest during testing (Roell and Bossema 1982). Reasons for these differences probably lie in nesting habits. Jackdaws nest in holes and do not need to be aggressive toward crows. Rooks nest in colonies and may thus rely on commotion as a deterrent. Magpies, on the other hand, nest in pairs and cannot rely on large numbers of conspecifics; instead they resort to direct attacks. Crows' nests are inconspicuous, and it may pay them to leave the area so that intruders do not locate young (see also Quinn et al. 2003).

Second, a given species may mob different species of predator in different ways and to differing extents (Buitron 1983; Griesser and Ekman 2005). Arabian babblers utter "tzwicks" when they first see an owl or a cat but then switch to trills with the former yet continue using "tzwicks" with the latter (Naguib et al. 1999). Possibly, call differences convey information about level of threat to conspecifics. Among nesting black-headed gulls at Ravenglass, Cumbria, UK, where red foxes pose a greater threat than stoats, and stoats a greater menace than European hedgehogs, adult gulls mob foxes at a higher rate than stoats but approach them less closely. Compared with stoats, gulls mob hedgehogs at a lower rate but get much closer to them, hitting them on most occasions (Kruuk 1964; see also Elliot 1985b).

Third, different calls are sometimes given to the same predator. Red-winged blackbirds give seven different calls to red-tailed hawk and raccoon mounts and to a model American carrion crow (table 11.1), although call rates

Table 11.1 Percentage of different types of calls used in nest defense by 30 pairs of red-winged blackbirds and proportions of individuals that gave the calls

Call types	Calls by males		Calls by females	
	Percentage of all calls	Proportion of birds giving call	Percentage of all calls	Proportion of birds giving call
Check	59.6	30/30	84.5	30/30
Chatter	0	0/30	0.9	15/30
Growl	2.8	11/30	0	0/30
Scream	0	0/30	13.7	8/30
Peet	1.2	1/30	0	0/30
Cheer	32.8	29/30	0	0/30
Seeet	3.6	9/30	0	0/30
Total calls given	2958	—	5979	—

Source: R. L. Knight and Temple 1988.
Note: Red-tailed hawk, American crow, and raccoon models were each presented to 10 different pairs of blackbirds when the nests contained eggs, 6–8 days after initiation of incubation.

Table 11.2 Ranks of mobbing intensity in Florida scrub jays to the presentation of a gopher snake (Florida pine snake) in the field

	Percentage of time mobbed	Latency	Closest approach	Bite snake	Other nonflight action patterns	Flight-derived action patterns	Sum of ranks
Male breeder	4	4	4	4	3.5	3	22.5
Female breeder	2.5	3	2	2	3.5	3	16.0
Helpers	2.5	2	3	2	2	3	14.5
Fledglings	1	1	1	2	1	1	7.0

Source: Francis, Hailman, and Woolfenden 1989.

do not differ between models. Screams, in particular, attract other blackbirds and elicit hovering over the models (R. L. Knight and Temple 1988).

Fourth, there is individual variation in mobbing responses (chapter 10). (i) In R. L. Knight and Temple's (1988) study, for instance, females gave more "checks" than males (table 11.1), whereas males approached the models more closely, and hovered over and struck the models more than females. (ii) Mobbing behavior varies considerably over the course of the reproductive cycle (chapter 10). Among birds such as barn swallows, which nest colonially, some individuals, termed active mobbers, emit mobbing calls, approach a stuffed eastern screech owl or a human to < 2 m, and even hit them; others (passive mobbers) are silent and circle at 2–10 m from the stimulus (Shields 1984). In tree swallows, parents of nestlings under threat are active mobbers, whereas neighbors mob passively (Winkler 1994). (iii) In cooperative breeders, helpers may mob less than adults (table 11.2). (iv) Young birds may mob less than adults, because young may be poor at recognizing predators, take time to develop mobbing responses, and be inept at flying from sources of danger (table 11.2; see also Griesser and Ekman 2005). Indeed, among Florida scrub jays, mobbing calls by fledglings are often ignored by adults (Francis, Hailman, and Woolfenden 1989).

Fifth, an individual's mobbing behavior varies according to conditions associated with risk. For example, within an age-sex class, larger, more-dominant birds are more likely to mob (Slagsvold 1985). Even during an approach toward a predator, the length of each advance decreases progressively to give prey greater opportunity to assess changing risk, whereas calling rate increases, perhaps to intimidate the predator more effectively (Curio and Regelmann 1985). Variation in the form that mobbing takes within individuals, between individuals, and across species makes an all-encompassing definition precarious.

11.4 Costs of mobbing

Presumably, mobbing is costly because it increases prey's proximity to the predator and possibly increases the length of time that prey stays in the predator's vicinity (Dugatkin and Godin 1992). In support of a supposed increase in risk, there are anecdotal accounts of predators capturing prey that mob them. For instance, a review reported 30 instances of different bird species being attacked, captured, or killed, or dying of stress while mobbing (Sordahl 1990), the most cited of which is Denson's (1979) report of a great horned owl capturing and flying away with one of several American crows that had been mobbing it. Nonetheless, the number of these anecdotes is surprisingly few. Moreover, we lack comparative information on the probability of prey being killed when it is and is not mobbing (Hennessy 1986). As a consequence, we have to resort to using the design features of mobbing to impute risk. These include approaching more-dangerous predators less frequently or closely than more-benign ones (Curio 1975; Coss and Owings 1985); greater likelihood of approach when mobbing calls are played far from a dangerous location, such as a raptor's nest (Forsman and Monkkonen 2001); more-cautious harassing where escape is impeded (Verbeek 1985); a closer approach when more than one mobber is involved (S. K. Robinson 1985; Curio and Regelmann 1986); a greater propensity to mob by larger and more-agile prey species (Forsman and Monkkonen 2001); and more caution shown toward predators in prey populations that have evolved in sympatry with dangerous predators than in those that have experienced only weak predation pressure over recent evolutionary time (Coss and Owings 1985; Towers and Coss 1990). Additional survivorship costs include reports of loud mobbing enticing predators to nest boxes (Krama and Krams 2005), attracting other predators to the area (McLean, Smith, and Stewart 1986), the clamor allowing predators to follow flocks through thick vegetation, mobbers colliding with each other or the ground during a melee

(Conover 1987), and young being left unprotected when parents leave to mob (Emlen et al. 1966; McNicholl 1973). These indirect data are extensive and therefore persuasive (but see Desrochers, Belisle, and Bourque 2002), but they indicate only that mobbing is risky and do not address extent of risk.

Circling a predator may be energetically costly, which could explain why more birds choose to mob a predator on windy days (Conover 1987); moreover, raptors such as sparrow hawks may be aerodynamically less stable on windy days, lowering risks of counterattack (Quinn and Cresswell 2004). Opportunity costs may involve reduction in time spent feeding (Winkler 1994; Gleason and Norconk 2002) and, for nesting birds, chilling of eggs or fledglings when the parent is off the nest (Emlen et al. 1966). Unfortunately, energetic and time costs of mobbing have yet to be quantified for any species.

11.5 Benefits of mobbing

That mobbing is an effective antipredator strategy comes from two sources: reduced predation in colonially nesting birds and increased breeding success as a result of nest defense by parents (chapter 10). In the former case, findings that predation rates are lower where birds nest close together has been coupled with observations that mobbing behavior reduces predation (table 11.3), although few of these studies show that it is mobbing per se, as opposed to other group-related benefits (chapter 8), that increases breeding success. In addition, neither area of inquiry is explicit about the evolutionary mechanism by which mobbing might raise fitness, because it might benefit the mobber or its relatives in many different ways (table 11.4). Such hypothesized benefits might accrue simultaneously, or be important only for some species or in some situations but not in others. At present, some of these hypotheses as to how mobbing confers fitness advantages have received considerable support, while others have received much less or have yet to be investigated.

11.5.a Direct benefits: lethal counterattack   The most obvious consequence of mobbing is a damaging attack on the predator that kills or injures it and thereby eliminates risk (Meinertzhagen 1959; Stanford 1998). There are rare accounts of a predator being killed in a mobbing melee (Humphries and Driver 1970; Cowlishaw 1994) and more-common reports of predators defending themselves against attack (Veen 1977). Obviously, this hypothesis is applicable only to interactions between certain prey species and certain predators, perhaps where differences in body size are small or prey are equipped with weaponry. Where these factors pertain, the effect of mobbing calls on recruiting conspecific or heterospecific prey to a mobbing site may have severe consequences for predators. If more than one prey individual is involved, benefits are seen as mutualistic: prey are only able to tackle the predator as a

Table 11.3 Indirect evidence that mobbing reduces predation in colonially nesting birds

Prey	Predator	Observed outcome measure	Reference[a]
Predation lower with colonial nesting			
Lapwing	Carrion crow	Predation rates on artificial nests lower within 50 m of lapwing nests than farther away	1, 2, 3
Common gull	Herring gull, great black-backed gull, carrion crow, Eurasian eagle owl, American mink, European badger	Experimental eggs disappear more rapidly close to solitary nests than colonial nests	4
Black-headed gull	Red fox	Breeding success of outlying nests lower than colony nests	5
Fieldfare	Carrion crow, black-billed magpie, Eurasian jay, great spotted woodpecker, Eurasian red squirrel, small mustelids	Experimental eggs disappear more rapidly on their own than near solitary nests, than near colony	6, 7, 8
Bank swallow	Blue jay	Tethered young swallow more likely to survive near nesting area than farther away	9
Mobbing reduces predation			
Black-headed gull	Herring gull, carrion crow	Higher rate of mobbing attacks reduces predation success	10
Eastern kingbird	Crows and hawks	Successful pairs are more aggressive than pairs that lost nests	11
Stonechat	Eurasian kestrel and other raptors and shrikes	Rates of whits higher in nests that were were not predated than those that were	12
American goldfinch	American kestrel and blue jay mounts	Higher female call rates at successful nests than unsuccessful mounts	13
7 species of forest birds	Powerful owl	Predation rate 8.75 times higher on nonmobbing than on mobbing species	14

[a]1. Goransson et al. 1975; 2. Elliott 1985a; 3. Baines 1990; 4. Gotmark and Andersson 1984; 5. I. J. Patterson 1965; 6. Andersson and Wiklund 1978; 7. Wiklund and Andersson 1980; 8. V. Haas 1985; 9. Hoogland and Sherman 1976; 10. Kruuk 1964; 11. Blancher and Robertson 1982; 12. Greig-Smith 1980; 13. R. L. Knight and Temple 1986; 14. Pavey and Smyth 1998.

group, since they are unable to do so alone. Risk of a damaging attack may underpin the second, move-on, hypothesis for mobbing.

11.5.b Direct benefits: the move-on hypothesis   Curio (1978) stated that a predator should leave an area sooner the more intensely and/or longer it is molested. There are two parts to this hypothesis: mobbing should deter the predator from hunting, and second, should increase the probability that the predator will not return to the area, assuming that it normally revisits hunting locales where it has been successful (table 11.4). Thus, it is in the interests of each mobber to drive the predator away. There is backing for the first idea but very little information for or against the second (Lima 2002).

Table 11.4 Curio's (1978) original table of predictions based on functional explanations for mobbing behavior by birds

	Behavior of	
Hypotheses	Prey bird	Predator
Move on		Predator leaves site and avoids it later on
Perception advertisement	A (stalking or) hunting predator elicits mobbing more readily than a resting one	(a) Predator gives up hunting or does not attempt to do so (b) One mobber suffices to persuade the predator that it is discovered
Selfish herd	Mobbers congregate so that their "danger zones" overlap	
Confusion	Mobbers move independently and erratically from each other	Predator fails more often when attacking a mobbing assembly. More specifically, its failures increase with increasing heterogeneity of the assembly
Attract the mightier	Small predators elicit stronger mobbing than larger ones	The most powerful predators are attracted by mobbing calls
Alerting others	Mobber lowers a receiver's threshold for flight-eliciting stimuli of (a) the predator mobbed, (b) all possible predators; (c) kin selection maintains warning	Predator fails more often when attacking a mobbing assembly
Silencing offspring	Mobbing fluctuates in parallel with vocal begging of young. One or both parents suffice; few mobbing calls suffice	Locates young by ear
Cultural transmission (of predator)	(a) A dangerous species not recognized innately comes to elicit mobbing through experience with it, (b) still naïve birds become conditioned to that species by mobbing of a knowledgeable tutor; (c) kin selection maintains cultural transmission	
Cultural transmission (of site)	See (a), (b), and (c) above. (d) Educated individuals avoid site of encounter later on	Predator returns to site where mobbed

Note: a tenth hypothesis concerned distress screaming rather than mobbing (see chapter 12).

At a proximate level, mobbing appears to distress some predators. Laboratory work demonstrates that hawks and owls are bothered by playbacks of mobbing vocalizations, as indicated by panic flights and attempts to hide (Shalter 1978; Flasskamp 1994). In the field, birds are more likely to mob a flying raptor than one that is perched, indirectly supporting the move-on hypothesis (Bildstein 1982; Pettifor 1990); and in a key study, free-living Eurasian kestrels were shown to move farther after being mobbed than when they were not mobbed (table 11.5). This is perhaps the strongest evidence yet for the move-on hypothesis.

At an ultimate level, indirect evidence shows that mobbing may deter predators because parental call rates are higher at nests that survive the breeding

Table 11.5 Distances (m) that European kestrels moved between perches or wind-hovering positions by either perch- or flight-hunting kestrels when they were or were not being mobbed

Distance travelled	Perch hunting		Flight hunting	
	Not mobbed	Mobbed	Not mobbed	Mobbed
Mean	145.4	384.7	80.9	552.9
SD	176.3	340.7	78.4	348.1
N	150	17	150	31

Source: Pettifor 1990.

Table 11.6 Relation between attack frequency of black-backed gulls and predation success of crows and herring gulls

	Crow			Herring gull		
	Success			Success		
Number of attacks per second	+	−	%	+	−	%
0	60	2	97	10	0	100
>0, <1/6	25	8	76	2	0	100
>1/6, <1	20	12	63	17	4	81
>1	2	15	13	52	59	47

Source: Kruuk 1964.
Note: Number of observations shown.

season than those that fail. Among American goldfinches, female vocalizations in response to a model blue jay were higher at successful nests (at 21 nests, 41.2 calls/min on average) than at unsuccessful nests (at 24 nests, 24.5 calls/min on average: R. L. Knight and Temple 1986b; see also Greig-Smith 1980). In a classic study, Kruuk (1964) showed that predation success on eggs of black-backed gulls by both carrion crows and herring gulls was reduced as they came under higher rates of attack from nesting black-backed gulls (table 11.6), providing direct evidence for mobbing increasing a component of reproductive success (see also Blancher and Robertson 1982).

A second, central assumption of the move-on hypothesis is that it drives predators away from an area either in the medium or long term. Folklore tells us that foxes or raccoons return to places (such as chicken coops) where they have been successful previously, but it is less clear if they avoid areas where they have been unsuccessful, and only a few studies have verified this (for example, Tinbergen, Impekoven, and Franck 1967). There is a little information on the medium-term effects of mobbing on predators. For example, in Brisbane Forest Park, Queensland, Australia, powerful owls, huge nocturnal owls, are mobbed if found by small birds. Although rainforest makes up only 12% of habitat within the park, owls were found roosting in the rainforest on 113 occasions and in open forest on 116 occasions. As roosting in rainforest was associated with a lower frequency of harassment by forest birds, it appears that owls sought out thick vegetation to avoid being found by mobbers. This had the effect of moving owls away from vegetation, where the majority of small birds foraged (Pavey and Smyth 1998). More generally, mobbing may actually be a driving force selecting for cryptic diurnal behavior in owls (Chandler and Rose 1988). Nonetheless, we urgently need information on where predators choose to hunt following successful and unsuccessful hunts and bouts of mobbing.

11.5.c Direct benefits: perception advertisement   A third hypothesis is that mobbing informs the predator that it has been discovered by prey. It is not clear how this hypothesis differs from the move-on hypothesis, as prey must have seen the predator if they are harassing it. Moreover, it is not clear why prey should take putative risks in approaching the predator when other species use less-risky forms of perception advertisement that require no approach (section 7.6). Nonetheless, in situations in which a predator does not move on following mobbing, perhaps because it is much larger than the mobbers, mobbing will at least have informed the predator that it has been seen (Flasskamp 1994).

11.5.d Direct benefits: selfish-herd effect and confusion effect   Curio (1978) suggested that groups of mobbers gain benefits through dilution of risk (W. D. Hamilton 1971). It is true that mobbing does have a rallying effect on conspecifics: screams of red-winged blackbirds to model predators attract other nesting parents, which then start to call and hover over the model (R. L. Knight and Temple 1988); and playbacks of "sweeet" calls of American goldfinches attract conspecifics that perch and call near the speaker (R. L. Knight and Temple 1986b). While mobbing does rally other birds, including heterospecifics (Forsman and Monkkonen 2001), dilution of risk is not sufficient by itself to explain why a prey animal should go out of its way to join a mobber if it puts the joiner in greater jeopardy. Curio also suggested that a group of erratically flying birds might confuse a predator, but currently there is no direct evidence that mobbed predators are confused.

11.5.e Direct benefits: attract the mightier   Mobbing a predator might attract other, more powerful predators that could kill the weaker predator or drive it off (Bourne 1977; Curio 1978). This hypothesis is very similar to the one that has been put forward for the adaptive significance of distress calls, or fear screams, given out by some birds and mammals when they are forcibly handled (section 12.9). A thin scattering of studies support the idea of fear screams attracting other predators that then usurp the first predator to obtain a relatively easy meal; occasionally the prey item escapes in the melee. There are anecdotes of mobbing attracting other predators to an area. For instance, northwestern crows are attracted to mobbing calls of American robins (McLean, Smith, and Stewart 1986). There are no reports of a mobbing individual moving out of the first predator's vicinity more quickly if other predators arrive, however. In short, it is difficult to envisage how attracting additional predators could be of benefit to a mobber that has not actually been captured.

11.5.f Indirect benefits: alerting others   The alerting-others hypothesis views mobbing as a type of warning signal, with beneficiaries principally, but not necessarily, being kin, as, for example, in Siberian jays (Griesser and Ekman 2005). An experiment with European blackbirds clearly shows that mobbing alerts other blackbirds. One "tutor" blackbird was presented with a stuffed little owl in a situation in which the experimental blackbird (the observer) could see and hear only the mobber but not the owl. The experimental bird was then presented with a frightening stimulus, the sudden appearance of a black plywood board. In the presence of a mobbing bird, the experimental subject took off twice as quickly as in situations when the other bird was not mobbing (Frankenberg 1981). Latencies to take off can be critical in escaping incoming raptors (Kenward 1978). In a second experiment, where the frightening stimulus was presented to the observer in the opposite direction from where the owl sat in relation to the tutor, the observer took off in a direction away from the owl toward the frightening stimulus, even though it could see only the mobber and not the owl being mobbed. Thus, mobbing provides directional information about the predator. These experiments show that alerting others about the presence, and sometimes location, of predators must be an important consequence of mobbing. It is not obvious, however, why prey should have to approach the predator in order to alert conspecifics, suggesting that alerting others is unlikely to be the primary function of mobbing.

Other researchers have viewed mobbing as a way of keeping conspecifics aware of the predator and monitoring its movements (Owings and Hennessy 1984; Stanford 1998). In this case, the initial mobber gains direct benefits by being able to resume feeding earlier or start to feed intermittently, because its companions are watching the predator. The recruited individuals benefit by being able to observe the threat themselves and share the burden of vigilance with others: a putative case of mutualism.

11.5.g Indirect benefits: silencing offspring   Mobbing may signal to offspring to be silent in the presence of a predator. While the hypothesis can apply only in the breeding season, during that period it could be an additional or alternative to other benefits on Curio's list. Some vocalizations cause offspring to change their behavior. As illustrations, calls made by bank swallows to an approaching American kestrel cause nestlings perched at burrow entrances to retreat tail first into their tunnels (Windsor and Emlen 1975); mobbing calls of herring gulls result in chicks hiding under rocks (Burger 1981); and alarm calls of moustached warblers actually cause chicks to jump out of the nests on the day of fledgling (Kleindorfer, Hoi, and Fessl 1996). Some species of birds do seem to have specific calls for silencing begging offspring: "whit" calls in stonechats (Greig-Smith 1980) and "BB" calls in American goldfinches (R. L. Knight and Temple 1986b), for instance. It is not always clear, however, whether such calls are mobbing or warning calls. In one unequivocal case, however, playbacks of mobbing screams of red-winged blackbirds caused young to stop begging within 1–5 seconds and crouch low in the nest, as well as attracted other blackbirds, which hovered near the speaker (R. L. Knight and Temple 1988). This example shows that silencing offspring may be an important beneficial consequence of mobbing vocalizations during the breeding season.

11.5.h Benefits unclear: cultural transmission   Under the cultural-transmission hypothesis, an individual learns to fear an object that it witnesses other birds mob. As a result, it avoids it later on, or begins to mob the object itself. In another version of the hypothesis (table 11.4), the receiver learns to avoid the place where the predator was mobbed; here the predator is explicitly assumed to return to the place where it encountered prey. There is considerable experimental evidence for the first part of the hypothesis but not for the second.

In a well-known experiment, a captive European blackbird (the "observer") was kept singly in an aviary that was separated from a conspecific "teacher." Between the two aviaries was a cardboard box with four radially and horizontally oriented chambers that could be rotated by the experimenter. The observer could be presented with a stuffed male noisy friarbird, an Australian honeyeater and therefore an ontogenetically and evolutionarily novel bird, while the teacher could be shown a stuffed little owl at the same time through judicious rotation of the box. In this experiment the teacher mobbed the owl, and the observer could see the other blackbird's mobbing behavior but presumably assumed that it was mobbing the friarbird. Observers quickly started to mob the honeyeater when the conspecific mobbed the little owl, and they maintained this response in the absence of teacher, mobbing it vociferously two hours later when tested (fig. 1.7). Further experiments showed that blackbirds could be conditioned to mob even a multicolored plastic bottle. Moreover, when observer birds were given the task of teacher, they could condition new observers to mob the novel friarbird. There was no discernible deterioration of information transfer through six presentations involving six observers that subsequently became teachers (Curio, Ernst, and Vieth 1978a, 1978b). Indeed, taped conspecific and alien species' mobbing calls sufficed by themselves to transmit enemy recognition of the friarbird as well (Vieth, Curio, and Ernst 1980).

Field studies lend indirect support to the cultural-transmission hypothesis. On Walney Island, Cumbria, Kruuk (1976) presented herring gulls and lesser black-backed gulls with a model of a fox, stoat, or European hedgehog with or without a dead gull. Gull flock sizes and alighting distances increased when the dead gull was present. Then, in a follow-up experiment, when gulls were shown the stoat presented alone they habituated to it, as judged by reduced flock sizes, but birds that had seen the stoat with the dead gull earlier continued to go on flocking (fig. 1.5). They appeared to have learned that the stoat was dangerous as a result of seeing it with a dead gull. In another, related study, Conover (1987) tested whether ring-billed gulls attracted from more-distant parts of the colony (passive mobbers) learned from experience. He knew that most passive mobbers were recruited from nearby, the majority from less than 30 m away. First he presented a model owl or human in one part of the colony; then he presented the same stimulus 50 m away but in a place that was obscured from the first site. At the second site, flock sizes of mobbers and flight distances were reduced. At 100 m away, however, flock sizes and flight distances were the same as at the initial site. He concluded that passive mobbers had learned something about the predator when it was first presented, perhaps via the behavior of conspecifics, and had become less interested and wary of the threat.

While observers can learn from others' mobbing behavior, it is unclear how this mechanism could benefit the teacher without knowing something about the social context under which cultural transmission occurs in the wild. Where mobbing is most prevalent in breeding pairs or among groups of related birds, kin selection is likely; where it is most prevalent in flocks, direct benefits or reciprocal altruism are imputed. Direct selection could operate through teaching others the characteristics of threatening stimuli, thereby increasing mobbing group size and predator harassment. Reciprocal altruism might operate if observers recognize future threats and warn the teacher of their approach, but this could only occur in stable groups where these individuals repeatedly encounter predators together. The extent to which flock composition of various species remains stable over time is likely to be very variable, however.

There is little support for mobbing conveying information about the site where the predator appears. While European starlings were more wary of an owl model after seeing it with a tethered starling than when the owl had been shown alone, they did not avoid the site at which the owl model had been presented, thereby failing to support the site-information hypothesis (Conover and Perito 1981). Similarly, on Walney Island, gulls' characteristically weak reactions to a stuffed hedgehog did not increase after gulls had mobbed a stuffed stoat and a dead gull in the same place (Kruuk 1976). The site-information hy-pothesis will only be relevant when predators' movements are localized or repetitious.

11.5.i Other hypotheses   One hypothesis not mentioned by Curio is that mobbing is a demonstration of fitness to conspecifics (Dugatkin and Godin 1992; Maklakov 2002). This hypothesis can help explain why birds sometimes mob an injured or dead conspecific (Slagsvold 1984b) and why certain species, such as the famous noisy miner, mob small and harmless heterospecifics so vigorously (K. E. Arnold 2000). For example, dominant birds often approach a hawk more closely than do subordinate birds (Moholt and Trost 1989), but it is not clear as to whether it is less costly for them to alight near a threat or whether they are demonstrating to conspecifics their ability to threaten and escape. More-aggressive mobbing of a dead hooded crow in flocking areas than on breeding territories supports the idea of self-advertisement, because several other crows are present nearby. That a stuffed Eurasian eagle owl was mobbed more on breeding territories than in flocking areas, however, hints at predator mobbing having an antipredator function but conspecific mobbing having another function in this species (Slagsvold 1984b).

A related hypothesis, that mobbing is a demonstration of prey's ability to escape predation (quality advertisement, section 7.6), has received no atten-tion (Dugatkin and Godin 1992).

11.6 Mobbing and group size

The move-on hypothesis for mobbing in birds has received the most support, and the idea is given further backing by three related pieces of information: breeding success increases in larger colonies of birds; in larger colonies, more birds mob predators; and mobs of birds thwart predatory attack. First, the as-sociation between egg survival and nesting in groups as opposed to nesting alone is well established in species such as fieldfares (Andersson and Wiklund 1978) and common gulls (Gotmark and Andersson 1984). Furthermore, breeding success is known to be higher with increasing number of close neighbors (for example, lapwings: Berg, Lindberg, and Kallebrink 1992) and for central versus peripheral birds (for example, black-headed gulls: I. J. Pat-terson 1965; barn swallows: Hoogland and Sherman 1976; sandwich terns: Fuchs 1977; see table 8.6). Similarly, nest survival increases with colony size in several species, including fieldfares (Wiklund and Andersson 1994; Haas 1985) and Eurasian jackdaws (Johnsson 1994). Note, however, that the dilution ef-fect is sufficient to explain these results. Second, the number of birds mobbing a predator increases with colony size in several species (fig. 11.1). Third, dur-ing unsuccessful predatory attacks, the number of birds mobbing a predator

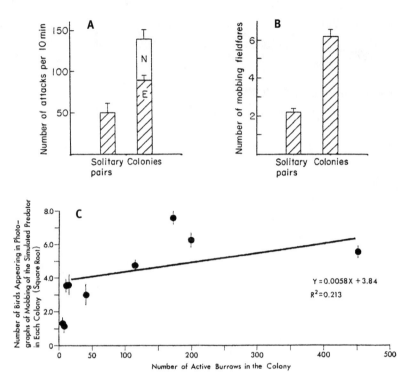

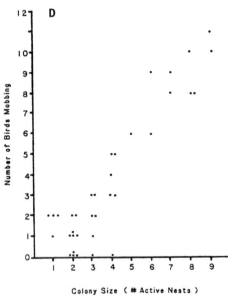

Figure 11.1 *A*, number of attacks on a magpie by fieldfares, and *B*, number of fieldfares attacking it near a nest of colonially breeding and a nest of solitary breeding fieldfares (Wiklund and Andersson 1994). The attack frequencies of the nearest experimental pairs (*E*) and neighbors (*N*) are shown. *C*, number of bank swallows appearing in photographs of mobs plotted against colony size (Hoogland and Sherman 1976). *D*, number of different individual yellow-rumped caciques mobbing predators during attacks by black caracaras against cacique colonies of different sizes (S. K. Robinson 1985).

Table 11.7 (a) Number (mean ± SE) of black-legged kittiwakes mobbing great skuas during attempts to predate their chicks (from Andersson 1976); (b) Number of yellow-rumped caciques mobbing Cuvier's toucans and black caracaras that began tearing open their nests (from S. K. Robinson 1985)

	Number of observations	Number mobbing
a		
Successful great skua attacks	7	5.6 ± 1.3
Unsuccessful great skua attacks	30	11.6 ± 1.1
b		
Successful at tearing open nest	23	2.2 ± 0.3
Chased away by caciques	20	5.3 ± 0.8

is greater than during successful ones, suggesting that bird numbers may be instrumental in reducing success (table 11.7). The possibility that breeding success in larger colonies increases simply as a result of either selfish-herd effects or dilution effects rather than nest defense may be eliminated in at least one species. Artificial nests placed next to red-winged blackbird colonies survived better the closer they were placed to the colony, but not because they were in central positions within the colony or because of colony size (Picman, Leonard, and Horn 1988). In summary, then, if studies from different species are all taken together, there is a reasonable link between mobbing and breeding success through its effect on driving predators away, although the full chain of causal events has yet to be demonstrated within a single population.

Mobbing is a poor method of moving predators on in all instances, however, because it necessitates prey noticing the predator in order to be able to respond to it, and the predator being deterred by calls, swoops, and strikes. At night, for instance, nesting or roosting birds seldom notice predators approaching. For example, fieldfares are relatively defenseless against nocturnally hunting owls (Wiklund and Andersson 1994), and least terns suffer high levels of depredation from nocturnally hunting black-crowned night herons (Brunton 1997). Very generally, mammalian predators hunt at night, whereas many avian predators are diurnal, so mammalian predators are less likely to be affected by mobbing than raptors. This may explain why in populations of red-winged blackbirds that suffer predation chiefly from birds (westerly populations), group mobbing can be beneficial, making synchronous nesting advantageous (Westneat 1992), whereas in more-easterly populations, where predation is chiefly from mammals, synchronous nesting is less prevalent (Weatherhead and Sommerer 2001). Indeed, we might expect colonial nesting to be a poorer option than cryptic solitary nesting where mammals are the chief predator, given that mammalian predators are often nocturnal and concentrate their search for food close to sites where they have been successful

(Tinbergen, Impekoven, and Franck 1967; Lemmetyinen 1971). An additional twist to the argument is found in fieldfares that nest alone and in colonies. Fieldfares eject feces at predators that destroy water-repellant properties of feathers but not of fur, leading one to suppose that antipredator benefits of colonial nesting would be higher where avian predators are abundant. In central Norway, the number of fieldfare colonies was positively correlated with small rodent density, because mustelids concentrated on small mammals; when the rodent population crashed, mustelids turned to fieldfare colonies, which benefited little from communal defense, with the result that solitary nesting now predominated (Hogstad 1995b).

Second, mobbing is only likely to be effective, and hence more likely to be manifested, when the predator is fearful of attacks by prey (C. R. Brown and Brown 1987). There are plenty of instances of birds tailoring their mobbing behavior to smaller predators (Kruuk 1976; see also Elliot 1985b). Indirect evidence supports this, too: colony size is negatively correlated with body size across five species of gull, possibly because small body size makes their defense against predators less efficient, forcing small gull species to nest together in larger colonies (Gotmark 1982).

11.7 Mobbing and mixed-species associations in birds

Nesting associations between timid and aggressive species have evolved independently on a number of different occasions. In these associations, one bird species gains antipredator advantages from nesting in association with another species as a consequence of the latter's mobbing or nest defense. The most common examples are birds breeding in open tundra habitats associating with birds of prey (for example, Summers et al. 1994) and diverse species nesting among tern and gull colonies (for example, Alberico, Reed, and Oring 1991), although other examples exist (table 11.8). In most of these cases, the timid (Erwin 1979) species has higher nesting success as a result of escaping predation when it nests close to the protector species (for example, table 11.9).

Associations do not occur by chance or by similar choice of nest sites. For example, red-billed choughs have different nest-site requirements from lesser kestrels, but they choose to nest close to these raptors, leaving many suitable nest sites unoccupied (Blanco and Tella 1997); and Bullock's orioles nest closer to yellow-rumped magpie nests than to conspecific nests (Richardson and Bolen 1999). In a detailed study of red-breasted geese nesting with raptors in the Pura and Pyasina basins in the Taymyr Peninsula, Siberia, Russia, Quinn and his coworkers (2003) found that geese began nesting 4–28 days later than protector raptor species, that they never nested at a cliff traditionally used by peregrine falcons unless the falcons were present, and that geese moved nests if falcons moved their aeries between years. Moreover, some species even time their breeding to take advantage of the period of peak nest defense shown by the protector species. In western Tokyo, Japan, azure-winged magpies breed most successfully when nesting near active nests of Japanese lesser sparrow hawks (Ueta 1994). Sparrow hawks chase predatory large-billed crows from the nesting area from the start of incubation until their young leave the nest. Most magpies started their own incubation 17–19 days after the start of incubation by hawks, coinciding with the peak level in nest defense by hawks (fig. 11.2). Indeed, magpies breeding within 100 m of a hawk's nest bred synchronously, but those nesting more than 1 km away showed greater variability in breeding phenology (Ueta 2001).

The cues that the more timid species uses to choose when, where, and with which species to nest are poorly understood (Larsen and Grundetjern 1997). Along the Taymyr Peninsula, red-breasted geese choose to nest with peregrines and snowy owls rather than rough-legged buzzards despite buzzard aeries being three to eight times more abundant along the rivers; but many also nest in association with herring gulls on islands that are free of arctic foxes. In an experiment in which a dog was taken to within 15–20 m of the nest, peregrines and snowy owls were found to be more assiduous in nest defense, as judged by the height of stoops and number of close stoops made at the decoy (fig. 11.3), suggesting that preferences were related to intensity of nest defense of the protector species in habitats in which foxes were present (Quinn et al. 2003).

A more sophisticated example comes from wood pigeons that nest close to Eurasian hobbies in poplar plantations in northern Italy. As an observer approached to < 1 m of the hobby nest tree, Bogliani, Sergio, and Tavecchia (1999) scored the behavior of hobbies on an eight-point scale ranging from being flushed and disappearing to "tight sitters" that also made several close stoops toward the intruder while continually uttering loud cries. They found that the probability of dummy wood pigeon nests being predated was lower if the protector hobbies were aggressive, if they were early nesters, and if the artificial nests were placed close to the hobby nest. Natural wood pigeon nests were found disproportionately close to hobbies that were tight-sitters on their

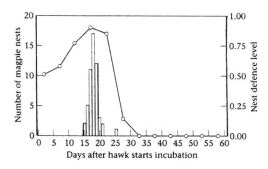

Figure 11.2 The level of nest defense by Japanese lesser sparrow hawks during each 5-day period (*open circles*) and the number of azure-winged magpies that started incubation each day (*histograms*) (Ueta 2001).

Table 11.8 Mobbing and mixed-species associations in birds

Beneficient species	Protector species	Observed outcome measure	Reference[a]
Silvery grebe, Rolland's grebe	Brown-hooded gull	Higher hatching success and lower predation rates on adults	1
Red-breasted goose	Peregrine falcon	Lower risk of predation near peregrine nest	2
	Snowy owl, rough-legged buzzard	Higher nesting success than at gull colonies in years of high predation pressure	3
Snow goose	Snowy owl	Nesting success declined with distance from snowy owl nest	4
Eider duck	Lesser snow goose	Nests not losing eggs prior to hatching are closer to goose nest	5
	Herring gull, common gull	Greater nest survival with increasing herring gull density	6
	Herring gull, common gull, greater and lesser black-backed gull	Lower proportion of nests destroyed within than outside gull colonies because hooded crows avoid gulls	7
Oldsquaw	Arctic tern	Arctic terns mob predators	8
Red-breasted merganser	Common tern	Lower rates of nest predation inside tern colony	9
Least bittern	Boat-tailed grackle	Higher reproductive success inside grackle colonies in one year	10
Merlin	Fieldfare	Higher breeding success close to fieldfare nests	11
Common snipe, common redshank, ruff	Black-tailed godwit, lapwing	Successful nests found closer to godwit nests	12
Bar-tailed godwit	Long-tailed skua, whimbrel, gray plover	Godwits nest close to skuas when rodents abundant but close to whimbrels and gray plovers when rodents scarce	13
Eurasian curlew	Eurasian kestrel	Artificial nests suffer lower predation close to kestrel nests	14
Spotted sandpiper	Common tern	Less predation by American mink than those nesting outside tern colonies	15
Turnstone	Arctic tern, common gull	Gulls chase crows and ravens away	16
Common gull	Arctic skua	Arctic skuas mob avian predators and domestic dogs	17
Sandwich tern	Black-headed gull	Experimental eggs have higher survival among tern nests	18
Black skimmer	Common tern	Terns chase predators away more often but no differences in nesting success	19
	Gull-billed tern	Greater response of terns to predator decoys	20
Wood pigeon	Eurasian hobby	Less predation on dummy nests close to aggressive hobbies	21
Sand-colored nighthawk	Yellow-billed tern, large-billed tern	Nighthawk greater hatching success near terns and skimmers that mob predators	22
Plain-fronted thornbird	Cattle tyrant, stripe-backed wren, saffron finch, troupial	Nesting success higher with associates	23
Azure-winged magpie	Japanese lesser sparrow hawk	Predation rate lower <50 m from sparrow hawk nest	24
Red-billed chough	Lesser kestrel	Breeding success higher within kestrel colonies	25
Fieldfare	Merlin	Higher breeding success close to merlin nests	26
Yellow wagtail, meadow pipit	Lapwing	Higher nest densities inside lapwing territories	27

Table 11.8 *(continued)*

Beneficient species	Protector species	Observed outcome measure	Reference[a]
Savannah sparrow	Herring gull	Gulls alert sparrows to presence of American crows and attack crows, resulting in lower predation rates	28
Brambling, redwing	Fieldfare	Strong nesting association with fieldfares that mirror fieldfare colony movements annually	29
Bullock's oriole	Yellow-billed magpie	Lower predation rate near magpie nests	30

[a] 1. Burger 1984; 2. Quinn and Korkorev 2002; 3. Quinn et al. 2003; 4. Bety et al. 2001; 5. G. J. Robertson 1995; 6. Gerell 1985; 7. Gotmark and Ahlund 1988; 8. R. M. Evans 1970; 9. Young and Titman 1986; 10. Post and Seals 1993; 11. Wiklund 1979; 12. Dyrcz, Witkowski, and Okulewicz 1981; 13. Larsen 2000; 14. Norrdahl et al. 1995; 15. Alberico, Reed, and Oring 1991; 16. Brearey and Hilden 1985; 17. Gotmark and Andersson 1980; 18. Fuchs 1977; Veen 1977; 19. Erwin 1979; 20. Pius and Leberg 1998; 21. Bogliani, Sergio, and Tavecchia 1999; 22. Groom 1992; 23. Lindell 1996; 24. Ueta 1994; 25. Blanco and Tella 1997; 26. Wiklund 1982; 27. Eriksson and Gotmark 1982; 28. Wheelwright, Lawler, and Weinstein 1997; 29. Slagsvold 1980; 30. D. S. Richardson and Bolen 1999.

Table 11.9 Distances (m) and fates of common snipe, and ruff and common redshank combined nests from occupied black-tailed godwit nests

	Snipe			Ruff and Redshank		
	Nest successful	Nest destroyed by predators	*p*-value	Nest successful	Nest destroyed by predators	*p*-value
Average distance from closest godwit nest	37	83.5	<0.01	39.5	95	<0.001
Average distance from 2 closest nests	53	100	<0.001	53	125	<0.001
Average distance from 3 closest nests	64	111	<0.001	64	153	<0.001

Source: Dyrez, Witkowski, and Okulewicz 1981.

nest and to hobbies that fledged more young. Some birds may use early nesting and aggressiveness of protector species as cues for where to place their nest.

Nonetheless, the nature of mixed-species associations is more complicated than one species simply obtaining antipredator benefits from the other, because the protector species is sometimes a formidable predator itself and may incorporate young and adults of the timid species into its diet (for example, Alberico, Reed, and Oring 1991; Wheelwright, Lawler, and Weinstein 1997). For example, along the Taymyr Peninsula again, where red-breasted geese nest in association with peregrine falcons, geese are also preyed on by peregrines. Geese nesting close to a peregrine aerie are more likely to be attacked by host falcons and desert their nests, whereas those nesting far away are more likely to be attacked by foxes. The optimal predicted distance that balanced these opposing pressures for a nesting goose was 46 m from an aerie, close to the actual average distance of 34 m. Risk sensitivity of decision making is highlighted by the finding that goose nests were found closer to peregrine's nests during years when arctic and brown lemming populations were low and hence predation pressure from arctic foxes was high (Quinn and Kokorev 2002; see also Larsen 2000)!

Across species the same principle applies. Eider ducks nest with larids, skuas, and even tethered sledge dogs in parts of their range, but the relationship with gulls is an uneasy one, because gulls are a heavy source of predation on eider eggs if they are left unguarded during the egg laying period. Once the eider female starts to incubate, however, gulls take few eggs (Gotmark 1989). Once the gulls themselves have started to breed, they drive hooded crows and minks away from the colony (Gerell 1985; Gotmark and Ahlund 1988), but they are of no help before this; for the eiders, correct timing is critical. The association between larids and 9 species of duck reflects this: mixed-species nesting associations were strong when ducks started to lay later or at the same time as gulls or terns, but associations were weak if ducks started to lay before the larids (table 11.10).

It is also too simplistic to view these mixed-species associations as commensal, with the more timid species obtaining antipredator benefits at no cost or benefit to the bolder. The association may be mutualistic if the timid species joins the protector during mobbing or gives alarm calls to approaching predators. For example, not only do fieldfares have higher breeding success when associating with merlins but merlins similarly have higher breeding success when associating with fieldfares, although the mechanism is not understood (Wiklund 1979, 1982). Undoubtedly, the protector species is likely to benefit from the dilution effect (K. L. Clark and Robertson 1979), amplified by the fact that a predator may be more likely to choose a timid bird's nest in event of attack; but there has been scant exploration of benefits for protector species.

Alternatively, some associations could be parasitic, as Groom (1992) has shown for sand-colored nighthawks that nest in mixed-species associations along beaches of the Manu River in southeastern Peru. There, four bird species, black skimmers, large-billed terns, yellow-billed terns, and sand-colored nighthawks, nest together, but only skimmers and terns attack the black caracaras, hawks, and falcons that take eggs and young off the beaches. Whereas nighthawk hatching success was positively correlated with numbers of terns and skimmers on the beach, tern, and skimmer hatching success was negatively correlated with numbers of nighthawks; and whereas nighthawk hatching success was greater < 10 m from a tern or skimmer nest, successful skimmer or tern nests had fewer nighthawk nests < 10 m away than unsuccessful nests. This was because rate of predation attempts increased with numbers of nighthawk but not tern and skimmer nests. Moreover, terns and skimmers devoted more time to vigilance and antipredator behavior and spent less time in parental care among groups of 90 or more nighthawks than among groups of 40 or fewer

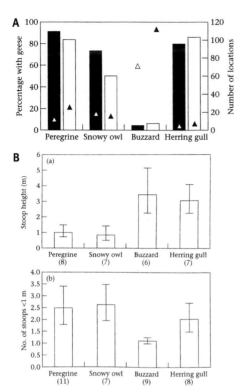

Figure 11.3 *A*, total number of raptor aeries and herring gull colonies along river valleys in the Pura and Pyasuna basins, Siberia (*open triangles*, 1996; *filled triangles*, 1999) and the percentage of these with which red-breasted geese were associated (*filled histograms*, 1996; *open histograms*, 1999). *B*, nest defense intensity of associate species in terms of (*a*), stoop height, and (*b*), stoop frequency (stoops < 1 m) during 2-minute trials on a surrogate arctic fox in 1999. Numbers in parentheses are sample sizes (pairs for raptors, colonies for gulls). Gulls responses are for multiple pairs. Histograms are means and back-transformed SEs (Quinn et al. 2003).

Table 11.10 Degree of nesting association between larids (including great black-backed, lesser black-backed, herring and common gulls) and common and arctic terns, and 9 species of duck in relation to their relative timing of egg laying

	Degree of association with larids			
	Strong	Moderate	Weak/absent	Start of laying
Tufted duck	X			Later or same time as larids
Greater scaup	X			Later or same time as larids
Velvet scoter		X		Later or same time as larids
Red-breasted merganser		X		Later or same time as larids
Mallard		X		About same time as larids (?)
Common teal		X		About same time as larids (?)
Northern shoveler		X		About same time as larids (?)
Goosander			X	Before larids
Common eider			X	Before larids

Source: Gotmark 1989.
Note: Data obtained from several Finnish archipelagoes in the Baltic Sea.

nighthawks. That late-nesting skimmers and terns always nested on beaches lacking nighthawks but nighthawks never nested on beaches where these birds were absent reinforces the idea that the association between nighthawks and the other species is parasitic (see also Brearey and Hilden 1985). In sum, mixed-species nesting associations driven by protector species' mobbing behavior can be mutualistic, commensal, or parasitic.

11.8 Group defense in mammals

In some groups of mammals, individuals bunch together in the face of predatory threat or physically attack a predator together, behaviors that are termed group defense or group attack depending on which behavior predominates, although the terms are often used interchangeably. In addition to these specific usages, it is recognized that instances of group defense or attack often more generally involve close inspection, scolding, and harassing predators, just as occurs in mobbing in birds. Therefore, "group defense" is also used as a catchall phrase embodying a suite of predator-directed activities. Informally, group defense, used here to include all these activities, is seen as a more forceful antipredator strategy than mobbing, however, because it can occur in response to predatory attack in some circumstances. For instance, banded mongooses have been seen to climb a tree in order to rescue a group member from the clutches of a martial eagle (Rood 1983). Information on group defense in mammals (in the general sense) is restricted almost exclusively to three taxonomic groups, sciurids, ungulates, and primates, although it occurs in African elephants and white rhinoceroses, too (Owen-Smith 1988). Continuous monitoring of the predator and attracting conspecifics to the site are common features in the three taxa; differences in the extent to which prey physically contact the predator must in part reflect relative differences in prey-predator weaponry and body size.

11.8.a Snake-directed behavior in sciurids   California ground squirrels show a slew of responses to northern Pacific rattlesnakes, an important predator. They may visually monitor snakes for extended periods of time from as close as 1–2 m; they may follow snakes for up to 5 minutes, advancing to within 30 cm; and they may approach a coiled snake and kick loose substrate toward it (Hennessy and Owings 1988). Similar behaviors are seen in rock squirrels (Owings et al. 2001) and in some species of kangaroo rat (Randall, Hatch, and Hekkala 1995; Randall and Boltas King 2001). Approaches are slow, deliberate advances interrupted with pauses, head bobbing, and vigorous tail flagging that may be conducted alone or with other ground squirrels. In the open, ground squirrels may also bite snakes and, in their burrow, kick soil as a prelude to trying to entomb them (Coss and Owings 1985). Some of these interactions end in strikes by snakes in which envenomation may or may not occur. Whether a subset of all of these ground squirrel behavior patterns is labeled harassment or mobbing is contingent on the predator's reaction and perhaps number of ground squirrels involved (Loughry 1987a), both of which are highly context dependent. Indeed, our attempts to understand antipredator defenses by breaking them down into inspection, harassment, and attack seem contrived when describing ground squirrel antipredator behavior (see section 7.6).

Despite little information on the consequences of different forms of snake-directed behavior for rattlesnakes or its costs for ground squirrels, some of the factors affecting snake-directed behavior are well documented. Ground squirrels recognize and respond to snakes visually, particularly when snakes are moving, but may overlook those that are coiled and immobile (Owings and Coss 1977); they react to them auditorily, using the sound of rattling to locate and harass snakes (Rowe and Owings 1978; Swaisgood, Rowe, and Owings 1999); and they respond to snake odor (Hennessy and Owings 1978). California ground squirrels raised in the laboratory react to snakes in a similar way to adults but more intensely, indicating some degree of innate recognition (Owings and Coss 1977). Mothers show more-intense snake-directed activities than other age-sex classes in California ground squirrels and Formosan squirrels, and in the former species differentiate large and small, and warm and cool snakes more assiduously than other age-sex classes (Tamura 1989; Swaisgood, Owings, and Rowe 1999; Swaisgood, Rowe, and Owings 2003). In contrast, males spend more time close to snakes in black-tailed prairie dogs (Loughry 1987a). Finally, ground squirrels from areas lacking poisonous snakes take greater risks during snake-directed activities than their counterparts from snake infested areas, although the reasons are unclear (Coss and Goldthwaite 1995).

Many of these careful studies have necessarily been performed in the laboratory (Loughry 1987b) using single individuals, so the extent to which harassment occurs in the wild and the number of animals involved in mobbing is not always well established (Hennessy and Owings 1988; Tamura 1989). While it seems probable that sciurids gather important information during mobbing and reduce the likelihood that snakes will remain in an area, both possibilities still need empirical backing.

11.8.b Protective behavior in ungulates   As in sciurids, ungulates may move toward, attack, flee, or stand and face a predator if they are approached by it. Most reported instances of ungulate attacks or standoffs are of mothers defending their offspring against predators (but see Marion and Sexton 1979). Group attacks in bovid and artiodactyls species are also associated with living in large groups, however, (Caro et al. 2004) with the most celebrated behavior perhaps being the circle or pinwheel formations that are taken up in response to single carnivores running around prey or to advances or attacks by a group of carnivores (for example, Miller and Gunn 1984). Propensity for ungulates to bunch together varies between populations and species. Wildebeest on the open Serengeti Plains do not bunch together in the face of attack, although they do form lines abreast; but in the thickly wooded Selous Game Reserve, Tanzania, they form a pinwheel (Creel and Creel 2002). White-tailed deer do not alter group size or formation in the face of coyote attack, but mule deer bunch together when encountering or fleeing from coyotes (fig. 11.4.a). In the second, cervid, example, coyotes were less likely to pursue and attack mule deer when individuals moved together than when they spread out (fig. 11.4.b), but this was not the case in attacks on white-tailed deer. Coyotes made short rushes at bunched groups of mule deer in an apparent attempt to cause an individual to bolt, but they did not enter groups, probably because of implicit or explicit threat of aggression. Ability to escape attack through flight is probably the explanation for these differences. Wildebeest in Selous are unable to detect wild dogs at long distances away and so take evasive action, whereas mule deer cannot run fast enough to outdistance coyotes.

What are the costs of group defense in its broad sense for ungulates? Costs of forming a defensive circle may be restricted principally to opportunity costs, whereas costs of group attack certainly involve energetic costs of pursuing the predator and possibly survivorship costs; only costs of group inspection have been documented thus far, where they were found to be quite substantial (FitzGibbon 1994). Benefits of these activities appear clear-cut: preventing a predator from making a successful attack (Shank 1977) or driving a predator away (for example, Berger 1979; Gese 1999), although we have no idea of the probabilities that predators would have quit hunting without such encouragement. Perception advertisement, moving the predator on, and learning about its characteristics may also be involved.

It seems unlikely that we will have to impute anything other than individual selection to explain group defense in its restricted sense. Given that there are examples of single ungulates standing their ground and driving predators away (Mech 1970) and of mammalian carnivores being able to bring down ungulate prey only after it has bolted (Caro 1994a, Creel and Creel 2002), it seems probable that group defensive formations can be explained simply in terms of individual selection. When offspring are present and mothers defend them alone or in a group, kin selection is a sufficient explanation. While there may be some mutualistic benefit from conspecifics covering each other's hindquarters, it seems unnecessary to invoke mutualism or reciprocal altruism in shaping defensive formations in ungulates.

The evolutionary forces underlying direct attack by groups of ungulates

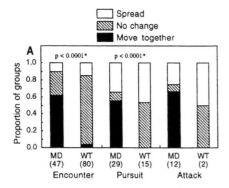

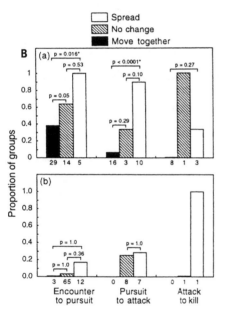

Figure 11.4 *A*, change in group formation in response to coyotes by mule deer (*MD*) and white-tailed deer (*WT*). Species, sample sizes, and hunt stage (encounter, pursuit, or attack) given below bars. *P*-values denote results of *G* tests used to compare frequency of response depending on species; there were insufficient data to compare species during attack. *B*, proportion of interactions that escalate depending on the formation of the group for (*a*), mule deer, and (*b*), white-tailed deer. Sample sizes are given below data bars. When data are absent, the sample size is shown (*n* = 0) to distinguish these cases from those in which data were available but no attempts escalated. * indicates significant, using a sequential Bonferroni correction (Lingle 2001).

may be more complicated, because individuals probably stand a better chance of deterring predators when they attack as a group than if they attack alone. Mutualistic coalitions of adults repulsing predatory attack may apply in stable groups, but some ungulate groups are composed of unrelated individuals whose composition is unstable. It would be interesting to relate the incidence of different forms of antipredator behavior to breeding systems in ungulates (see Caro et al. 2004). Relative body size and weaponry are clearly important additional factors in whether ungulates attack predators. Large ungulates such as moose, buffalo, and eland are notable in driving predators away whether prey are alone or in groups and whether or not offspring are present (Mech

1970; Kruuk 1972). Smaller species do not have this ability; instead, they freeze or hide in vegetation (Estes 1974; Caro and FitzGibbon 1992).

11.8.c Group attacks in primates   The majority of reports of mobbing, group defense, or group attacks in primates consist of attacks against predators and comprise screaming, running along the ground or through trees toward predators, and physical contact, including slapping, biting, trying to grab the talons of eagles, and even stoning dangerous intruders (for example, Hamilton, Buskirk, and Buskirk 1975; Baenninger, Estes, and Baldwin 1977; Chapman 1986; Hiraiwa-Hasegawa et al. 1986; Bartecki and Heymann 1987; Gautier-Hion and Tutin 1988; Eason 1989; Struhsaker and Leakey 1990; Ross 1993; Passamani 1995; Iwamato et al. 1996). Solitary primates may exhibit similar behavior, but this is poorly documented. As usual, costs of group attacks for individuals have rarely been examined; an exception is Cowlishaw's (1994) review, in which he documented individuals of various baboon species being killed in 1 out of 33 attacks on predators. Many of these attacks were successful in driving predators away, however, and some resulted in the death of the predator (4 out of 11 documented cases: table 11.11).

In most group attacks, males take a predominant role. For example, male red colobus respond to attacks of chimpanzees by communally and aggressively defending other group members, interposing themselves between the apes and immature and female colobus, and even leaping onto, biting, and chasing chimpanzees (Stanford 1995; see also Struhsaker and Leakey 1990). Data show that the more male colobus that are actively involved in confronting chimpanzees, the lower probability of chimpanzees hunting successfully (fig. 11.5). Similar to coyotes attacking mule deer, chimpanzees usually kill when red colobus immatures and females scatter or flee. From a comparative perspective, increased male representation in primate groups is associated with risk of heightened predation (van Schaik and Horstermann 1994; Hill and Lee 1998), but it is difficult to establish whether this relationship is associated with benefits obtained from predator mobbing or from other group-size effects such as corporate vigilance (chapter 4), or dilution of predation risk (chapter 8).

Among group-living primates, by no means do all species or populations mob or defend themselves, in the broad sense, against predators. In Gombe Stream National Park, Tanzania, for example, blue monkeys and red-tailed monkeys flee immediately on hearing or seeing chimpanzees and are rarely hunted by them; red colobus, on the other hand, stand their ground, possibly because they are larger and less fleet than cercopithecoids (Stanford 1995). Even among red colobus, the Tai Forest population is less aggressive toward chimpanzees than those at Gombe (table 11.12). Boesch (1994) believes that forest structure underscores these population differences (but see Stanford 1998). At

Table 11.11 Predator deterrence by different baboon species

Predator species	Number of predators	Retaliators	Outcome of retaliations	Number of cases
Python	1	group	unknown	9
Chimpanzee	U	male/s	P foiled	U
Jackal	2	1 male	P deterred	1
Dog	U	20 males	P fled	U
	U	1 male	P maimed/killed	U
Cheetah	3	1 male	P bypass group	1
	1	2 males	P bypass group	1
Lion	4	group	unknown	1
	3	group	P leaves	1
	3	group	P leaves	?
Leopard	1	2 males	P and 1D killed	1
	1	1 male	P chased	1
	1	male/s	P wounded	1
	1	group	P fled/killed	1
	1	unknown	P killed	1
	1	3 males	P chased	1
	1	males	unknown	1
	1	group	P mobbed	1
	1	group	P chased	2
	1	male/s	P killed	1
	1	group	P mobbed	1
Leopard/cheetah	1	group	P fled	1
	1	male/s	P mobbed	2
Hyena	1	1 female	No effect	1

Source: See Cowlishaw 1994 for references.
Note: P, predator; D, baboon defender; U, data unavailable.

Gombe, both red colobus and chimpanzees forage in areas where the canopy is low and irregular, bringing the colobus close to chimpanzees. As a consequence, colobus repeatedly harass chimpanzees whether the latter are hunting or not in order to deter them from attacking. In Tai, however, the closed canopy is 40 – 60 m high and the colobus simply ascend silently on seeing chimpanzees, or else move toward and mingle with a troop of Diana monkeys, noted for their extreme vigilance. While one explanation for Gombe colobus' belligerent behavior is that the forest offers them no other alternatives, another is that Gombe red colobus weigh more and Gombe chimpanzees weigh less than those in Tai, making a mobbing ploy more formidable. Alternative behavioral options that operate on minute-to-minute, ecological, and evolutionary timescales are a useful but still relatively neglected way of conceptualizing antipredator options open to prey facing specific predators under specific environmental circumstances (Bshary and Noe 1997b).

In conclusion, mobbing in birds and group defense in mammals have been addressed by different groups of researchers. Taxonomic xenophobia has stifled attempts to relate these two types of antipredator defense to each other,

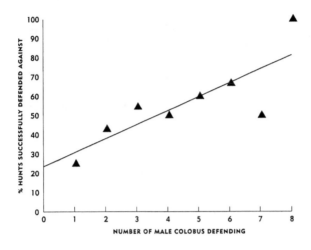

Figure 11.5 Number of male red colobus counterattacking chimpanzees, plotted against the percentage of chimpanzee hunts which ended in the colobus group being defended successfully at Gombe Stream National Park, Tanzania, 1991–95 (Stanford 1998).

Table 11.12 Percentage of different reactions of red colobus monkeys to visual presence of chimpanzees in two different populations when colobus could not (No) or could (Yes) see a human observer at the same time

	Tai		Gombe	
	No	Yes	No	Yes
Panic[a]	0	0	2	30
Flee/none[b]	93	100	42	20
Threat[c]	0	0	13	10
Mobbing[d]	7	0	2	40
Total	71	6	45	20

Source: Boesch 1994.
[a] Panic indicates a high level of response caused by the presence of chimpanzees and humans together.
[b] Flight and no reaction are combined, since they are regularly observed in the same group and shown by the same individual during the encounter.
[c] A threat occurs when the colobus advance in the trees toward the chimpanzees, calling aggressively.
[d] Mobbing is an attack against the chimpanzees with physical contact.

but the behavior patterns, approach, inspection, scolding, harassment, standing ground, and physical contact are extremely similar in both classes. Thus, there are merits in bringing these areas together; for example, using rules about primate group attack derived from relative group and body sizes of different prey and predator species to make predictions about propensity to mob in birds; or extrapolating growing support for the move-on hypothesis developed in birds over to primates and so examine predators' reactions to harassment. Although we do not expect the relative importance of different functions of mobbing or defense to be the same across classes or even between species within classes, concepts and experimental protocols are readily transferable.

11.9 Summary

When a predator approaches and before it has launched an attack, prey may mob it. Mobbing consists of several behavior patterns that include approaching a predator, vocalizing loudly, and physical attack. Between and within species, the propensity to mob varies enormously and differs by sex, age, changes in the reproductive cycle, type of predator, and degree of risk. Many aspects of mobbing indicate that mobbing places prey at a heightened risk of predation; occasionally, mobbers are captured by a predator.

There are many potential consequences of mobbing for the predator itself, for the mobber, and for recruits. Occasionally, predators are killed by the mob, but more commonly the predator is sufficiently harassed that it leaves the immediate area. In comparison with other hypotheses, this "move-on" hypothesis has received most support. Mobbing necessarily informs the predator that it has been detected. It can also alert conspecifics and heterospecifics to the presence of a predator. Members of a mob may gain benefits from the dilution and confusion effects. In some species, offspring are silenced on hearing mobbing calls. Laboratory experiments show that mobbing conveys information about a predator to conspecifics, causing them to mob it themselves. Cultural transmission of fearful responses can pass between several tutors and observers.

In nature, colonially nesting birds, especially seabirds, benefit from forming mobs. Breeding success is higher in colonies, the number of birds in mobs increases with colony size, and the number of birds in a mob is associated with an increased likelihood of a predatory attempt being unsuccessful. In a small number of species, birds of one species nest together with a pugnacious species that drives predators away from all of their nests. Protector species may also gain benefits from mixed-species nesting associations, but the degree to which these associations are commensal, mutualistic, or parasitic varies considerably on a case-by-case basis.

Group defense in mammals has parallels with mobbing in birds, because it may consist of inspecting, scolding, and harassing a predator, although it can additionally include joint attack and defense. Aspects of group defense have been studied in three groups of mammals. In ground squirrels, individuals approach and follow snakes, monitoring their movements. In ungulates, individuals in some populations form tight circles in the face of predatory attack. In some primate populations, individuals physically attack predators and drive them off. Differences in willingness to approach predators closely may be related to relative body size of prey and predator and the ability to use other antipredator options, such as flight.

12 Flight and Behaviors of Last Resort

12.1 Introduction

Even when a prey animal perceives that a predator has oriented toward it or is approaching in its general direction, it still has a surprising number of options left open for escape. This chapter discusses these methods and, for convenience, is ordered according to increasing proximity of predator to prey. Depending on the prey species, prey can remain where it is and attempt to avoid detection through crypsis (section 12.2) or, alternatively and when the predator is very close, suddenly expose a conspicuous part of its body and attempt to startle the predator, which may temporarily delay final contact and enable it to escape (section 12.3). Another possibility is to stand its ground and attack the predator (section 12.4). Usually prey are smaller and less well armed than predators, however, forcing them to execute an escape. There are numerous means by which homeotherms escape from a predator, including flying or running away, jumping out of the way, climbing into or dropping out of trees, burrowing, or entering water. These differ among prey species, among environments in which the prey lives, and according to species of predator (section 12.5). Whatever the means of quitting the area, however, the distance from the predator at which the prey decides to flee is influenced by a variety of factors, including risk of capture where it stands, costs of flight, and alternative means of escape (Ydenberg and Dill 1986; section 12.6). In birds, which usually fly from predators, risk of capture is particularly influenced by body weight, resulting in complex relationships between mass of birds and their takeoff ve-

Figure 12.0 (*Facing page*) Canadian lynx pursuing snowshoe hare through deep snow. In this situation in which other antipredator defenses have failed, the hare must rely on speed, sharp turns, and leaps to escape with its life (reproduced by kind permission of Sheila Girling).

locity and angle of ascent; in turn, this influences the way in which birds manage their weight gain and loss in relation to predation risk (section 12.7).

Whether a prey animal chooses to stand its ground or to flee, it may nonetheless be captured. Now it can struggle and attempt to injure the predator and wriggle free, or even autotomize a part of its body that its tormentor has secured. Though found principally in invertebrates and poikilotherms, autotomy occurs in a few mammals (section 12.8). A captured prey animal may also utter fear screams, the function of which is still improperly understood (section 12.9); or it can even feign death, seemingly making a coup de grace unnecessary (section 12.10). While no single species or individual animal demonstrates all of these behavior patterns, indeed some are incompatible with each other, it is instructive to analyze the circumstances in which these behaviors of last resort are manifested.

12.2 Freezing and immobility

It has long been known that immobility is a critical component of remaining cryptic. For example, L. de Ruiter (1952) drew attention to the importance of stick caterpillars resembling twigs provided that they remained completely immobile. Among homeotherms, certain species freeze when they become aware of a predator in their vicinity (as distinguished from remaining cryptic for long periods of time in the absence of immediate threat: section 3.5). In birds, species such as willow grouse become motionless when a person or dog approaches (for example, Gabrielsen, Blix, and Ursin 1985), whereas in mammals, species including woodchucks and common voles freeze when threatened (E. N. Smith and Woodruff 1980; Gerkema and Verhulst 1990; Sundell and Ylonen 2004). In even-toed ungulates, freezing (among adults) is associated with living in dense forests, being small, and having striped or spotted coats, all of which are likely to enhance crypticity (Caro et al. 2004). Absence or reduction of movement lowers threat of predation in relation to raptors and owls hunting small rodents (Metzgar 1967; Ambrose 1972; Snyder 1975; Snyder, Jenson, and Cheney 1976), although it is less clear whether immobility reduces predation risk from mammalian or snake predators. Immobility is a common response in avian offspring faced with danger. Mothers give characteristic warning calls (Starkey and Starkey 1973) that elicit rapid crouching on the ground in precocial young or in the nest in altricial offspring, and fear bradycardia in which the heart rate plummets during the period when threat is imminent (R. D. Thompson et al. 1968; fig. 12.1). Predisposition toward freezing differs both by sex (Balph 1977) and by age (Gochfeld 1981). For example, in the semiprecocial black skimmer, young nestlings under 3 weeks of age remain still in scrapes as a predator approaches, but older chicks run off to a new hiding place before crouching and making a new scrape; the change coincides with a

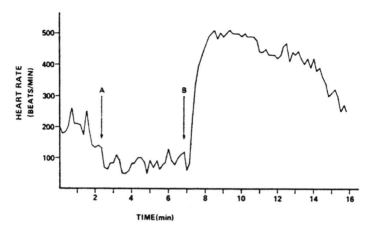

Figure 12.1 Heart rate responses in an incubating willow ptarmigan when approached by one person. A, hen can see the approaching person; B, the same person turns and leaves the area (Gabrielsen, Blix, and Ursin 1985).

plumage transformation from cryptic sand-colored down to black and brown juvenile plumage. Freezing is potentiated in some species by an observer looking directly at prey or, in experimental situations, by the presence of prominent eyes on a model (Gallup, Cummings, and Nash 1972; Scaife 1976a, 1976b).

How can we explain the pattern of freezing responses across taxa? Among lizards, where most of the work on cryptic antipredator behavior in vertebrates has been carried out, some species, usually sit-and-wait predators, show cryptic antipredator behavior; species that forage widely flee from predators (Vitt and Congdon 1978). Nevertheless, within species, individuals shift from flight to immobility according to a variety of environmental and reproductive factors that may reduce flight speed, including a lowering of temperature and whether females are gravid (Bauwens and Thoen 1981; Hertz, Huey, and Nero 1982; D. G. Smith 1997). The decision over which defense to use may also be governed closely by the probability of escape in homeotherms. For example, red colobus cease calling and hide high up in trees when playbacks of predatory chimpanzee calls are played to them; but if a group of Diana monkeys is nearby, they will simply move directly and quite openly toward the Dianas; chimpanzees usually refrain from attacking Diana monkeys because of their heightened vigilance (Boesch 1994; Bshary and Noe 1997b). Nonetheless, social and environmental factors affecting immobility, and indeed the whole topic of immobility, has received very little notice.

12.3 Defense calls and flash coloration
If prey stand their ground while a predator approaches, they may suddenly make a (often abrupt) noise (unfortunately named a defense call, given that this term can be used for so many other calls given in the presence of a pred-

ator: section 1.3.b) or expose a brightly colored part of their body (termed flash coloration). Reluctantly, I will continue to use the term *defense calls* here to avoid introducing more jargon. Defense calls differ from distress calls (section 12.9) in that the former are given prior to contact with the predator, the latter after. Very little research indeed has been carried out on defense calls or flash coloration in homeotherms, but there is an informal consensus that defense calls are quite common in mammals but less common in birds, whereas flash coloration occurs sporadically in birds but is absent in mammals. There are no data that address these suppositions systematically; nevertheless, there are plenty of anecdotes, as follows.

Most species of carnivores growl and hiss at predators when in jeopardy, whereas collared peccaries give prolonged snorts and gnash their teeth (Elder 1956). Some lagomorphs and the duck-billed platypus surprisingly growl when threatened (Collias 1960), which may constitute mimicry of dangerous carnivores. In birds, defense calls include hissing sounds given by parids and tetraonids (Klump and Shalter 1984), growling in American coots (Gullion 1952), jaguarlike growls emitted by crested guans (Kilham 1978), and rattling sounds made by burrowing owls that sound like a rattlesnake (D. J. Martin 1973). Sometimes auditory signals are accompanied by visual displays. For example, while snorting or growling, European badgers piloerect hair over their whole body, caudally piloerect, exposing white underfur, and head flag, thereby emphasizing their striped face mask (Butler and Roper 1994). Olfaction may play a role, too, as when animals defecate or urinate in the presence of a predator (McManus 1970). More-directed spraying of urine or anal gland secretions, as seen in mustelids, or vomiting, as seen in procellariiformes, could similarly shock a predator.

Flash marks, that is, conspicuously colored feathers that are normally hidden, are found in many taxa, but the focus of research has been on epaulets used in sexually selected displays. Some waders flash brightly colored rump patches at approaching predators (Woodland, Jaafar, and Knight 1980; Brooke 1998), and shorebirds expose bright white undersides when they take off en masse (Buchanan et al. 1988), but these have been respectively interpreted as pursuit-deterrent signals and possible attempts to confuse rather than startle the predator. In practice, it may be very difficult to tease apart these functions. On the basis of verbal argument only, white rump patches in mammals that are exposed while fleeing are considered not to startle the predator but to be conspecific warning or pursuit-deterrent signals.

Some of these behaviors and morphological traits are assumed to startle the predator, causing it to halt its attack for an instant and thus buy extra time for escape (but see Jablonski 1999); others threaten the predator and perhaps stop it from attacking at all. It is difficult to distinguish between these possibilities, because the difference hinges, in part, on whether prey are capable of effective counterattack. In circumstances in which prey are relatively defenseless (young snowshoe hares growling, for example) or mimicry appears to be involved (burrowing owls), the predator's startle response (also called startle effect) is probably triggered by novelty, oddity, conspicuousness, or anomaly of the call or display, as occurs in insect-avian predator interactions (Schlenoff 1985). Where prey are formidable (as in peccaries), threat of injury may be more important than surprise in halting predatory attack. Conceptually, the former requires no prior experience of prey, whereas the latter may demand some.

In a similar vein, it is difficult to distinguish between defense calls and flash coloration on the one hand, and aposematic displays on the other. Both may startle the predator, but whereas the first involves simply shocking and need not be backed up by honest threat (for example, coots), the second involves persuading the predator not to attack because of adverse consequences that are arguably honest except in the case of Batesian mimicry (Maynard Smith and Harper 2003). Interestingly, the predicted outcome of defense against naïve younger predators is different in the two cases. When faced with aposematic prey, inexperienced predators may be unaware of the strength of toxic defenses and more likely to attack than older predators, but conversely be more surprised by flash coloration and less likely to attack than experienced predators. In nature, these differences may be more apparent than real. For example, while mustelids characteristically produce musky anal secretion, there is a range of smells, with some species having more-putrid odors than others, so that the repugnant species might be termed aposematic but the more-benign species classed as eliciting a startle response (Macdonald 1985).

12.4 Counterattack

In some instances, prey may attack the predator. Counterattack is probably more prevalent in prey that possess weapons, such as sharp teeth in shrews but not in mice, sharp beaks in parrots but not in toucans, but there are no systematic data. Most famously, species with noxious scent glands, such as skunks, are very willing to stand their ground and spray their attacker. Other mustelids, including badgers and ratels, have reputations for dogged defense. Among birds, the albatrosses, petrels, and storm petrels, all of them ground nesters, regurgitate stomach oil that is made up of fatty acids, fatty alcohols, glyceryl, and wax esters (Warham 1996) principally derived from food (Clarke and Prince 1976; Imber 1976). While there are a number of possible reasons that these procellariiformes produce oil, regurgitation by both adults and young is a very effective means of defense, because feathers of predatory birds, such as gulls, become waterlogged and can rapidly lead to drowning (Swennen 1974), whereas the fur of carnivores, such as foxes, becomes foul smelling

(J. Fisher 1952). Regurgitation in response to threat also occurs in other bird species, including turkey vultures (Vogel 1950). Various mammals, such as guanacoes, show parallel but less antagonistic behavior when they spit mucous secretions at adversaries (W. H. Hudson 1892).

Factors that determine whether a prey animal counterattacks must include risks of injury and death and alternative options. The former two will depend on relative size of prey and predator, relative group sizes of prey and predator, and whether the prey has armor or weaponry. For instance, across artiodactyls, species weighing > 20 kg are more likely to attack a predator than smaller species (Caro et al. 2004). In lizards, the extent to which alternative methods of escape are available, including crypticity and flight (Heatwole 1968; Losos et al. 2002), are influenced by prey morphology and environmental factors such as cover and refuges, but the work that has been conducted in this area is confined principally to poikilotherms.

12.5 Methods of escape

Surveying homeotherms in general, categories of escape are incredibly varied, ranging from ponderous movement to extremely fast dashes. Aside from flat-out sprints, fast turns can be used to great effect in escaping predators. Fast turns demand a relatively slow speed and low body mass and can therefore be used advantageously by prey, because predators are commonly heavier than prey and have often accelerated to high speed before they attack (Howland 1974). Furthermore, flight along the ground or in the air is often erratic, involving zigzagging, looping, wild bouncing, and sudden twisting (termed protean behavior: Humphries and Driver 1967, 1970), and may prevent a predator from making an accurate attack on a prey individual or delay its attack by startling or confusing it (section 8.4.a). Little is understood about the benefits of protean behavior. Across bovids, zigzagging is seen principally in open-country species (Caro et al. 2004), intimating that it is prominent where other avenues of escape are limited. In open habitat, FitzGibbon (1990a) found that cheetahs lost more distance from their quarry when Thomson's gazelles started sharp turns or zigzags closer to them in the course of the chase than if they started to zigzag farther away, so it is used to best effect as a last resort.

Some species have prominent color patches that they display principally during flight (Cott 1940; Edmunds 1974; Humphries and Driver 1967) such as European rabbits, but again, the function of these flash marks is still an open question (for example, Stoner, Bininda-Emonds, and Caro 2003). Among waders, taxa that flock have flashier backs and coverts than nonflocking relatives, but whether flash marks confuse the predator or serve some other function such as coordinating flight is unknown (Brooke 1998). After flight has ceased, animals often cover their flash marks, causing the previously conspicuous prey to vanish suddenly, at least to the human eye (for example, white-tailed deer: Caro et al. 1995), but the effect that this has on predators is not known.

Intraspecific variation in escape tactics requires additional explanation, but little can be offered at present. Some very general rules are that different patterns of escape often vary by age, because younger animals have shorter legs or wings and less stamina than adults, making it risky to rely simply on speed (Walther 1969); moreover, they are less able to defend themselves because of their small size. Females may suffer an impaired ability to escape because of physiological costs of reproduction or mass changes associated with egg production or pregnancy (for example, Kullberg, Houston, and Metcalfe 2002; Kullberg, Metcalfe, and Houston 2002). Additionally, where more than one topographic escape route is available, individuals of the same species will flee in the more open and dangerous habitat but stand their ground in steep terrain (Bleich 1999).

12.5.a Birds   In a review of escape tactics of North American birds informed by natural history, Lima (1993) attempted to make generalizations about fleeing based on escape destinations. He classified patterns of escape as follows: into woody vegetation (mostly passerines), into herbaceous vegetation (marsh and grassland birds), around the back of tree trunks (woodpeckers, creepers, and nuthatches), or into the air, here distinguishing between speed-based tactics (doves and Galliformes), aerial dodging (larger gulls, owls, and corvids), and socially coordinated escape (Charadriiformes and Passeriformes: for example, Buchanan et al. 1988). In addition, he recognized that some species are difficult to categorize, such as speckled mousebirds, which drop like rocks out of trees into herbaceous vegetation under aerial attack. Some birds, including waterfowl and kingfishers, also plunge into water (Johnson 1925; Skinner 1928); some, including mourning doves and European starlings, even dive straight at the ground!

Why do different species of birds escape in different ways? Focusing on 43 Emberizine finch species, Lima (1993) tried to document escape destinations according to species' level of sociality and vegetation density in which they lived, and from this extracted four suites of commonly correlated traits based on number of species. Type I birds (see table 12.1) live in exposed habitats, where they can be detected by predators rather easily and therefore rely on either group-related vigilance or group-related evasion in the air to escape. Type II birds live in exposed areas in habitats rich in woody vegetation. They rely on group-enhanced vigilance to detect predators but escape into woody vegetation. Type III finches live alone in dense habitats and make their escape into woody vegetation. Type IV birds are cryptic, furtive, rather solitary spe-

Table 12.1 Distribution of Emberizine finch species among the 36 life history "combinations" defined by 3 escape destinations, 4 levels of sociality, and 3 types of habitat vegetational density

	Escape destination								
	Air			Herbaceous vegetation			Woody vegetation		
	Habitat			Habitat			Habitat		
Sociality	E	M	D	E	M	D	E	M	D
Solitary						5[IV]			5[III]
Low					2	1[IV]		5[II]	2[III]
Moderate								15[II]	
High	7[I]							1	

Source: Lima 1993.
Note: E denotes exposed; M, moderate density of vegetation; and D, dense vegetation. Entries are the number of species placed in a given category. Roman numerals refer to possible basic suites of life history traits for Emberizine finches (see text).

cies that skulk in herbaceous cover (see also Lima 1990b, 1992b). While descriptive, the study is ambitious because it considers two ecological and one social factor simultaneously in classifying four different sorts of escape tactics.

A different category of study used first principles to predict escape gambits of prey by measuring differences in body mass, wingspan, and wing area of eight candidate prey species (Hedenstrom and Rosen 2001). This study considered three alternatives in relation to an aerial attack by an Eleanora's falcon: outclimbing, outmaneuvering the predator in a turning gambit, and diving away. Hedenstrom and Rosen knew that small size, long wings, large flight muscles, and a high wing-beat frequency enhanced climbing flight, which enabled them to predict correctly that only the dunlin, and possibly the common swift, had a faster velocity than Eleanora's falcon in their sample of species and could escape in this way. They also knew that only species with relatively large but short wings can manage a small turning radius, allowing them to predict that only the arctic tern could escape using a turning gambit. High diving speed is attained by large size, and is seen in plunge-diving water birds but also in passerines that can pull out of the dive at the last second.

Categorizations of escape responses are necessarily based on the most common responses to the most common predator, but within species, individuals sometimes exhibit different forms of escape in response to different types of attack. For instance, when a bird is attacked it has to make a split-second decision as to the angle that it will take off and the speed that it will fly, because there is a trade-off between maximizing linear acceleration and climb rate when flying (Witter and Cuthill 1993). Blue tits suddenly approached by a model merlin take off at a steep angle and in 65% of escape flights roll and

Table 12.2 Escape responses of common redshanks on different types of attack

Response	Ground/crouch	Creek dive	Fly
Sparrowhawk			
Surprise	1	28	364
Nonsurprise	0	6	37
Stoop	1	5	10
Percentage (*N*)*	90.5% (21)	19.6% (51)	7.5% (428)
Peregrine			
Surprise	20	18	25
Nonsurprise	3	15	10
Stoop	17	18	12
Percentage (*N*)*	2.4% (41)	1.9% (52)	14.3% (56)
Merlin			
Surprise	7	6	29
Nonsurprise	0	0	0
Stoop	0	0	1
Percentage (*N*)*	22.2% (9)	0% (6)	0% (32)

Source: Cresswell 1993.
*Percentage of responses that resulted in captures (and *N*).

loop within the first meter of flight, probably trying to outmaneuver the raptor by flying above it and in the opposite direction. The angle of ascent is steeper when the model is flown toward them at a low angle of attack, although predator approach speed does not affect takeoff angle. Nonetheless, rapid approach does cause the tits to dodge sideways more often: approach at 2 m/sec results in 0 out of 10 dodges, versus 13 out of 29 when approached at 5 m/sec and at a low angle of attack (Lind, Kaby, and Jakobsson 2002). Similar alterations in angles of ascent have been reported for great tits (Kullberg, Jakobsson, and Fransson 1998) but not for European robins (Lind et al. 1999).

Prey also modify their form of escape to different predators. In some instances we know that intraspecific variation in escape is geared to the probability of escaping the predator. For example, in Cresswell's common redshank study, flying from sparrow hawks resulted in a capture rate of only 8% and was performed most, whereas crouching on the ground resulted in a 91% capture rate and was performed least. In relation to peregrines, flying resulted in a 14% capture rate, whereas crouching and creek diving taken together resulted in a 2% capture rate and was performed about twice as often. For merlins, flying away was always successful and was performed most often (table 12.2; see also Cresswell 1996). As yet there are precious few other studies that match escape pattern to probability of escape in this fashion.

12.5.b Mammals Mammals also demonstrate many forms of escape, including outright rapid flight from a predator (for example, white-tailed deer: Lingle 1992), running in zigzags (for example, kangaroo rat: Djawdan and Garland 1988), intermittent locomotion (for example, degu: McAdam and Kramer 1998), jumping sideways (for example, California ground squirrel: Owings and Coss 1977), jumping vertically (for example, African ground squirrel: Ewer 1966), dropping to the ground during flight (for example, Thomson's gazelle: Walther 1969), gliding between trees (for example, prehensile-tailed mammals: Emmons and Gentry 1983), direct flight to cover (for example, marmots: Barash 1973), running to a burrow (for example, Belding's ground squirrel: L. W. Turner 1973), digging a burrow (for example, armadillos: Harrison Matthews 1971), retreating to crags (for example, dall sheep: Murie 1944) or vertical cliff faces (for example, mountain goats: Fox and Streveler 1986), climbing up grass stems (for example, voles: Jedrzejewski and Jedrzejewska 1990), climbing up trees (for example, red colobus: Bshary and Noe 1997b), climbing onto terminal branches (for example, yellow baboons: Altmann and Altmann 1970), climbing down trees (for example, tamarins: Heymann 1990), hiding behind trees (for example, de Brazza's monkey: Wahome, Rowell, and Tsingalia 1993), running into water (for example, capybara: Macdonald 1984), and even descending from trees into water (for example, sloths: Hingston 1932). The extent to which some of these lesser known antipredator behavior patterns are associated with particular environmental variables or predators is barely understood, other than some obvious points that terrestrial species do not drop out of trees and arboreal species do not burrow! Nonetheless, we know that small species of artiodactyl and those inhabiting rocky habitats use refuges or burrows, whereas species that are pursued by coursing predators are more likely to enter water (Caro et al. 2004).

Within mammalian species, only a few of the many factors that influence these patterns of escape have been investigated. One of them is habitat, which clearly has an important influence on flight speed (Lima and Dill 1990; Lima 1998). As illustrations, in the degu, a diurnal rodent from Chile, flight velocity is faster across open than shrub habitats (Vasquez, Ebensperger, and Bozinovic 2002). Conversely, in juvenile Townsend's ground squirrels, running speeds are slower in shrub habitat than in open areas (Schooley, Sharpe, and van Horne 1996; see also Carey 1985). In the first case, researchers argue that predation risk is higher in open habitat, because prey can be seen; in the second they argue that risk is higher in shrubs because of ambush! Other factors that affect running speed under natural conditions include sex, distance run, substrate, and incline (Blumstein 1992).

Some attention has been paid to the decision as to when a prey animal

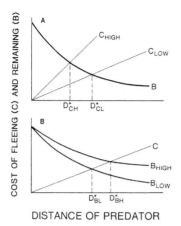

Figure 12.2 A simple economic model of flight distance. The cost of remaining (confusingly denoted as B) when a predator is at a particular distance is directly proportional to risk of capture, which increases to a maximum when $D = 0$. Costs of flight (C), such as lost foraging opportunity, increase with distance (time to attack) in a linear fashion. The animal should choose flight over continued feeding when $B > C$. The crossover of the B and C curves thus defines an optimal flight distance (D^*). A, increasing the cost of flight decreases the optimal flight distance ($D^*_{CH} < D^*_{CL}$). (B). Increasing risk (cost of remaining) has the opposite effect ($D^*_{BH} > D^*_{BL}$) (Ydenberg and Dill 1986).

should emerge from a refuge after retreating or fleeing from a predator. These models assume that the prey cannot monitor the predator from the confines of its bolt-hole (Hugie 2003). If a predator waits for the prey to emerge, prey waiting a fixed length of time will quickly be superseded by that of the predator waiting for just a little longer. The evolutionarily stable strategy is for the predator's waiting times to follow a negative exponential function, randomly selecting a length of time in which to wait. The prey's distribution of waiting times is more positively skewed, rapidly increasing after the predator is likely to have departed and then gradually decreasing with increasing waiting time. The models predict that predators will rarely outwait prey.

12.6 Flight distance

Whatever the means of escape, the distance at which a prey animal flees from a predator is determined by both the cost of fleeing and the cost of remaining at the site (Ydenberg and Dill 1986). Costs of flight include leaving a rich patch of food; flight distance will be reduced if these are high (fig. 12.2a); costs of remaining include having to run a long distance to a refuge; these will increase flight distance if they are high (fig. 12.2b). Historically, recognizing that flight distance (also, and sometimes less confusingly, known as flight reaction distance, flight initiation distance, approach distance, flush distance, or escape flight distance) is the outcome of a cost-benefit analysis was important, because up until 1986, when Ydenberg and Dill published their review, most of the influential studies on flight from predators had assumed that the distance at which prey detected a predator was the same as the distance at which it fled, and this obscured consideration of any cost-benefit decision making (G. V. N. Powell 1974; Siegfried and Underhill 1975; Kenward 1978; Barnard 1980; but see Lazarus 1979; Greig-Smith 1981b). These studies were all of birds, in which

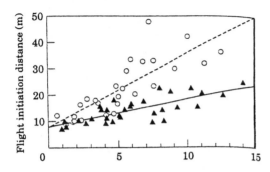

Figure 12.3 The relationship between flight initiation distances and distance to burrow for woodchucks approached by an observer on the same side of the burrow as the woodchuck (*solid triangles, solid line*) or on the opposite side of the burrow from the woodchuck (*open circles, dashed lines*) (Kramer and Bonenfant 1997).

flight often follows detection closely in time; early (and less influential) studies of mammals had disregarded detection distance and focused on flight distance only!

Ydenberg and Dill (1986) predicted that flight distance would decrease with increasing cost of flight. Costs of flight can derive from the mechanical effort put into running or flying, as where gravid females flee at greater distances than nonreproductives (Bauwens and Thoen 1981), or from the risk of leaving a high-quality feeding patch. House sparrows, for example, show a negative correlation between flight distance and seed density in patches (Barnard 1980).

That flight distances increase with risk of capture is supported by many studies. Thus, flight distances tend to be greater when there is a paucity of cover, or refuges are difficult to attain (Grant and Noakes 1987; Dill and Houtman 1989). For example, when approached by a person, free-living woodchucks not only show increased flight distances the farther they are from their burrow (Bonenfant and Kramer 1996), but the rate of increase is steeper when the observer approaches the woodchuck from the opposite side of the burrow, because the animal will now have to run toward the person to escape (fig. 12.3). Nevertheless, across species, the relationship between cover and flight distance may be complicated by cover offering different kinds of feeding opportunities to different species, changing the cost-benefit decision of when to flee (Fernandez-Juricic, Jiminez, and Lucas 2002). Also, some species may feed close to cover, because it affords obstruction from being spotted by predators, but do not flee into cover if the predator proceeds with an attack, whereas other species use cover as an escape destination (Lima 1990b, 1992b).

Risk of capture may also be modified by behavior and condition of both predator and prey. For example, flight distances of African ungulates are greater when they are approached by predators that are running, not walking; by predators approaching in groups, not alone; and by a predator approaching directly toward them (Estes and Goddard 1967; Ewer 1968; Walther 1969; Burger and Gochfeld 1991). Also, flight distances increase when females are ac-

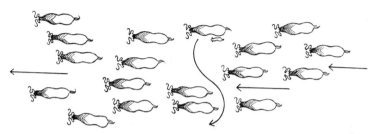

Figure 12.4 Herd of wildebeest passing a sleeping spotted hyena. One wildebeest cow has a calf and avoids the hyena, whereas the others ignore it (Kruuk 1972).

companied by newborn offspring, which cannot flee quickly (Altmann 1958; Rowe-Rowe 1974; FitzGibbon and Lazarus 1995; fig. 12.4).

Flight distance may also be affected by prey group size irrespective of its effects on detection distance (chapter 4). On the one hand, a small flock of common redshanks is more likely to fly instantly upon a sparrow hawk appearing than is a large flock (Cresswell 1994a). On the other hand, sanderlings in larger flocks do not take off at shorter distances from a person than those in smaller flocks (Roberts and Evans 1993). More generally, there is considerable variation across studies of birds, with flight distance increasing with group size in some species (for example, brent geese: Owens 1977), whereas in others, the relationship is flat or concave at intermediate group sizes (for example, barred ground doves: Greig-Smith 1981b). The reasons that individual predation risk might decline in larger groups include dilution and confusion (chapter 8) but will vary according to predator hunting style and prey escape patterns, making generalizations difficult.

Finally, flight distance may also be influenced by alternative methods of escape when a predator appears. Thus, animals relying on crypsis have a lower cost of remaining at a site than conspicuous animals and should have shorter flight distances. Some support for this comes from lizards (Heatwole 1968; D. G. Smith 1997). Similarly, well-armored animals should flee at shorter distances from those lacking armor. Data, again from lizards, show that armored cordylids run shorter distances than unarmored species and quickly enter a refuge because they are short-legged, bulky, and move slowly (Losos et al.

2002). Although this study did not record flight distance per se, it reiterates the point that armor reduces but does not eliminate threat of predation and that it constrains flight rather than liberates the animal from the necessity to flee.

While a number of factors affect flight distance, flight distance itself can have consequences on the form of escape responses, because shorter flight distances may necessitate more-drastic escape measures. For example, both blue tits and great tits take off at a steeper angle in response to a model merlin flown at them when they detect danger at a shorter than a farther distance. In addition, blue tits are more likely to dodge sideways: this occurred in 20 out of 34 flights when they detected a predator at 1 m, but 6 out of 20 dodged when detecting it at 2.3 m away; great tits employed this escape measure far less frequently. A shallower takeoff angle permits higher acceleration, but at short flight distances the blue tits sacrifice this for maximizing height or increasing maneuverability (Lind et al. 2003). In general, there have been few attempts to match flight or detection distance to escape responses, but it is glaringly obvious that these are related.

12.7 Flight and weight gain in birds

When energetic requirements cannot be met directly from feeding, lipid deposits can be metabolized—during periods of inclement weather or during migration, for instance. Fat storage carries costs, however, the most notable of which is enhanced risk of predation, although others are involved (Witter and Cuthill 1993). Predation risk falls into two categories, the necessity to feed and its associated increased risk of predation as a result of reduced vigilance or foraging in an exposed habitat (for example, McNamara 1990; Bednekoff and Houston 1994; McNamara, Houston, and Lima 1994; Cuthill and Houston 1997; Brodin 2001; chapter 4), and mass-dependent impaired ability to take off at an appropriate angle, climb, accelerate, attain maximum velocity, maneuver appropriately, or fly a long distance (Pennycuick 1975; Hedenstrom 1992; Howland 1974; Marden 1987). These are problems faced principally by birds, although they also apply to bats (Aldridge and Brigham 1988; Hughes and Rayner 1991). Unfortunately, empirical data on mass-dependent flight performance is complicated, because different studies focus on different flight measures, and some of these measures, such as velocity and angle of ascent, may be traded off against each other; experimental or natural alterations in weight vary among studies, and there may be adverse threshold effects of weight gain; and voluntary flights differ from those in response to threat. In general, however, small increases in weight, as, for example, between dawn, when birds are hungry, and dusk, when they have fed, do not alter flight velocity, acceleration, or angle of ascent in response to a stimulus threat (table 12.3). Interestingly, this suggests birds may not take off at maximum flight output at dawn. None-

Table 12.3 Effects of increased body mass on measures of flight velocity and ascent in birds listed according to mass increase

Prey	Predator or other alarm	Alarm	Variation in mass	Increase in mass (%)	Reduction in velocity (%)	Reduction in ascent (%)	Reference[a]
Common starling	Shout	Yes	Gravid	7[b]	0[c]	29[b]	1
Pied flycatcher	—[d]	No	Incubation	7	10	—	2
Greenfinch	Merlin	Yes	Diurnal	7	0	0	3
Zebra finch	Tapping	Yes	Diurnal	c.7[b]	0[c]	—	4
	—[d]	No	Diurnal	7	20	—	5
Great tit	Merlin	Yes	Diurnal	8	0	0	6
				8	0	0	6
Willow tit	Merlin	Yes	Diurnal	8	0	0	7
				8	0	0	7
Yellowhammer	Merlin	Yes	Diurnal	8	0	0	3
Common starling	—[d]	No	Weights	10[b]	0[c]	40–50[b]	8
Blue tit	—[d]	No	Incubation	14	20	—	9
European robin	Merlin	Yes	Migratory	27	0	17	10
Blackcap	Merlin	Yes	Migratory	60	17	32	11
Sedge warbler	Merlin	Yes	Migratory	67	26	0	12

[a] 1. Lee et al. 1996; 2. Kullberg, Metcalfe, and Houston 2002; 3. van der Veen and Lindstrom 2000; 4. Veasey, Metcalfe, and Houston 1998; 5. Metcalfe and Ure 1995; 6. Kullberg, Jakobsson, and Fransson 1998; 7. Kullberg 1998; 8. Witter, Cuthill, and Bonser 1994; 9. Kullberg, Houston, and Metcalfe 2002; 10. Lind et al. 1999; 11. Kullberg, Fransson, and Jakobsson 1996; 12. Kullberg, Jakobsson, and Fransson 2000.
[b] Estimated from original study.
[c] No effect denoted as 0.
[d] Not measured.

theless, this effect is not seen in all studies. Where weight increase was similar in magnitude but was caused by being gravid, female starlings showed a reduction in angle of ascent (Lee et al. 1996), resulting perhaps from changes in the center of body mass.

When birds make routine flights as opposed to alarm flights, mass-dependent effects on flight performance are manifest. For example, in zebra finches there is a strong relationship between mass and flight velocity in flights when birds were not alarmed, but this is hardly evident in alarmed flights (Metcalfe and Ure 1995; Veasey, Metcalfe, and Houston 1998). Apparently, birds performing routine flights reduce speed to save energy when they are fatter, but under threat they sacrifice energetic considerations and attain maximal speed. In other species such as willow tits and great tits, however, there was no effect of mass increase on routine flight velocity (Kullberg 1998; Kullberg, Jakobsson, and Fransson 1998).

Across studies, birds sacrifice angle of ascent before they sacrifice flight speed (table 12.3). For example, adding increasingly heavy weights to the base of the tail feathers of starlings reduced the angle of ascent (and aerial maneu-

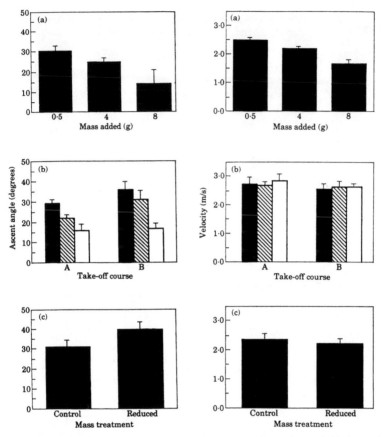

Figure 12.5 (a), effects of adding mass, as shown, to the backs of European starlings between the wings; birds had to ascend 1.25 m (course B) and were tested immediately. (b), effects of adding mass to the base of tail feathers of starlings 3 days before takeoff trials, where birds had to ascend either 0.75 m (course A) or 1.25 m (course B) to reach cover; black bars denote 0 g; hatched bars, 3.4 g; open bars 6.8 g. (c), effects of depriving starlings of food for 4 hours prior to the experiment (Reduced). Left-hand panels show angle of ascent; right-hand panels, velocity (Witter, Cuthill, and Bonser 1994).

verability) but not the flight velocity; a parallel experiment in which body mass was reduced by 4 hours' food deprivation confirmed these results (fig. 12.5). As a low angle of takeoff allows for the most rapid acceleration, birds face a trade-off between a steep climb and horizontal speed, so it is surprising that they opt for the latter, given that it is known that attack success of raptors is greatly reduced once its prey becomes airborne (Page and Whitacre 1975; Cresswell 1993, 1996) (and that it was models of raptors that were used in many of these studies). Nonetheless, if alternative methods of escape such as flying to cover (Lima 1993) are considered, they may tip the balance in favor of flight velocity, as demonstrated by current data.

In situations in which increases in body mass are more marked, as when birds fuel up prior to migration, flight from an alarm stimulus is affected more strongly. Thus, when body mass increases by 25% or more, either velocity or angle of ascent or both may be reduced (table 12.3), perhaps because birds are no longer able to trade off speed for angle of ascent. This might explain why extensive fuel loads are rare and are seen only prior to passage over areas where feeding is difficult (Kullberg, Fransson, and Jakobsson 1996).

Given that maintaining high levels of fat entails at least two forms of predation risk (increased feeding and impaired flight), one might expect birds to manage their fat reserves according to risk of predation (Lima 1986; Mc-Namara, Houston, and Lima 1994). In the short term, birds do stop feeding when exposed to a predator and may defecate, both of which reduce mass and make takeoff easier (Lilliendahl 2000). Over the longer term, birds become lighter in the presence of predators. As illustrations, yellowhammers and greenfinches show reductions in body mass as a result of seeing a stuffed sparrow hawk or through handling by a person, in the former case because they waited longer before resuming feeding (van der Veen 1999; van der Veen and Sivars 2000), in the latter because they made a decision to lose weight despite ample time for feeding (Lilliendahl 1997; see also Carrascal and Polo 1999). More strikingly, over a period of 45 years in England when sparrow hawk numbers fluctuated greatly as a consequence of organochlorine pesticide poisoning, first declining, then being absent, then increasing, and finally stabilizing as pesticide application was curtailed, the body mass of great tits, a principal prey item for sparrow hawks, increased, remained high, declined, and then stabilized (fig. 12.6). Furthermore, in counties where sparrow hawks never declined (western England and Wales) and where they never recovered (East Anglia), tit body weights remained stable over time, while in those where sparrow hawks showed recovery (southern England), tit weights declined sharply by an average of 9.7%. Wrens, a species taken rarely by sparrow hawks, showed no change in any of the three areas (Gosler, Greenwood, and Perrins 1995). Repeated capture of individual tits showed that these changes stemmed from individuals altering their body mass rather than from selection by hawks. In addition, in seven studies of great tits and blue tits, the shape of the relationship between mass and fledgling survival hinged on the presence of sparrow hawks, with heaviest chicks surviving best in study sites where sparrow hawks were very rare, but heaviest chicks not having highest survival rates in sites where sparrow hawks had become reestablished (see Adriaensen et al. 1998 for a review). Even cover can affect management of fat reserves, because common starlings with access to cover show higher fat scores, although no change in body mass or wing loading, implying birds with lower perceived predation risk maintain greater fat reserves (Witter, Cuthill, and Bonser 1994).

Conversely, some species of birds gain weight under threat of predation.

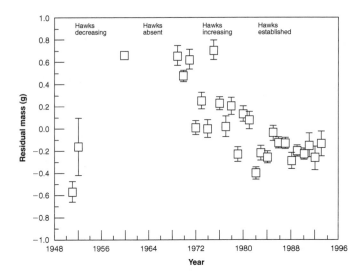

Figure 12.6 Changes in winter (October–March) great tit mass in Wytham Woods, Oxordshire, UK, between 1951 and 1993 (mean + SE). Mass was corrected by regression for body size, time of day, and ambient temperature (mean temperature at Oxford on day of capture) by using the equation: mass $= -22.7 + 0.264$ wing length $+ 1.43$ time -0.389 temperature ($F_{1,5091} = 815.4$, $P < 10^{-6}$), which explained 32.5% of the original variance. Great tits were markedly heavier during the years when sparrow hawks were absent, and declined in mass after their return (Gosler, Greenwood, and Perrins 1995).

For example, yellowhammers showed faster increases in weight over the course of a day when a sparrow hawk had been shown to them in the morning (Lilliendahl 1998), and tufted titmice exhibited a more rapid weight gain after exposure to a sharp-shinned hawk (Pravosudov and Grubb 1998b). In a migratory context, likely to be different from resident birds balancing ordinary energy requirements, blackcaps increased the amount of food that they ate when exposed to a sparrow hawk. Blackcap night activity (suggesting that birds were keen to leave a stopover site) was higher under risk of predation, so blackcaps may have increased feeding to be able to leave the dangerous site earlier (Fransson and Weber 1997). The relationship between fat reserves and predation risk is complex, because birds that are capable of high foraging rates, such as dominant birds, carry lower energy reserves than subordinates in some situations (for example, Ekman and Lilliendahl 1993; Gentle and Gosler 2001; Cresswell 2003) but larger energy reserves in others (Verhulst and Hogstad 1996). Likely, the discrepancy hinges on whether dominance has a greater effect on predation or starvation in a given circumstance (Hake 1996; Verhulst and Hogstad 1996).

The whole issue of management of fat reserves under risk of predation in birds is under active discussion, because studies have used differing methods to portray predation risk and have observed different age-sex classes that fly in

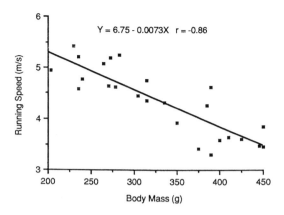

Figure 12.7 Relationship between body mass and maximum running speed in Belding's ground squirrels. Only the 5 fastest squirrels in each 50-g interval were used in the analysis ($n = 25$) (Trombulak 1989).

rather different ways under similar weight changes (J. G. Burns and Ydenberg 2002; Krams 2002). Additionally, discrepancies may stem from some birds perceiving a simulated predation event as more an interruption of foraging, which they make up for later in the day. Where interruption and predation risk have been separated experimentally (by depriving birds of food without a threat being present), yellowhammers and blue tits responded to interruption by increasing food intake and to predation risk by reducing it (van der Veen and Sivars 2000; Rands and Cuthill 2001). From a biological rather than a methodological standpoint and therefore of more interest, species differences in forms of escape and use of cover may be responsible for differences in management of fat reserves, but these still await investigation.

A negative relationship between body fat and flight speed has also been reported for mammals, although data here are far less extensive. Belding's ground squirrels that were weighed, placed in a cloth sack, and then released to the sound of yelling and striking the ground ran more slowly the heavier they were (fig. 12.7), and heavier animals found it more difficult to maintain their speed over the first two seconds of flight (Trombulak 1989). If such a relationship is found more generally in other mammals, it raises questions as to how mammals engineer pregnancy to reduce its impact on flight speeds: think of prey species that frequently resort to rapid flight, and predators, such as cheetahs, that sprint to catch their prey.

12.8 Autotomy and deflection of attack

Under predatory attack, certain vertebrates, namely lepidosaurian lizards, some salamanders, and a few rodents, autotomize, or shed, their tail; among invertebrates, autotomy of appendages is widespread. In lizards, tails are lost through both intervertebral and intravertebral autotomy, although the latter is more common, and tails are often regenerated later after breakage. Tails are usually shed after the tail has been seized, and may subsequently act as a distraction that occupies the predator's attention, since tails may thrash for 5 minutes after separation. Irrespective of the type of autotomy, the rest of the lizard can escape (E. N. Arnold 1988). While the effectiveness of tail loss in thwarting further pursuit is likely to vary according to its owner's speed, its proximity to a refuge, size of lizard, and the extent to which the tail can be consumed easily, autotomy demonstrably reduces mortality in some species (for example, ground skink: Dial and Fitzpatrick 1983; *Eumeces* sp.: Vitt and Cooper 1986). Some species of skink have a brightly colored tail that is thrashed in the presence of a snake, apparently to draw the snake's strike; it is then autotomized (Cooper and Vitt 1985). Lizards that have lost their tail and that are unable to employ autotomy adopt cryptic behavior instead (Formanowicz, Brodie, and Bradley 1990).

Tail loss occurs in some rodents (table 12.4) where it has evolved several times, occurring in eight out of 29 families. Similarities with reptiles are superficial, because tails never move following autotomy. Two forms of tail loss occur: loss of the epidermal sheath of the tail, by far the most common, and breakage across vertebrae. In the former, sometimes all but often just the distal third of the skin of the tail separates from the underlying muscles and vertebral column, with only slight loss of blood (Michener 1976). The remainder of the tail, bone and tissue is consumed later by the owner, leaving a short stump or none at all. In the latter form, found almost exclusively in a handful of American rodents, the whole tail, including skin, vertebral axis, muscles, and tendons, is lost during breakage. This usually occurs when the owner turns around its tail axis or makes a violent jump, breaking the tail at any point (Dubost and Gasc 1987).

Incidence of tail loss in rodents can be surprisingly high—63% of male golden spiny mice, for example—suggesting that seizure by the tail must be relatively commonplace. While researchers who have live-trapped and handled small rodents can attest to losing subjects through tail autotomy, no systematic research into its antipredator benefits has been conducted. We do not know the response of different predators to autotomy, whether the predator is normally startled, whether it consumes the epidermal sheath, or the proportion of escapes that result from tail loss, nor do we understand the ecological circumstances in which autotomy is beneficial.

Tail autotomy in rodents does not involve deflection of attack, because, as far as we know, tails are never waved in anticipation of attack and do not wriggle once discarded. R. A. Powell (1982) has suggested that the black tail tip of some species of weasel may deflect predatory attack by raptors, however. He trained three red-tailed hawks to attack models of weasels in the laboratory. Three models were long (40 cm) and three were short (17 cm) cylinders of white, artificial fur plus a "tail." Each set of models had a black mark either on

Table 12.4 Rodent species that undergo tail loss

Species and family
Funisciurus substriatus (Sciuridae)
Tamias striatus (Sciuridae)
Liomys pictus (Heteromyidae)
Perognathus fallax (Heteromyidae)*
Perognathus panamintinus (Heteromyidae)*
Neotoma lepida (Muridae)
Peromyscus boylii (Muridae)
Peromyscus floridanus (Muridae)
Sigmodon hispidus (Muridae)
Apodemus sylvaticus (Mus sylvaticus) (Muridae)
Apodemus flavicollis (Muridae)
Apodemus agrarius (Muridae)
Rattus norvegicus (Muridae)
Rattus rattus (Muridae)
Zyzomys woodwardi (Muridae)
Zyzomys pedunculatus (Muridae)
Lophuromys sp. (Muridae)
Acomys russatus (Muridae)
Acomys cahirinus (Muridae)
Acomys wilsoni (Muridae)
Acomys percivali (Muridae)
Mus musculus (Muridae)
Glis glis and other dormice (Gliridae)
Eliomys quercinus (Gliridae)
Muscardinus avellanarius (Gliridae)
Dryomys nitedula (Gliridae)
Graphiurus crassi (Gliridae)
Graphiurus sp. (Gliridae)
Lagidium peruanum (Chinchillidae)
Mysateles melanurus (Capromys melanurus) (Capromidae)*
Mysateles nana (Capromys nanus) (Capromyidae)*
Mysateles prehensilis (Capromyidae)
Octodon degus (Octodontidae)*
Proechimys sp. (and other Echimyidae)*
Proechimys longicaudatus (Echimidae)

Source: Shargal et al. 1999.
*Species with tail breaks across vertebrae.

the "back," at the end of the "tail," or else had no black mark at all. Hawks consistently missed the long-tailed models that had a tail spot more often than other types of long-tailed models, but missed the short-tailed models having no spot most often. In the case of the long-tailed models, hawks appeared to check their attack and miss at the last moment, as if surprised by some aspect of the model. Powell suggested that it would be advantageous for large weasel species, such as long-tailed weasels, to have black tail tips but for short weasels, such as least weasels, to have no black mark. This work has never been pursued, but it has the potential to help explain the function of black tail tips in numerous mammals, from slender mongooses to banner-tail kangaroo rats.

There are many striking cases of deflective devices in adult butterflies, including eye spots and markings on wings, and in caterpillars, including false heads, that have been shown to deflect strikes of birds to less important parts of the body (for example, Cott 1940; Wourms and Wasserman 1985), but the study of deflection in vertebrates has lagged far behind. In some lizards, such as broad-headed skinks, whose tails are autotomous, tails are undulated before flight from a predator and then again after flight in the absence of a predator. W. E. Cooper (1998a, 1998b) interprets this second display as being anticipatory and given in order to deflect any attack from an as-yet-undetected predator. While the idea has yet to be substantiated, it provides a new way of thinking about why many mammals hold the prominent distal ends of their tail high when walking or running.

Mammals and birds may release fur and feathers when held, but little is known about this phenomenon (Ratner 1975).

12.9 Fear screams

When caught by a predator or handled by a person, birds but also mammals and anurans (for example, Hodl and Gollman 1986) utter harsh, loud vocalizations called distress calls or fear screams. These are often accompanied by a violent struggle (Perrone 1980), although not necessarily by attempts to bite or peck their handler (Conover 1994). Distress calls in birds are of high intensity and cover a wide frequency range composed of noise or a strong harmonic structure given in repeated and prolonged bursts (Aubin 1987; Aubin and Bremond 1989; Neudorf and Sealy 2002); common pipistrelle bats give distress calls at low frequencies (Russ, Racey, and Jones 1998). Consequently, such calls can be heard over long distances and are easy to localize (Bremond and Aubin 1990); they seem designed to attract other animals. Conspecifics are attracted to such calls in many species, but in others they are repelled by them (Frings and Jumber 1954). Distress calls have similar properties in many species, suggesting convergent signaling to heterospecific prey or to common predators; indeed, heterospecifics are attracted to each other's distress calls

Table 12.5 Predictions (columns) of the 5 hypotheses (rows) for the function of distress calls in birds

Hypotheses	Members of consp. groups call most	Members of mixed flocks call more than solitary or pairs only	Captured birds call more in dense habitats	Larger birds call more	Distress calls are paired with struggle	Birds respond to calls by mobbing predator
Warning kin	Yes	No	No	No	No	No
Calling for help	Yes	No	No	No	No	Yes
Mobbing	No	Yes	No	No	No	Yes
Startle predator	No	No	No	Yes	Yes	No
Predator manipulation	No	No	Yes	Yes	No	No

Source: Neudorf and Sealy 2002.

(Stefanski and Falls 1972b; Jurisevic and Sanderson 1998; J. M. Ryan, Clark, and Lackey 1985; Chu 2001; Russ et al. 2004), as are predators (Klump and Shalter 1984). Incidence of fear screaming varies within species, as shown by the same individual screaming to differing extents depending on the handling procedure (Perrone and Paulson 1979), and between species; for example, a greater percentage of individuals call in larger species (Perrone 1980; Jurisevic and Sanderson 1998; Neudorf and Sealy 2002).

Five closely related hypotheses have been proposed to explain the function of distress calls: calling for help, warning kin, mobbing, startling the predator, and attracting other, secondary predators (Conover 1994; table 12.5). Several of these hypotheses have received some support, implying fear screams have multiple functions.

Rohwer, Fretwell, and Tuckfield (1976) suggested that distress calls are given to warn conspecifics, particularly kin, of danger. The most persuasive support for this hypothesis comes from studies of nest mates. For instance, distress calling by black skimmer chicks uttered when they are handled at the nest results in their siblings running off (Gochfeld 1981). More indirect evidence shows that across species, permanent residents, which are more likely to have kin present, scream more than winter residents when handled in mist nets (Rohwer, Fretwell, and Tuckfield 1976). Similarly, across species from North America and Europe, incidence of distress calling is associated with increasing flock size (Greig-Smith 1982; Inglis et al. 1982) and being sedentary (Perrone 1980; Greig-Smith 1982), both of which are proxies for kin being in the vicinity, although not with stability of groups (Greig-Smith 1982). In contrast, distress calling in birds from a lower montane rainforest in Costa Rica do not show associations with flocking (Neudorf and Sealy 2002).

A second, closely related hypothesis is that birds call conspecifics for help

(Rohwer, Fretwell, and Tuckfield 1976). Here conspecifics are expected to re-spond to calls by approaching the distressed bird. That diurnal winter mi-grants scream more than nocturnal migrants (Rohwer, Fretwell, and Tuckfield 1976) and, that when they have young nestlings, white-throated, song, and swamp sparrow males fly directly at a speaker playing (albeit adult) distress calls (Stefanski and Falls 1972a), support the call-for-help idea. None-theless, in a direct test of the call-for-help and warning hypotheses, G. E. Hill (1986) found that when a hawk model was presented in association with tufted titmouse distress calls, tufted titmice approached the speaker very cautiously and to only 7 m of the model, both of which are at odds with aiding a captured conspecific. When the hawk model was presented without accompanying dis-tress calls, some titmice approached the model very closely and appeared not to notice it, which favors a warning function for distress calls instead. Most damning, playbacks of distress calls are routinely used by wildlife managers to keep some bird species away from an area (Frings and Jumber 1954)!

A third hypothesis, that distress calls attract mobbers, provides a mecha-nism as to how other birds might aid a conspecific through driving the pred-ator away. This hypothesis views distress calling as a type of mobbing call that recruits others which may themselves benefit from moving the predator on (section 11.5.b). Phainopeplas produce loud screams and mimic other species' vocalizations when they are in distress, suggesting that heterospecifics may be attracted by hearing a cacophony of species' calls (Chu 2001). Nonetheless, this hypothesis has received little support thus far. For example, there was no evidence of more distress calling occurring in neotropical rainforest birds when they were in mixed-species flocks than when they were solitary, despite mobbing by heterospecifics being common in other contexts (Neudorf and Sealy 2002). In contrast, a study of common pipistrelles documented a twenty- to eightyfold increase in numbers of emerging bats passing over a cage to which bats giving distress calls were confined. Researchers interpreted this be-havior as mobbing, believing that such numbers would be sufficient to drive a predator away (Russ, Racey, and Jones 1998).

Fourth, fear screams could startle a predator, causing it to drop its prey for an instant and allowing escape (Driver and Humphries 1969). That larger spe-cies fear-scream more than smaller ones and that birds struggle at the same time as screaming suggest that they try to take advantage of a momentary weakening of grip (Conover 1994; but see Neudorf and Sealy 2002). More-over, when taken out of mist nets, birds give distress calls at higher rates if they are held by their wings or legs, a situation in which they might break free, than by the head or body (table 12.6). There is some evidence that fear screams do startle naïve coyotes, as indicated by flinches, jumping back, and other re-sponses to playbacks of distress calls, but these responses wane quite fast when

Table 12.6 Mean number of distress calls given by different passerines while held for 15 seconds in each position by their head, body, leg, or wing

Species	N	Head	Body	Leg	Wing	p-value
Blue jay	7	0.3	0.1	7.6	8.7	0.01
American robin	7	1.3	0.3	2.4	17.6	0.002
Gray catbird	11	1.5	0	3.2	7.8	0.01
Northern mockingbird	16	7.3	9.5	14.0	8.4	0.03
Common starling	24	2.9	0.1	8.4	11.5	<0.0001
Song sparrow	8	0.1	0.1	3.1	4.6	0.02
White-throated sparrow	10	0	0	9.8	8.3	0.0003
House finch	17	0.1	0	1.6	2.5	0
Chipping sparrow	5	0	0	1.2	5.6	0.02
Red-winged blackbird	43	0.1	0.1	1.6	8.2	<0.0001
Northern oriole	10	2.0	0.6	6.2	6.4	0.0009

Source: Conover 1994.

screams are replayed to them (Wise, Conover, and Knowlton 1999). Also, experiments with captive raccoons and opossums found that 20% were startled by a distress call (Conover 1994). Nonetheless, if screams serve to startle a predator, they should be brief and explosive; but, au contraire, they are usually repetitive.

A final hypothesis states that fear screams attract other predators to the prey, and that during an ensuing dispute the first predator may drop the prey and allow it to escape. Predators are certainly attracted to distress calls. For example, when varied thrush and brown towhee calls were played in the field at Point Reyes Bird Observatory, California, in seven trials, playbacks attracted a Cooper's hawk once; two sharp-shinned hawks once, the first one chasing the second one; and a great horned owl twice. The accipters and owls appeared within as little as 10 seconds and 30 seconds, respectively, and approached to within 6 m of the speaker (Perrone 1980; see also Koenig et al. 1991); rapid approach of other predators would be key to enabling the prey to escape before death. More systematically, Hogstedt (1983) in Sweden found that larger predators, goshawks and red foxes for instance, had a greater probability of approaching broadcast fear screams than smaller predators, such as kestrels. This could be advantageous for the prey, because a large predator is more likely to usurp a small predator than vice versa. In a sample of 30 species of birds that were mist netted by Hogstedt, larger species were more likely to scream, consistent with the idea that they take longer to disable and that screaming is more effective in larger than in smaller species because their screams are louder and carry farther. Moreover, birds occupying open habitats screamed less frequently than those in moderately exposed habitats; in turn, the latter screamed less than those living in dense cover, where visibility is low and the need to attract attention by vocalizing is greater. Against the hypothesis, when distress calls were played to two coyotes, the one in possession of prey intensified its attack rather than relinquished hold of it (Wise, Conover, and Knowlton 1999). At present, there is mixed or relatively strong support for four of the five hypotheses, suggesting that fear screams serve several functions.

12.10 Death feigning

In stark contrast with uttering distress calls when captured, some species feign death; the juxtaposition of these behaviors provides a showcase example of the incompatibility of certain antipredator defenses. Death feigning, also referred to as death shamming, akinesis, thanatosis, tonic immobility or animal hypnosis, represents a catatonia-like paralysis that initially takes the form of muscular rigidity but may also be followed by hypotonicity. Vocal behavior and heart and respiratory rates are suppressed, and while animals usually have their eyes open, they may intermittently close them, appearing asleep or dead (table 12.7). This state may last from a few seconds to several hours. Death feigning is surprisingly widespread in birds and mammals as well as reptiles and amphibians (Dodd and Brodie 1976; Caldwell, Thorp, and Jervey 1980; Greene 1988).

In birds, death feigning has been documented in many species (Armstrong 1947), ranging from eider ducks (Perry 1938) to turkey vultures (Vogel 1950) to Eurasian bullfinches (Tinbergen 1962), and its neurobiological basis is reasonably well understood (for example, Gallup and Maser 1977). Immobility in birds can be induced either by a firm grasp or by gentle stroking. Several factors prolong its duration. In chickens, these include warning calls (R. B. Jones 1986), social rearing (Rovee-Collier, Kaufman, and Farina 1980), close proximity of the experimenter, staring at the chicken (Gallup, Cummings, and Nash 1972), and lack of opportunity to escape (Arduino and Gould 1984). Birds are therefore aware of their surroundings and opportunities for leaving

Table 12.7 Comparison of behavioral differences between sleep and feigning death in common opossums

Sleep	Feigned death
1. Eyes and mouth closed	1. Eyes and mouth open
2. Dorsal aspect upward	2. Lateral aspect upward
3. Feet tucked under	3. Feet visible and toes usually flexed
4. No response to sharp sounds	4. Ears twitch at sharp sound
5. Arouses with a start and turns head toward source of prodding or blowing	5. No response to prodding or blowing; may retract lips slightly

Source: Francq 1969.

while feigning death. Death feigning develops at around 7–10 days in chickens (Ratner and Thompson 1960), and in other avian species requires a few trials before it is performed properly. It rarely occurs in adults with chicks (Armstrong 1965), presumably because it precludes nest defense.

In mammals, death feigning has been described in several species, including the pampas fox (Hudson 1892), African ground squirrel (Ewer 1966), laboratory rat (Ratner 1967), zorilla (Estes 1991), European rabbit (Ewell 1981), and, of course, opossums (McManus 1970). Death feigning in opossums is induced by grabbing and shaking the animal and fails to decline with repeated attempts to induce it. Opossums lie on their side with their eyes open and monitor their surroundings while immobile, taking longer to recover if a threat is present (Sentell and Compton 1987).

Surprisingly little work has been conducted on the relationship between death feigning and surviving predatory attack despite a consensus that this is its function. The principal study is over 25 years old. It showed that of 50 ducks that became immobile after being seized by foxes in small outdoor pens, 29 survived the ordeal. This was because rather than killing the ducks, some foxes mouthed, cached, and then left them, allowing the ducks time to recover and escape (Sargeant and Eberhardt 1975). Interestingly, the study indicated that death feigning provided most protection against naïve foxes; experienced foxes were more likely to kill the birds or chew off their legs. Death feigning is probably adaptive against predators that show a temporal separation between capturing and killing their prey and where the latter is precipitated by movement of the prey (R. K. R. Thompson et al. 1981). In addition, Ruxton, Sherratt, and Speed (2004) suggest that death feigning may be prevalent in prey that live in groups, circumstances in which the predator can attempt multiple kills. That death feigning occurs in so many vertebrates suggests that these conditions pertain quite widely; nonetheless, the phylogenetic distribution and environmental conditions, especially predator identity and pressure, associated with death feigning are poorly documented, as are the correlates of its incompatible alternative, fear screaming.

In conclusion, birds and mammals use several different ways to halt a predator's advance at the last moment, to flee, and to save themselves after contact has been made, although each species employs only one or two. On a broad scale, defense options open to each species are constrained by its phenotype. Thus, a cryptic species can freeze or show flash colors, whereas a gaudy species cannot use these defenses to good effect; a digitigrade species can run fast or perhaps zigzag, but a fossorial one cannot; and ostentatious behavior, tail waving or screaming, are at odds with tonic immobility. That said, there are many closely related species with similar phenotypes that use different types of flight and behaviors of last resort which cannot be interpreted so easily using constraints. To explain the distribution of such defenses, comparative, phylogenetically controlled analyses that match antipredator defenses to ecological and social variables within clades can make headway (Caro et al. 2004), but they are necessarily limited by quality of data, recourse to broad categorization of independent variables, vagaries of the phylogenetic reconstruction, and inability to verify associations without direct observation or better experimentation. Nonetheless, they do at least move toward explaining the evolutionary patterning of bizarre escape and defenses in nature.

12.11 Summary

Just before or during a predatory attack, prey animals can employ a battery of antipredator defenses to escape being killed. As the predator approaches, they can attempt to be cryptic by remaining immobile. Once close at hand, prey can startle the predator by giving a sudden vocalization, some of which apparently mimic dangerous animals, or expose brightly colored parts of their body. Prey may also launch a counterattack using noxious secretions, strength, weaponry, armor, or strength of numbers.

Both birds and mammals show many diverse forms of escape, including zigzagging flight (termed protean behavior), displaying color patches during flight, and quickly covering these patches at its termination. Flight in birds can be categorized by the type of destination (into vegetation types or air) and by aerial maneuvering. Forms of flight may be tailored to the most promising way to escape from a particular predator. Mammals show a bewildering array of escape tactics as well.

The distance at which prey flees from a predator depends on a balance between the costs of flight, including lost opportunities for feeding, and costs of remaining. Both types of costs are influenced by multiple factors, including cover, type of predator approach, prey group size, and alternative means of escape.

Body weight in birds affects flight performance. Whereas small increases in weight gain do not alter flight parameters, larger ones reduce flight ability. Normally, angle of ascent is sacrificed before flight velocity. Some birds lower their fat reserves under heightened predation risk so as to reduce time spent foraging and counteract impaired flight performance, but empirical results are not entirely consistent.

Like lizards, some rodent species shed their tail if contacted by predators. There is some evidence that tails may be used to deflect attack in weasels. If captured, some birds and mammals utter loud fear screams that seem designed to attract attention. Five hypotheses have been put forward to explain the function of fear screaming: warning kin, calling for help, inciting mobbing, startling the predator, and attracting secondary predators that may displace the original predator, allowing prey to escape. There is mixed support for most of them, suggesting that fear screaming has multiple functions. Some species pretend to be dead when captured and so elude being killed. Death feigning and fear screaming are good examples of antipredator behaviors that are incompatible.

13 Framing Questions about Antipredator Defenses

13.1 Introduction

The list of antipredator defenses chronicled in this book is long and daunting, and although I have been at pains to separate the mechanisms by which prey mount defenses against predators at different stages of the predatory sequence, this should not be viewed as artificially inflating the variety of defenses shown by birds and mammals. For example, group size of prey differentially affects encounter rate with predators, attack probability, likelihood of capture, and dilution of risk in different ways between a single prey-predator pair (Creel and Creel 2002). Nonetheless, even if we disregard defenses that operate at several stages of the predatory sequence, such as grouping, the number of qualitatively different defenses is still incredibly impressive. One just has to consider the variety of shorebirds' distraction displays (table 10.4) or the plethora of antipredator traits by which mammals avoid predation by wolves (table 13.1).

Before trying to understand why animals exhibit so many defenses, we can make the task easier by distilling the list a little, because many defenses constitute parts of morphological and behavioral complexes that only work effectively when these traits are expressed together (section 13.2). Tight associations between morphological and behavioral traits are one of the most important and long-recognized aspects of antipredator defenses. In addition, we can dismiss the idea that large numbers of defenses are simply the total sum of different prey species showing a restricted number of defenses to different predator species (section 13.3); every prey species exhibits more than one antipredator defense to a given predator.

Figure 13.0 (*Facing page*) Giraffes have dappled coats that are assumed to help them blend in with patchy light experienced in woodland habitats. Despite their large size and relatively few predators (principally lions and people), crypticity is still apparently necessary for both young and adults (reproduced by kind permission of Sheila Girling).

Table 13.1 Antipredator characteristics and behavior of wolf prey species taken from numerous studies

Species:	H	AH	Bv	Wb	G	S	Sa	Bb	D	P	C	E	Mo	B	M	FU	MU	AU	Sp
Physical traits																			
Size													X	X	X				
Weapons																			
Antlers/horns																X[a]	X		
Hooves																		X	
Cryptic coloration																		X[b]	
Speed/agility	X					X			X										
Lack of scent							X[b]												
Behavior																			
Birth synchrony																		X	
Hiding									X[b]	X[b]									
Following			X[b]	X[b]						X[b]				X[b]					
Aggressiveness																		X	
Grouping	X				X	X			X	X	X	X	X	X					
Vigilance																			X
Vocalizations						X			X										
Visual signals		X				X					X	X							
Landscape use																			
Migration									X	X	X								X
Nomadism							X		X		X	X							
Spacing																			
Away									X	X					X				
Out									X	X	X								
Escape features																			
Water			X						X		X	X							
Steepness					X	X													
Shorelines											X								
Burrows				X															

Source: Mech and Peterson 2003.
Note: Column headings are as follows: H, hare; AH, arctic hare; Bv, American beaver; Wb, wild boar; G, goat; S, sheep; Sa, saiga; Bb, blackbuck; D, deer; P, pronghorn; C, caribou; E, elk; Mo, musk ox; B, bison; M, moose; FU, female ungulates; MU, male ungulates; AU, all ungulates; Sp, all species.
[a] Some females.
[b] Young or neonates.

At this juncture, it may be profitable to fathom why so many defenses are seen in nature by asking three different questions that treat numbers of prey and numbers of predator species separately. Parsing out such questions concatenates the list and makes it more tractable. First, why does a single species show different antipredator defenses to different predators; that is, why does it have predator species–specific defenses, not a set of generalized defenses that are effective against all predators (section 13.4)? Second, why do different prey species use different defenses to thwart the success of the same predator species (section 13.5)? Third, why does a single prey species show a number of antipredator defenses to a single predator species, each of which is designed to curtail a different step of the predatory sequence (section 13.6)? The answers to these questions are not yet solved (Pearson 1989); that is, we do not know the extent to which defenses are redundant, the degree to which defenses are mutually incompatible alternatives to each other, or why some species target their defense efforts at early stages of the predatory sequence, others at later stages. Nonetheless, there is an emerging consensus that within a single prey species, antipredator defenses against even one predator can best be interpreted correctly when we understand how a prey species responds to other predator species in its environment. In section 13.7, I discuss the extent to which prey defenses and predator attack mechanisms can be thought of as a product of coevolution, or a kind or arms race between a particular prey and a particular predator. Pressing my luck, perhaps, I then take a step back and outline questions about antipredator defenses that are of general significance (section 13.8). I restrict myself to an arbitrary round number of ten, five of which center on prey and five on predators; other authors would doubtless settle on other issues. I end with some leads as to why defenses are far from perfect (section 13.9).

My task in this chapter, therefore, is to demonstrate ways in which it is productive to ponder the diversity of antipredator defenses, reemphasize the importance of considering prey and predator behavior together, deal with the issue of coevolution, and offer suggestions as to how the study of antipredator defenses can be advanced.

13.2 Synergism between morphology and behavior

For years, biologists have recognized that effective crypsis necessitates animals living in certain habitats and resting in particular attitudes. Cott (1940) was assiduous in repeatedly drawing attention to synergisms between morphological and behavioral traits. Later, N. Tinbergen (1965) listed four behavioral correlates of crypsis: diurnal immobility, living on a background that matches the animal's coloration, adopting a position that provides maximum concealment, and living well spaced out (beyond the distance at which predators detect prey easily). Examples of these phenomena are plentiful in arthropods: the melanistic form of the peppered moth must rest on dark tree trunks, leaf-mimicking tettigonids must rest facing downward on stems so that their abdomen looks like budding leaves, and stick insects wave gently when the wind picks up (M. H. Robinson 1969; see also Chai and Srygley 1990). It is equally easy to find parallels in birds and mammals: dull-colored nightjars remain immobile on roads until danger is very close (Cott 1940), sloths move languidly along the undersides of branches (Hingston 1932), and pottos ponderously clutch branches in a hesitant fashion in the breeze. Cott (1940) also commented that aposematic animals are sluggish and diurnal and are often gregarious. While there are few birds and mammals that are aposematic, many draw attention to parts of their body using exaggerated movements and bright patches of color simultaneously: tail flicking in birds that have white undersides to their tail, hissing and teeth baring in carnivores under threat, and rattling and erecting quills in porcupines. In short, anecdotal evidence indicates that behavior and morphology are intricately linked in most defenses, although these associations have yet to be investigated systematically in homeotherms. Such analyses will involve showing that selection favors combinations of traits being expressed together and that such correlational selection drives functional integration of traits and possibly genetic integration as well.

In North American snakes, some species are striped, others are blotched or banded. The former are diurnal, use open habitats, and flight as their primary defense; the latter are secretive, showing aggressive or cryptic antipredator behavior. Through optical illusion, striped snakes appear stationary when crawling slowly and by extension appear to be moving slower than they actually are. Within garter snakes, a similar pattern is found, with striped individuals showing direct flight and unmarked individuals or those with broken patterns showing stereotyped reversals or sudden changes in direction during escape that is thought to foster crypsis after initial detection by a predator. Within one population, individuals with the highest probabilities of survival performed uninterrupted flight if they were striped but fled evasively if unstriped or spotted, presumably as a result of differential predation from visually hunting predators (Brodie 1992). It would be instructive to carry out similar analyses in homeotherms, although to date, very few examples have been identified of single species with alternative antipredator morphologies or coloration related to defense.

Dewitt, Sih, and Hucko (1999) have drawn attention to the idea that there may be four classes of relationship between morphological and behavioral antipredator defenses, of which Brodie's snake example is only one. Their first distinction is codependent traits, which are mechanically linked. An example would be leg musculature and bipedal locomotion in heteromyid rodents. Their second distinction is complementary traits, in which morphology and behavior are mechanically independent but are used together for greater effect. An example would be larger species being more likely to counterattack. The third category is cospecialization, in which morphology and behavior are positively correlated. This would apply to striped and blotched garter snakes showing flight and crypticity, respectively. Their fourth class is compensatory, in which morphology and behavior are negatively correlated. An example would be slow-running juvenile marmots foraging close to a rocky talus that provides refuge, but adults, which rely on speed, foraging farther away. The first two axes pertain to species-specific traits, whereas the second two usually address differences between individuals of a particular species. Cott's attention focused on species (almost nothing was known about interindividual variation in defenses then) and refers principally to the complementary distinction. Indeed, cospecialization and compensatory relationships between antipredator morphology and behavior usually demand knowledge of individual differences in defenses (Rundle and Bronmark 2001; but see Mikolajewski and Johansson 2004) and are made more credible by repeated observations of the same individual, neither of which have been given a great deal of attention in relation to homeotherms as yet. That said, the issue deserves thorough study (section 13.8.c).

13.3 Defenses shown by different prey to different predators

One trivial explanation for the plethora of defenses seen in nature is that every prey species principally employs one or two types of defense to all predators and that the variety of defenses that biologists have documented merely reflect the number of prey species that have been studied! This idea can be quickly dismissed. Certainly, some prey appear to have very little opportunity to use anything more than one or a few antipredator defenses to avoid predation. For instance, against ornate hawk eagles, a sit-and-wait predator, red-cap moustached tamarins principally rely on three behaviors only: early warning through vigilance or alarm calls and dropping to the forest floor. Hawk eagles, perched high in trees, watch patiently for monkeys and then use a rapid diving descent lasting no more than a very few seconds to attack them. Tamarins have so little time that they can rely only on personal vigilance, and alarm calls of conspecifics and heterospecifics, to drop and so avoid being struck and carried off (Peres 1993). In some circumstances, therefore, prey may have a very limited set of defenses, but other defenses such as shifts in diurnal activity or use of space may also be in play; and, generally, it is far more common to find single prey using a menu of defenses against one predator (see table 13.1). There are numerous examples in this book such as Thomson's gazelles exercising vigilance, inspection, snorting, stotting, flight, and occasionally immobility against cheetah attack (Caro 1994b; FitzGibbon and Lazarus 1995). That well-studied species seem to possess a greater number of defenses than less-studied species implies that multiple defenses are likely to be the rule. It would therefore be erroneous to ponder the variety of defenses that biologists have described as simply a large constellation of a limited set of responses shown by many different prey species.

13.4 Prey employ different defenses against different predators

It is common knowledge that a given avian or mammalian prey species modulates its response to different predators both quantitatively and qualitatively. In most cases, differences in antipredator defenses are interpreted as being adaptive based solely on their design characteristics. Thus, American pikas in the Rocky Mountains of Colorado, USA, delay calling more and call less frequently to long-tailed weasels than to American martens (Ivins and Smith

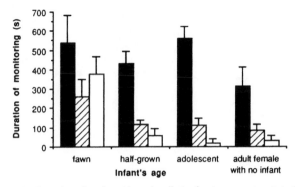

Figure 13.1 The mean time Thomson's gazelle mothers with attendant offspring of varying ages spent monitoring 3 predator species, compared with an adult female with no offspring. Bars indicate SE. Predator species: cheetah, *solid histograms*; spotted hyena, *hatched histograms*; golden jackal, *open histograms*. There was no significant effect of infant's age on the time that mothers stared at cheetahs, but there was on the duration of staring at hyenas and jackals (FitzGibbon and Lazarus 1995).

1983). The explanation is that weasels are more dangerous than martens, because they are smaller and can follow pikas more easily into the interstices of the talus, where pikas hide. Similarly, gazelle mothers in the Serengeti Plains spend a greater amount of time monitoring cheetahs than spotted hyenas and jackals; the former are stealth predators that use surprise, the latter are coursers that usually approach prey openly. Also, gazelle mothers monitor jackals for far longer when they have fawns; these canids pose a threat primarily to fawns and rarely attack older gazelles (fig. 13.1). These explanations invoke differences in hunting style and predation pressure in order to interpret quantitative differences in antipredator defenses.

Qualitative differences in responses to different predators are commonly reported, too. Research on gulls, for example, describes disparate defensive actions taken against different predators (Veen 1977; table 13.2). Again, these are presumed to constitute adaptive responses, not only to the hunting style of each predator but also to the threat that they pose. For example, in gulls it is argued that each antipredator defense is an adaptive response to whether the threat is to adults, to chicks on the ground, or to both. To peregrines, which are a threat to adults, black-headed gulls flee; to herring gulls and European coots, which threaten chicks, they mount attacks; and to red foxes and stoats, which threaten both, they show a mix of sometimes ineffectual behaviors (Kruuk 1964). The more predators there are, therefore, the more defenses will be manifested, because there will be greater variance in hunting technique and perhaps different types of threat. In support of this, positive correlations have been noticed between the number of antipredator defenses shown by bank voles to different predator species and the contribution that bank voles play in that predator's diet (fig. 3.12); also between the diversity of escape responses in

Table 13.2 Antipredator behavior of Franklin's gulls

Predator	Occurrence of predation in gull colony	Actual predation	Reaction of gulls	Effect on predator
Human	Normally rare	None directly, causes death due to chicks getting lost	Silent panic	None
Mink	Common in one of the years	Killed young often in excess	None observed	?
Moose	Rare in colony during breeding season	None directly, may destroy some nests	Mobbing	Colony avoidance
Marsh hawk	More than 1/day	Infrequently seen to take adults and chicks	Silent panic	Some avoidance of colony
American coot	Approximately 35/day	Observed to eat eggs; competition for nest sites	Attack	Sometimes effective
Great horned owl	Approximately 1/day	Kills adults and juveniles nightly	None	None
Crow, herring gull, ring-billed gull	Sometimes present	None	None	None

Source: Burger 1974.

populations of whiptail lizards and the extent to which their tail was broken, an index of predation pressure (Schall and Pianka 1980).

Unfortunately, interpretations based on the apparent efficacy of various responses are rather unsatisfactory, and it is more convincing if antipredator defenses are matched to predatory success and failure rates. Cresswell (1993) observed that common redshanks principally fly from sparrow hawks, creek dive or crouch on the ground from peregrines, and fly from merlins. He showed that flight resulted in the lowest rate of capture from sparrow hawks; dives and crouches, in the lowest success rate from peregrines; and creek dives and flight, in the lowest deaths from merlins (table 12.2). Studies such as this are more persuasive in demonstrating that antipredator defenses are indeed tailored to the methods that each predator uses to pursue and capture prey.

Defenses in multipredator environments may be subject to additional selection pressures not encountered in single-predator systems, because a given defense may be beneficial against one predator but be costly when used against another. In the simplest sense, to avoid predation in one microhabitat, prey may shift to another, where they increase their exposure to a different predator species (Kerfoot and Sih 1987); but the principle applies to both behavior and morphology. Fieldfares nesting close together benefit in relation to avian nest predators, because parents mob them together, but clumped nests are vulnerable to terrestrial predators that practice area-searching behavior (M. Andersson and Wiklund 1978). Three-spine sticklebacks benefit in relation to piscivorous bird predators, because they are awkward to handle, but they fall prey to predatory odonate nymphs, which find them easier to grasp; this promotes loss of spines in some populations (Reimchen 1980). Antipredator defense practiced against one predator may make prey more vulnerable to another predator if the predators hunt in different ways and the two defenses are incompatible (for example, armor and flight speed), or two types of defensive behavior compete for a prey individual's time (for example, vigilance and hiding). On the other hand, if predators hunt in the same way, or if a given defense can be used to thwart danger imposed by predators that hunt in different ways (for example, using vigilance to perceive sit-and-wait predators or spot rapidly approaching predators), then defenses against two predators may be entirely compatible.

Degree of specificity of antipredator defenses against each of two predators affects the way in which the defenses interact. When prey show nonspecific defenses that are effective against several predators, such as hiding in a refuge, the presence of one predator probably reduces the ability of the second predator to be successful. If, however, prey exhibit antipredator defenses that are specific to each predator species, then compromises may be in order, rendering the prey less effective against either threat. In such cases, for example, vigilance against aerial and terrestrial predators, prey may be best off feeding as quickly as possible to minimize exposure time, a risky proposition in the short term (Lima 1992a, 1992b; Matsuda, Abrams, and Hori 1993; Sih, Englund, and Wooster 1998). Unfortunately, we do not really understand why some defenses are specific to certain predators, whereas others can be used to thwart attack by a range of predators.

In sum, there is a consensus that knowledge of the number of predators that pose a danger to a prey species, relative predation pressure, and antipredator defenses to each are all necessary to interpret the adaptive significance of defenses to a predator. Studies that focus on single predator-prey interactions in multipredator systems are sadly inadequate for understanding prey defenses, because antipredator defense to one may be a compromise to many.

13.5 Different prey use different defenses against the same predator

Although many antipredator defenses are seen again and again among particular taxa (such as stotting in ungulate species and alarm calling in primates), providing some common comparative themes, it has also long been recognized that different prey species mount different defenses to the same predator. Over the last 100 years, for example, the idea that there are alternative and often incompatible defenses in nature has been repeatedly brought to attention in the antipredator literature (table 13.3). Many of these alternatives focus on the inherent contradiction between crypsis and using force to drive a predator away,

Table 13.3 Antipredator defenses that are incompatible

Antipredator defenses		Reference[a]
Countershading	Size or weaponry	1
Turning white in winter	Defense using force	2
Spotted young	Using sheltered retreats or defending young	2
Hiding or freezing	Flight or counterattack	3
Inconspicuous parental behavior around nests	Aggressive nest defense	4
Flight	Confrontation	5
Maneuverability in birds	Size in birds	6
Death feigning	Distress calls	7

[a] 1. G. H. Thayer 1909; 2. Cott 1940; 3. Kruuk 1972; Estes 1974; Jarman 1974; 4. Murphy, Cummings, and Palmer 1997; 5. Lingle 2002; 6. Andersson and Norberg 1981; 7. Gallup and Maser 1977; Hogstedt 1983.

but others point out the difficulties in attempting different types of flight, or between feigning death and calling loudly. These alternative defense mechanisms stem from logical argument and empirical data. In general, large animals use force and flight; intermediate-sized prey employ a wide array of antipredator devices, including signaling to predators; and small prey species rely on crypsis (Caro and FitzGibbon 1992). For instance, in the Bandipur tiger reserve, India, smaller species remain motionless, whereas larger prey species attack predators (table 13.4). As a rather mundane generality, therefore, one can state that body size, at least, drives interspecific differences in antipredator defenses.

Nonetheless, similar-sized species also show different defenses to the same predator. For instance, mule deer and white-tailed deer living sympatrically in southern Alberta, Canada, react to coyotes in different ways (Lingle and Pellis 2002). Whereas white-tailed deer are more likely to flee than mule deer at all stages of the hunt sequence, mule deer are more likely to stand their ground and be aggressive. These different defense mechanisms stem from white-tailed deer being able to run faster than mule deer, and for even white-tailed fawns to outdistance coyotes, but for adult mule deer to be outdistanced by coyotes. Indeed, mule deer that flee are more likely to be attacked than those that stand their ground. Differences between these two ungulates stem from the timing and form of their gait, with the longer suspension of the mule deer gallop allowing them to clear obstacles and be more maneuverable at a cost of being slower (Lingle 1992, 1993), although physical differences do not appear to be involved. It is not clear why mule deer can afford to be more aggressive than white-tails. This study represents one of the few where we have a reasonable causal understanding of interspecific differences in defenses (see also Hakkarainen et al. 1992); but, in general, we do not know why different prey sometimes use the same or different defenses to thwart attack by the same predator.

Table 13.4 Antipredator behavior seen in large mammal prey in Bandipur tiger reserve, India

Species	Preferred habitat	O	C	B	Ac	Dd	Tf	F	Dc	Cr	M	D	Sc	Pg	W	At
Asian elephant	Scrub forest, grassland	+	+	+	?	−	−	+	−	−	−	−	−	−	−	+
Wild pig	Scrub forest and short grass with water	+	+	+	G	−	−	+	+	−	−	+	+	?	−	+
Chital	Scrub, short grass ecotone	+	+	+	+	+	+	+	+	?	+	+	+	+	+	−
Sambar	Scrub forest, tall grass	+	+	+	+	+	−	+	+	−	+	+	+	?	+	+
Indian muntjac	Forests with dense undergrowth	+	+	−	+	?	+	+	+	?	+	−	−	?	−	+
Gaur	Forest grassland	+	+	+	S	−	−	+	−	−	−	−	−	−	−	+
Indian hare	Scrub, short grass	+	?	−	?	?	+	+	+	+	+	−	−	−	−	−

Source: Johnsingh 1983.
Note: Column headings are as follows: O, watching; C, curiosity attraction; B, bunching; Ac, alarm call; Dd, distraction display; Tf, tail flashing; F, flight; Dc, distress call in flight; Cr, abrupt crouching while running; M, staying motionless in cover; D, abrupt dispersion; Sc, starting call; Pg, deposition of pedal gland secretion; W, seeking refuge in water; At, attack.
G, grunt; S, snort; +, present; −, absent; ?, unknown.

That antipredator defenses to the same predator differ between closely related prey species imply that morphological, social, or environmental variables must differ between prey species. These differences might arise from differential defense capabilities at any stage in the predatory sequence, such as competence in detecting predators, in fleeing, or in self-defense; from differences in group size that would additionally influence these defenses; from habitat preferences that constrain particular defense plans; from differences in predation pressure from the predator in question; or even indirectly from other predator species that have to be avoided. This makes the number of factors to be considered virtually unmanageable. As Kruuk wrote as far back as 1972,

> If there is indeed such a connection between the direct antipredator responses on the one hand and indirect antipredator mechanisms like group size and structure on the other, and if the indirect antipredator mechanisms are important factors in other aspects of the species' ecology, as discussed before, one must then, when comparing the reactions of different prey species to one predator, also take into account the whole ecology of these various species. The indirect antipredator behavior of the *one* prey species may be very well adapted to the ecological importance and behavior of *different* predators, but it is difficult, if not impossible, to make this kind of simple comparison between antipredator behavior of *different* prey species to *one* predator. (p. 207; emphasis in the original)

A simpler research gambit, then, is to examine the diversity of defenses of a single prey species to a single species of predator.

13.6 Prey summon several defenses against the same predator

13.6.a Repeated use of the same defense   One of the difficulties in characterizing antipredator defenses is that the same defense can be called upon several times during the predatory sequence: vigilance to see if predators are in the vicinity, vigilance to monitor predators as they approach; hiding to escape detection, hiding after failed pursuit. While one might be forgiven for thinking that a given antipredator defense would affect the outcome of a predation attempt in the same way whenever it was exhibited, defenses sometimes yield different benefits depending on the stage of predation and may even result in costs at one stage of the sequence that are overshadowed by net benefit when other stages of the sequence are considered. Grouping is an antipredator defense that operates repeatedly during predation attempts on prey, but it may have mixed blessings, as shown for wildebeest and impala being hunted by wild dogs.

In the wooded environment of the northern Selous Game Reserve of Tanzania, larger groups of both wildebeest and impala are more likely to be encountered by wild dogs than smaller groups. Intermediate-sized groups of wildebeest are more likely to be attacked by wild dogs than large or small herds, but there is no effect of group size on the probability of impala being attacked. Subsequently, the probability of a kill being made is more likely when herd size is large for both wildebeest and impala, findings that run contrary to prevailing dogma on the benefits of grouping. Even when dilution risk is taken into account, the probability of a wildebeest dying increases with herd size whether it is hunted by larger or small packs, although in different ways (fig. 13.2). Across the wildebeest population, the most common herd size is 21–30 individuals, suggesting that herd size is shaped by other factors, most probably defense against other, more common predators in the area, such as spotted hyenas, or by food competition. For impala, however, the probability of dying declines with herd size once dilution is taken into account (fig. 13.3). Here, the most common herd size corresponds closely to herd size that minimizes predation by wild dogs, suggesting that impala are responding principally to predation by this species and little to pressure from lions or spotted hyenas (Creel and Creel 2002). This study illustrates the ways in which a single defense has stage-specific effects on predation risk.

13.6.b Different defenses   There are at least four circumstances in which single species can exhibit more than one antipredator defense to a single species of predator. First, prey from one population may show one type of defense in one environment, but prey from another population exhibit a different type else-

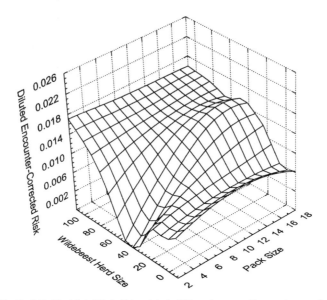

Figure 13.2 The diluted risk of death for individual wildebeest in herds of different size, corrected for nonrandom rates of encounter between wild dogs and herds of different size (Creel and Creel 2002).

where. Thus, red colobus ascend into high closed canopy and become cryptic when approached by chimpanzees in the Tai Forest, but are forced to confront chimpanzees in the low and broken canopy of Gombe (Bshary and Noe 1997b). The extent to which this is common in bird and mammal prey is improperly understood at present, because there have been too few antipredator studies of the same species in different habitats. Second, different individuals, even of the same age-sex class, may show different defenses at the same stage of the predatory sequence, as documented earlier for garter snakes, but this phenomenon appears relatively rare in homeotherms. Third, a single prey individual may demonstrate different defenses at different stages of the predatory sequence; for example, vigilance as a predator approaches, flight when pursued, and feigning death when the predator is closing in. This is by far the most common sort of observation of multiple defenses in birds and mammals. Fourth, the same individual may exhibit different defenses at the same stage of the predatory sequence. For instance, if terrestrial predators approach closely, ungulate mothers defend newborn offspring against medium or large predators, but the same female will show little aggressive defense during other stages of her reproductive cycle (see Lingle and Pellis 2002). Our current comprehension of homeotherm antipredator defenses therefore reveals a single prey animal showing a diversity of defenses to a single predator over the whole predatory sequence and sometimes within a specific stage of the sequence. Why should this be?

The most obvious explanation for multiple defenses to a single predator is

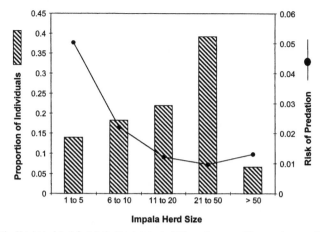

Figure 13.3 The diluted risk of death for individual impala in herds of different size, corrected for nonrandom rates of encounter between wild dogs and herds of different size. The histograms depict the proportion of individuals found in herds of different size in the population (Creel and Creel 2002).

that prey needs to mount a hierarchy of barricades; that is, increasingly potent lines of defense are used as a predator overcomes initial ones. Characteristically, less costly behaviors are employed first, with earlier defenses, such as crypsis and vigilance, being energetically inexpensive but later ones, such as flight, being energetically dear. One contributing factor to this trend must be that later defenses are employed less frequently than earlier ones, allowing the individual to incur greater costs using them. Defenses often become increasingly specialized toward the end of the predatory sequence as well, possibly because earlier defenses are used to detect a great many potentially dangerous animals in the environment, whereas later defenses involve more-intimate contact with a fewer number of threats which need to be tailored to the predator's particular characteristics more closely.

Why do some prey species invest more heavily in leaving the predatory sequence early on by showing extreme vigilance, for example, and others apparently disregard early opportunities for predator evasion and concentrate on thwarting later phases (for example, animals with spines and dermal shields) (Endler 1986, 1991a)? Constraints on defense capabilities related to other aspects of a species' biology must play a role, but it is difficult to make more than general statements at this juncture. For instance, species that are unable to run fast, perhaps because they have large claws used for breaking open termite nests, are forced to rely on armor; and species that cannot fly rapidly are forced to use vigilance. In addition, aspects of species' ecology, such as living in a burrow, must limit opportunities for early warning. Furthermore, certain types of predation pressure, such as that mounted by sit-and-wait predators, must similarly limit possibilities of advanced warning. At present, we cannot piece these different parts of the puzzle together. Testable predictions (conducted in a taxonomically restricted set of species that have similar morphologies) that pit aspects of species' ecology against the most important defense shown by each species are needed.

Viewing multiple defenses hierarchically leads one into thinking that the prey's only goal is to get away from the predator, but this is not necessarily so. It is possible that the prey animal may want not only to escape but also to influence the predator's subsequent movements or to condition it not to attack the prey individual again. Aposematism is the paradigmatic morphological defense in this instance; behaviorally, prey may drive the predator away by mobbing, but at another time may attack it, or slink off in a third. More information on both predators' behavior following interactions with prey (Lima 2002) and on predators' learning abilities are necessary before we can make headway with these ideas.

One reason that a single individual may show more than one defense at the same stage of the predatory sequence is that a given predator species may catch and subdue the same prey animal in different ways, through surprise attack, extended chase, or cornering, each of which demands different defenses. Furthermore, the predator may attempt repeated attacks, initially ambushing the prey, but if that fails, pursuing it over long distances (Treves 1999b). Predators do not hunt using just one simple method, or at least rarely.

13.7 Predator-prey coevolution

Many antipredator defenses in animals are notable for being specialized, and several examples can be found of predators having specific counteradaptations, although most of these examples are found in invertebrates and few in homeotherms. For example, rabbits disappear into burrows when in danger, and weasels have an elongate body shape that allows them to enter burrows. This raises the issue of whether these defenses and counterattack mechanisms have arisen as a result of coevolutionary change. Coevolution, defined as reciprocal evolutionary change in interacting species (J. N. Thompson 1994), demands that the two interacting partners are intimately involved with each other during some part of their life cycle. To demonstrate coevolution, differential reciprocal evolutionary change must be shown to be operating over different parts of a geographic landscape. Usually, interactions must occur between extreme specialists; that is, the number of other species with which one species interacts needs to be limited in order for evolutionary change in one party to have sufficient impact on the other party that it, too, undergoes evolutionary change. Thus, species-specific internal parasites and their hosts are likely candidates for witnessing coevolution in attacking and counterattacking traits; brood parasite-host systems also fit this mold.

In contradistinction, predator-prey interactions over evolutionary time are not a promising forum for coevolutionary change, because most predators attack more than one prey species and can therefore switch to attacking another prey item rather than evolving a counterattack if a new defense evolves. Similarly, prey are usually subject to predation from more than one predator species (Sih, Englund, and Wooster 1998), putting a brake on the evolution of a predator-specific defense mechanism. From the outset, lack of intimacy between one prey and one predator, whatever their taxonomic affiliation, makes coevolutionary change improbable (Brodie and Brodie 1999). One promising exception may be of transient and resident killer whales living off British Columbia and Alaska. There, residents that are salmon feeders use pulsed calls to communicate; but transients, which specialize in killing marine mammals, refrain from using pulsed calls, except just after making a kill; marine mammals but not fish detect killer whale calls and respond with antipredator behavior (Seecke, Ford, and Slater 2005). Perhaps another candidate for coevolution in homeotherm prey is between ground squirrels and rattlesnakes. There, resistance to snake venom is closely matched to snake predation pressure across geographically separated squirrel populations (table 9.5), but there is no evidence, as yet, of a counter-response in snakes, such as venom toxicity.

While coevolution is usually thought of as occurring between a highly specialized prey and predator, it is also recognized in taxonomic groups of prey and suites of predators. For instance, families of marine shell-breaking predators rose in number over evolutionary time, while the numbers of families of well-protected, tightly coiled and sculptured shells increased over evolutionary time as well (Vermeij 1994). This is called diffuse coevolution (Jansen 1980). Nonetheless, there are still relatively few examples of diffuse coevolution in predator-prey systems, and none in homeotherms.

Endler (1991a) listed six speculative reasons for absence of either type of coevolutionary change in predator-prey interactions. Predators attack several rather than one species of prey; apostatic selection by predators will promote shifts to additional prey species; specialized defenses occur later in the predatory sequence, usually resulting in predators experiencing generalized defenses (early on) more frequently; stronger selection on defenses than on attack mechanisms, because one involves loss of life, the other loss of food; lower densities of predators than prey, resulting in lower effective population size and opportunities for directional selection in predators; and faster evolution of prey, because of more-rapid generation times. There has been little attempt to test the veracity of these ideas (but see Brodie and Brodie 1999), but, that said, the few putative examples of coevolution do appear to have characteristics that circumvent these concerns (for example, Hieber et al. 2002).

From the outset, then, the likelihood of discovering strong cases of coevolutionary change in predator-prey interactions is limited and is even more constrained in birds and mammals, where single prey–single predator interactions are most uncommon. Perhaps the most profitable areas to look could be in simple predator-prey systems such as in the Arctic or on islands, where predator species richness is low.

13.8 Ten pressing questions

13.8.a How important is coloration in antipredator defense?

Brightness and hue of cuticle, scales, feathers, and hair have been shaped by many selective forces that can be divided into three very broad categories: concealment, communication, and physiological processes. Concealment itself may operate through background matching, obliterative shading, disruptive coloration, or countershading (Cott 1940), any of which may be involved in helping prey avoid being detected by predators. Unfortunately, evidence that they do help is surprisingly weak, although often given superficial weight by certitude. (Who would argue that speckled down in precocial fledglings is not for camouflage?) Tests of the adaptive significance of coloration still principally consist of anecdote, observations, and comparative analyses of selected taxa that seek to answer rather general questions of whether particular principles of coloration pertain to the species or group under study.

More detailed questions, such as why is this particular part of the body conspicuous or why is the end of the tail of this species black, not red, have yet to be tackled in earnest. This will demand knowledge of trade-offs of different selection pressures, asking questions such as Does gaudy plumage involved in mate choice actually increase predation risk? From an antipredator standpoint, we will need to know how different predators react to the same patch of color. For example, black and white striped skunks, apparently repellent to carnivores, are an important prey of great horned owls (Johnsgard 2002). Answers to detailed questions about significance of coloration in defense will necessitate experimental study, in the tradition of Gotmark's mounts; but countering the criticism that models exclude behavioral differences will necessitate using artificially marked live animals, which raises ethical difficulties.

13.8.b How can we explain patterns of morphological and physiological defenses across taxa?

In general, morphological and physiological defenses in homeotherms have been granted little attention by biologists other than regarding them as scientific curiosities. Given that repeated observations show that antipredator behavior is tied so intricately to morphology, one would expect that the behavior of armored species should be quite different from those of undefended species. There is little evidence for this, however: to the untrained eye, porcupines and turkeys appear just as cautious and risk averse as do capybaras and tinamous (but see Ruxton, Sherratt, and Speed 2004). This raises the evolutionary question of why armor arose in the first place, and, related to this, why quills, spines, and armor plates are so rare in mammals and absent in extant birds, First forays into these questions might be studies of the eco-correlates of morphological defense structures, or attempting to match defenses to predation pressure, as carried out across populations of California ground squirrels and Pacific rattlesnakes in relation to antivenom (Poran, Coss, and Benjamini 1987).

In addition, there is a hint that armored and aposematic mammals are of intermediate body weight (Lovegrove 2001), but the argument that small species can rely on crypsis, while large ones are protected by size seems unsatisfactory until we can verify it with comparative analyses at a small scale, comparing predation rates on armored versus nonarmored or aposematic versus inconspicuous species pairs living sympatrically.

13.8.c How do antipredator morphology and behavior interact? It has long been appreciated that behavioral and morphological defenses complement each other, the effectiveness of dull coloration being promoted by immobility being a classic example. To date, there are only a handful of studies that have examined whether morphology precedes behavior over evolutionary time or vice versa, as has been carried out for aposematism and aggregation (Sillen-Tullberg 1988). Nor do we know whether complementarity leads to codependence, in which utilization of defensive structures necessitates behavioral adjustment. While it will be easy to score presence or absence of morphological traits, good natural history observations are still required before we can map presence or absence of most defense behaviors onto phylogenetic trees.

Across individuals, behavior and morphology may be positively correlated or negatively correlated (Mikolajewski and Johansson 2004). The latter phenomenon (compensation) is extremely common in nature (consider defenseless young having to remain hidden in a nest), but the former (cospecialization), which implies some redundancy in defenses, may be just as common, and superficially describes the battery of defenses mounted by a prey species against a single predator (section 13.6.b). That individuals are willing to incur the evolutionary or developmental costs of exhibiting two or more defenses implies that they face greater predation risk than those that restrict themselves to one, but once again, we have no data.

13.8.d Do prey recognize individual predators? We know that some homeotherms utter conspecific warning calls when they see a predator, others signal extent of danger, and others signal the type of threat. There is currently no consensus as to why signals of different species carry different meanings. As primate signals are the most sophisticated, it is easy to say that cognition is key, but ecological factors may be involved. Different avenues of predator-specific escape may force sophisticated signaling about direction of threat. Extrapolating further, there may be good ecological reasons for recognizing, if not signaling about, particular individual predators. Given that prey animals often have a home range or foraging area that overlaps the range of only a few individual predators that they see time and again (for example, Cresswell and Quinn 2004), it would pay prey to recognize whether the predator is hunting or resting (possibly quite easy), is hungry or satiated (cues available from its demeanor), and even predator identity (good hunter or poor hunter having watched it previously). Knowledge of any of these factors would allow prey to tailor their foraging and vigilance more sensitively to predation risk and thus reduce opportunity costs. At present, however, we simply assume that prey cannot do this, but some evidence suggests that certain animals are capable of sophisticated discrimination (Slobodchikoff et al. 1991); this has never been followed up using natural predators.

13.8.e How common are multifunctional defenses?   A major impediment to understanding the adaptive significance of antipredator defenses and hence the study of defenses in general is that a single behavior often has several beneficial consequences, either at the same stage in a predatory sequence or at different stages (section 1.3.a). As illustrations, it is reasonable to argue that an individual may be vigilant to look out for predators, monitor the proximity of other group members, and gauge its distance to cover; warning calls may signal to conspecifics and to the predator; and mobbing may drive a predator away, alert conspecifics, and silence offspring. Accumulating empirical evidence supports the idea that behavioral defenses have several functions, either because different studies lend weight to different hypotheses or one study finds it impossible to discriminate between two or more predictions.

Behavioral ecologists are trained to be interested in the primary function of a behavior, by definition that outcome through which natural selection principally promotes reproductive success; but this carries the danger of disregarding other, almost equally important consequences. For some behavior patterns involved in antipredator defenses, we should not be concerned about uncovering multiple functions. Instead, it may be profitable to discern the circumstances in which multiple consequences are to be expected. For example, living in groups will facilitate signaling to conspecifics and mobbing involving conspecifics, while living in open country might promote antipredator vigilance and vigilance that monitors escape departures of conspecifics.

13.8.f How do predators respond to interactions with prey over time?   Many antipredator defenses appear selected to influence the behavior of predators during the prey-predator interaction, but there has been little attempt to determine whether defenses continue to exert an effect following the interaction. For example, inspection behavior and mobbing result in moving the predator on, but it is not clear how quickly different predator species return to that site. More generally, we do not understand how an unsuccessful or a successful attack affects the subsequent behavior of any given predator. Prey do make some antipredator decisions based on location, such as moving nest sites following loss of a previous nest to predators (Marzluff 1988), indicating that some predators return to places where they have been successful. We do not know, however, which species of predators move into an area, mine it for prey, and then move elsewhere and do the same again, and which predators repeatedly visit a series of sites regularly or randomly (Lima 2002). If prey could gauge this, it might enable them to reduce vigilance in the days that follow an unsuccessful attack.

It seems likely that there are certain locations within a predator's home range or territory where topography is suited to a surprise attack (trees bordering an estuary, or a steep rise allowing approach from below) that predators use time after time, but it is not clear whether prey recognize these.

13.8.g How common are multiple attacks on grouped prey?   The dilution effect assumes that one predator makes a single attack on a group of prey and that the probability of being the victim is the reciprocal of group size, but in some instances this does not describe what we see in nature. Single predators have been reported as making multiple attacks on groups: they make a kill, feed, and then, while still in view, attack the group again when it has quieted down. In cases of surplus killing, a predator may amble through an unsuspecting group, repeatedly killing without eating (Kruuk 1972). Members of a pack of predators may fan out and take several prey animals from the group (Packer, Scheel, and Pusey 1990; Creel and Creel 2002). Therefore, the extent to which a group of prey faces predation threat by predator species that embark on killing sprees, and the frequency with which predators make multiple kills, will impact the benefits of the dilution effect on prey. Current models and empirical investigation disregard multiple kills or acknowledge them as aberrant or outlying data points.

13.8.h Do predators select prey on the basis of condition?   There are conflicting hypotheses as to whether predators preferentially target prey according to its physical condition. Some argue that coursing predators target prey individuals in poor condition, whereas stalking predators do not use this criterion but attack the closest individual (section 7.6). Another argument is that birds in poor condition are more prone to attack (Kenward 1978), or, alternatively, birds with high body reserves have an impaired ability to escape (Gosler,

Greenwood, and Perrins 1995). Doubtless, there are no glib answers to these debates, because prey's condition will be more or less important under different circumstances. As examples, condition is likely to have little effect for birds attacked while still on the ground (Whitfield et al. 1999) or for raptors that find it very easy to capture a particular prey species, but in other situations in which predators chase prey, it may be important (Temple 1987). Relative predation risk from predator species hunting in different ways may affect prey's foraging decisions, since they affect condition. Moreover, differences in selectivity among predators raise the possibility that the same prey uses condition-dependent signals against some predators but not others.

Assessing prey's condition is difficult, because it requires finding and measuring carcasses and comparing them to a comparable live control population, but it is important because it makes predictions about intraspecific variation in vigilance, risk-sensitive foraging, and quality advertisement, and population consequences of predation.

13.8.i Do individual predators vary in hunting style?   Antipredator defense models and observational studies assume that every individual of a given predator species hunts in the same way and has the same prey preferences. Long-term knowledge of individual predators shows that this is incorrect. For example, some individual cheetahs routinely run into herds of Thomson's gazelles rather than stalk them, and some specialize on capturing wildebeest (Caro 1994a). Falconers readily attest to individual differences among members of their mews. If intraspecific variation in predator hunting behavior is the norm, it will affect antipredator defenses in at least three ways. First, without individual recognition of predators, prey may not be able to take appropriate evasive action against two predators with different hunting techniques, or even against one if it varies its style. Second, even in simple systems in which prey are attacked by a single predator species, specialization on certain antipredator defenses will be selected against. Third, in relation to prey preferences, predator-prey ratios will not predict predation pressure on a particular prey species or prey morph if some predators are predisposed to attacking one prey type but other predators to attacking another (see also Sherratt and MacDougall 1995). Given that marked individual differences in hunting behavior are evident, the consequences for antipredator defenses and prey population dynamics need attention.

13.8.j How do predator learning mechanisms affect antipredator defenses?
Hunting behavior develops gradually in most carnivores and raptors, being influenced by genetic variables, maternal factors, and trial-and-error learning. Naïve predators impose different and less predictable selection pressures on

prey populations than do adult predators. For instance, there has been great debate about the origins of aposematism when all predators are naïve to a conspicuous prey morph (chapter 7). Difficulties of origin go beyond the evolution of aposematism, however, because it is not clear how pursuit-deterrent signals (or even armor or venom resistance, if only partially effective) could evolve in these circumstances either. Even after these antipredator defenses have appeared in the course of evolution, prey still have to cope with a proportion of the predator population that are naïve either because they are juveniles or because they have yet to encounter the defense as adults. These individuals may attempt to hunt in the face of signals of unprofitability and exhaust prey or occasionally capture them. Thus, the mechanisms by which predators learn about prey (such as trial-and-error learning) and the speed with which they learn (number of predation attempts), as well as the proportion of juveniles in the predator population (which will be affected by reproductive rates) will impact the relative benefits of antipredator signals emitted by prey. To take an extreme, if all predators were juveniles, the signals would have little effect on the probability of attack, and costs of signal production would select against them. While these issues are well appreciated in the study of aposematism (Beatty, Beirinckx, and Sherratt 2004; Ruxton, Sherratt, and Speed 2004), they have not been considered in relation to other antipredator defenses.

13.9 Why are defenses imperfect?

While it has been noted that predators are often unsuccessful in their attempts to capture prey (review in Vermeij 1982) and that antipredator defenses therefore operate relatively well (Abrams 1986), it is less often explicitly acknowledged that animals continue to be captured by predators. The reasons that no animal is immune from predation can be answered at several levels. Over an evolutionary timescale, there may be insufficient genetic variation to enable prey to evolve perfect defenses, or there may have been insufficient time for prey to have evolved effective defenses since a new form of predation or a new predator appeared. Alternatively, pressure from multiple predators may prevent adequate response to any one if defenses conflict. Predators may even be evolving new means of subduing prey rapidly, not because of interactions between predators and their prey but because of interactions between predators and their own predators, to which prey cannot respond appropriately (Vermeij 1994). At present, these form little more than a list of possibilities.

Over an ecological timescale, the energetic or time cost of a defense may be sufficiently large that an individual cannot achieve the desired end point; trade-offs between antipredator behavioral defenses and foraging activities may mean that balancing the risk of predation against starvation results in prey behaving suboptimally. Alternatively, behavior that is beneficial in one context may carry over into another context and be maladaptive (Sih, Kats, and Maurer 2003). Fearlessness in intrasexual combat may be unwise against a formidable predator. Developmental costs may also play a role in the appearance of morphological defenses if these are expensive and reduce the chances of achieving a desired body size. The reproductive value of an individual may even be sufficiently low that it is not worth investing in elaborate defense anymore (C. W. Clark 1994), although this idea would gain greater credence if old animals were selectively taken by predators.

Over the timescale of an interaction between predator and prey, constraints may again be important, as when prey are unable to see approaching predators at night (Kruuk 1964); or there are perceptual limitations, such as an inability to assess the motivational state of a predator (Abrams 1994), or when prey have poor knowledge of refuge topography to which to bolt. Finally, defenses may fail, simply because an animal makes a mistake, slipping on loose stones or losing balance in a steep turn. At a proximate level, luck must play a large role in whether the prey lives or dies.

13.10 Summary

Many of the morphological defenses observed in nature require an animal to restrict itself to certain habitats or to rest in particular attitudes; this is especially true of crypsis and aposematism. Thus, defenses can be thought of as morphological and behavioral complexes in many instances.

Antipredator defenses are numerous, not because each prey species has a single line of defense and there are so many species of prey, but because one prey species shows different antipredator defenses to both the same and different predators. Defenses aimed at various predators differ quantitatively and qualitatively and appear adjusted to varying predator hunting styles and types of threat that they pose. In multipredator environments, generalized defenses may be effective against several predators, whereas specific defenses against one predator may interfere with those against another.

Different prey species show different defenses toward the same predator because of differences in prey morphology or environmental factors, making it difficult to tease out the underlying reasons for interspecific variation in defenses.

Prey frequently use the same defense at different stages of the predatory sequence as well as different defenses at different stages; these often become more costly and more specific as the sequence progresses. The reasons that some species invest more heavily in early defenses but others more in later defenses are poorly understood.

There is little reason to think that predator-prey interactions are a product of reciprocal evolutionary change, because they are insufficiently intimate and they do not occur between unique combinations of prey and predator species. Depending on the time frame, antipredator defenses may sometimes fail because of insufficient time for prey to have evolved defenses against a new form of predation, trade-offs in time or resource allocation between defenses and other activities, or perceptual constraints.

Ten questions of general significance to the study of antipredator defenses are raised near the end of the chapter.

Scientific names of vertebrates mentioned in the text

(based on Schmidt and Inger 1957; Clements 1991; Robins et al. 1991; D. E. Wilson and Cole 2000)

FISH
 ORDER CLUPEIFORMES
  Flat-iron herring   *Harengula thrissina*
  Pacific sardine   *Sardinops sagax*
 ORDER CYPRINIFORMES
  Mississippi silvery minnow   *Hybognathus nuchalis*
  Chub   *Leuciscus cephalus*
  European minnow   *Phoxinus phoxinus*
  Fathead minnow   *Pimephales promelas*
  Roach   *Rutilus rutilus*
  Trahira   *Hoplias malabaricus*
 ORDER SALMONIFORMES
  Northern pike   *Esox lucius*
  Rainbow trout   *Oncorhynchus mykiss*
 ORDER SILURIFORMES
  Atlantic salmon   *Salmo satar*
 ORDER GADIFORMES
  Burbot   *Lota lota*
 ORDER ANTHERINIFORMES
  Giant rivulus   *Rivulus rivulus*
  Banded killifish   *Fundulus diaphinus*
  Guppy   *Poecilia reticulata*
  Atlantic silverside   *Menidia menidia*

Figure A.0 (*Facing page*) The brown kiwi is flightless and relies on cryptic coloration and skulking behavior to avoid detection. Introduced terrestrial predators have had a devastating impact on their populations (reproduced by kind permission of Sheila Girling).

ORDER GASTEROSTEIFORMES
  Brook stickleback   *Culaea inconstans*
  Three-spined stickleback   *Gasterosteus aculeatus*
  White perch   *Morone americana*
ORDER PERCIFORMES
  Largemouth bass   *Micropterus salmoides*
  Cichlid   *Crenicichla alta*
  African jewel fish   *Hemichronus bimaculatus*
  Perch   *Perca fluviatilis*
  Damselfish   *Pomacentrus coelestis*
  Blue acara cichlid   *Aequideus pulcher*
  Striped parrotfish   *Scarus iserti*
  Spotlight parrotfish   *Sparisoma viride*

REPTILES
ORDER CROCODILIA
  Black caiman   *Melanosuchus niger*
ORDER SAURIA
  Leopard gecko   *Eublepharis macularius*
  Zebra-tailed lizard   *Callisaurus draconoides*
  Puerto Rican anole   *Anolis cristatellus*
  Puerto Rican giant anole   *Anolis cuvieri*
  Curly-tailed lizard   *Leiocephalus carinathus*
  Ground skink   *Scincella lateralis*
  Broad-headed skink   *Eumeces laticeps*
  Whiptail lizard   *Cnemidophorus tigris*
  Greater earless lizard   *Cophosaurus texanus*
ORDER OPHIDIA
  Rat snake   *Bogertophis triapsis*
  Gopher snake or bull snake   *Pituophis melanoleucus*
  Hognose snake   *Heterodon platyrhinos*
  Western terrestrial garter snake   *Thamnophis elegans*
  Indian cobra   *Naja naja*
  Green mamba   *Dendroaspis angusticeps*
  Black mamba   *Dendroaspis polylepis*
  European adder   *Vipera berus*
  Puff adder   *Bitis arietans*
  Copperhead   *Agkistrodon contortrix*
  Eastern diamondback rattlesnake   *Crotalus adamanteus*
  Western diamondback rattlesnake   *Crotalus atrox*
  Pacific rattlesnake   *Crotalus viridis*

MAMMALS

ORDER MONOTREMATA
  Short-beaked echidna   *Tachyglossus aculeatus*
  Duck-billed platypus   *Ornithorhynchus anatinus*
ORDER DIDELPHIMORPHIA
  Common opossum   *Didelphis marsupialis*
  Virginia opossum   *Didelphis virginianus*
ORDER DASYUROMORPHIA
  Thylacine   *Thylacinus cynocephalus*
  Tasmanian devil   *Sarcophilus laniarius*
ORDER DIPROTODONTIA
  Silver-gray brushtail possum   *Trichosurus vulpecula*
  Red kangaroo   *Magaleia rufa*
  Tammar wallaby   *Macropus eugenii*
  Western gray kangaroo   *Macropus fushiginosus*
  Eastern gray kangaroo   *Macropus giganteus*
  Quokka   *Setonix brachyurus*
  Red-necked pademelon   *Thylogale thetis*
  Swamp wallaby   *Wallabia bicolor*
ORDER XENARTHRA
  Pale throated three-toed sloth   *Bradypus tridactylus*
  Pink fairy armadillo   *Chlamyphorus truncatus*
  Nine-banded armadillo   *Dasypus novemcinctus*
  Giant armadillo   *Priodontes maximus*
  Southern three-banded armadillo   *Tolypeutes matacus*
  Giant or Brazilian anteater   *Myrmecophaga jubata*
ORDER INSECTIVORA
  Lesser hedgehog tenrec   *Echinpos telfairi*
  Common or tailless tenrec   *Tenrec ecaudatus*
  Streaked tenrec   *Hemicentetes semispinosus*
  Greater hedgehog tenrec   *Setifer setosus*
  European hedgehog   *Erinaceus europaeus*
  Moonrat   *Echinosorex gymnurus*
  Northern short-tailed shrew   *Blarina brevicauda*
  Common or Eurasian shrew   *Sorex araneus*
ORDER SCANDENTIA
  Pen-tailed shrew   *Ptilocercus lowi*
ORDER CHIROPTERA
  Common pipistrelle   *Pipistrellus pipistrellus*
ORDER PRIMATES
  Red-fronted or brown lemur   *Eulemur fulvus*
  Ring-tailed lemur   *Lemur catta*
  Ruffed lemur   *Variegata variegata*
  Verreaux's or white sifaka   *Propithecus verreauxi*
  Potto   *Perodicticus potto*

Common or white-tufted-ear marmoset	*Callithrix jacchus*
Saddleback tamarin	*Saguinus fuscicollis*
Red bellied tamarin or red-chested moustached tamarin	*Saguinus labiatus*
Black-chested moustached tamarin or red-cap moustached tamarin	*Saguinus mystax*
Mantled howler monkey	*Alouatta palliata*
Mexican black howler monkey	*Alouatta pigra*
Central American spider monkey	*Ateles geoffreyi*
White-fronted capuchin	*Cebus albifrons*
Brown capuchin	*Cebus appella*
White-faced capuchin	*Cebus capucinus*
Wedge-capped capuchin	*Cebus nigrivittatus*
Weeping capuchin	*Cebus olivaceus*
Squirrel monkey	*Saimiri sciureus*
Sooty mangabey	*Cercocebus atys*
Vervet monkey	*Cercopithecus aethiops*
Red-tailed or redtail monkey	*Cercopithecus ascinus*
Campbell's monkey or guenon	*Cercopithecus campbelli*
Moustached monkey	*Cercopithecus cephus*
Diana monkey	*Cercopithecus diana*
Blue monkey or samango	*Cercopithecus mitis*
de Brazza's monkey	*Cercopithecus neglectus*
White-nosed guenon or putty-nosed monkey	*Cercopithecus nictitans*
Crowned guenon	*Cercopithecus pogonias*
Lesser white-nosed monkey	*Cercopithecus petaurista*
Gray-cheeked mangabey	*Lophocebus albigena*
Long-tailed macaque	*Macaca fascicularis*
Pig-tailed macaque	*Macaca nemestrina*
Rhesus macaque	*Macaca mulatta*
Bonnet macaque	*Macaca radiata*
Barbary macaque	*Macaca sylvanus*
Talapoin monkey	*Miopithecus talapoin*
Olive baboon	*Papio anubis*
Yellow baboon	*Papio cynocephalus*
Hamadryas baboon	*Papio hamadryas*
Black and white colobus	*Colobus guereza*
Western black and white or king colobus	*Colobus polykomos*
Nilgiri langur	*Presbytis johni*
Thomas's langur	*Presbytis thomasi*
Red colobus	*Procolobus badius*
Hanuman langur	*Semnopithecus entellus*
Kloss's gibbon	*Hylobates klossii*

Gorilla	*Gorilla gorilla*
Human	*Homo sapiens*
Chimpanzee	*Pan troglodytes*
Orangutan	*Pongo pongo*

ORDER CARNIVORA

Arctic fox	*Alopex lagopus*
Golden jackal	*Canis aureus*
Black-backed jackal	*Canis mesomelas*
Coyote	*Canis latrans*
Wolf	*Canis lupus*
Domestic dog or dingo	*Canis familiaris*
African wild dog	*Lycaon pictus*
Pampas fox	*Pseudalopex gymnocercus*
Red fox	*Vulpes vulpes*
Cheetah	*Acinonyx jubatus*
Caracal	*Caracal caracal*
Ocelot	*Leopardus pardalis*
Canadian lynx	*Lynx canadensis*
Bobcat	*Lynx rufus*
Leopard cat	*Prionailurus bengalensis*
Mountain lion, cougar or puma	*Puma concolor*
Lion	*Panthera leo*
Jaguar	*Panthera onca*
Leopard	*Panthera pardus*
Tiger	*Panthera tigris*
Domestic cat	*Felis catus*
Ring-tailed mongoose	*Galidia elegans*
Slender mongoose	*Galerella sanguinea*
Dwarf mongoose	*Helogale parvula*
Meerkat	*Suricata suricatta*
Indian gray mongoose	*Herpestes edwardsii*
Egyptian mongoose	*Herpestes ichneumon*
Banded mongoose	*Mungos mungo*
Spotted hyena	*Crocuta crocuta*
Striped hyena	*Hyaena hyaena*
Aardwolf	*Proteles cristatus*
Giant otter	*Pteronura brasiliensis*
European badger	*Meles meles*
Ratel or honey badger	*Mellivora capensis*
Hog-nosed skunk	*Conepatus leuconotus/mesoleucus*
Hooded skunk	*Mephitis macroura*
Striped skunk	*Mephitis mephitis*
Spotted skunk	*Spilogale putorius*
Greater grison	*Galictis vittata*

Wolverine	*Gulo gulo*
Zorilla or striped polecat	*Ictonyx striatus*
Stoat or ermine	*Mustela erminea*
Polecat	*Mustela putorius*
American mink	*Mustela vision*
American marten	*Martes americana*
European pine marten	*Martes martes*
Fisher	*Martes pennanti*
European polecat or ferret	*Mustela putoris/Putoris furo*
Long-tailed weasel	*Mustela frenata*
European or least weasel	*Mustela nivalis*
Striped weasel	*Poecilogale albinucha*
White-nosed coati	*Nasua narica*
Raccoon	*Procyon lotor*
American badger	*Taxidea taxus*
South American sea lion	*Otaria byronia*
Gray seal	*Halichoerus grypus*
Weddell seal	*Leptonychotes wedellii*
Ringed seal	*Phoca hispida*
Harbor seal	*Phoca vitulina*
American black bear	*Ursus americanus*
Grizzly bear	*Ursus arctos*
Polar bear	*Ursus maritimus*

ORDER CETACEA

Killer whale	*Orcinus orca*

ORDER PROBOSCIDEA

Asian elephant	*Elephas maximus*
African elephant	*Loxondata africana*

ORDER PERISSODACTYLA

Burchell's zebra	*Equus burchelli*
Grevy's zebra	*Equus grevyi*
White rhinoceros	*Ceratotherium simum*
Black rhinoceros	*Diceros bicornis*
Indian rhinoceros	*Rhinoceros unicornis*

ORDER TUBILIDENTATA

Aardvark	*Orycteropus afer*

ORDER ARTIODACTYLA

Warthog	*Phacochoerus africanus/aethiopicus*
Red river hog	*Potamocherus porcus*
Wild pig or wild boar	*Sus scrofa*
Collared peccary	*Pecari angulatus*
Guanaco	*Lama guanicoe*
Giraffe	*Giraffa camelopardalis*
Chital	*Axis axis*

Red deer or elk	*Cervus elaphus*
Japanese sika deer	*Cervus nippon*
Sambar	*Cervus unicolor*
Fallow deer	*Dama dama*
Indian muntjac	*Muntiacus muntjak*
Moose	*Alces alces*
Reeve's muntjac	*Muntiacus reevesi*
White-tailed deer	*Odocoileus virginianus*
Black-tailed or Mule deer	*Odocoileus hemionus*
Caribou	*Rangifer tarandus*
Pronghorn	*Antilocapra americana*
Impala	*Aepyceros melanopus*
Coke's hartebeest	*Alcelaphus buselaphus*
Wildebeest	*Connochaetes taurinus*
Topi or Tiang	*Damiliscus korrigum*
Tsessbe	*Damiliscus lunatus*
Springbok	*Antidorcas marsupialis*
Blackbuck	*Antilope cervicapra*
Grant's gazelle	*Gazella granti*
Thomson's gazelle	*Gazella thomsoni*
Klipspringer	*Oreotragus oreotragus*
Oribi	*Ourebia ourebi*
Steinbok or steenbok	*Raphicerus campestris*
Saiga antelope	*Saiga tatarica*
American bison	*Bison bison*
Gaur	*Bos frontalis*
African buffalo	*Synerus caffer*
Eland	*Taurotragus oryx*
Bushbuck	*Tragelaphus scriptus*
Greater kudu	*Tragelaphus strepsiceros*
Domestic goat	*Capra hircus*
Spanish ibex	*Capra pyrenaica*
Mountain goat	*Oreamnos americanus*
Musk ox	*Ovibos moschatus*
Mountain, bighorn or Dall sheep	*Ovis canadensis/dalli*
Mouflon	*Ovis orientalis*
Common rhebok	*Pelea capreolus*
Waterbuck	*Kobus ellipsiprymnus/defessa*
White-eared or Uganda kob	*Kobus kob*
Beisa oryx	*Oryx beisa*
Gemsbok	*Oryx gazella*
Sable antelope	*Hippotragus niger*
Southern reedbuck	*Redunca arundinum*

ORDER PHOLIDOTA
Giant pangolin   *Manis gigantea*
ORDER RODENTIA
Formosan or Pallas's squirrel   *Callosciurus erythraeus*
Mountain beaver   *Aplodontia rufa*
Gunnison's prairie dog   *Cynomys gunnisoni*
White-tailed prairie dog   *Cynomys leucurus*
Black-tailed prairie dog   *Cynomys ludovicianus*
Alpine marmot   *Marmota marmota*
Hoary marmot   *Marmota caligata*
Golden marmot   *Marmota caudate*
Yellow-bellied marmot   *Marmota flaviventris*
Woodchuck   *Marmota monax*
Gray squirrel   *Sciurus carolinensis*
Eastern fox squirrel   *Sciurus niger*
Eurasian red squirrel   *Sciurus vulgaris*
Uinta ground squirrel   *Spermophilus armatus*
California ground squirrel   *Spermophilus beecheyi*
Belding's ground squirrel   *Spermophilus beldingi*
Columbian ground squirrel   *Spermophilus columbianus*
Golden-mantled ground squirrel   *Spermophilus lateralis*
Arctic ground squirrel   *Spermophilus parryii*
Richardson's ground squirrel   *Spermophilus richardsonii*
Round-tailed ground squirrel   *Spermophilus tereticaudus*
Townsend's ground squirrel   *Spermophilus townsendii*
Thirteen-lined ground squirrel   *Spermophilus tridecemlineatus*
Rock squirrel   *Spermophilus variegatus*
Yellow-pine chipmunk   *Tamias amoenus*
Gray-collared or gray-neck chipmunk   *Tamias cinereicollis*
Siberian chipmunk   *Tamias sibiricus*
Sonoma chipmunk   *Tamias/Eutamias sonomae*
Eastern chipmunk   *Tamias striatus*
Red squirrel   *Tamiasciurus hudsonicus*
African ground squirrel   *Xerus erythropus*
Canadian or American beaver   *Castor canadensis*
Valley pocket gopher
  or Botta's pocket gopher   *Thomomys bottae*
Desert kangaroo rat   *Dipodomys deserti*
Merriam's kangaroo rat   *Dipodomys merriami*
Chisel-toothed kangaroo rat   *Dipodomys microps*
Banner-tail kangaroo rat   *Dipodomys spectabilis*
Pale kangaroo mouse   *Microdipodops pallidus*
Arizona pocket mouse   *Perognathus amphus*
Apache pocket mouse   *Perognathus apache*

Little pocket mouse	*Perognathus longimembris*
Lesser Egyptian jerboa	*Jaculus jaculus*
European water vole	*Arvicola terrestris*
Bank vole	*Clethrionomys glareolus*
Red-backed vole	*Clethrionomys rutilus*
Arctic lemming	*Dicrostonyx torquatus*
Brown lemming	*Lemmus sibiricus*
Field or short-tailed vole	*Microtus agrestis*
Common vole	*Microtus arvalis*
Prairie vole	*Microtus ochrogaster*
Meadow vole	*Microtus pennsylvanicus*
Social vole	*Microtus socialis*
Black-bellied hamster	*Cricetus cricetus*
Golden hamster	*Mesocricetus auratus*
Dune hairy-footed gerbil	*Gerbillurus tytonis*
Allenby's gerbil	*Gerbillus allenbyi*
Lesser Egyptian gerbil	*Gerbillus gerbillus*
Greater Egyptian gerbil	*Gerbillus pyramidum*
Fat sand rat	*Psammomys obesus*
Great gerbil	*Rhombomys opimus*
Cairo spiny mouse	*Acomys cahirinus*
Maned or crested rat	*Lophiomys imhausi*
Norway or brown rat	*Rattus norvegicus*
Roof rat	*Rattus rattus*
Kimberly rock rat	*Zyzomys woodwardi*
White-throated wood rat	*Neotoma albigula*
Gray wood rat	*Neotoma floridana*
Golden or common spiny mouse	*Acomys cahirinus*
Darwin's leaf-eared mouse	*Phyllotis darwini*
House mouse	*Mus musculus*
Northern grasshopper mouse	*Onychomys leucogaster*
Deer mouse	*Peromyscus maniculatus*
Oldfield mouse	*Peromyscus polionotus*
Hispid cotton rat	*Sigmodon hispidus*
Spring hare	*Pedetes capensis*
Naked mole rat	*Heterocephalus glaber*
Asian garden dormouse	*Eliomys melanurus*
African brush-tailed porcupine	*Atherus africanus*
Cape porcupine	*Hystrix africaeaustralis*
Indian crested porcupine	*Hystrix indica*
Brazilian tree porcupine	*Coendou prehensilis*
North American porcupine	*Erethizon dorsatum*
Plains viscacha	*Lagostomus maximus*
Brazilian guinea pig	*Cavia aperea*

Patagonian mara	*Dolichotis patagonium*
Pacarana	*Dinomys branickii*
Capybara	*Hydrochaeris hydrochaeris*
Degu	*Octodon degus*
ORDER LAGOMORPHA	
Collared pika	*Ochotona collaris*
American pika	*Ochotona princeps*
Snowshoe or varying hare	*Lepus americanus*
Arctic hare	*Lepus arcticus*
Black-tailed jackrabbit	*Lepus californicus*
Brown hare	*Lepus europeanus*
Black-naped or Indian hare	*Lepus nigricollis*
Mountain hare	*Lepus timidus*
European rabbit	*Oryctolagus cuniculus*

BIRDS

ORDER RHEIFORMES	
Greater rhea	*Rhea americana*
ORDER STRUTHIONIFORMES	
Ostrich	*Struthio camelus*
ORDER CASUARIIFORMES	
Emu	*Dromalus novaehollandiae*
ORDER DINORNITHIFORMES	
Brown kiwi	*Apteryx australis*
ORDER PODICIPEDIFORMES	
Rolland's or white-tufted grebe	*Rollandia rolland*
Western grebe	*Aechmophorus occidentalis*
Silvery grebe	*Podiceps occipitalis*
ORDER SPHENISCIFORMES	
Adelie penguin	*Pygoscelis adeliae*
Chinstrap penguin	*Pygoscelis antarctica*
Antarctic giant petrel	*Macronectes giganteus*
ORDER PROCELLARIIFORMES	
Herald's or Trindade petrel	*Pterodroma arminjoniana*
Kermadec petrel	*Pterodroma neglecta*
Blue petrel	*Halobaena caerula*
Thin or slender-billed prion	*Pachyptila belcheri*
Antarctic prion	*Pachyptila desolata*
Leach's storm petrel	*Oceanodroma leucorhoa*
Wedge-rumped storm petrel	*Oceanodroma tethys*
Common diving-petrel	*Pelecanoides urinatrix*
ORDER PELECANIFORMES	
Double-crested cormorant	*Phalacrocorax auritus*
Pelagic cormorant	*Phalacrocorax pelagicus*
White pelican	*Pelecanus erythrorhynchos*

ORDER ANSERIFORMES

Brant or brent goose	*Branta bernicla*
Canada goose	*Branta canadensis*
Barnacle goose	*Branta leucopsis*
Red-breasted goose	*Branta ruficollis*
Common shelduck	*Tadorna tadorna*
Spur-winged goose	*Plectropterus gambensis*
White-fronted goose	*Anser albifrons*
Pink-footed goose	*Anser brachyrhynchus*
Egyptian goose	*Alopochen aegyptiacus*
Wigeon	*Anas Americana/penelope*
Northern shoveler	*Anas clypeata*
Snowgoose	*Anser caerulescens*
Common or green-winged teal	*Anas crecca*
Blue-winged teal	*Anas discors*
Mallard	*Anas platyrhynchos*
Gadwall	*Anas strepera*
Lesser scaup	*Aythya affinis*
Tufted duck	*Anthya fuligula*
Greater scaup	*Anthya marila*
Common eider or eider duck	*Somateria mollissima*
Oldsquaw or long-tailed duck	*Clangula hyemalis*
Velvet or white-winged scoter	*Melanitta fusca*
Goosander or Common merganser	*Mergus merganser*
Red-breasted merganser	*Mergus serrator*

ORDER PHOENICOPTERIFORMES

Andean flamingo	*Phoenicoparrus audinus*

ORDER CICONIIFORMES

Little blue heron	*Egretta caerulea*
Gray heron	*Ardea cinerea*
Great blue heron	*Ardea herodias*
Buff-backed heron	*Ardea ibis*
Black-crowned night-heron	*Nycticorax nycticorax*
Least bittern	*Ixobrychus exilis*
Little bittern	*Ixobrychus minutus*
Marabou stork	*Leptoptilos crumeniferus*

ORDER FALCONIFORMES

Turkey vulture	*Cathartes aura*
Bald eagle	*Haliaeetus leucocephalus*
African fish eagle	*Haliaeetus vocifer*
Black-chested snake eagle	*Circaetus pectoralis*
Bateleur	*Terathopius ecaudatus*
Western marsh harrier	*Circus aeruginosus*
Hen harrier, northern harrier, or marsh hawk	*Circus cyaneus*

Madagascar harrier hawk	*Gymnogenys radiatus*
Cooper's hawk	*Accipiter cooperii*
Northern goshawk	*Accipiter gentilis*
Japanese lesser sparrow hawk	*Accipiter gularis*
Sparrow hawk	*Accipiter nisus*
Sharp-shinned hawk	*Accipiter striatus*
African goshawk	*Accipiter tachiro*
Harris's hawk	*Parabuteo unicinctus*
Great black-hawk	*Buteogallus urubitinga*
Red-shouldered hawk	*Buteo lineatus*
Red-tailed hawk	*Buteo jamaicensis*
Rough-legged buzzard	*Buteo lagopus*
Harpy eagle	*Harpia harpyja*
Spanish imperial eagle	*Aquila adalberti*
Tawny eagle	*Aquila rapax*
African hawk eagle	*Hieraaetus pennatus*
Ornate hawk eagle	*Spizaetus ornatus*
Martial eagle	*Polemaetus bellicosus*
Crowned hawk-eagle	*Stephanoaetus coronatus*
Secretary bird	*Sagittarius serpentarius*
Merlin	*Falco columbarius*
Eleanora's falcon	*Falco eleonorae*
Lesser falcon	*Falco naumanni*
Peregrine falcon	*Falco peregrinus*
American kestrel	*Falco sparverius*
Eurasian hobby	*Falco subbuteo*
Eurasian kestrel	*Falco tinnunculus*
Crowned hawk eagle	*Stephanoaetus coronatus*
Black caracara	*Daptrius ater*

ORDER GRUIFORMES

Eastern swamphen or purple gallinule	*Porphyrio porphyrio*
Common moorhen	*Gallinula chloropus*
American coot	*Fulica americana*
European coot	*Fulica atra*

ORDER GALLIFORMES

Crested guan	*Penelope purpurascens*
Red or willow grouse or ptarmigan	*Lagopus lagopus*
White-tailed ptarmigan	*Lagopus leucurus*
Rock ptarmigan	*Lagopus mutus*
Sage grouse	*Centrocercus urophasianus*
Red-legged partridge	*Alectoris rufa*
Gray partridge	*Perdix perdix*
Eurasian quail	*Cotornix cotornix*
Japanese quail	*Cotornix japonica*

Domestic chicken or red jungle fowl	*Gallus gallus*
Domestic or wild turkey	*Meleagris gallopavo*
Northern bobwhite or Bobwhite quail	*Colinus virginianus*

ORDER CHARADRIIFORMES

Eurasian woodcock	*Scolopax rusticola*
Common snipe	*Gallinago gallinago*
Surfbird	*Aphriza virgata*
Bar-tailed godwit	*Limosa lapponica*
Black-tailed godwit	*Limosa limosa*
Whimbrel	*Numenius phaeopus*
Eurasian curlew	*Numenius arquata*
Spotted sandpiper	*Tringa macularia*
Common redshank	*Tringa totanus*
Ruddy turnstone	*Arenaria interpres*
Sanderling	*Calidris alba*
Dunlin	*Calidris alpina*
Baird's sandpiper	*Calidris bairdii*
Red knot	*Calidris canutus*
Purple sandpiper	*Calidris maritina*
Western sandpiper	*Calidris mauri*
Least sandpiper	*Calidris minutilla*
Spotted thick-knee or stone curlew	*Burhinus capensis*
Eurasian oystercatcher	*Haemotopus ostralegus*
Ruff	*Philomachus pugnax*
European golden plover	*Pluvialis apricaria*
Gray plover	*Pluvialis squatarola*
Kentish plover	*Charadrius alexandrius*
Piping plover	*Charadrius melodus*
Three-banded plover	*Charadrius tricollaris*
Killdeer	*Charadrius vociferous*
Thick-billed plover	*Charadrius wilsonia*
Eurasian dotterel	*Eudromias morinellus*
Blacksmith plover	*Vanellus armatus*
Crowned lapwing	*Vanellus coronatus*
Long-toed lapwing	*Vanellus crassirostris*
Lapwing	*Vanellus vanellus*
Laughing gull	*Larus atricilla*
Herring gull	*Larus argentatus*
California gull	*Larus californicus*
Common or mew gull	*Larus canus*
Ring-billed gull	*Larus delawarensis*
Kelp gull	*Larus dominicianus*
Lesser black-backed gull	*Larus fuscus*
Glaucous-winged gull	*Larus glaucescens*

Heermann's gull	*Larus heermanni*
Brown-hooded gull	*Larus maculipennis*
Great black-backed gull	*Larus marinus*
Franklin's gull	*Larus pipixcan*
Common black-headed gull	*Larus ridibundus*
Black-legged kittiwake	*Rissa tridactyla*
Large-billed tern	*Phaetusa simplex*
Least tern	*Sterna antillarum*
Little tern	*Sterna albifrons*
Great crested tern	*Sterna bergii*
Cayenne tern	*Sterna eurygnatha*
Forster's tern	*Sterna forseteri*
Common tern	*Sterna hirundo*
Royal tern	*Sterna maxima*
Gull-billed tern	*Sterna/Gelochelidon nilotica*
Arctic tern	*Sterna paradisaea*
Sandwich tern	*Sterna sandvicensis*
Yellow-billed tern	*Sterna superciliaris*
Brown skua	*Catharacta lonnbergi/antarctica*
South polar skua	*Catharacta maccormicki*
Long-tailed skua or jaeger	*Stercocarius longicaudus*
Arctic skua or parasitic jaeger	*Stercocarius parasiticus*
Pomarine jaeger	*Stercocarius pomarinus*
Great skua	*Stercocarius/Catharacta skua*
Black skimmer	*Rynchos niger*
Guillemot or thick-billed murre	*Uria lomvia*
Rhinoceros auklet	*Cerorhinca monocerata*
ORDER PTEROCLIDIFORMES	
Yellow-throated sandgrouse	*Pterocles gutturalis*
ORDER COLUMBIFORMES	
Wood pigeon	*Columba palumbus*
Spotted dove	*Streptopelia chinensis*
Barbary dove	*Streptopelia risoria*
Laughing dove	*Streptopelia senegalensis*
Zebra or barred ground dove	*Geopelia striata*
Mourning dove	*Zenaida macroura*
ORDER COLIIFORMES	
Speckled mousebird	*Colias striatus*
ORDER STRIGIFORMES	
Eastern screech owl or long-eared owl	*Otus asio*
Snowy owl	*Nyctea scandiaca*
Eurasian eagle owl	*Bubo bubo*
Verreaux's eagle owl	*Buteo lacteus*
Great horned owl	*Bubo virginianus*

Tawny owl	*Strix aluco*
Ferruginous pigmy owl	*Glaucidium brasilianum/nanum*
Pigmy owl	*Glaucidium passerinum*
Pearl-spotted owlet	*Glaucidium perlatum*
Little owl	*Athene noctua*
Burrowing owl	*Athene cunicularia*
Tengmalm's owl	*Aegolius funereus*
Short-eared owl	*Asio flammeus*
Powerful owl	*Ninox strenua*

ORDER CAPRIMULGIFORMES

Sand-colored nighthawk	*Chordeiles nupestris*
Whip-poor-will	*Caprimulgus vociferus*

ORDER APODIFORMES

Common swift	*Apus apus*

ORDER TROGONIFORMES

Long-tailed hermit	*Phaethornis superciliosus*

ORDER CORACIIFORMES

Common kingfisher	*Alcedo atthis*
Turquoise-browed motmot	*Eumomota superciliosa*
Lilac-breasted roller	*Coracias caudate*
Eurasian hoopoe	*Upupa epops*
Black and white casqued hornbill	*Ceratogymna subcylindricus*
Abyssinian ground hornbill	*Bucorvus abyssinicus/leadbeateri*

ORDER PICIFORMES

Cuvier's toucan	*Ramphastos cuvieri*
Great spotted woodpecker	*Dendrocopus major*
Red-cockaded woodpecker	*Picoides borealis*
Downy woodpecker	*Picoides pubescens*
Hairy woodpecker	*Picoides villosus*

ORDER PASSERIFORMES

Red-faced spinetail	*Cranioleuca erythrops*
Plain-fronted or common thornbird	*Phacellodomus rufifrons*
Bluish-slate antshrike	*Thamnomanes schistogynus*
Guianan cock-of-the-rock	*Rupicola rupicola*
Scale-crested pygmy-tryant	*Lophotriccus pileatus*
Eye-ringed flatbill	*Rhynchocyclus brevirostris*
Black-tailed flycatcher	*Myiobius atricaudus*
Dusky flycatcher	*Empidonax oberholseri*
Cattle tyrant	*Machetornis rixosus*
Eastern kingbird	*Tyrannus tyrannus*
Great kiskadee	*Pitangus sulphuratus*
Noisy friarbird	*Philemon corniculatus*
Noisy miner	*Manorina melanocephala*
New Zealand bellbird	*Anthornis melanura*

New Zealand robin	*Petroica australis*
Hooded pitohui	*Pitohui kirhocephalus*
Variable pitohui	*Pitohui dichrous*
White-winged chough	*Corcorax melanorhamphos*
Blue-capped ifrita	*Ifrita kowaldi*
Rufous-backed fantail	*Rhipidura rufidorsa*
Fork-tailed drongo	*Dicrurus adsimilis*
Scrub jay	*Aphelocoma californica*
Florida scrub jay	*Aphelocoma coerulescens*
Mexican scrub jay	*Aphelocoma ultramarina*
Blue jay	*Cyanocitta cristata*
Steller's jay	*Cyanocitta stelleri*
Eurasian jay	*Garrulus glandarius*
Siberian jay	*Perisoreus infaustus*
Azure-winged magpie	*Cyanopica cyana*
Yellow-billed magpie	*Pica nuttali*
Black-billed magpie	*Pica pica*
Pinyon jay	*Gymnorhinus cyanocephalus*
Red-billed or red-winged chough	*Pyrrhocorax pyrrhocorax*
American crow	*Corvus brachyrhynchos*
Northwestern crow	*Corvus caurinus*
Raven	*Corvus corax*
Carrion or hooded crow	*Corvus corone*
Rook	*Corvus frugilegus*
Large-billed crow	*Corvus macrorhynchos*
Eurasian jackdaw	*Corvus monedula*
Great gray or northern shrike	*Lanius excubitor*
Loggerhead shrike	*Lanius ludovicianus*
Phainopepla	*Phainopepla nitens*
Western bluebird	*Sialia mexicana*
Finsch's flycatcher thrush or Finsch's rufous flycatcher	*Neocossyphus/Stizorhina finschii*
Rufous flycatcher-thrush	*Neocossyphus/Stizorhina fraseri*
White-tailed ant-thrush	*Neocossyphus poensis*
Red-tailed ant-thrush	*Neocossyphus rufus*
Varied thrush	*Zoothera naevia*
Wood thrush	*Catharus mustelina*
Hermit thrush	*Catharus guttatus*
Redwing	*Turdus iliacus*
European blackbird	*Turdus merula*
American robin	*Turdus migratorius*
Song thrush	*Turdus philomelos*
Fieldfare	*Turdus pilaris*
Gray catbird or warbler	*Dumetella carolinensis*

Northern mockingbird	*Mimus polyglottus*
Superb starling	*Lamprotornis/Spreo superbus*
Brown thrasher	*Toxostoma rufum*
Common or European starling	*Sturnus vulgaris*
Southern black flycatcher	*Melaenornis pammelaina*
Collared flycatcher	*Fidecula albicollis*
Pied flycatcher	*Fidecula hypoleuca*
European robin	*Erithacus rubecula*
Common redstart	*Phoenicurus phoenicurus*
Stonechat	*Saxicola torquata*
Blackstart	*Cercomela melanura*
Red-breasted nuthatch	*Sitta canadensis*
White-breasted nuthatch	*Sitta carolinensis*
Stripe-backed wren	*Campylorhynchus nuchalis*
Rufous-naped wren	*Campylorhynchus rufinucha*
House wren	*Troglodytes aedon*
Wren	*Troglodytes troglodytes*
Tree swallow	*Tachycineta bicolor*
Carolina wren	*Thyrothorus ludovicianus*
Bush tit	*Psaltriparus minimus*
Bank swallow or sand martin	*Riparia riparia*
Cliff swallow	*Hirundo pyrrhonota*
Barn swallow	*Hirundo rustica*
Ruby-crowned kinglet	*Regulus calendula*
Silver-eye	*Zosterops lateralis*
Eurasian bullfinch	*Pyrrhula pyrrhula*
Moustached warbler	*Acrocephalus melanopogon*
Aquatic warbler	*Acrocephalus paludicola*
Sedge warbler	*Acrocephalus schoenobaenus*
Seychelles warbler	*Acrocephalus sechellensis*
Willow warbler	*Phylloscopus trochilus*
Blackcap	*Silvia atricapilla*
Arabian babbler	*Turdoides squamiceps*
Jungle babbler	*Turdoides striatus*
Coal tit	*Parus ater*
Black-capped chickadee	*Parus atricapillus*
Tufted titmouse	*Parus bicolor*
Carolina chickadee	*Parus carolinensis*
Blue tit	*Parus caeruleus*
Crested tit	*Parus cristatus*
Great tit	*Parus major*
Willow tit	*Parus montanus*
Marsh tit	*Parus palustris*
Mexican chickadee	*Parus sclateri*

Cape or southern penduline tit	*Anthoscopus minutus*
Desert lark	*Ammomanes deserti*
Crested lark	*Galerida cristata*
Skylark or Eurasian lark	*Alauda arvensis*
House sparrow	*Passer domesticus*
White wagtail	*Motacilla alba*
Yellow wagtail	*Motacilla flava*
White-browed sparrow-weaver	*Plocepasser mahali*
Red-billed quelea	*Quelea quelea*
Seychelles fody	*Foudia sechellarum*
Common waxbill	*Estrilda astrild*
Zebra finch	*Taeniopyga guttata*
Nutmeg manikin or scaly-breasted munia	*Lonchura punctulata*
Meadow pipit	*Anthus pratensis*
Alpine accentor	*Prunella collaris*
Reed bunting	*Emberiza schoeniclus*
Yellowhammer	*Emberiza citrinella*
Swamp sparrow	*Melospiza georgiana*
Song sparrow	*Melospiza melodia*
White-throated sparrow	*Zonotrichia albicollis*
White-crowned sparrow	*Zonotrichia leucophrys*
Dark-eyed or northern junco	*Junco hyemalis*
Yellow-eyed junco	*Junco phaeonotus*
White-winged shrike-tanager	*Lanio versicolor*
Savannah sparrow	*Passerculus sandwichensis*
Chaffinch	*Fringilla coelebs*
Brambling	*Fringilla montifringilla*
European goldfinch	*Cardeuelis carduelis*
Greenfinch	*Carduelis chloris*
Pine siskin	*Carduelis pinus*
American goldfinch	*Carduelis tristis*
House finch	*Carpodacus mexicanus*
Purple finch	*Carpodacus purpureus*
Evening grosbeak	*Coccothraustes vespertinus*
Orange-crowned warbler	*Vermivora celata*
Virginia warbler	*Vermivora virginiae*
Yellow warbler	*Dendroica petechia*
Prairie warbler	*Dendroica discolor*
Red warbler	*Ergaticus ruber*
Black and white warbler	*Mniotilta varia*
American tree sparrow	*Spizella arborea*
Chipping sparrow	*Spizella passerina*
MacGillivray's warbler	*Oporornis tolmiei*

Brown towhee	*Pipilo fuscus*
Red-winged blackbird	*Agelaius phoeniceus*
Saffron finch	*Sicalis flaveola*
Dickcissel	*Spiza americana*
Lazuli bunting	*Passerina amoena*
Indigo bunting	*Passerina cyanea*
Chestnut-headed oropendola	*Psarocolius wagleri*
Yellow-rumped cacique	*Cacicus cela*
Red-rumped cacique	*Cacicus haemorrhus*
Bullock's or Northern oriole	*Icterus galbula*
Troupial	*Icterus icterus*
Northern cardinal	*Cardinalis cardinalis*
Brewer's blackbird	*Euphagus cyanocephalus*
Boat-tailed grackle	*Quiscalus major*
Common grackle	*Quiscalus quiscula*

REFERENCES

Abensperg-Traun, M. 1991. Survival strategies of the echidna *Tachyglossus aculeatus* Shaw 1792 (Monotremata: Tachyglossidae). *Biol Conserv* 58:317–28.

Abrahams, M. V. 1995. The interaction between antipredator behaviour and antipredator morphology: Experiments with fathead minnows and brook sticklebacks. *Can J Zool* 73:2209–15.

Abrahams, M. V., and L. M. Dill. 1989. A determination of the energetic equivalence of the risk of predation. *Ecology* 70:999–1007.

Abrams, P. A. 1986. Adaptive responses of predators to prey and prey to predators: The failure of the arms-race analogy. *Evolution* 40:1229–47.

———. 1994. Should prey overestimate the risk of predation? *Am Nat* 144:317–28.

Abramsky, Z., M. L. Rosenzweig, and A. Subach. 2002. The costs of apprehensive foraging. *Ecology* 83:1330–40.

Ackerman, J., A. L. Blackmer, and J. M. Eadie. 2004. Is predation on waterfowl nests density dependent?—Tests at three spatial scales. *Oikos* 107:128–40.

Ackerman, J., and J. M. Eadie. 2003. Current versus future reproduction: An experimental test of parental investment decisions using nest desertion by mallards (*Anas platyrhynchos*). *Behav Ecol Sociobiol* 54:264–73.

Ackers, S. H., and C. N. Slobodchikoff. 1999. Communication of stimulus size and shape in alarm calls of Gunnison's prairie dogs, *Cynomys gunnisoni*. *Ethology* 105:149–62.

Adriaensen, A., A. A. Dhondt, S. Van Dongen, L. Lens, and E. Matthysen. 1998. Stabilizing selection on blue tit fledgling mass in the presence of sparrowhawks. *Proc R Soc Lond Ser B Biol Sci* 265:1011–16.

Alados, C. L. 1985. An analysis of vigilance in the Spanish ibex (*Capra pyrenaica*). *Z Tierpsychol* 68:58–64.

Alados, C. L., and J. Escos. 1988. Alarm calls and flight behaviour in Spanish ibex (*Capra pyrenaica*). *Biol Behav* 13:11–21.

Alatalo, R. V., and P. Helle. 1990. Alarm calling by individual willow tits, *Parus montanus*. *Anim Behav* 40:437–42.

Alatalo, R. V., and J. Mappes. 1996. Tracking the evolution of warning signals. *Nature* 382:708–10.

Figure R.0 (*Facing page*) A piping plover incubating its eggs. Despite cryptic coloration of both parent and eggs, the parents have still chosen a nest site near, albeit minimal, cover (reproduced by kind permission of Sheila Girling).

Alberico, J. A. R., J. M. Reed, and L. W. Oring. 1991. Nesting near a common tern colony increases and decreases spotted sandpiper nest predation. *Auk* 108:904–10.

Alberts, S. C. 1994. Vigilance in young baboons: Effects of habitat, age, sex and maternal rank on glance rate. *Anim Behav* 47:749–55.

Alberts, S. C., and J. Altmann. 1995. Balancing costs and opportunities: Dispersal in male baboons. *Am Nat* 145:279–306.

Albrecht, T., and P. Klvana. 2004. Nest crypsis, reproductive value of a clutch and escape decisions in incubating female mallards *Anas platyrhynchos*. *Ethology* 110:603–13.

Aldridge, H. D. J. N., and R. M. Brigham. 1988. Load carrying and maneuverability in an insectivorous bat: A test of the 5% "rule" of radio-telemetry. *J Mammal* 69:379–82.

Alexander, R. D. 1974. The evolution of social behaviour. *Annu Rev Ecol Syst* 5:325–383.

Alkon, P. U., and D. Saltz. 1988. Influence of season and moonlight on temporal activity patterns of Indian crested porcupines (*Hystrix indica* Kerr). *J Mammal* 69:71–81.

Allan, J. R., and T. J. Pitcher. 1986. Species segregation during predator evasion in cyprinid fish shoals. *Freshw Biol* 16:653–59.

Allen, J. A. 1874. On geographical variation in color among north American squirrels; with a list of the species and varieties of the American squirridae occurring north of Mexico. *Proc Bost Soc Nat Hist* 16:276–94.

Allen, J. A. 1988. Frequency-dependent selection by predators. *Philos Trans R Soc Lond B Biol Sci* 319:485–503.

Altmann, M. 1958. The flight distance in free-ranging big game. *J Wildl Manag* 22:207–09.

Altmann, S. A. 1956. Avian mobbing behavior and predator recognition. *Condor* 58:241–53.

Altmann, S. A., and J. Altmann. 1970. *Baboon ecology*. Chicago: University of Chicago Press.

Alvarez, F. 1993. Alertness signaling in two rail species. *Anim Behav* 46:1229–31.

Amat, J. A., R. M. Fraga, and G. M. Arroyo. 1999. Replacement clutches by Kentish plovers. *Condor* 101:746–51.

Amat, J. A., and J. A. Masero. 2004. Predation risk on incubating adults constrains the choice of thermally favourable nest sites in a plover. *Anim Behav* 67:293–300.

Ambrose, H. W. III. 1972. Effect of habitat familiarity and toe-clipping on rate of owl predation in *Microtus pennsylvanicus*. *J Mammal* 53:909–12.

Andelt, W. F., and T. D. I. Beck. 1998. Effect of black-footed ferret odors on behavior and reproduction of prairie dogs. *Southwestern Naturalist* 43:344–51.

Andersen, D. E. 1990. Nest-defense behavior of red-tailed hawks. *Condor* 92:991–97.

Anderson, C. M. 1986. Predation and primate evolution. *Primates* 27:15–39.

Anderson, J. R. 1984. Ethology and ecology of sleep in monkeys and apes. *Adv Study Behav* 14:165–229.

Andersson, M. 1976. Predation and kleptoparasitism by skuas in a Shetland seabird colony. *Ibis* 118:208–17.

———. 1981. On optimal predator search. *Theor Popul Biol* 19:58–86.

———. 1983. On the functions of conspicuous seasonal plumages in birds. *Anim Behav* 31:1262–64.

———. 1986. Evolution of condition-dependent sex ornaments and mating preferences: Sexual selection based on viability differences. *Evolution* 40:804–16.

———. 1994. *Sexual selection*. Princeton, NJ: Princeton University Press.

Andersson, M., and R. A. Norberg. 1981. Evolution of reversed sexual size dimorphism and sex role partitioning among birds of prey with a size scaling of flight performance. *Biol J Linn Soc* 15:105–30.

Andersson, M., and C. G. Wiklund. 1978. Clumping versus spacing out: Experiments on nest predation in fieldfares (*Turdus pilaris* L.). *Anim Behav* 26:1207–12.

Andersson, M., C. G. Wiklund, and H. Rundgren. 1980. Parental defence of offspring: A model and an example. *Anim Behav* 28:536–42.

Andersson, S., and C. G. Wiklund. 1987. Sex role partitioning during offspring protection in the rough-legged buzzard *Buteo lagopus*. *Ibis* 129:103–07.

Andraso, G. M., and J. N. Barron. 1995. Evidence for a trade-off between defensive morphology and startle-response performance in the brook stickleback (*Culaea inconstans*). *Can J Zool* 73:1147–53.

Andren, H. 1991. Predation: An overrated factor for over-dispersion of birds' nests? *Anim Behav* 1063–69.

———. 1995. Effects of landscape composition on predation rates at habitat edges. In *Mosaic landscapes and ecological processes*, ed. L. Hansson, L. Fahrig, and G. Merriam, 225–55. London: Chapman and Hall.

Andren, H., and P. Angelstam. 1988. Elevated predation rates as an edge effect in habitat islands: Experimental evidence. *Ecology* 69:544–47.

Arduino, P. J. Jr., and J. L. Gould. 1984. Is tonic immobility adaptive? *Anim Behav* 32:921–23.

Arenz, C. L., and D. W. Leger. 1997a. Artificial visual obstruction, antipredator vigilance, and predator detection in the thirteen-lined ground squirrel (*Spermophilus tridecemlineatus*). *Behaviour* 134:1101–14.

———. 1997b. The antipredator vigilance of adult and juvenile thirteen-lined ground squirrels (Sciuridae: *Spermophilus tridecemlineatus*): Visual obstruction and simulated hawk attacks. *Ethology* 103:945–53.

———. 1999a. Thirteen-lined ground squirrel (Sciuridae: *Spermophilus tridecemlineatus*) antipredator vigilance: Monitoring the sky for aerial predators. *Ethology* 105:807–816.

———. 1999b. Thirteen-lined ground squirrel (Sciuridae: *Spermophilus tridecemlineatus*) antipredator vigilance decreases as vigilance cost increases. *Anim Behav* 57:97–103.

———. 2000. Antipredator vigilance of juvenile and adult thirteen-lined ground squirrels and the role of nutritional need. *Anim Behav* 59:535–41.

Armitage, K. B., and C. M. Chiesura. 1994. Time and wariness in yellow-bellied marmots. *J Mt Ecol* 2:1–8.

Armstrong, E. A. 1947. *Bird display and behaviour: An introduction to the study of bird psychology*. New York: Oxford University Press.

———. 1949. Diversionary display.—Part 1. Connotation and terminology. *Ibis* 91:88–97.

———. 1954. The ecology of distraction display. *Anim Behav* 2:121–35.

———. 1956. Distraction display and the human predator. *Ibis* 98:641–54.

———. 1965. *Bird display and behavior*. New York: Dover.

Arnold, E. N. 1988. Caudal autotomy as a defense. In *Biology of the reptilia*, ed. C. Gans and R. B. Huey, 16:235–73. New York: Alan R. Liss.

Arnold, K. E. 2000. Group mobbing behaviour and nest defence in a cooperatively breeding Australian bird. *Ethology* 106:385–93.

Arnold, S. J., and Wassersug, R. J. 1978. Differential predation on metamorphic anurans by garter snakes (*Thamnophis*): Social behavior as a possible defense. *Ecology* 59:1014–22.

Artiss, T., W. M. Hochachka, and K. Martin. 1999. Female foraging and male vigilance in white-tailed ptarmigan (*Lagopus leucurus*): Opportunism or behavioural coordination? *Behav Ecol Sociobiol* 46:429–34.

Artiss, T., and K. Martin. 1995. Male vigilance in white-tailed ptarmigan, *Lagopus leucurus*: Mate guarding or predator detection? *Anim Behav* 49:1249–58.

Ashmole, N. P. 1963. The biology of the Wideawake or Sooty Tern *Sterna fuscata* on Ascension Island. *Ibis* 103b:297–364.

Atkeson, T. D., R. L. Marchinton, and K. V. Miller. 1988. Vocalizations of white-tailed deer. *Am Midl Nat* 120:194–200.

Aubin, T. 1987. Respective parts of the carrier and of the frequency modulation in the semantic of distress calls. An experimental study of *Sturnus vulgaris* by means of digital synthesis methods. *Behaviour* 100:123–33.

Aubin, T., and J. C. Bremond. 1989. Parameters used for recognition of distress calls in two species: *Larus argentatus* and *Sturnus vulgaris*. *Bioacoustics* 2:22–33.

Baack, J. K., and P. V. Switzer. 2000. Alarm calls affect foraging behavior in eastern chipmunks (*Tamias striatus*, Rodentia: Sciuridae). *Ethology* 106:1057–66.

Bachman, G. C. 1993. The effect of body condition on the trade-off between vigilance and foraging in Belding's ground squirrels. *Anim Behav* 46:233–44.

Badyaev, A. V., W. J. Etges, J. D. Faust, and T. E. Martin. 1998. Fitness correlates of spur length and spur asymmetry in male wild turkeys. *J Anim Ecol* 67:845–52.

Badyaev, A. V., T. E. Martin, and W. J. Etges. 1996. Habitat sampling and habitat selection by female wild turkeys: Ecological correlates and reproductive consequences. *Auk* 113:636–46.

Baenninger, R., R. D. Estes, and S. Baldwin. 1977. Anti-predator behaviour of baboons and impalas toward a cheetah. *East Afr Wildl J* 15:327–29.

Baines, D. 1990. The roles of predation, food and agricultural practice in determining the breeding success of the lapwing *Vanellus vanellus* on upland grasslands. *J Anim Ecol* 59:915–29.

Baker, R. R. 1985. Bird colouration: In defence of unprofitable prey. *Anim Behav* 33:1387–88.

Baker, R. R., and M. V. Hounsome. 1983. Bird coloration: Unprofitable prey model supported by ringing data. *Anim Behav* 31:614–15.

Baker, R. R., and G. A. Parker. 1979. The evolution of bird colouration. *Philos Trans R Soc Lond B Biol Sci* 287:63–130.

Bakken, G. S., V. C. Vanderbilt, W. A. Buttemer, and W. R. Dawson. 1978. Avian eggs: Thermoregulatory value of very high near-infrared reflectance. *Science* 200:321–23.

Balda, R. P., G. C. Bateman, and G. F. Foster. 1972. Flocking associates of the pinion jay. *Wilson Bull* 84:60–76.

Baldellou, M., and P. Henzi. 1992. Vigilance, predator detection and the presence of supernumeray males in vervet monkey troops. *Anim Behav* 43:451–61.

Balmford, A., and M. Turyaho. 1992. Predation risk and lek-breeding in Uganda kob. *Anim Behav* 44:117–27.

Balph, D. M., and D. F. Balph. 1966. Sound communication of Uinta ground squirrels. *J Mamm* 47:440–50.

Balph, M. H. 1977. Sex differences in alarm responses of wintering evening grosbeaks. *Wilson Bull* 89:325–27.

Banks, P. B. 2001. Predation-sensitive grouping and habitat use by eastern grey kangaroos: A field experiment. *Anim Behav* 61:1013–21.

Barash, D. 1973. The social biology of the Olympic marmot. *Anim Behav Monogr* 6:173–245.

Barash, D. P. 1975a. Marmot alarm-calling and the question of altruistic behavior. *Am Midl Nat* 94:468–70.

———. 1975b. Evolutionary aspects of parental behavior: Distraction behavior of the alpine accentor. *Wilson Bull* 87:367–73.

Barber, H. N. 1954. Genetic polymorphism in the rabbit in Tasmania. *Nature* 173:1227–29.

Barbosa, A. 1995. Foraging strategies and their influence on scanning and flocking behaviour of waders. *J Avian Biol* 26:182–86.

Barlow, G. W. 1972. The attitude of fish eye-lines in relation to body shape and to stripes and bars. *Copeia* 1972:4–12.

———. 1974. Extraspecific imposition of social grouping among surgeon-fishes (Pisces: Acanthuridae). *J Zool (Lond)* 174:333–40.

Barnard, C. J. 1980. Flock feeding and time budgets in the house sparrow (*Passer domesticus* L.). *Anim Behav* 28:295–309.

———, ed. 1984. *Producers and scroungers.* Beckenham, UK: Croom Helm.

Barnard, C. J., and H. Stephens. 1981. Prey size selection by lapwings in lapwing/gull associations. *Behaviour* 77:1–22.

———. 1983. Costs and benefits of single and mixed species flocking in fieldfares (*Turdus pilaris*) and redwings (*T. iliacus*). *Behaviour* 84:91–123.

Barnard, C. J., and D. B. A. Thompson. 1985. *Gulls and plovers: The ecology and behaviour of mixed-species feeding groups.* New York: Columbia University Press.

Barnard, C. J., D. B. A. Thompson, and H. Stephens. 1982. Time budgets, feeding efficiency and flock dynamics in mixed species flocks of lapwings, golden plovers and gulls. *Behaviour* 8:44–69.

Barreto, G. R., and D. W. Macdonald. 1999. The response of water voles, *Arvicola terrestris*, to the odours of predators. *Anim Behav* 57:1107–12.

Barta, Z., R. Flynn, and L-A. Giraldeau. 1997. Geometry for a selfish foraging group: A genetic algorithm approach. *Proc R Soc Lond Ser B Biol Sci* 264:1233–1238.

Bartecki, U., and E. W. Heymann. 1987. Field observation of snake-mobbing in a group of saddleback tamarins, *Saguinus fuscicollis nigrifrons. Folia Primatol* 48:199–202.

Bartholomew, G. A., and H. H. Caswell Jr. 1951. Locomotion in kangaroo rats and its adaptive significance. *J Mammal* 32:155–69.

Bartle, J. A., and P. M. Sagar. 1987. Intraspecific variation in the New Zealand Bellbird *Anthornis melanura. Notornis* 34:253–306.

Batary, P., and A. Baldi. 2004. Evidence of an edge effect on avian nest success. *Cons Biol* 18:389–400.

Bauwens, D., and C. Thoen. 1981. Escape tactics and vulnerability to predation associated with reproduction in the lizard *Lacerta vivipara. J Anim Ecol* 50:733–43.

Bayne, E. M., and K. A. Hobson 1997. Comparing the effects of landscape fragmentation by forestry and agriculture on predation of artificial nests. *Conserv Biol* 11:1418–1429.

Beani, L., F. Dessi-Fulgheri. 1995. Mate choice in the grey partridge, *Perdix perdix*: role of physical and behavioural male traits. *Anim Behav* 49:347–356.

Beatty, C. D., K. Beirinckx, and T. N. Sherratt. 2004. The evolution of mullerian mimicry in multispecies communities. *Nature* 431:63–67.

Beauchamp, G. 1998. The effect of group size on mean food intake rate in birds. *Biol Rev Camb Philos Soc* 73:449–472.

———. 2001. Should vigilance always decrease with group size? *Behav Ecol Sociobiol* 51:47–52.

Beauchamp, G., and B. Livoreil. 1997. The effect of group size on vigilance and feeding rate in spice finches (*Lonchura punctulata*). *Can J Zool* 75:1526–31.

Becker, P. H. 1995. Effects of coloniality on gull predation on common tern (*Sterna hriundo*) chicks. *Colon Waterbirds* 18:11–22.

Beddard, F. E. 1892. *Animal coloration: An account of the principal facts and theories related to the colours and markings of animals.* London: Swan Sonnenschein & Co.

Bednekoff, P. A. 1997. Mutualism among safe, selfish sentinels: A dynamic game. *Am Nat* 150:373–92.

————. 2001. Coordination of safe, selfish sentinels based on mutual benefits. *Ann Zool Fen* 38:5–14.

Bednekoff, P. A., and A. I. Houston. 1994. Dynamic models of mass-dependent predation, risk-sensitive foraging, and premigratory fattening in birds. *Ecology* 75:1131–40.

Bednekoff, P. A., and S. L. Lima. 1998a. Randomness, chaos and confusion in the study of antipredator vigilance. *Trends Ecol Evol* 13:284–87.

————. 1998b. Re-examining safety in numbers: Interactions between risk dilution and collective detection depend upon predator targeting behaviour. *Proc R Soc Lond Ser B Biol Sci* 265:2021–26.

Bednekoff, P. A., and R. Ritter. 1994. Vigilance in Nxai Pan springbok, *Antidorcas marsupialis*. *Behaviour* 129:1–11.

Begg, R. J., and C. R. Dunlop. 1980. Security eating, and diet in the large rock-rat, *Zyzomys woodwardi* (Rodentia: Muridae). *Aust Wildl Res* 7:63–70.

Bekoff, M. 1995. Vigilance, flock size, and flock geometry: Information gathering by western evening grosbeaks (Aves, Fringillidae). *Ethology* 99:150–61.

Beletsky, L. D. 1989. Alert calls of male red-winged blackbirds: Do females listen? *Behaviour* 111:1–12.

Beletsky, L. D., B. J. Higgins, and G. H. Orians. 1986. Communication by changing signals: Call switching in red-winged blackbirds. *Behav Ecol Sociobiol* 18:221–29.

Belk, M. C., and M. H. Smith. 1996. Pelage coloration in oldfield mice (*Peromyscus polionotus*): Antipredator adaptation? *J Mammal* 77:882–90.

Belt, T. W. 1874. *The naturalist in Nicaragua.* London: Longmans.

Bengston, S. A. 1972. Reproduction and fluctuations in size of duck populations at lake Myvatn, Iceland. *Oikos* 23:35–58.

Bennett, A. T. D., I. C. Cuthill, J. C. Partridge, and K. Lanan. 1997. Ultra-violet colours predict mate preferences in starlings. *Proc Natl Acad Sci USA* 94:8618–21.

Bennett, M. B. 1987. Fast locomotion of some kangaroos. *J Zool* 212:457–64.

Benson, S. B. 1936. Concealing coloration among some desert rodents of the southwestern United States. *Univ Calif Publ Zool* 40:1–69.

Benson, S. V. 1952. *The observer's book of birds.* London: Frederick Warne.

Berg, A., T. Lindberg, and K. G. Kallebrink. 1992. Hatching success of lapwings on farmland: Differences between habitats and colonies of different sizes. *J Anim Ecol* 61:469–76.

Berger, J. 1978a. Group size, foraging, and antipredator ploys: An analysis of bighorn sheep decisions. *Behav Ecol Sociobiol* 4:91–99.

————. 1978b. Maternal defensive behavior in bighorn sheep. *J Mammal* 59:620–21.

————. 1979. "Predator harassment" as a defensive strategy in ungulates. *Am Midl Nat* 102:197–99.

————. 1991. Pregnancy incentives, predation constraints and habitat shifts: Experimental and field evidence for wild bighorn sheep. *Anim Behav* 41:61–77.

————. 1998. Future prey: Some consequences of the loss and restoration of large mammalian carnivores on prey. In *Behavioral ecology and conservation biology*, ed. T. Caro, 80–100. New York: Oxford University Press.

————. 1999. Anthropogenic extinction of top carnivores and interspecific animal behaviour: Implications of the rapid decoupling of a web involving wolves, bears, moose and ravens. *Proc R Soc Lond Ser B Biol Sci* 266:2261–67.

Berger, J., and C. Cunningham. 1988. Size-related effects on search times in North American grassland female ungulates. *Ecology* 69:177–83.

————. 1994. Phenotypic alterations, evolutionarily significant structures, and rhino conservation. *Cons Biol* 8:833–40.

Berger, J., J. E. Swenson, and I-L. Persson. 2001. Recolonizing carnivores and naïve prey: Conservation lessons from Pleistocene extinctions. *Science* 291:1036–39.

Bergerud, A. T. 1985. Antipredator strategies of caribou: Dispersion along shorelines. *Can J Zool* 63:1324–29.

Bergerud, A. T., H. E. Butler, and D. R. Miller. 1984. Antipredator tactics of calving caribou: Dispersion in mountains. *Can J Zool* 62:1566–75.

Bergerud, A. T., R. Ferguson, and H. E. Butler. 1990. Spring migration and dispersion of woodland caribou at calving. *Anim Behav* 39:360–68.

Bergerud, A. T., and R. E. Page. 1987. Displacement and dispersion of parturient caribou at calving as antipredator tactics. *Can J Zool* 65:1597–1606.

Bergstrom, C. T., and M. Lachmann. 2001. Alarm calls as costly signals of antipredator vigilance: The watchful babbler game. *Anim Behav* 61:535–42.

Bertram, B. C. R. 1978a. Living in groups: Predators and prey. In *Behavioural ecology: An evolutionary approach*, ed. J. R. Krebs and N. B. Davies, 64–96. Oxford: Blackwell Scientific Publications.

———. 1978b. *Pride of lions.* London: J. M. Dent & Sons.

———. 1980. Vigilance and group size in ostriches. *Anim Behav* 28:278–86.

Bertram, B. C. R., and F. A. Burger. 1981. Are ostrich *Struthio camelus* eggs the wrong colour? *Ibis* 123:207–10.

Best, L. B. 1978. Field sparrow reproductive success and nesting ecology. *Auk* 95:9–22.

Best, L. B., and D. F. Stauffer. 1980. Factors affecting nesting success in riparian bird communities. *Condor* 82:149–58.

Betts, B. J. 1976. Behaviour in a population of Columbian ground squirrels, *Spermophilus columbianus columbianus. Anim Behav* 24:652–80.

Bety, J., G. Gauthier, J-F. Giroux, and E. Korpimaki. 2001. Are goose nesting success and lemming cycles linked? Interplay between nest density and predators. *Oikos* 93:388–400.

Beveridge, F. M., and J. M. Deag. 1987. The effects of sex, temperature and companions on looking up and feeding in single and mixed species flocks of house sparrows (*Passer domesticus*), chaffinches (*Fringilla coelebs*) and starlings (*Sturnus vulgaris*). *Behaviour* 100:303–20.

Biermann, G. C., and R. J. Robertson. 1981. An increase in parental investment during the breeding season. *Anim Behav* 29:487–89.

———. 1983. Residual reproductive value and parental investment. *Anim Behav* 31:311–12.

Bildstein, K. L. 1982. Responses of northern harriers to mobbing passerines. *J Field Ornithol* 53:7–14.

———. 1983. Why white-tailed deer flag their tails. *Am Nat* 121:709–15.

Bjorklund, M. 1991. Coming of age in fringillid birds: Heterochrony in the ontogeny of secondary sexual characters. *J Ecol Biol* 4:83–92.

Black, J. M., C. Carbone, R. L. Wells, and M. Owen. 1992. Foraging dynamics in goose flocks: The cost of living on the edge. *Anim Behav* 44:41–50.

Black, J. M., and M. Owen. 1989. Agonistic behaviour in barnacle goose flocks: Assessment, investment and reproductive success. *Anim Behav* 37:199–209.

Blancher, P. J., and R. J. Robertson. 1982. Kingbird aggression: Does it deter predation? *Anim Behav* 30:929–45.

———. 1985. Predation in relation to spacing of kingbird nests. *Auk* 102:654–58.

Blanco, G., and J. L. Tella. 1997. Protective association and breeding advantages of choughs nesting in lesser kestrel colonies. *Anim Behav* 54:335–42.

Bleich, V. C. 1999. Mountain sheep and coyotes: Patterns of predator evasion in a mountain ungulate. *J Mammal* 80:283–89.

Bleich, V. C., R. T. Bowyer, and J. D. Wehausen. 1997. Sexual segregation in mountain sheep: Resources or predation? *Wildl Monogr* 134:1–50.

Blumstein, D. T. 1992. Multivariate analysis of golden marmot maximum running speed: A new method to study MRS in the field. *Ecology* 73:1757–67.

———. 1995a. Golden-marmot alarm calls. I. The production of situationally specific vocalizations. *Ethology* 100:113–25.

———. 1995b. Golden-marmot alarm calls. II. Asymmetrical production and perception of situationally specific vocalizations? *Ethology* 101:25–32.

———. 1996. How much does social group size influence golden marmot vigilance? *Behaviour* 133:1133–51.

———. 1998. Quantifying predation risk for refuging animals: A case study with golden marmots. *Ethology* 104:501–16.

Blumstein, D. T., and K. B. Armitage. 1997a. Alarm calling in yellow-bellied marmots: I. The meaning of situationally variable alarm calls. *Anim Behav* 53:143–71.

———. 1997b. Does sociality drive the evolution of communicative complexity? A comparative test with ground-dwelling scuirid alarm calls. *Am Nat* 150:179–200.

Blumstein, D. T., and W. Arnold. 1995. Situational specificity in Alpine-marmot alarm communication. *Ethology* 100:1–13.

Blumstein, D. T., and J. C. Daniel. 2002. Isolation from mammalian predators differentially affects two congeners. *Behav Ecol* 13:657–63.

———. 2004. Yellow-bellied marmots discriminate between the alarm calls of individuals and are more responsive to calls from juveniles. *Anim Behav* 68:1257–65.

Blumstein, D. T., J. C. Daniel, A. S. Griffin, and C. S. Evans. 2000. Insular tammar wallabies (*Macropus eugenii*) respond to visual but not acoustic cues from predators. *Behav Ecol* 11:528–35.

Blumstein, D. T., J. C. Daniel, and I. G. McLean. 2001. Group size in quokkas. *Austr J Zool* 49:641–49.

Blumstein, D. T., J. C. Daniel, and B. P. Springelt. 2004. A test of the multi-predator hypothesis: Rapid loss of antipredator behavior after 130 years of isolation. *Ethology* 110:919–34.

Blumstein, D. T., C. S. Evans, and J. C. Daniel. 1999. An experimental study of behavioural group size effects in tammar wallabies, *Macropus eugenii. Anim Behav* 58:351–60.

Blumstein, D. T., E. Fernandez-Juricic, O. LeDee, E. Larsen, I. Rodriguez-Prieto, and C. Zugmeyer. 2004. Avian risk assessment: Effects of perching height and detectability. *Ethology* 110:273–85.

Blumstein, D. T., M. Mari, J. C. Daniel, J. G. Ardon, A. S. Griffin, and C. S. Evans. 2002. Olfactory predator recognition: Wallabies may have to learn to be wary. *Anim Cons* 5:87–93.

Blumstein, D. T., J. Steinmetz, K. B. Armitage, J. C. Daniel. 1997. Alarm calling in yellow-bellied marmots: II. The importance of direct fitness. *Anim Behav* 53:173–84.

Boag, D. A., S. G. Reebs, and M. A. Schroeder. 1984. Egg loss among spruce grouse inhabiting lodgepole pine forests. *Can J Zool* 62:1034–37.

Boesch, C. 1991. The effects of leopard predation on grouping patterns in forest chimpanzees. *Behaviour* 117:220–42.

———. 1994. Chimpanzee-red colobus monkeys: A predator-prey system. *Anim Behav* 47:1135–48.

Bogliani, G., F. Sergio, and G. Tavecchia. 1999. Woodpigeons nesting in association with hobby falcons: Advantages and choice rules. *Anim Behav* 57:125–31.

Boland, C. R. J. 2003. An experimental test of predator detection rates using groups of free-living emus. *Ethology* 109:209–22.

Bond, A. B., and A. C. Kamil. 2002. Visual predators select for crypticity and polymorphism in virtual prey. *Nature* 415:609–13.

Bonenfant, M., and D. L. Kramer. 1996. The influence of distance to burrow on flight initiation distance in the woodchuck, *Marmota monax*. *Behav Ecol* 7:299–303.

Booth, C. L. 1990. Evolutionary significance of ontogenetic colour change in animals. *Biol J Linn Soc* 40:125–63.

Borowski, Z. 1998. Influence of weasel (*Mustela nivalis* Linnaeus, 1766) odour on spatial behaviour of root voles (*Microtus oeconomus* Pallas, 1776). *Can J Zool* 76:1799–1804.

Bosque, C., and M. T. Bosque. 1995. Nest predation as a selective factor in the evolution of developmental rates in altricial birds. *Am Nat* 145:234–60.

Boucher, D. H. 1977. On wasting parental investment. *Am Nat* 111:786–88.

Bourne, W. R. P. 1977. The function of mobbing. *Br Birds*70:266–68.

Bouskila, A. 1995. Interactions between predation risk and competition: A field study of kangaroo rats and snakes. *Ecology* 76:165–78.

———. 2001. A habitat selection game of interactions between rodents and their predators. *Ann Zool Fenn* 38:55–70.

Bowers, M. A. 1988. Seed removal experiments on desert rodents: The microhabitat by moonlight effect. *J Mammal* 69:201–4.

Bowman, G. B., and L. D. Harris. 1980. Effect of spatial heterogeneity on ground-nest depredation. *J Wildl Manag* 44:806–13.

Bowyer, R. T., J. G. Kie, and V. Van Ballenberghe. 1998. Habitat selection by neonatal black-tailed deer: Climate, forage, or risk of predation? *J Mammal* 79:415–25.

Bowyer, R. T., V. Van Ballenberghe, J. G. Kie, and J. A. K. Maier. 1999. Birth-site selection by Alaskan moose: Maternal strategies for coping with a risky environment. *J Mammal* 80:1070–83.

Boyko, A. R., R. M. Gibson, and J. R. Lucas. 2004. How predation risk affects the temporal dynamics of avian leks: Greater sage grouse versus golden eagles. *Am Nat* 163:154–65.

Bradbury, J. W., and S. L. Vehrencamp. 1998. *Principles of animal communication.* Sunderland, MA: Sinauer Assn.

Branch, L. C. 1993. Social organization and mating system of the plains viscacha (*Lagostomus maximus*). *J Zool (Lond)* 229:473–91.

Brandon, R., G. Labanick, and J. Huheey. 1979. Relative palatability, defensive behavior, and mimetic relationships of red salamanders (*Psuedotriton rubber*), mud salamanders (*Psuedotriton montanus*), and red efts (*Notopthalmus vridescens*). *Herpetologica* 35:289–303.

Brashares, J. S., T. Garland Jr., and P. Arcese. 2000. Phylogenetic analysis of coadaptation in behavior, diet, and body size in the African antelope. *Behav Ecol* 11:452–63.

Braude, S., D. Ciszek, N. E. Berg, and N. Shefferly. 2001. The ontogeny and distribution of countershading in colonies of the naked mole-rat (*Heterocephalus glaber*). *J Zool (Lond)* 253:351–57.

Brearey, D., and O. Hilden 1985. Nesting and egg-predation by turnstones *Arenaria interpres* in land colonies. *Ornis Scand* 16:283–92.

Breitwisch, R. 1988. Sex differences in defence of eggs and nestlings by northern mockingbirds, *Mimus polyglottos. Anim Behav* 36:62–72.

Breitwisch, R., N. Gottlieb, and J. Zaias. 1989. Behavioral differences in nest visits between male and female northern mockingbirds. *Auk* 106:659–65.

Bremond, J-C., and T. Aubin. 1990. Responses to distress calls by black-headed gulls, *Larus ridibundus*: The role of non-degraded features. *Anim Behav* 39:503–11.

Brightsmith, D. J. 2000. Use of arboreal termiteria by nesting birds in the Peruvian Amazon. *Condor* 102:529–38.

Brillhart, D. B., and D. W. Kaufman. 1991. Influence of illumination and surface structure on space use by prairie deer mice (*Peromyscus maniculatus bairdii*). *J Mammal* 72:764–68.

Broadbent, D. E. 1971. *Decision and stress*. London: Academic Press.

Brock, V. E., and R. H. Riffenburgh. 1960. Fish schooling: A possible factor in reducing predation. *J Cons Int Explor Mer* 25:307–17.

Brodie, E. D. Jr. 1977. Hedgehogs use toad venom in their own defence. *Nature* 268:627–28.

Brodie, E. D. Jr., B. J. Ridenhour, and E. D. Brodie III. 2002. The evolutionary response of predators to dangerous prey: Hotspots and coldspots in the geographic mosaic of coevolution between garter snakes and newts. *Evolution* 56:2067–82.

Brodie, E. D. III. 1992. Correlational selection for color pattern and antipredator behavior in the garter snake *Thamnophis ordinoides*. *Evolution* 46:1284–98.

Brodie, E. D. III, and E. D. Brodie Jr. 1999. Predator-prey arms races. *BioScience* 49:557–68.

Brodin, A. 2001. Mass-dependent predation and metabolic expenditure in wintering birds: Is there a trade-off between different forms of predation? *Anim Behav* 62:993–99.

Brooke, M. de L. 1998. Ecological factors influencing the occurrence of "flash marks" in wading birds. *Funct Ecol* 12:339–46.

Brooke, M. de L., S. Hanley,and S. B. Laughlin. 1999. The scaling of eye size with body mass in birds. *Proc R Soc Lond Ser B Biol Sci* 266:405–12.

Brown, C. H. 1982. Ventroloquial and locatable vocalizations in birds. *Z Tierpsychol* 59:338–50.

———. 1984. The vocal range of alarm calls in thirteen-lined ground squirrels. *Z Tierpsychol* 65:273–88.

Brown, C. R., and B. M. Brown. 1987. Group-living in cliff swallows as an advantage in avoiding predators. *Behav Ecol Sociobiol* 21:97–108.

Brown, E. D.1985. Functional interrelationships among the mobbing and alarm caws of common crows (*Corvus brachyrhynchos*). *Z Tierpsychol* 67:17–33.

Brown, G. E., J-G. J. Godin, and J. Pedersen. 1999. Fin-flicking behaviour: A visual antipredator alarm signal in a charachin fish, *Hemigrammus erythrozonus*. *Anim Behav* 58:469–75.

Brown, J. H., and G. B. West. 2000. *Scaling in biology*. New York: Oxford University Press.

Brown, J. S. 1988. Patch use as an indicator of habitat preference, predation risk, and competition. *Behav Ecol Sociobiol* 22:37–47.

———. 1989. Desert rodent community structure: A test of four mechanisms of coexistence. *Ecol Monogr* 59:1–20.

———. 1992. Patch use under predation risk: I. Models and predictions. *Ann Zool Fenn* 29:301–9.

———. 1999. Vigilance, patch use and habitat selection: Foraging under predation risk. *Evol Ecol Res* 1:49–71.

———. 2000. Foraging ecology of animals in response to heterogeneous environments. In *Ecological consequences of environmental heterogeneity*, ed. J. Hutchins and A. Stewart, 181–214. Oxford: Blackwell Science.

Brown, J. S., and P. U. Alkon. 1990. Testing values of crested porcupine habitats by experimental food patches. *Oecologia* 83:512–18.

Brown, J. S., B. P. Kotler, R. J. Smith, and W. O. Wirtz II. 1988. The effects of owl predation on the foraging behavior of heteromyid rodents. *Oecologia* 76:408–15.

Brown, L. N. 1965. Selection in a population of house mice containing mutant individuals. *J Mammal* 46:461–65.

Brown, R. E., and D. W. Macdonald. 1985. *Social odours in mammals*. Oxford: Clarendon Press.

Brunton, D. H. 1986. Fatal antipredator behavior of a killdeer. *Wilson Bull* 98:605–7.

———. 1990. The effects of nesting stage, sex, and type of predator on parental defense by killdeer (*Charadrius vociferous*): Testing models of avian parental defense. *Behav Ecol Sociobiol* 26:181–90.

———. 1997. Impacts of predators: Center nests are less successful than edge nests in a large nesting colony of least terns. *Condor* 99:372–80.

Bshary, R. 2001. Diana monkeys, *Cercopithecus diana*, adjust their anti-predator response behaviour to human hunting strategies. *Behav Ecol Sociobiol* 50:251–56.

Bshary, R., and R. Noe. 1997a. Red colobus and Diana monkeys provide mutual protection against predators. *Anim Behav* 54:1461–74.

———. 1997b. Anti-predation behaviour of red colobus monkeys in the presence of chimpanzees. *Behav Ecol Sociobiol* 41:321–33.

Buchanan, J. B., C. T. Schick, L. A. Brennan, and S. G. Herman. 1988. Merlin predation on wintering dunlins: Hunting success and dunlin escape tactics. *Wilson Bull* 100:108–18.

Buckley, P. A., and F. G. Buckley. 1977. Hexagonal packing of royal tern nests. *Auk* 94:36–43.

Buitron, D. 1983. Variability in the responses of black-billed magpies to natural predators. *Behaviour* 87:209–36.

Buler, J. J., and R. B. Hamilton. 2000. Predation of natural and artificial nests in a southern pine forest. *Auk* 117:739–47.

Bumann, D., J. Krause, and D. Rubenstein. 1997. Mortality risk of spatial positions in animal groups: The danger of being in the front. *Behaviour* 134:1063–76.

Bures, S., and V. Pavel. 1997. The effect of nestling condition on risk-taking in meadow pipits. *Anim Behav* 54:1531–34.

Burger, J. 1974. Breeding adaptations of Franklin's gull (*Larus pipixcan*) to a marsh habitat. *Anim Behav* 22:521–67.

———. 1981. A model for the evolution of mixed-species colonies of Ciconiiformes. *Q Rev Biol* 56:143–67.

———. 1984. Grebes nesting in gull colonies: Protective associations and early warning. *Am Nat* 123:327–37.

Burger, J., and M. Gochfeld. 1991. Human disturbance and birds: Tolerance and response distances of resident and migrant species in India. *Env Cons* 18:158–65.

———. 1992. Effect of group size on vigilance while drinking in the coati, *Nasua narica* in Costa Rica. *Anim Behav* 44:1053–57.

———. 1994. Vigilance in African mammals: Differences among mothers, other females, and males. *Behaviour* 131:153–69.

Burhans, D. E., and F. R. Thompson III. 2001. Relationship of songbird nest concealment to nest fate and flushing behavior of adults. *Auk* 118:237–42.

Burke, D. M., K. Elliott, L. Moore, W. Dunford, E. Nol, J. Phillips, S. Holmes, and K. Freemark. 2004. Patterns of nest predation on artificial and natural nests in forests. *Conserv Biol* 18:381–88.

Burnett, L., and G. R. Hosey. 1987. Frequency of vigilance behaviour and group size in rabbits (*Oryctolagus cuniculus*). *J Zool (Lond)* 212:367–68.

Burns, J. G., and R. C. Ydenberg. 2002. The effects of wing loading and gender on the escape flights of least sandpipers (*Calidris minutilla*) and western sandpipers (*Calidris mauri*). *Behav Ecol Sociobiol* 52:128–36.

Burns, K. J. 1998. A phylogenetic perspective on the evolution of sexual dichromatism in tanagers (Thraupidae): The role of female versus male plumage. *Evolution* 52:1219–24.

Burrell, H. 1927. *The platypus: Its discovery, zoological position, form and characteristics, habits, life history, etc.* Sydney: Angus & Robertson.

Burtt, E. H. Jr. 1979. Tips on wings and other things. In *The behavioral significance of color*, ed. E. H. Burtt Jr., 75–110. New York: Garland STPM Press.

———. 1981. The adaptiveness of animal colors. *BioScience* 31:721–29.

Buskirk, W. H. 1976. Social systems in a tropical forest avifauna. *Am Nat* 110:293–310.

Butcher, G. S., and S. Rohwer. 1989. The evolution of conspicuous and distinctive coloration for communication in birds. *Curr Ornithol* 6:51–108.

Butler, J. M., and T. J. Roper. 1994. Escape tactics and alarm responses in badgers *Meles meles*: A field experiment. *Ethology* 99:313–22.

Butler, M. A., and J. J. Rotella. 1998. Validity of using artificial nests to assess duck-nest success. *J Wildl Manag* 62:163–71.

Butynski, T. M. 1984. Nocturnal ecology of the springhare, *Pedetes capensis*, in Botswana. *Afr J Ecol* 22:7–22.

Buxton, P. A. 1923. *Animal life in deserts: A study of the fauna in relation to the environment.* London: Edward Arnold.

Byers, J. A. 1997. *American pronghorn: Social adaptations and the ghosts of predators past.* Chicago: University of Chicago Press.

Byers, J. A., and Byers, K. Z. 1983. Do pronghorn mothers reveal the locations of their hidden fawns? *Behav Ecol Sociobiol* 13:147–56.

Byrkjedal, I. 1987. Antipredator behavior and breeding success in greater golden-plover and Eurasian dotterel. *Condor* 89:40–47.

Cahalane, V. H. 1961. *Mammals of North America.* New York: Macmillan.

Caine, N. G. 1990. Unrecognized anti-predator behaviour can bias observational data. *Anim Behav* 39:195–96.

Caine, N. G., and S. L. Marra. 1988. Vigilance and social organization in two species of primates. *Anim Behav* 36:897–904.

Caine, N. G., and P. J. Weldon. 1989. Responses by red-bellied tamarins (*Saguinus labiatus*) to fecal scents of predatory and non-predatory neotropical mammals. *Biotropica* 21:186–89.

Calder, C. J., and M. L. Gorman. 1991. The effects of red fox *Vulpes vulpes* faecal odours on the feeding behaviour of Orkney voles *Microtus arvalis*. *J Zool* 224:599–606.

Caldwell, G. S. 1986. Predation as a selective force on foraging herons: Effects of plumage color and flocking. *Auk* 103:494–505.

Caldwell, G. S., and R. W. Rubinoff. 1983. Avoidance of venomous sea snakes by naïve herons and egrets. *Auk* 100:195–98.

Caldwell, J. P., J. H. Thorp, and T. O. Jervey. 1980. Predator-prey relationships among larval dragonflies, salamanders, and frogs. *Oecologia* 46:285–89.

Caraco, T. 1979. Time budgeting and group size: A test of theory. *Ecology* 60:618–27.

Caraco, T., and M. C. Bayham. 1982. Some geometric aspects of house sparrow flocks. *Anim Behav* 30:990–96.

Caraco, T., S. Martindale, and H. R. Pulliam. 1980. Avian time budgets in the presence of a predator. *Nature* 285:400–401.

Carey, H. V. 1985. The use of foraging areas by yellow-bellied marmots. *Oikos* 44:273–79.

Carey, H. V., and P. Moore. 1986. Foraging and predation risk in yellow-bellied marmots. *Am Midl Nat* 116:267–75.

Carl, G. R., and C. T. Robbins. 1988. The energetic cost of predator avoidance in neonatal ungulates: Hiding versus following. *Can J Zool* 66:239–46.

Carlisle, T. R. 1982. Brood success in variable environments: Implications for parental care allocation. *Anim Behav* 30:824–36.

———. 1985. Parental response to brood size in a cichlid fish. *Anim Behav* 33:234–38.

Caro, T. M. 1986a. The functions of stotting: A review of the hypotheses. *Anim Behav* 34:649–62.

———. 1986b. The functions of stotting in Thomson's gazelles: Some tests of the predictions. *Anim Behav* 34:663–84.

———. 1987. Cheetah mothers' vigilance: Looking out for prey or for predators? *Behav Ecol Sociobiol* 20:351–61.

———. 1994a. *Cheetahs of the Serengeti Plains: Group living in an asocial species.* Chicago: University of Chicago Press.

———. 1994b. Ungulate antipredator behaviour: Preliminary and comparative data from African bovids. *Behaviour* 128:189–228.

———. 1995. Pursuit-deterrence revisited. *Trends Ecol Evol* 10:500–503.

———. 2005. The adaptive significance of coloration in mammals. *BioScience* 55:125–36.

Caro, T. M., and C. D. FitzGibbon. 1992. Large carnivores and their prey: The quick and the dead. In *Natural enemies: The population biology of predators, parasites and diseases,* ed.: M. J. Crawley, 117–42. Oxford: Blackwell Scientific Publications.

Caro, T. M., C. M. Graham, C. J. Stoner, and M. M. Flores. 2003. Correlates of horn and antler shape in bovids and cervids. *Behav Ecol Sociobiol* 55:32–41.

Caro, T. M., C. M. Graham, C. J. Stoner, and J. K. Vargas. 2004. Adaptive significance of antipredator behaviour in artiodactyls. *Anim Behav* 67:205–28.

Caro, T. M., and M. D. Hauser. 1992. Is there teaching in nonhuman animals? *Q Rev Biol* 67:151–74.

Caro, T. M., L. Lombardo, A. W. Goldizen, and M. Kelly. 1995. Tail-flagging and other antipredator signals in white-tailed deer: New data and synthesis. *Behav Ecol* 6:442–50.

Carrascal, L. M., and E. Moreno. 1992. Proximal costs and benefits of heterospecific social foraging in the great tit, *Parus major. Can J Zool* 70:1947–52.

Carrascal, L. M., and V. Polo. 1999. Coal tits, *Parus ater,* lose weight in response to chases by predators. *Anim Behav* 58:281–85.

Carrier, D. R. 1996. Ontogenetic limits on locomotor performance. *Physiol Zool* 69:467–88.

Carrillo, J., and J. M. Aparicio. 2001. Nest defence behaviour of the Eurasian kestrel (*Falco tinnunculus*) against human predators. *Ethology* 107:865–75.

Case, T. J. 1978. On the evolution and adaptive significance of postnatal growth rates in the terrestrial vertebrates. *Q Rev Biol* 53:243–82.

Cassini, M. H. 1991. Foraging under predation risk in the wild guinea pig Cavia aperea. *Oikos* 62:20–24.

Catterall, C. P., M. A. Elgar, and J. Kikkawa. 1992. Vigilance does not covary with group size in an island population of silvereyes (*Zosterops lateralis*). *Behav Ecol* 3:207–10.

Cawthorn, J. M., D. L. Morris, E. D. Ketterson, and V. Nolan Jr. 1998. Influence of experimentally elevated testosterone on nest defence in dark-eyed juncos. *Anim Behav* 56:617–21.

Chai, P. 1986. Field observation and feeding experiments on the responses of rufous-tailed jacamars (*Galbula ruficanda*) to free-flying butterflies in a tropical rainforest. *Biol J Linn Soc* 29:161–89.

Chai, P., and R. B. Srygley. 1990. Predation and the flight, morphology, and temperature of Neotropical rain-forest butterflies. *Am Nat* 135:748–65.

Chalfoun, A. D., M. J. Ratnaswamy, and F. R. Thompson III. 2002. Songbird nest predators in forest-pasture edge and forest interior in a fragmented landscape. *Ecol Appl* 12:858–67.

Chandler, C. R., and R. K. Rose. 1988. Comparative analysis of the effects of visual and auditory stimuli on avian mobbing behavior. *J Field Ornithol* 59:269–77.

Chapman, C. 1985. The infuence of habitat on behaviour in a group of St. Kitts green monkeys. *J Zool (Lond)* 206:311–20.

———. 1986. *Boa constrictor* predation and group response in white-faced Cebus monkeys. *Biotropica* 18:171–72.

Chapman, C. A., and L. J. Chapman. 1996. Mixed-species primate groups in the Kibale forest: Ecological constraints on association. *Int J Primatol* 17:31–50.

————. 2000. Interdemic variation in mixed-species association patterns: Common diurnal primates of Kibale National Park, Uganda. *Behav Ecol Sociobiol* 47:129–39.

Chapman, C. A., L. J. Chapman, and L. Lefebvre. 1990. Spider monkey alarm calls: Honest advertisement or warning calls? *Anim Behav* 39:197–98.

Chapman, D. M., and U. Roze. 1997. Functional histology of quill erection in the porcupine, *Erethizon dorsatum*. *Can J Zool* 75:1–10.

Charnov, E. L. 1993. *Life history invariants: Some explorations of symmetry in evolutionary ecology.* Oxford: Oxford University Press.

Charnov, E. L., and J. R. Krebs. 1975. The evolution of alarm calls: Altruism or manipulation? *Am Nat* 109:107–12.

Charnov, E. L., G. H. Orians, and K. Hyatt. 1976. Ecological implications of resource depression. *Am Nat* 110:247–59.

Chase, I. D. 1980. Cooperative and non-cooperative behavior in animals. *Am Nat* 115:827–57.

Cheesman, R. E. 1926. *In unknown Arabia.* London: MacMillan & Co.

Cheney, D. L., and R. M. Seyfarth. 1981. Selective forces affecting the predator alarm calls of vervet monkeys. *Behaviour* 76:25–61.

————. 1985a. Social and non-social knowledge in vervet monkeys. *Philos Trans R Soc Lond B Biol Sci* 308:187–201.

————. 1985b. Vervet monkey alarm calls: Manipulation through shared information? *Behaviour* 94:150–66.

————. 1988. Assessment of meaning and the detection of unreliable signals by vervet monkeys. *Anim Behav* 36:477–86.

————. 1991. Truth and deception in animal communication. In *Cognitive ethology: The minds of other animals*, ed. C. A. Ristau, 127–51. Hillsdale, NJ: Lawrence Erlbaum Associates.

Cheney, D. L., and R. W. Wrangham. 1987. Predation. In *Primate societies*, ed. B. B. Smuts, D. L. Cheney, R. M. Seyfarth, R. W. Wrangham, and T. T. Struhsaker, 227–39. Chicago: University of Chicago Press.

Childress, M. J., and M. A. Lung. 2003. Predation risk, gender and the group size effect: Does elk vigilance depend upon the behaviour of conspecifics? *Anim Behav* 66:389–98.

Chu, M. 2001. Heterospecific responses to scream calls and vocal mimicry by phainopeplas (Phainopepla nitens) in distress. *Behaviour* 138:775–87.

Churchfield, S. 1990. *The natural history of shrews.* London: Christopher Helm.

Clark, C. W. 1994. Antipredator behavior and the asset-protection principle. *Behav Ecol* 5:159–70.

Clark, C. W., and M. Mangel. 1986. The evolutionary advantages of group foraging. *Theor Popul Biol* 30:45–79.

Clark, K. L., and R. J. Robertson. 1979. Spatial and temporal multi-species nesting aggregation in birds as anti-parasite and anti-predator defenses. *Behav Ecol Sociobiol* 5:359–371.

Clark, R. G., and T. D. Nudds. 1991. Habitat patch size and duck nesting success: The crucial experiments have not been performed. *Wildl Soc Bull* 19:534–43.

Clark, R. G., and B. K. Wobeser. 1997. Making sense of scents: Effects of odour on survival of simulated duck nests. *J Avian Biol* 28:31–37.

Clarke, A., and P. A. Prince. 1976. The origin of stomach oil in marine birds: Analyses of the stomach oil from six species of subantarctic procellariiform birds. *J Exp Mar Biol Ecol* 23:15–30.

Clarke, B. C. 1962. Natural selection in mixed populations of two polymorphic snails. *Heredity* 17:319–45.

————. 1969. The evidence for apostatic selection. *Heredity* 24:347–52.

Clarke, J. A. 1983. Moonlight's influence on predator/prey interactions between short-eared owls (*Asio flammeus*) and deermice (*Peromyscus maniculatus*). *Behav Ecol Sociobiol* 13:205–9.

Clements, J. F. 1991. *Birds of the world: A check list*. Vista, CA: Ibis Publishing Company.

Clode, D. 1993. Colonially breeding seabirds: Predators or prey? *Trends Ecol Evol* 8:336–38.

Cloudsley-Thompson, J. L. 1980. *Tooth and claw: Defensive strategies in the animal world*. London: J. M. Dent & Sons Ltd.

Clutton-Brock, T. H. 1982. The functions of antlers. *Behaviour* 79:108–25.

———. 1984. Reproductive effort and terminal investment in iteroparous animals. *Am Nat* 123:212–29.

———. 1991. *The evolution of parental care*. Princeton, NJ: Princeton University Press.

Clutton-Brock, T. H., S. D. Albon, and P. H. Harvey. 1980. Antlers, body size and breeding group size in the Cervidae. *Nature* 285:565–67.

Clutton-Brock, T. H., D. Gaynor, G. M. McIlrath, A. D. C. MacColl, R. Kansky, P. Chadwick, M. Manser, J. D. Skinner, and P. N. M. Brotherton. 1999. Predation, group size and mortality in a cooperative mongoose, *Suricata suricatta*. *J Anim Ecol* 68:672–83.

Clutton-Brock, T. H., F. E. Guinness, and S. D. Albon. 1982. *Red deer: Behavior and ecology of two sexes*. Chicago: University of Chicago Press.

Clutton-Brock, T. H., and P. H. Harvey. 1977a. Primate ecology and social organization. *J Zool* 183:1–39.

———. 1977b. Species differences in feeding and ranging behaviour in primates. In *Primate ecology*, ed. T. H. Clutton-Brock, 557–79. London: Academic Press.

———. 1979. Comparison and adaptation. *Proc R Soc Lond Ser B Biol Sci* 205:547–65.

Clutton-Brock, T. H., M. J. O'Riain, P. N. M. Brotherton, D. Gaynor, R. Kansky, A. S. Griffin, and M. Manser. 1999. Selfish sentinels in cooperative mammals. *Science* 284:1640–44.

Cody, M. L. 1973. Coexistence, coevolution and convergent evolution in seabird communities. *Ecology* 54:31–44.

Colagross, A. M. L., and A. Cockburn. 1993. Vigilance and grouping in the eastern grey kangaroo, *Macropus giganteus*. *Austr J Zool* 41:325–34.

Cole, L. C. 1954. The population consequences of life history phenomena. *Q Rev Biol* 29:103–37.

Coleman, R. M., and M. R. Gross. 1991. Parental investment theory: The role of past investment. *Trends Ecol Evol* 6:404–6.

Collias, N. E. 1960. An ecological and functional classification of animal sounds. In *Animal sounds and communication*, ed. W. E. Lanyon and W. N. Tavolga, 368–91. Washington, DC: American Institute of Biological Sciences.

Collias, N. E., and E. C. Collias. 1978. Cooperative breeding behavior in the white-browed sparrow weaver. *Auk* 95:472–84.

———. 1984. *Nest building and bird behavior*. Princeton, NJ: Princeton University Press.

Collins, D. A. 1984. Spatial pattern in a troop of yellow baboons (*Papio cynocephalus*) in Tanzania. *Anim Behav* 32:536–53.

Collins, L. R., and J. F. Eisenberg. 1980. Notes on the behavior and breeding of pacaranas in captivity. *Int Zoo Yearb* 12:108–14.

Colwell, M. A., and L. W. Oring. 1990. Nest-site characteristics of prairie shorebirds. *Can J Zool* 68:297–302.

Conover, M. R. 1987. Acquisition of predator information by active and passive mobbers in ringed-billed gull colonies. *Behaviour* 102:41–57.

———. 1994. Stimuli eliciting distress calls in adult passerines and response of predators and birds to their broadcast. *Behaviour* 131:19–37.

Conover, M. R., and J. J. Perito. 1981. Response of starlings to distress calls and predator models holding conspecific prey. *Z Tierpsychol* 57:163–72.

Cooke, F., R. F. Rockwell, and D. B. Lank. 1995. *The snow geese of La Perouse Bay*. Oxford: Oxford University Press.

Coolen, I., and L-A. Giraldeau. 2003. Incompatability between antipredator vigilance and scrounger tactic in nutmeg manikins, *Lonchura punctulata*. *Anim Behav* 66:657–64.

Coolen, I., L-A. Giraldeau, and M. Lavoie. 2001. Head position as an indicator of producer and scrounger tactics in a ground feeding bird. *Anim Behav* 61:895–903.

Cooper, J. M., and J. A. Allen. 1994. Selection by wild birds on artificial dimorphic prey on varied backgrounds. *Biol J Linn Soc* 51:433–46.

Cooper, W. E. Jr. 1998a. Conditions favoring anticipatory and reactive displays deflecting predatory attack. *Behav Ecol* 9:598–604.

———. 1998b. Reactive and anticipatory display to deflect predatory attack to an autotomous lizard tail. *Can J Zool* 76:1507–10.

———. 2000. Pursuit deterrence in lizards. *Saudi J Bio Sci* 7:15–29.

———. 2001. Multiple roles of tail display by the curly-tailed lizard *Leiocephalus carinatus*: Pursuit deterrent and deflective roles of a social signal. *Ethology* 107:1137–49.

Cooper, W. E. Jr., and L. J. Vitt. 1985. Blue tails and autotomy: Enhancement of predation avoidance in juvenile skinks. *Z Tierpsychol* 70:265–76.

Coppinger, R. 1970. The effect of experience and novelty on avian feeding behavior with references to the evolution of warning coloration in butterflies. II. Reactions of naïve birds to novel insects. *Am Nat* 104:323–35.

Cords, M. 1990. Vigilance and mixed-species association of some East African forest monkeys. *Behav Ecol Sociobiol* 26:297–300.

———. 1995. Predator vigilance costs of allogrooming in wild blue monkeys. *Behaviour* 132:559–69.

Coss, R. G. 1991. Context and animal behavior III: The relationship between early development and evolutionary persistence of ground squirrel antisnake behavior. *Ecol Psychol* 3:277–315.

———. 1999. Effects of relaxed natural selection on the evolution of behavior. In *Geographic variation in behavior: Perspectives on evolutionary mechanisms*, ed. S. A. Foster and J. A. Endler, 180–208. New York: Oxford University Press, New York.

Coss, R. G., and R. O. Goldthwaite. 1995. The persistence of old designs for perception. In *Perspectives in ethology*, vol. 11, ed. N. S. Thompson, 83–148. New York: Plenum Press.

Coss, R. G., K. L. Guse, N. S. Poran, and D. G. Smith. 1993. Development of antisnake defenses in California ground squirrels (*Spermophilus beecheyi*): II. Microevolutionary effects of relaxed selection from rattlesnakes. *Behaviour* 124:137–64.

Coss, R. G., and D. H. Owings. 1978. Snake-directed behavior by snake naïve and experienced California ground squirrels in a simulated burrow. *Z Tierpsychol* 48:421–35.

———. 1985. Restraints on ground squirrel antipredator behavior: Adjustments over multiple time scales. In *Issues in the ecological study of learning*, ed. T. D. Johnston and A. T. Pietrewicz, 167–200. Hillsdale, NJ: Lawrence Erlbaum Associates.

Coss, R. G., and U. Ramakrishnan. 2000. Perceptual aspects of leopard recognition by wild bonnet macaques (*Macaca radiata*). *Behaviour* 137:315–35.

Cote, S. D., A. Peracino, and G. Simard. 1997. Wolf, *Canis lupus*, predation and maternal defensive behavior in mountain goats, *Oreamnos americanus*. *Can Field-Nat* 111:389–92.

Cott, H. B. 1940. *Adaptive coloration in animals*. London: Metheun & Co.

———. 1946/7. The edibility of birds: Illustrated by five years' experiments and observations (1941–1946) on the food preferences of the hornet, cat and man: and considered with special reference to the theories of adaptive coloration. *Proc Zool Soc Lond* 116:371–524.

Cott, H. B., and C. W. Benson. 1970. The palatability of birds, mainly based upon observations of a tasting panel in Zambia. *Ostrich Suppl* 8:357–84.

Coulson, J. C. 1968. Differences in the quality of birds nesting in the center and on the edges of a colony. *Nature* 217:478–79.

Cowlishaw, G. 1994. Vulnerability to predation in baboon populations. *Behaviour* 131:293–304.

———. 1997a. Trade-offs between foraging and predation risk determine habitat use in a desert baboon population. *Anim Behav* 53:667–86.

———. 1997b. Refuge use and predation risk in a desert baboon population. *Anim Behav* 54:241–53.

———. 1998. The role of vigilance in the survival and reproductive strategies of desert baboons. *Behaviour* 135:431–52.

Cowlishaw, G., M. J. Lawes, M. Lightbody, A. Martin, R. Pettifor, and J. M. Rowcliffe. 2004. A simple rule for the costs of vigilance: Empirical evidence from a social forager. *Proc R Soc Lond Ser B Biol Sci* 271:27–33.

Crabtree, R. L., L. S. Broome, and M. L. Wolfe. 1989. Effects of habitat characteristics on gadwall nest predation and nest-site selection. *J Wildl Manag* 53:129–37.

Craig, J. L. 1982. On the evidence for a "pursuit deterrent" function of alarm signals of swamphens. *Am Nat* 119:753–55.

Creel, S., and N. M. Creel. 2002. *The African wild dog: Behavior, ecology and conservation*. Princeton, NJ: Princeton University Press.

Creel, S., G. Spong, and N. Creel. 2001. Interspecific competition and the population biology of extinction-prone carnivores. In *Conservation of carnivores*, ed. J. L. Gittleman, S. M. Funk, D. Macdonald, and R. K. Wayne, 35–60. Cambridge: Cambridge University Press.

Cresswell, W. 1993. Escape response by redshanks, *Tringa totanus*, on attack by avian predators. *Anim Behav* 46:609–11.

———. 1994a. Flocking is an effective anti-predation strategy in redshanks, *Tringa totanus*. *Anim Behav* 47:433–42.

———. 1994b. The function of alarm calls in redshanks, *Tringa totanus*. *Anim Behav* 47:736–38.

———. 1994c. Song as a pursuit-deterrent signal, and its occurrence relative to other anti-predation behaviours of skylark (*Alauda arvensis*) on attack by merlins (*Falco columbarius*). *Behav Ecol Sociobiol* 34:217–23.

———. 1996. Surprise as a winter hunting strategy in sparrowhawks *Accipter nisus*, peregrines *Falco peregrinus* and merlins *F. columbarius*. *Ibis* 138:684–92.

———. 1997a. Nest predation: The relative effects of nest characteristics, clutch size and parental behaviour. *Anim Behav* 53:93–103.

———. 1997b. Nest predation rates and nest detectability in different stages of breeding in blackbirds *Turdus merula*. *J Avian Biol* 28:296–302.

———. 2003. Testing the mass-dependent predation hypothesis: In European blackbirds poor foragers have higher overwinter body reserves. *Anim Behav* 65:1035–44.

Cresswell, W., G. M. Hilton, and G. D. Ruxton. 2000. Evidence for a rule governing the avoidance of superfluous escape flights. *Proc R Soc Lond Ser B Biol Sci* 267:733–37.

Cresswell, W., J. Lind, U. Kaby, J. L. Quinn, and S. Jakobsson. 2003. Does an opportunistic predator preferentially attack nonvigilant prey? *Anim Behav* 66:643–48.

Cresswell, W., and J. L. Quinn. 2004. Faced with a choice, sparrowhawks more often attack the more vulnerable prey group. *Oikos* 104:71–76.

Cresswell, W., J. L. Quinn, M. J. Whittingham, and S. Butler. 2003. Good foragers can also be good at detecting predators. *Proc R Soc Lond Ser B Biol Sci* 270:1069–76.

Crook, J. H. 1965. The adaptive significance of avian social organization. *Symp Zool Soc Lond* 14:181–218.

Crook, J. H., and J. S. Gartlan. 1966. Evolution of primate societies. *Nature* 210:1200–1203.

Croze, H. 1970. Searching image in carrion crows. *Z Tierpsychol* Suppl 5:1–86.

Cullen, A. 1969. *Window onto wilderness*. Nairobi: East African Publishing House.

Cullen, J. M. 1960. Some adaptations in the nesting behaviour of terns. *Proc Int Ornithol Congr* 12:153–57.

Curio, E. 1970. Validity of the selective coefficient of a behaviour trait in hawkmoth larvae. *Nature* 228:382.

———. 1975. The functional organization of anti-predator behaviour in the pied flycatcher: A study of avian visual perception. *Anim Behav* 23:1–115.

———. 1976. *The ethology of predation*. New York: Springer-Verlag.

———. 1978. The adaptive significance of avian mobbing: I. Teleonomic hypotheses and predictions. *Z Tierpsychol* 48:175–83.

———. 1980. An unknown determinant of sex-specific altruism. *Z Tierpsychol* 53:139–52.

———. 1988a. Cultural transmission of enemy recognition by birds. In *Social learning: Psychological and biological perspectives*, ed. T. R. Zentall and B. G. Galef Jr., 75–97. Hillsdale, NJ: Lawrence Erlbaum Associates.

———. 1988b. Relative realized lifespan and delayed cost of parental care. *Am Nat* 131:825–36.

———. 1993. Proximate and developmental aspects of antipredator behavior. *Adv Study Behav* 22:135–238.

Curio, E., U. Ernst, and W. Vieth. 1978a. Cultural transmission of enemy recognition: One function of mobbing. *Science* 202:899–901.

———. 1978b. The adaptive significance of avian mobbing. II. Cultural transmission of enemy recognition in blackbirds: effectiveness and some constraints. *Z Tierpsychol* 48:184–202.

Curio, E., G. Klump, and K. Regelmann. 1983. An anti-predator response in the great tit (*Parus major*): Is it tuned to predator risk? *Oecologia* 60:83–88.

Curio, E., and H. Onnebrink. 1995. Brood defense and brood size in the great tit (*Parus major*): A test of a model of unshared parental investment. *Behav Ecol* 6:235–41.

Curio, E., and K. Regelmann. 1985. The behavioural dynamics of great tits (*Parus major*) approaching a predator. *Z Tierpsychol* 69:3–18.

———. 1986. Predator harassment implies a real deadly risk: a reply to Hennessy. *Ethology* 72:75–78.

———. 1987. Do great tit *Parus major* parents gear their brood defence to the quality of their young? *Ibis* 129:344–52.

Curio, E., K. Regelmann, and U. Zimmermann. 1984. The defence of first and second broods by great tit (*Parus major*) parents: A test of predictive sociobiology. *Z Tierpsychol* 66:101–27.

———. 1985. Brood defence in the great tit (*Parus major*): The influence of life-history and habitat. *Behav Ecol Sociobiol* 16:273–83.

Cushing, B. S. 1985. Estrous mice and vulnerability to weasel predation. *Ecology* 66:1976–78.

Cuthill, I. C., and A. I. Houston. 1997. Managing time and energy. In *Behavioural ecology*, ed. J. R. Krebs, and N. B. Davies, 4th ed., 97–120. Oxford: Blackwell.

Cuthill, I. C., M. Stevens, T. Maddocks, C. A. Parraga, and T. S. Troscianko. 2005. Disruptive coloration and background pattern matching. *Nature* 434:72–74.

Cuyler, W. K. 1924. Observations on the habits of the striped skunk (*Mephitis mesomelas varians*). *J Mammal* 5:180–89.

Daan, S., and J. Tinbergen. 1979. Young guillemots (*Uria lomvia*) leaving their arctic breeding cliffs: A daily rhythm in numbers and risk. *Ardea* 67:96–100.

D'Agostino, G. M., L. E. Giovinazzo, and S. W. Eaton 1981. The sentinel crow as an extension of parental care. *Wilson. Bull* 93:394–95.

Dahlgren, J. 1990. Females choose vigilant males: An experiment with monogamous grey partridge, *Perdix perdix*. *Anim Behav* 39:646–51.

Dale, S., R. Gustavsen, and T. Slagsvold. 1996. Risk taking during parental care: A test of three hypotheses applied to the pied flycatcher. *Behav Ecol Sociobiol* 39:31–42.

Dall, S. R. X., B. P. Kotler, and A. Bouskila. 2001. Attention, "apprehension" and gerbils searching in patches. *Ann Zool Fen* 38:15–23.

Daly, M., P. R. Behrends, M. I. Wilson, and L. F. Jacobs. 1992. Behavioural modulation of predation risk: Moonlight avoidance and crepuscular compensation in a nocturnal desert rodent, *Dipodomys merriami*. *Anim Behav* 44:1–9.

Daly, M., M. Wilson, P. R. Behrends, and L. F. Jacobs. 1990. Characteristics of kangaroo rats, *Dipodomys merriami*, associated with differential predation risk. *Anim Behav* 40:380–89.

Damme, R. van, and T. J. M. van Dooren. 1999. Absolute versus per unit body length speed of prey as an estimator of vulnerability to predation. *Anim Behav* 57:347–52.

Daniel, J. C., and D. T. Blumstein. 1998. A test of the acoustic adaptation hypothesis in four species of marmots. *Anim Behav* 56:1517–28.

Darling, F. F. 1938. *Bird flocks and the breeding cycle, a contribution to the study of avian sociality.* Cambridge: Cambridge University Press.

Darst, C. R., P. A. Menendez-Guerrero, L. A. Coloma, and D. C. Cannatella. 2005. Evolution of dietary specialization and chemical defense in poison frogs (Dendrobatidae): A comparative analysis. *Am Nat* 165:56–69.

Darwin, C. 1871. *The descent of man and selection in relation to sex.* 2nd ed. London: John Murray.

Davenport, C. M. 1908. Elimination of self-coloured birds. *Nature* 78:101.

Davidson, G. W. H. 1985. Avian spurs. *J Zool* 206:353–66.

Davies, N. B. 2000. *Cuckoos, cowbirds and other cheats.* London: T&AD Poyser.

Davis, L. S. 1984. Alarm calling in Richardson's ground squirrels (*Spermophilus richardsonii*). *Z Tierpsychol* 66:152–64.

Davison, W. B., and E. Bollinger. 2000. Predation rates on real and artificial nests of grassland birds. *Auk* 117:147–53.

Dawkins, M. 1971. Perceptual changes in chicks: Another look at the "search image" concept. *Anim Behav* 19:566–74.

Dawkins, M. S., and T. Guilford. 1995. An exaggerated preference for simple neural network models of signal evolution. *Proc R Soc Lond Ser B Biol Sci* 261:357–60.

Dawkins, R. 1976. *The selfish gene.* Oxford: Oxford University Press.

———. 1979. Twelve misunderstandings of kin selection. *Z Tierpsychol* 51:184–200.

Dawkins, R., and T. R. Carlisle. 1976. Parental investment, mate desertion and a fallacy. *Nature* 262:131–33.

Dawson, T. J., and C. R. Taylor. 1973. Energetic cost of locomotion in kangaroos. *Nature* 246:313–14.

Dayan, T., and D. Simberloff. 1996. Patterns of size separation in carnivore communities. In *Carnivore behavior, ecology and evolution, II*, ed. J. L. Gittleman, 243–66. Ithaca, NY: Cornell University Press.

Dayan, T., D. Simberloff, E. Tchernov, and Y. Yom-Tov. 1990. Feline canines: Community-wide character displacement among the small cats of Israel. *Am Nat* 136:39–60.

Deecke, V. B., J. K. B. Ford, and P. J. B. Slater. 2005. The vocal behaviour of mammal-eating killer whales: Communicating with costly calls. *Anim Behav* 69:395–405.

Deecke, V. B., P. J. B. Slater, and J. K. B. Ford. 2002. Selective habituation shapes acoustic predator recognition in harbour seals. *Nature* 420:171–73.

Dehn, M. M. 1990. Vigilance for predators: Detection and dilution effects. *Behav Ecol Sociobiol* 26:337–42.

Denson, R. D. 1979. Owl predation on a mobbing crow. *Wilson Bull* 91:133.

Despland, E., and S. J. Simpson. 2005. Food choices of solitarious and gregarious locusts reflect cryptic and aposematic antipredator strategies. *Anim Behav* 69:471–79.

Desrochers, A., M. Belisle, and J. Bourque. 2002. Do mobbing calls affect the perception of predation risk by forest birds? *Anim Behav* 64:709–14.

Dewitt, T. J., A. Sih, and J. A. Hucko. 1999. Trait compensation and cospecialization in a freshwater snail: Size, shape and antipredator behaviour. *Anim Behav* 58:397–407.

Dial, B. E. 1983. Lizard tail autotomy: Function and energetics of postautotomy tail movement in *Scincella lateralis*. *Science* 219:391–93.

———. 1986. Tail display in two species of iguanid lizards: A test of the "predator signal" hypothesis. *Am Nat* 127:103–11.

Dial, B. E., and L. C. FitzPatrick. 1983. Lizard tail autotomy: Function and energetics of postautotomy tail movement in *Scincella lateralis*. *Science* 219:391–93.

Diamond, J. M. 1982. Mimicry of friarbirds by orioles. *Auk* 99:187–96.

Diaz, M. 1992. Rodent seed predation in cereal crop areas of central Spain: Effects of physiognomy, food availability, and predation risk. *Ecography* 15:77–85.

Dice, L. R. 1947. Effectiveness of selection by owls of deermice (*Peromyscus maniculatus*) which contrast in color with their background. *Contrib Lab Vertebr Biol Univ Mich* 34:1–20.

Dice, L. R., and P. M. Blossom. 1937. *Studies of mammalian ecology in southwestern North America with special attention to colors of desert mammals.* Washington, DC: Carnegie Institution.

Dickman, C. R. 1992. Predation and habitat shift in the house mouse, *Mus domesticus*. *Ecology* 73:313–22.

Dickman, C. R., and C. P. Doncaster. 1984. Responses of small mammals to red fox (*Vulpes vulpes*) odour. *J Zool* 204:521–31.

Dijak, W. D., and F. R. Thompson III. 2000. Landscape and edge effects on the distribution of mammalian predators in Missouri. *J Wildl Manag* 64:209–16.

Dill, L. M. 1987. Animal decision making and its ecological consequences: The future of aquatic ecology and behaviour. *Can J Zool* 65:803–11.

Dill, L. M., and A. H. G. Fraser. 1984. Risk of predation and the feeding behavior of juvenile coho salmon (*Oncorhynchus kisutch*). *Behav Ecol Sociobiol* 16:65–71.

Dill, L. M., and R. Houtman. 1989. The influence of distance to refuge on flight initiation distance in the gray squirrel (*Sciurus carolinensis*). *Can J Zool* 67:233–35.

Dion, N., K. A. Hobson, and S. Lariviere. 2000. Interactive effects of vegetation and predators on the success of natural and simulated nests of grassland songbirds. *Condor* 102:629–34.

Djawdan, M. 1993. Locomotor performance of bipedal and quadrupedal heteromyid rodents. *Funct Ecol* 7:195–202.

Djawdan, M., and T. Garland Jr. 1988. Maximal running speeds of bipedal and quadrupedal rodents. *J Mammal* 69:765–72.

Dodd, C. K., and E. D. Brodie Jr. 1976. Defensive mechanisms of neotropical salamanders with an experimental analysis of immobility and the effect of temperature on immobility. *Herpetologia* 32:269–90.

Dolby, A. S., and T. C. Grubb Jr. 1998. Benefits to satellite members in mixed-species foraging groups: An experimental analysis. *Anim Behav* 56:501–09.

———. 2000. Social context affects risk taking by a satellite species in a mixed-species foraging group. *Behav Ecol* 11:110–14.

Doligez, B., E. Danchin, and J. Clobert. 2002. Public information and breeding habitat selection in a wild bird population. *Science* 297:1168–70.

Donovan, T. M., P. W. Jones, E. M. Annand, and F. R. Thompson III. 1997. Variation in local-scale edge effects: Mechanisms and landscape context. *Ecology* 78:2064–75.

Doucet, G. J., and J. R. Bider. 1969. Activity of *Microtus pennsylvanicus* related to moon phase and moonlight revealed by the sand transect technique. *Can J Zool* 47:1183–86.

Dow, H., and S. Fredga. 1983. Breeding and natal dispersal of the goldeneye, *Bucephala clangula*. *J Anim Ecol* 52: 681–95.

Driver, P. M., and D. A. Humphries. 1969. The significance of the high-intensity alarm call in captured passerines. *Ibis* 111:243–44.

Dubost, G., and J-P Gasc. 1987. The process of total tail autotomy in the South-American rodent, *Proechimys*. *J Zool (Lond)* 212:563–72.

Duebbert, H. F., and H. A. Kantrud. 1974. Upland duck nesting related to land use and predator reduction. *J Wildl Manag* 38:257–65.

Duebbert, H. F., and J. T. Lokemoen. 1976. Duck nesting in fields of undisturbed grass-legume cover. *J Wildl Manag* 40:39–49.

Duebbert, H. F., J. T. Lokemoen, and D. E. Sharp 1983. Concentrated nesting of mallards and gadwalls on Miller Lake Island, North Dakota. *J Wildl Manag* 47:729–40.

Duffey, E., N. Creasey, and K. Williamson. 1950. The rodent-run distraction behaviour of certain waders. *Ibis* 92:27–33.

Dugatkin, L. A., and J-G. J. Godin. 1992. Prey approaching predators: A cost-benefit perspective. *Ann Zool Fen* 29:233–52.

Dukas, R., and C. W. Clark. 1995. Sustained vigilance and animal performance. *Anim Behav* 49:1259–67.

Dukas, R., and A. C. Kamil. 2000. The cost of limited attention in blue jays. *Behav Ecol* 11:502–6.

Dumbacher, J. P., B. M. Beehler, T. F. Spande, H. M. Garraffo, and J. W. Daly. 1992. Homobatrachotoxin in the genus *Pitohui*: Chemical defense in birds? *Science* 258:799–801.

Dumbacher, J. P., and R. C. Fleischer. 2001. Phylogenetic evidence for colour pattern convergence in toxic pitohuis: Mullerian mimicry in birds? *Proc R Soc Lond Ser B Biol Sci* 268:1971–76.

Dumbacher, J. P., and S. Pruett-Jones. 1996. Avian chemical defense. *Curr Ornithol* 13, ed. V. Nolan Jr. and E. D. Ketterson, 137–74.

Dumbacher, J. P., T. F. Spande, and J. W. Daly. 2000. Batrachotoxin alkaloids from passerine birds: A second toxic bird genus (*Ifrita kowaldi*) from New Guinea. *Proc Natl Acad Sci USA* 97:12970–75.

Dunbar, R. I. M. 1988. *Primate social systems*. London: Croom Helm.

Dunbar, R. I. M., and E. P. Dunbar. 1974. Social organization and ecology of the klipspringer (*Oreotragus oreotragus*) in Ethiopia. *Z Tierpsychol* 35:481–93.

Dunford, C. 1977. Kin selection for ground squirrel alarm calls. *Am Nat* 111:782–85.

Dunn, E. 1977. Predation by weasels (*Mustela nivalis*) on breeding tits (*Parus* spp.) in relation to the density of tits and rodents. *J Anim Ecol* 46:634–52.

Durant, S. M. 1998. Competition refuges and coexistence: An example from Serengeti carnivores. *J Anim Ecol* 67:370–86.

———. 2000a. Predator avoidance, breeding experience and reproductive success in endangered cheetahs, *Acinonyx jubatus*. *Anim Behav* 60:121–30.

———. 2000b. Living with the enemy: Avoidance of hyenas and lions by cheetahs in the Serengeti. *Behav Ecol* 11:624–32.

Dyawdan, M. 1993. Locomotor performance of bipedal and quadrupedal heteromyid rodents. *Funct Ecol* 7:195–202.

Dyrcz, A., J. Witkowski, and J. Okulewicz. 1981. Nesting of "timid" waders in the vicinity of "bold" ones as an antipredator adaptation. *Ibis* 123:542–45.

Eadie, W. R. 1938. The dermal glands of shrews. *J Mammal* 19:171–74.

Eason, P. 1989. Harpy eagle attempts predation on adult howler monkey. *Condor* 91:469–70.

East, M. 1981. Alarm calling and parental investment in the robin *Erithacus rubecula. Ibis* 123:223–30.

East, M., H. Hofer, and A. Turk. 1989. Functions of birth dens in spotted hyaenas (*Crocuta crocuta*). *J Zool* 219:690–97.

Eaton, R. L. 1976. A possible case of mimicry in larger mammals. *Evolution* 30:853–56.

Eckardt, W., and K. Zuberbuhler. 2004. Cooperation and competition in two forest monkeys. *Behav Ecol* 15:400–11.

Edmunds, M. 1974. *Defence in animals.* New York: Longman.

Edmunds, M., and R. A. Dewhirst. 1994. The survival value of countershading with wild birds as predators. *Biol J Linn Soc* 51:447–52.

Edstrom, A. 1992. *Venomous and poisonous animals.* Malabar, FL: Kreiger Publishing.

Edwards, J. 1983. Diet shifts in moose due to predator avoidance. *Oecologia* 60:185–89.

Eggers, S., M. Griesser, and J. Ekman. 2005. Predator-induced plasticity in nest visitation rates in the Siberian jay (*Perisoreus infaustus*). *Behav Ecol* 16:309–15.

Ehrlich, P. R., and A. H. Ehrlich. 1973. Coevolution: Heterotypic schooling in Caribbean reef fishes. *Am Nat* 107:157–60.

Eilam, D., T. Dayan, S. Ben-Eliyahu, I. Schulman, G. Shefer, and C. A. Hendrie. 1999. Differential behavioural and hormonal responses of voles and spiny mice to owl calls. *Anim Behav* 58:1085–93.

Eisenberg, J. F., and E. Gould. 1970. *The Tenrecs: A study in mammalian behavior and evolution.* Smithsonian Contributions to Zoology 27. Washington, DC: Smithsonian Institution Press.

Eisner, T., J. Conner, J. E. Carrel, J. P. McCormick, A. J. Slagle, C. Gans, and J. C. O'Reilly. 1990. Systemic retention of ingested cantharidin by frogs. *Chemoecology* 1:57–62.

Ekman, J. 1987. Exposure and time use in willow tit flocks: The cost of subordination. *Anim Behav* 35:445–52.

Ekman, J. B., and K. Lilliendahl. 1993. Using priority to food access: Fattening strategies in dominance-structured willow tit (*Parus montanus*) flocks. *Behav Ecol* 4:232–38.

Elcavage, P., and T. Caraco. 1983. Vigilance behaviour in house sparrow flocks. *Anim Behav* 31:303–4.

Elder, J. B. 1956. Watering patterns of some desert game animals. *J Wildl Manag* 20:368–78.

Elgar, M. A. 1986. House sparrows establish foraging flocks by giving chirrup calls if resources are divisible. *Anim Behav* 34:169–74.

———. 1987. Food intake and resource availability: Flocking decisions in house sparrows. *Anim Behav* 35:1168–76.

———. 1989. Predator vigilance and group size in mammals and birds: A critical review of the empirical evidence. *Biol Rev* 64:13–33.

Elgar, M. A., P. J. Burren, and M. Posen. 1984. Vigilance and perception of flock size in foraging house sparrows (*Passer domesticus* L.). *Behaviour* 90:215–23.

Elgar, M. A., and C. P. Catterall. 1981. Flocking and predator surveillance in house sparrows: Test of an hypothesis. *Anim Behav* 29:868–72.

Elgar, M. A., H. McKay, and P. Woon. 1986. Scanning, pecking and alarm flights in house sparrows (*Passer domesticus* L.) *Anim Behav* 34:1892–94.

Elliot, R. D. 1985a. The exclusion of avian predators from aggregations of nesting lapwings (*Vanellus vanellus*). *Anim Behav* 33:308–14.

————. 1985b. The effects of predation risk and group size on the anti-predator responses of nesting lapwings *Vanellus vanellus*. *Behaviour* 92:168–87.

Emerson, S. B., H. W. Greene, and E. L. Charnov. 1994. Allometric aspects of predator-prey interactions. In *Ecological morphology: Integrative organismal biology*, ed. P. C. Wainwright and S. M. Reilly, 123–39. Chicago: University of Chicago Press.

Emlen, J. T. Jr., D. E. Miller, R. M. Evans, and D. H. Thompson. 1966. Predator-induced parental neglect in a ring-billed gull colony. *Auk* 83:677–79.

Emmons, L. H. 1978. Sound communication among African rainforest squirrels. *Z Tierpsychol* 47:1–49.

————. 1983. A field study of the African brush-tailed porcupines (*Atherus africanus*), by radiotelemetry. *Mammalia* 47:183–94.

Emmons, L. H., and A. H. Gentry. 1983. Tropical forest structure and the distribution of gliding and prehensile-tailed vertebrates. *Am Nat* 121:513–24.

Emslie, S. D., N. Karnovsky, and W. Trivelpiece. 1995. Avian predation at penguin colonies on King George Island, Antarctica. *Wilson Bull* 107:317–27.

Endler, J. A. 1978. A predator's view of animal color patterns. *Evol Biol* 11:319–64.

————. 1980. Natural selection on color patterns in *Poecilia reticulata*. *Evolution* 34:76–91.

————. 1981. An overview of the relationships between mimicry and crypsis. *Biol J Linn Soc* 16:25–31.

————. 1983. Natural and sexual selection on colour patterns in Poeciliid fishes. *Environ Biol Fishes* 9:173–90.

————. 1984. Progressive background matching in moths, and a quantitative measure of crypsis. *Biol J Linn Soc* 22:187–231.

————. 1986. Defense against predators. In *Predator-prey relationships*, ed. M. E. Feder and G. V. Lauder, 109–34. Chicago: University of Chicago Press.

————. 1987. Predation, light intensity and courtship behaviour in *Poecilia reticulata* (*Pisces: Poeciliidae*). *Anim Behav* 35:1376–85.

————. 1988. Frequency-dependent predation, cypsis and aposematic coloration. *Philos Trans R Soc Lond B Biol Sci* 319:505–23.

————. 1990. On the measurement and classification of colour in studies of animal colour patterns. *Biol J Linn Soc* 41:315–52.

————. 1991a. Interactions between predators and prey. In *Behavioural ecology: An evolutionary approach*, ed. J. R. Krebs and N. B. Davies, 169–96. Oxford: Blackwell Scientific Publications.

————. 1991b. Variation in the appearance of guppy colour patterns to guppies and their predators under different visual conditions. *Vision Res* 31:587–608.

————. 1992. Signals, signal conditions, and the direction of evolution. *Am Nat* 139:S125–S153.

Endler, J. A., and J. Mappes. 2004. Predator mixes and the consequences of aposematic signals. *Am Nat* 163:532–47.

Endler, J. A., and M. Thery. 1996. Interacting effects of lek placement display behaviour, ambient light and colour patterns in three neotropical forest-dwelling birds. *Am Nat* 148:421–52.

Engen, S., T. Jarvi, and C. Wiklund. 1986. The evolution of aposematic coloration by individual selection: A life-span survival model. *Oikos* 46:397–403.

Epple, G., J. R. Mason, D. L. Nolte, and D. L. Campbell. 1993. Effects of predator odors on feeding in the mountain beaver (*Aplodontia rufa*). *J Mammal* 74:715–22.

Eriksson, M. O., and F. Gotmark. 1982. Habitat selection: Do passerines nest in association with lapwings Vanellus vanellus as defence against predators? *Ornis Scand* 13:189–92.

Erpino, M. J. 1968. Nest-related activities of black-billed magpies. *Condor* 70:154–65.

Erwin, R. M. 1979. Species interactions in a mixed colony of common terns (*Sterna hirundo*) and black skimmers (*Rynchops niger*). *Anim Behav* 27:1054–62.

Escalante, P., and J. W. Daly. 1994. Alkaloids in extracts of feathers of the red warbler. *J Ornithol* 135:410.

Eshel, I. 1978. On a prey-predator nonzero-sum game and the evolution of gregarious behavior of evasive prey. *Am Nat* 112:787–95.

Esler, D., and J. B. Grand. 1993. Factors influencing depredation of artificial duck nests. *J Wildl Manag* 57:597–601.

Estes, R. D. 1974. Social organization of the African bovidae. In *The behaviour of ungulates and its relationship to management*, ed. V. Geist and F. Walther, 166–205. Morges, Switzerland: IUCN.

———. 1976. The significance of breeding synchrony in the wildebeest. *E Afr Wildl J* 14:135–52.

———. 1991. *The behavior guide to African mammals*. Berkeley and Los Angeles: University of California Press.

Estes, R. D., and R. K. Estes. 1979. The birth and survival of wildebeest calves. *Z Tierpsychol* 50:45–95.

Estes, R. D., and J. Goddard. 1967. Prey selection and hunting behavior of the African wild dog. *J Wildl Manag* 31:52–70.

Evans, C. S. 1997. Referential signals. *Perspectives in Ethology* 12:99–143.

Evans, C. S., L. Evans, and P. Marler 1993. On the meaning of alarm calls: Functional reference in an avian vocal system. *Anim Behav* 46:23–38.

Evans, C. S., J. M. Macedonia, and P. Marler. 1993. Effects of apparent size and speed on the response of chickens, *Gallus gallus*, to computer-generated simulations of aerial predators. *Anim Behav* 46:1–11.

Evans, C. S., and P. Marler. 1995. Language and animal communication: Parallels and contrasts. In *Comparative approaches to cognitive science. Complex adaptive systems*, ed. H. L. Roitblat and J. A. Meyer, 341–82. Cambridge, MA: MIT Press.

Evans, R. M. 1970. Oldsquaws nesting in association with arctic terns at Churchill, Manitoba. *Wilson Bull* 82:383–90.

Ewell, A. H. 1981. Tonic immobility as a predator defense in the rabbit *Oryctolagus cuniculus*. *Behav Neural Biol* 31:483–89.

Ewer, R. F. 1966. Juvenile behaviour in the African ground squirrel, *Xerus erythropus* (E. Geoff.). *Z Tierpsychol* 23:190–216.

———. 1968. *Ethology of mammals*. London: Elek Science.

Farren, W. 1908. The crouching habit of the stone curlew. *Br Birds* 1:301–8.

Feekes, F. 1981. Biology and colonial organization of two sympatric caciques, *Cacicus c. cela* and *Cacicus h. haemorrhous* (Icteridae, Aves) in Suriname. *Auk* 69:83–107.

Fenn, M. G. P., and D. W. Macdonald. 1995. Use of middens by red foxes: Risk reverses rhythms of rats. *J Mammal* 76:130–36.

Fenton, M. B., N. G. H. Boyle, T. M. Harrison, and D. J. Oxley. 1977. Activity patterns, habitat use, and prey selection by some African insectivorous bats. *Biotropica* 9:73–85.

Fenton, M. B., I. L. Rautenbach, S. E. Smith, C. M. Swanepoel, J. Grosell, and J. van Jaarsveld. 1994. Raptors and bats: Threats and opportunities. *Anim Behav* 48:9–18.

Ferguson, J. W. H. 1987. Vigilance behaviour in white-browed sparrow-weavers *Plocepasser mahali*. *Ethology* 76:223–35.

Ferguson, S. H., A. T. Bergerud, and R. Ferguson. 1988. Predation risk and habitat selection in the persistence of a remnant caribou population. *Oecologia* 76:236–45.

Fernandez, G. J., A. F. Capurro, and J. C. Reboreda. 2003. Effect of group size on individual and collective vigilance in greater rheas. *Ethology* 109:413–25.

Fernandez-Juricic, E., J. T. Erichsen, and A. Kacelnik. 2004. Visual perception and social foraging in birds. *Trends Ecol Evol* 19:25–31.

Fernandez-Juricic, E., M. D. Jiminez, and E. Lucas. 2002. Factors affecting intra- and inter-specific variations in the difference between alert distances and flight distances for birds in forested habitats. *Can J Zool* 80:1212–20.

Fernandez-Juricic, E., B. Kerr, P. A. Bednekoff, and D. W. Stephens. 2004. When are two heads better than one? Visual perception and information transfer affect vigilance coordination in foraging groups. *Behav Ecol* 15:898–906.

Fernandez-Juricic, E., S. Siller, and A. Kacelnik. 2004. Flock density, social foraging, and scanning: An experiment with starlings. *Behav Ecol* 15:371–79.

Fernandez-Juricic, E., R. Smith, and A. Kacelnik. 2005. Increasing the costs of conspecific scanning in socially foraging starlings affects vigilance and foraging behaviour. *Anim Behav* 69:73–81.

Ferrer, M., L. Garcia, and R. Cadenas. 1990. Long-term changes in nest defence intensity of the Spanish imperial eagle. *Ardea* 78:395–98.

Festa-Bianchet, M. 1988. Seasonal range selection in bighorn sheep: Conflicts between forage quality, forage quantity, and predator avoidance. *Oecologia* 75:580–86.

Fichtel, C. 2004. Reciprocal recognition of sifaka (*Propithecus verreauxi verreauxi*) and redfronted lemur (*Eulemur fulvus rufus*) alarm calls. *Anim Cogn* 7:45–52.

Fichtel, C., and K. Hammerschmidt. 2002. Responses of redfronted lemurs to experimentally modified alarm calls: Evidence for urgency-based changes in call structure. *Ethology* 108:763–77.

———. 2003. Responses of squirrel monkeys to their experimentally modified mobbing calls. *J Acoust Soc Am* 113:2927–32.

Fichtel, C., and P. M. Kappeler. 2002. Anti-predator behavior of group-living Malagasy primates: Mixed evidence for a referential alarm call system. *Behav Ecol Sociobiol* 51:262–75.

Ficken, M. S. 1990. Acoustic characteristics of alarm calls associated with predation risk in chickadees. *Anim Behav* 39:400–401.

———. 2000. Call similarities among mixed species flock associates. *Southwest Nat* 45:154–58.

Ficken, M. S., and J. Popp. 1996. A comparative analysis of passerine mobbing calls. *Auk* 113:370–80.

Ficken, R. W., P. E. Matthiae, and R. Horwich. 1971. Eye marks in vertebrates: Aids to vision. *Science* 173:936–38.

Filliater, T. S., R. Breitwisch, and P. M. Nealen. 1994. Predation on northern cardinal nests: Does choice of nest site matter? *Condor* 96:761–68.

Fischer, J. 1998. Barbary macaques categorize shrill barks into two call types. *Anim Behav* 55:799–807.

Fischer, J., K. Hammerschmidt, D. L. Cheney, and R. M. Seyfarth. 2001. Acoustic structure of female chacma baboon barks. *Ethology* 107:33–54.

Fisher, J. 1952. *The fulmar*. London: Collins.

Fisher, R. A. 1930. *The genetical theory of natural selection*. Oxford: Oxford University Press (Clarendon).

Fishman, M. A. 1999. Predator inspection: Closer approach as a way to improve assessment of potential threats. *J Theor Biol* 196:225–35.

Fitch, H. S. 1949. Study of snake populations in central California. *Am Midl Nat* 41:513–79.

FitzGibbon, C. D. 1989. A cost to individuals with reduced vigilance in groups of Thomson's gazelles hunted by cheetahs. *Anim Behav* 37:508–510.

———. 1990a. Anti-predator strategies of immature Thomson's gazelles hiding and the prone response. *Anim Behav* 40:846–55.

————. 1990b. Mixed-species grouping in Thomson's and Grant's gazelles: The antipredator benefits. *Anim Behav* 39:1116–26.

————. 1990c. Why do hunting cheetahs prefer male gazelles? *Anim Behav* 40:837–45.

————. 1993. Antipredator strategies of female Thomson's gazelles with hidden fawns. *J Mammal* 74:758–62.

————. 1994. The costs and benefits of predator inspection behaviour in Thomson's gazelles. *Behav Ecol Sociobiol* 34:139–48.

FitzGibbon, C. D., and J. H. Fanshawe. 1988. Stotting in Thomson's gazelles: An honest signal of condition. *Behav Ecol Sociobiol* 23:69–74.

————. 1989. The condition and age of Thomson's gazelles killed by cheetahs and wild dogs. *J Zool* 218:99–107.

FitzGibbon, C. D., and J. Lazarus. 1995. Anti-predator behaviour of Serengeti ungulates: Individual differences and population consequences. In *Serengeti II: Dynamics, management and conservation of an ecosystem*, ed. A. R. E. Sinclair and P. Arcese, 274–96. Chicago: University of Chicago Press.

Flasskamp, A. 1994. The adaptive significance of avian mobbing. V. An experimental test of the "move-on" hypothesis. *Ethology* 96:322–33.

Forbes, M. R. L., R. G. Clark, P. J. Weatherhead, and T. Armstrong. 1994. Risk-taking by female ducks: Intra- and interspecific tests of nest defense theory. *Behav Ecol Sociobiol* 34:79–85.

Ford, E. B. 1964. *Ecological genetics.* London: Metheun.

Formanowicz, D. R. Jr., E. D. Brodie Jr., and P. J. Bradley. 1990. Behavioural compensation for tail loss in the ground skink, *Scincella lateralis. Anim Behav* 40:782–84.

Forslund, P. 1993. Vigilance in relation to brood size and predator abundance in the barnacle goose, *Branta leucopsis. Anim Behav* 45:965–73.

Forsman, A., and S. Appelqvist. 1998. Visual predators impose correlational selection on prey color pattern and behavior. *Behav Ecol* 9:409–13.

Forsman, A., and J. Herrstrom. 2004. Asymmetry in size, shape, and color impairs the protective value of conspicuous color patterns. *Behav Ecol* 15:141–47.

Forsman, J. T., and M. Monkkonen. 2001. Responses by breeding birds to heterospecific song and mobbing call playbacks under varying predation risk. *Anim Behav* 62:1067–73.

Foster, W. A., and J. E. Treherne. 1981. Evidence for the dilution effect in the selfish herd from fish predation on a marine insect. *Nature* 293:466–67.

Fox, J. L., and G. P. Streveler. 1986. Wolf predation on mountain goats in southeastern Alaska. *J Mammal* 67:192–95.

Fragaszy, D. M. 1990. Sex and age differences in the organization of behaviour in wedge-capped capuchins, *Cebus olivaceus. Behav Ecol* 1:81–94.

Francis, A. M., J. P. Hailman, and G. E. Woolfenden. 1989. Mobbing by Florida scrub jays: Behaviour, sexual asymmetry, role of helpers and ontogeny. *Anim Behav* 38:795–816.

Francq, E. N. 1969. Behavioral aspects of feigned death in the opossum Didelphis marsupialis. *Am Midl Nat* 81:556–68.

Frankenberg, E. 1981. The adaptive significance of avian mobbing. IV. "Alerting others" and "perception advertisement' in blackbirds facing an owl. *Z Tierpyschol* 55:97–118.

Fransson, T., and T. P. Weber. 1997. Migratory fuelling in blackcaps (*Sylvia atricapilla*) under perceived risk of predation. *Behav Ecol Sociobiol* 41:75–80.

Frederick, P. C., and M. W. Collopy. 1989. The role of predation in determining reproductive success of colonially nesting wading birds in the Florida everglades. *Condor* 91:860–67.

Fretwell, S. D. 1972. *Populations in a seasonal environment.* Princeton, NJ: Princeton University Press.

Frid, A. 1997. Vigilance by female Dall's sheep: Interactions between predation risk factors. *Anim Behav* 53:799–808.

Frings, H., and J. Jumber. 1954. Preliminary studies on the use of a specific sound to repel starlings (*Sturnus vulgaris*) from objectionable roosts. *Science* 119:318–19.

Fryxell, J. M., J. Greever, and A. R. E. Sinclair. 1988. Why are migratory ungulates so abundant? *Am Nat* 131:781–98.

Fryxell, J. M., and A. R. E. Sinclair. 1988. Migration by large ungulates. *Trends Ecol Evol* 3:237–41.

Fuchs, E. 1977. Predation and anti-predator behaviour in a mixed colony of terns *Sterna* sp. and black-headed gulls *Larus ridibundus* with special reference to the sandwich tern *Sterna sandvicensis*. *Ornis Scand* 8:17–32.

Fulk, G. W. 1972. The effects of shrews on the space utilization of voles. *J Mammal* 53:461–78.

Gabrielsen, G. W., A. S. Blix, and H. Ursin. 1985. Orienting and freezing responses in incubating ptarmigan hens. *Physiol Behav* 34:925–34.

Gaddis, P. 1980. Mixed flocks, accepters, and antipredator behavior. *Condor* 82:348–49.

Gagliardo, A., and T. Guilford. 1993. Why do warning-coloured prey live gregariously? *Proc R Soc Lond Ser B Biol Sci* 251:69–74.

Gaioni, S. J., and C. S. Evans 1986. Perception of the frequency characteristics of distress calls by mallard ducklings (*Anas platyrhynchos*). *Behaviour* 99:250–74.

Galbraith, H. 1988. Effects of agriculture on the breeding ecology of lapwing *Vanellus vanellus*. *J Appl Ecol* 25:487–503.

Gallup, G. G. Jr., W. H. Cummings, and R. F. Nash. 1972. The experimenter as an independent variable in studies of animal hypnosis in chickens (*Gallus gallus*). *Anim Behav* 20:166–69.

Gallup, G. G. Jr., and J. D. Maser. 1977. Tonic immobility: Evolutionary underpinnings of human catalepsy and catatonia. In *Psychopathology: Experimental models*, ed. J. M. Maser and M. E. P. Seligman, 334–57. San Francisco: W. H. Freeman.

Galton, F. 1871. Gregariousness in cattle and men. *MacMillan's Mag* 23:353–57.

Gamberale, G., and B. S. Tullberg. 1996a. Evidence for a peak-shift in predator generalization among aposematic prey. *Proc R Soc Lond Ser B Biol Sci* 263:1329–34.

———. 1996b. Evidence for a more effective signal in aggregated aposematic prey. *Anim Behav* 52:597–601.

———. 1998. Aposematism and gregariousness: The combined effect of group size and coloration on signal repellence. *Proc R Soc Lond Ser B Biol Sci* 265:889–94.

Gamberale-Stille, G. 2000. Decision time and prey gregariousness influence attack probability in naïve and experienced predators. *Anim Behav* 60:95–99.

———. 2001. Benefit by contrast: An experiment with live aposematic prey. *Behav Ecol* 12:768–72.

Gamberale-Stille, G., and T. Guilford. 2003. Contrast versus colour in aposematic signals. *Anim Behav* 65:1021–26.

Gamberale-Stille, G., and B. S. Tullberg. 2001. Fruit or aposematic insect? Context-dependent colour preferences in domestic chicks. *Proc R Soc Lond Ser B Biol Sci* 268:2595–29.

Garcia, J. E., and F. Braza. 1993. Sleeping sites and lodge trees of the night monkey (Aotus azarae) in Bolivia. *Int J Primatol* 14:467–76.

Garland, T. Jr. 1983a. The relation between maximal running speed and body mass in terrestrial mammals. *J Zool* 199:157–70.

———. 1983b. Scaling the ecological cost of transport to body mass in terrestrial mammals. *Am Nat* 121:571–87.

———. 1985. Ontogenetic and individual variation in size, shape and speed in the Australian agamid lizard *Amphibolurus nuchalis*. *J Zool* 207:425–39.

Garland, T. Jr., F. Geiser, and R. V. Baudinette. 1988. Comparative locomotor performance of marsupial and placental mammals. *J Zool* 215:505–22.

Garland, T. Jr., and C. M. Janis. 1993. Does metatarsal/femur ratio predict maximal running speed in cursorial mammals? *J Zool* 229:133–51.

Garland, T. Jr., and J. B. Losos. 1994. Ecological morphology of locomotor performance in squamate reptiles. In *Ecological morphology: Integrative organismal biology*, ed. P. C. Wainwright and S. M. Reilly, 240–302. Chicago: University of Chicago Press.

Garner, G. W., and J. A. Morrison. 1980. Observations of interspecific behavior between predators and white-tailed deer in southwestern Oklahoma. *J Mammal* 61:126–30.

Garrett, M. G., and W. L. Franklin. 1988. Behavioral ecology of dispersal in the black-tailed prairie dog. *J Mammal* 69:236–50.

Gaston, A. J. 1977. Social behaviour within groups of jungle babblers (*Turdoides striatus*). *Anim Behav* 25:828–48.

Gates, J. E., and L. W. Gysel. 1978. Avian nest dispersion and fledging success in field-forest ecotones. *Ecology* 59:871–83.

Gauthier-Clerc, M., A. Tamisier, and F. Cezilly. 1998. Sleep-vigilance trade-off in green-winged teals (*Anas crecca crecca*). *Can J Zool* 76:2214–18.

Gautier-Hion, A., R. Quris, and J-P. Gautier. 1983. Monospecific vs polyspecific life: A comparative study of foraging and antipredatory tactics in a community of *Cercopithecus* monkeys. *Behav Ecol Sociobiol* 12:325–35.

Gautier-Hion, A., and C. E. G. Tutin. 1988. Simultaneous attack by adult males of a polyspecific troop of monkeys against a crowned hawk eagle. *Folia Primatol* 51:149–51.

Gavish, L., and B. Gavish. 1981. Patterns that conceal a bird's eye. *Z Tierpsychol* 56:193–204.

Geist, V. 1974. On the relationship of social evolution and ecology in ungulates. *Am Zool* 14:205–20.

Gendron, R. P. 1986. Searching for cryptic prey: Evidence for optimal search rates and the formation of search images in quail. *Anim Behav* 34:898–912.

Gendron, R. P., and J. E. R. Staddon. 1983. Searching for cryptic prey: The effect of search rate. *Am Nat* 121:172–86.

Gentle, L. K., and A. G. Gosler. 2001. Fat reserves and perceived predation risk in the great tit, *Parus major*. *Proc R Soc Lond Ser B Biol Sci* 268:487–91.

Gerell, R. 1985. Habitat selection and nest predation in a common eider population in southern Sweden. *Ornis Scand* 16:129–39.

Gerkema, M. P., and S. Verhulst. 1990. Warning against an unseen predator: A functional aspect of synchronous feeding in the common vole, *Microtus arvalis*. *Anim Behav* 40:1169–78.

Gershenson, S. 1945. Evolutionary studies on the distribution and dynamics of melanism in the hamster (*Cricetus cricetus* L.). II. Seasonal and annual changes in the frequency of black hamsters. *Genetics* 30:1233–51.

Gese, E. M. 1999. Threat of predation: Do ungulates behave aggressively towards different members of a coyote pack? *Can J Zool* 77:499–503.

Getty, T. 2002. The discriminating babbler meets the optimal diet hawk. *Anim Behav* 63:397–402.

Ghalambor, C. K., and T. E. Martin. 2000. Parental investment strategies in two species of nuthatch vary with stage-specific predation risk and reproductive effort. *Anim Behav* 60:263–67.

———. 2001. Fecundity-survival trade-offs and parental risk-taking in birds. *Science* 292:494–97.

———. 2002. Comparative manipulation of predation risk in incubating birds reveals variability in the plasticity of responses. *Behav Ecol* 13:101–8.

Gil-da-Costa, A. Palleroni, M. D. Hauser, J. Touchton, and P. J. Kelley. 2003. Rapid acquisition of an alarm response by a neotropical primate to a newly introduced avian predator. *Proc R Soc Lond Ser B Biol Sci* 270:605–10.

Gilchrist, H. G. 1999. Declining thick-billed murre *Uria lomvia* colonies experience higher gull predation rates: An inter-colony comparison. *Biol Conserv* 87:21–29.

Gill, S. A., and S. G. Sealy. 1996. Nest defence by yellow warblers: Recognition of a brood parasite and an avian nest predator. *Behaviour* 133:263–82.

Gilliam, J. F., and D. F. Fraser. 1987. Habitat selection under predation hazard: Test of a model with foraging minnows. *Ecology* 68:1856–62.

Gingerich, P. D. 1975. Is the aardwolf a mimic of the hyaena? *Nature* 253:191–92.

Gittleman, J. L., and M. E. Gompper. 2001. The risk of extinction—What you don't know will hurt you. *Science* 291:997–99.

Gittleman, J. L., and P. H. Harvey. 1980. Why are distasteful prey not cryptic? *Nature* 286:149–50.

Gittleman, J. L., P. H. Harvey, and P. J. Greenwood. 1980. The evolution of conspicuous coloration: Some experiments in bad taste. *Anim Behav* 28:897–99.

Gleason, T. M., and M. A. Norconk. 2002. Predation risk and antipredator adaptations in white-faced sakis, Pithecia pithecia. In *Eat or be eaten: Predator sensitive foraging among primates*, ed. L. E. Miller, 169–84. Cambridge: Cambridge University Press.

Gloger, C. W. L. 1833. *Das Abandern der Vogel durch Einfluss des Klimas*. Breslau: A. Schulz.

Glueck, E. 1987. An experimental study of feeding, vigilance and predator avoidance in a single bird. *Oecologia* 71:268–72.

Gochfeld, M. 1980. Mechanisms and adaptive value of reproductive synchrony in colonial seabirds. In *Behavior of marine animals*, ed. J. Burger, B. J. Olla, and H. E. Winn, 4:207–70. New York: Plenum Press.

———. 1981. Responses of young black skimmers to high-intensity distress notes. *Anim Behav* 29:1137–45.

———. 1984. Antipredator behavior: Aggressive and distraction displays of shorebirds. In *Behavior of marine animals*, ed. J. Burger and B. J. Olla, 5:289–377. New York: Plenum Press.

Godin, J-G. J., and S. L. Crossman. 1994. Hunger-dependent predator inspection and foraging behaviours in the threespine stickleback (*Gasterosteus aculeatus*) under predation risk. *Behav Ecol Sociobiol* 34:359–66.

Godin, J-G.J., and S. A. Davis. 1995a. Who dares, benefits: Predator approach behaviour in the guppy (*Poecilia reticulata*) deters predator pursuit. *Proc R Soc Lond Ser B Biol Sci* 259:193–200.

———. 1995b. Boldness and predator deterrence: A reply to Milinski & Boltshauser. *Proc R Soc Lond Ser B Biol Sci* 262:107–12.

Godin, J-G. J., and J. Morgan. 1985. Predator avoidance and school size in a cyprinodontid fish, the banded killifish (*Fundulus diaphanous* Lesueur). *Behav Ecol Sociobiol* 16:105–10.

Godin, J-G, J., and S. A. Smith. 1988. A fitness cost of foraging in the guppy. *Nature* 333:69–71.

Goldman, E. A. 1935. Pocket gophers of the *Thomomys bottae* groups in the United States. *Proc Biol Soc Wash* 48:153–58.

Goldsmith, A. E. 1990. Vigilance behavior of pronghorns in different habitats. *J Mammal* 71:460–62.

Goldthwaite, R. O., R. G. Coss, and D. H. Owings. 1990. Evolutionary dissipation of an antisnake system: Differential behavior by California and Arctic ground squirrels in above- and belowground contexts. *Behaviour* 112:246–69.

Goodhart, C. G. 1975. Does the aardwolf mimic a hyaena? *Zool J Linn Soc* 57:349–56.

Goodwin, D. 1953. Observations on voice and behaviour of the red-legged partridge *Alectoris rufa*. *Ibis* 95:581–614.

Goransson, G., J. Karlsson, S. G. Nilsson, and S. Ulfstrand. 1975. Predation on birds' nest in relation to antipredator aggression and nest density: An experimental study. *Oikos* 26:117–20.

Gorman, M. L. 1984. The response of prey to stoat (*Mustela erminea*) scent. *J Zool* 202:419–23.

Gosler, A. G., J. J. D. Greenwood, and C. Perrins 1995. Predation risk and the cost of being fat. *Nature* 377:621–23.

Gosse, P. H. 1847. *Birds of Jamaica*. London: John Van Voorst.

Gotmark, F. 1982. Coloniality in five Larus gulls: A comparative study. *Ornis Scand* 13:211–24.

———. 1987. White underparts in gulls function as hunting camouflage. *Anim Behav* 35:1786–92.

———. 1989. Costs and benefits to eiders nesting in gull colonies: A field experiment. *Ornis Scand* 20:283–88.

———. 1992a. Anti-predator effect of conspicuous plumage in a male bird. *Anim Behav* 44:51–55.

———. 1992b. Blue eggs do not reduce nest predation in the song thrush, *Turdus philomelos*. *Behav Ecol Sociobiol* 30:245–52.

———. 1993. Conspicuous coloration in male birds is favoured by predation in some species and disfavoured in others. *Proc R Soc Lond Ser B Biol Sci* 253:143–46.

———. 1994a. Does a novel bright colour patch increase or decrease predation? Red wings reduce predation risk in European blackbirds. *Proc R Soc Lond Ser B Biol Sci* 256:83–87.

———. 1994b. Are bright birds distasteful? A re-analysis of H.B. Cott's data on the edibility of birds. *J Avian Biol* 25:184–97.

———. 1995. Black-and-white plumage in male pied flycatchers (*Ficedula hypoleuca*) reduces the risk of predation from sparrowhawks (*Accipter nisus*) during the breeding season. *Behav Ecol* 6:22–26.

———. Simulating a colour mutation: Conspicuous red wings in the European blackbird reduce the risk of attacks by sparrowhawks. *Funct Ecol* 10:355–59.

———. 1997. Bright plumage in the magpie: Does it increase or reduce the risk of predation? *Behav Ecol Sociobiol* 40:41–49.

Gotmark, F., and M. Ahlund. 1988. Nest predation and nest site selection among eiders *Somateria mollussima*: The influence of gulls. *Ibis* 130:111–23.

Gotmark, F., and M. Andersson. 1980. Breeding associations between common gull *Larus canus* and Arctic skua *Stercorarius parasiticus*. *Ornis Scand* 11:121–24.

———. 1984. Colonial breeding reduces nest predation in the common gull (*Larus canus*). *Anim Behav* 32:485–92.

Gotmark, F., D. Blomqvist, O. C. Johansson, and J. Bergkvist. 1995. Nest site selection: A trade-off between concealment and view of the surroundings? *J Avian Biol* 26:305–12.

Gotmark, F., and A. Hohlfalt. 1995. Bright male plumage and predation risk in passerine birds: Are males easier to detect than females? *Oikos* 74:475–84.

Gotmark, F., and J. Olsson. 1997. Artificial colour mutation: Do red-painted great tits experience increased or decreased predation? *Anim Behav* 53:83–91.

Gotmark, F., and P. Post. 1996. Prey selection by sparrowhawks, *Accipter nisus*: Relative predation risk for breeding passerine birds in relation to their size, ecology, and behaviour. *Philos Trans R Soc Lond B Biol Sci* 351:1559–77.

Gotmark, F., P. Post, J. Olsson, and D. Himmelmann. 1997. Natural selection and sexual dimorphism: Sex-biased sparrowhawk predation favours crypsis in female chaffinches. *Oikos* 80:540–48.

Gotmark, F., and U. Unger. 1994. Are conspicuous birds unprofitable prey? Field experiments with hawks and stuffed prey species. *Auk* 111:251–262.

Gottfried, B. M. 1979. Anti-predator aggression in birds nesting in old field habitats: An experimental analysis. *Condor* 81:251–57.

Gottfried, B. M., and C. F. Thompson. 1978. Experimental analysis of nest predation in an old-field habitat. *Auk* 95:304–12.

Gould, E. 1978. The behavior of the moonrat, *Echinosorex gymnurus* (Erinaceidae) and the pentail shrew, *Ptilocercus lowi* (Tupaiidae) with comments on the behavior of other insectivora. *Z Tierpsychol* 48:1–27.

Gould, L., L. M. Fedigan, and L. M. Rose. 1997. Why be vigilant? The case of the alpha animal. *Int J Primatol* 18:401–14.

Gould, S. J. 1966. Allometry and size in ontogeny and phylogeny. *Biol Rev* 41:587–640.

Grafen, A. 1990. Biological signals as handicaps. *J Theor Biol* 144:517–46.

Grand, T. C. 2000. Risk-taking by threespine sticklebacks (*Gasterosteus aculeatus*) pelvic phenotypes: Does morphology predict behaviour? *Behaviour* 137:889–906.

Grand, T. I. 1991. Patterns of muscular growth in the African bovidae. *Appl. Anim Behav Sci.* 29:471–82.

Grant, J. W. A., and D. L. G. Noakes. 1987. Escape behaviour and use of cover by young-of-the-year brook trout, *Salvelinus fontinalis*. *Can J Fish Aquat Sci* 44:1390–96.

Gray, D. R. 1987. *The muskoxen of Polar Bear Pass*. Markham, ON: Fitzhenry and Whiteside.

Green, W. C. H., and A. Rothstein 1993. Asynchronous parturition in bison: Implications for the hider-follower dichotomy. *J Mammal* 74:920–25.

Greene, E., and T. Meagher. 1998. Red squirrels, *Tamiasciurus hudsonicus*, produce predator-class specific alarm calls. *Anim Behav* 55:511–18.

Greene, H. W. 1977. The aardwolf as hyaena mimic: An open question. *Anim Behav* 25:245.

———. 1983. Dietary correlates of the origin and radiation of snakes. *Am Zool* 23:431–41.

———. 1988. Antipredator mechanisms in reptiles. In *Biology of the reptilia*, ed. C. Gans and R. B. Huey, 16:1–152. New York: Alan R. Liss.

Greene, H. W., and R. W. McDiarmid. 1981. Coral snake mimicry: Does it occur? *Science* 213:1207–12.

Greenwood, J. J. D. 1984. The functional basis of frequency-dependent food selection. *Biol J Linn Soc* 23:177–99.

Greig-Smith, P. W. 1980. Parental investment in nest defence by stonechats (*Saxicola torquata*). *Anim Behav* 28:604–19.

———. 1981a. The role of alarm responses in the formation of mixed-species flocks of heathland birds. *Behav Ecol Sociobiol* 8:7–10.

———. 1981b. Responses to disturbance in relation to flock size in foraging groups of barred ground doves *Geopelia striata*. *Ibis* 123:103–6.

———. 1982. Distress calling by woodland birds. *Anim Behav* 30:299–301.

Griesser, M. 2003. Nepotistic vigilance behavior in Siberian jay parents. *Behav Ecol* 14:246–50.

Griesser, M., and J. Ekman. 2004. Nepotistic alarm calling in the Siberian jay, *Perisoreus infaustus*. *Anim Behav* 67:933–39.

———. 2005. Nepotistic mobbing behaviour in the Siberian jay, *Perisoreus infaustus*. *Anim Behav* 69:345–52.

Griffin, A. S., and C. S. Evans. 2003. Social learning of antipredator behaviour in a marsupial. *Anim Behav* 66:485–92.

Griffin, A. S., C. S. Evans, and D. T. Blumstein. 2001. Learning specificity in acquired predator recognition. *Anim Behav* 62:577–89.

———. 2002. Selective learning in a marsupial. *Ethology* 108:1103–14.

Griffith, S. C., I. P. F. Owens, and K. A. Thuman. 2002. Extra pair paternity in birds: A review of interspecific variation and adaptive function. *Molec Ecol* 11:2195–2212.

Griffiths, M. 1978. *The biology of the monotremes*. New York: Academic Press.

Grinnell, J. 1903. Call notes of the bush-tit. *Condor* 5:85–87.

Groom, M. J. 1992. Sand-colored nighthawks parasitize the antipredator behavior of three nesting bird species. *Ecology* 73:785–93.

Gross, M. R. 1996. Alternative reproductive strategies and tactics: Diversity within sexes. *Trends Ecol Evol* 11:92–98.

Gross, M. R., and A. M. MacMillan. 1981. Predation and the evolution of colonial nesting in bluegill sunfish (*Lepomis macrochirus*). *Behav Ecol Sociobiol* 8:163–74.

Guiler, E. R. 1953. Distribution of the brush possum in Tasmania. *Nature* 172:1091–93.

Guilford, T. 1985. Is kin selection involved in the evolution of warning coloration? *Oikos* 45:31–36.

———. 1986. How do "warning colours" work? Conspicuousness may reduce recognition errors in experienced predators. *Anim Behav* 34:286–88.

———. The evolution of conspicuous coloration. *Am Nat* 131:S7–S21.

———. 1990. The evolution of aposematism. In *Insect defenses*, ed. D. L. Evans and J. O. Schmidt, 23–61. Albany: State University of New York Press.

———. 1992. Predator psychology and the evolution of prey coloration. In *Natural enemies: The population biology of predators, parasites and diseases*, ed. M. J. Crawley, 377–94. Oxford: Blackwell Scientific Publications.

———. Go-slow signaling and the problem of automimicry. *J Theor Biol* 170:311–16.

Guilford, T., and M. S. Dawkins. 1987. Search images not proven: A reappraisal of recent evidence. *Anim Behav* 35:1838–45.

———. 1988. Search image versus search rate: A reply to Lawrence. *Anim Behav* 37:160–62.

Guillemain, M., G. R. Martin, and H. Fritz. 2002. Feeding methods, visual fields and vigilance in dabbling ducks (*Anatidae*). *Funct Ecol* 16:522–29.

Gullion, G. W. 1952. The displays and calls of the American coot. *Wilson Bull* 64:83–98.

Gunness, M. A., and P. J. Weatherhead. 2002. Variation in nest defense in ducks: Methodological and biological insights. *J Avian Biol* 33:191–98.

Gustafson, E. J., and L. W. VanDruff. 1990. Behavior of black and gray morphs of *Sciurus carolinensis* in an urban environment. *Am Midl Nat* 123:186–92.

Guthrie, R. D. 1967. Fire melanism among mammals. *Am Midl Nat* 77:227–30.

Gyger, M., P. Marler, and R. Pickert. 1987. Semantics of an avian alarm call system: The male domestic fowl, *Gallus domesticus. Behaviour* 102:15–40.

Haas, C. S. 1997. Seasonality of births in bighorn sheep. *J Mammal* 78:1251–60.

Haas, C. S., and D. Valenzuela. 2002. Anti-predator benefits of group living in white-nosed coatis (*Nasua narica*). *Behav Ecol Sociobiol* 51:570–78.

Haas, V. 1985. Colonial and single breeding in fieldfares, *Turdus pilaris* L.: A comparison of nesting success in early and late broods. *Behav Ecol Sociobiol* 16:119–24.

Hackmann, N., C. S. Zamora, and E. Stauber. 1990. The white eye secretion in Aplodontia. In *Chemical signals in vertebrates 5*, ed. D. W. Macdonald, D. Muller-Schwarze, and S. E. Natynczuk, 139–46. Oxford: Oxford University Press.

Haftorn, S. 2000. Contexts and possible functions of alarm calling in the willow tit, Parus montanus; the principle of "better safe than sorry." *Behaviour* 137:437–49.

Hager, M. C., and G. S. Helfman. 1991. Safety in numbers: Shoal size choice by minnows under predatory threat. *Behav Ecol Sociobiol* 29:271–76.

Hagman, M., and A. Forsman. 2003. Correlated evolution of conspicuous coloration and body size in poison frogs (Dendrobatidae). *Evolution* 57:2904–10.

Hailman, J. P. 1977. *Optical signals: Animal communication and light.* Bloomington: Indiana University Press.

Hailman, J. P., K. J. McGowan, and G. E. Woolfenden. 1994. Role of helpers in the sentinel behaviour of the Florida scrub jay (*Aphelocoma c. coerulescens*). *Ethology* 97:119–40.

Hake, M. 1996. Fattening strategies in dominance-structured greenfinch (*Carduelis chloris*) flocks in winter. *Behav Ecol Sociobiol* 39:71–76.

Hakkarainen, H., and E. Korpimaki. 1994. Nest defence of Tengelmam's owls reflects offspring survival prospects under fluctuating food conditions. *Anim Behav* 48:843–49.

Hakkarainen, H., E. Korpimaki, T. Mappes, and P. Palokangas. 1992. Kestrel hunting behaviour towards solitary and grouped *Microtus agrestis* and *M. epiroticus*—A laboratory experiment. *Ann Zool Fen* 29:279–84.

Haland, A. 1989. Nest spacing and antipredator behaviour of the fieldfare *Turdus pilaris* in alpine habitats. *Fauna Norv Ser C Cinclus* 12:11–20.

Hall, D. J., and E. E. Werner. 1977. Seasonal distribution and abundance of fishes in the littoral zone of a Michigan lake. *Trans Am Fish Soc* 106:545–55.

Halpin, Z. 1983. Naturally-occurring encounters between black-tailed prairie dogs (*Cynomys ludovicianus*) and snakes. *Am Midl Nat* 109:50–54.

Halupka, K., and L. Halupka. 1997. The influence of reproductive season stage on nest defence by meadow pipits (*Anthus pratensis*). *Ethol Ecol Evol* 9:89–98.

Halupka, L. 1999. Nest defence in an altricial bird with uniparental care: The influence of offspring age, brood size, stage of the breeding season and predator type. *Ornis Fenn* 76:97–105.

Hamilton, W. D. 1964. The genetical evolution of social behaviour. *J Theor Biol* 7:1–52.

———. 1971. Geometry for the selfish herd. *J Theor Biol* 31:295–311.

Hamilton, W. J. III. 1973. *Life's color code.* New York: McGraw-Hill.

Hamilton, W. J. III, R. E. Buskirk, and W. H. Buskirk. 1975. Defensive stoning by baboons. *Nature* 256:488–89.

Hamlin, K. L., and L. L. Schweitzer. 1979. Cooperation by coyote pairs attacking mule deer fawns. *J Mammal* 60:849–50.

Hansell, M. 1996. The function of lichen flakes and white spider cocoons on the outer surface of birds' nests. *J Nat Hist* 30:303–11.

———. 2000. *Bird nests and construction behaviour.* Cambridge: Cambridge University Press.

Hanson, M. T., and R. G. Coss. 1997. Age differences in the response of California ground squirrels (*Spermophilus beecheyi*) to avian and mammalian predators. *J Comp Psychol* 111: 174–84.

———. 2001. Age differences in the response of California ground squirrels (*Spermophilus beecheyi*) to conspecific alarm calls. *Ethology* 107:259–75.

Hare, J. F. 1998. Juvenile Richardson's ground squirrels, *Spermophilus richardsonii*, discriminate among individual alarm callers. *Anim Behav* 55:451–60.

Hare, J. F., and B. A. Atkins. 2001. The squirrel that cried wolf: Reliability detection by juvenile Richardson's ground squirrels (*Spermophilus richardsonii*). *Behav Ecol Sociobiol* 51:108–12.

Harfenist, A., and R. C. Ydenberg. 1995. Parental provisioning and predation risk in rhinoceros auklets (*Cerorhinca monocerata*): Effects on nestling growth and fledging. *Behav Ecol* 6:82–86.

Harkin, E. L., W. F. D. van Dongen, M. E. Herberstein, and M. A. Elgar. 2000. The influence of visual obstructions on the vigilance and escape behaviour of house sparrows, *Passer domesticus*. *Austr J Zool* 48:259–63.

Harper, S. J., and G. O. Batzli. 1996. Effects of predators on structure of the burrows of voles. *J Mammal* 77:1114–21.

Harris, M. A., J. O. Murie, and J. A. Duncan. 1983. Responses of Columbian ground squirrels to playback of recorded calls. *Z Tierpsychol* 63:318–30.

Harris, M. P. 1969. The biology of storm petrels in the Galapagos Islands. *Proc Calif Acad Sci* 37:95–166.

Harrison, C. 1985. *A field guide to the nests, eggs and nestlings of British and European birds.* London: Collins.

Harrison Matthews, L. 1971. *The life of mammals.* Vol. 2. New York: Universe Books, New York.

Hart, A., and S. W. Lendrem. 1984. Vigilance and scanning patterns in birds. *Anim Behav* 32:1216–24.

Hart, B. L., L. A. Hart, M. S. Mooring, and L. Olubayo. 1992. Biological basis of grooming behaviour in antelope: The body-size, vigilance and habitat principles. *Anim Behav* 44:615–31.

Hartley, M. J., and M. L. Hunter Jr. 1998. A meta-analysis of forest cover, edge effects, and artificial nest predation rates. *Cons Biol* 12:465–69.

Hartley, P. H. T. 1950. An experimental analysis of interspecific recognition. *Symp Soc Exp Biol* 4:313–36.

Harvey, P. H., J. J. Bull, M. Pemberton, and R. J. Paxton. 1982. The evolution of aposematic coloration in distasteful prey: A family model. *Am Nat* 119:710–19.

Harvey, P. H., and P. J. Greenwood. 1978. Anti-predator defence strategies: Some evolutionary problems. In *Behavioural ecology: An evolutionary approach*, ed. J. R. Krebs and N. B. Davies, 129–51. Oxford: Blackwell Scientific Publications.

Harvey, P. H., M. Kavanagh, and T. H. Clutton-Brock. 1978. Sexual dimorphism in primate teeth. *J Zool* 186:475–86.

Harvey, P. H., and M. D. Pagel. 1991. *The comparative method in evolutionary biology*. Oxford: Oxford University Press.

Harvey, P. H., and R. J. Paxton. 1981. The evolution of aposematic coloration. *Oikos* 37:391–93.

Harvey, P. H., D. E. L. Promislow, and A. F. Read. 1989. Causes and correlates of life history differences among mammals. In *Comparative socioecology: the behavioural ecology of humans and other mammals*, ed. V. Standen and R. A. Foley, 305–18. Oxford: Blackwell Scientific Publications.

Haskell, D. 1994. Experimental evidence that nestling begging behaviour incurs a cost due to nest predation. *Proc R Soc Lond Ser B Biol Sci* 257:161–64.

———. 1999. The effect of predation on begging-call evolution in nestling wood warblers. *Anim Behav* 57:893–901.

Hasson, O. 1991. Pursuit-deterrent signals: Communication between prey and predator. *Trends Ecol Evol* 6:325–29.

———. 1994. Cheating signals. *J Theor Biol* 167:223–38.

Hasson, O., R. Hibbard, and G. Ceballos. 1989. The pursuit deterrent function of tail-wagging in the zebra-tailed lizard (*Callisaurus draconoides*). *Can J Zool* 67:1203–9.

Hauber, M. E., and P. W. Sherman. 1998. Nepotism and marmot alarm calling. *Anim Behav* 56:1049–52.

Hauser, M. D. 1986. Male responsiveness to infant distress calls in free-ranging vervet monkeys. *Behav Ecol Sociobiol* 19:65–71.

———. 1988a. Variation in maternal responsiveness in free-ranging vervet monkeys: A response to infant mortality risk? *Am Nat* 131:573–87.

———. 1988b. How infant vervet monkeys learn to recognize starling alarm calls: The role of experience. *Behaviour* 105:187–201.

———. 1996. *The evolution of communication*. Cambridge, MA: MIT Press.

Hauser, M. D., and C. Caffrey. 1994. Anti-predator response to raptor calls in wild crows, *Corvus brachyryynchos hesperis*. *Anim Behav* 48:1469–71.

Hauser, M. D., and R. W. Wrangham. 1990. Recognition of predator and competitor calls in non-human primates and birds: A preliminary report. *Ethology* 86:116–30.

Hay, M. E., and P. J. Fuller. 1981. Seed escape from heteromyid rodents: The importance of microhabitat and seed preference. *Ecology* 62:1395–99.

Heard, D. C. 1992. The effect of wolf predation and snow cover on musk-ox group size. *Am Nat* 139:190–204.

Heathcote, C. F. 1987. Grouping in eastern grey kangaroos in open habitat. *Austr Wildl Res* 14:343–48.

Heatwole, H. 1968. Relationship of escape behavior and camouflage in anoline lizards. *Copeia* 1968:109–13.

Hebblewhite, M., and D. H. Pletscher. 2002. Effects of elk group size on predation by wolves. *Can J Zool* 80:800–809.

Hedenstrom, A. 1992. Flight performance in relation to fuel load in birds. *J Theor Biol* 158:535–37.

Hedenstrom, A., and T. Alerstam. 1992. Climbing performance of migrating birds as a basis for estimating limits of fuel-carrying capacity and muscle. *J Exp Biol* 164:19–38.

———. 1994. Optimal climbing flight in migrating birds: Predictions and observations of knots and turnstones. *Anim Behav* 48:47–54.

———. 1995. Optimal flight speed in birds. *Philos Trans R Soc Lond B Biol Sci* 348:471–87.

———. 1996. Skylark optimal flight speeds for flying nowhere and somewhere. *Behav Ecol* 7:121–26.

Hedenstrom, A., and M. Rosen. 2001. Predator versus prey: On aerial hunting and escape strategies in birds. *Behav Ecol* 12:150–56.

Hegner, R. E. 1985. Dominance and anti-predator behaviour in blue tits (*Parus caeruleus*). *Anim Behav* 33:762–68.

Heikkila, J., K. Kaarsalo, O. Mustonen, and P. Pekkarinen. 1993. Influence of predation risk on early development and maturation in three species of *Clethrionomys* voles. *Ann Zool Fenn* 30:153–61.

Heinsohn, R. G. 1987. Age-dependent vigilance in winter aggregations of cooperatively breeding white-winged choughs (*Corcorax melanorhamphos*). *Behav Ecol Sociobiol* 20:303–6.

Heller, R., and M. Milinski. 1979. Optimal foraging of sticklebacks on swarming prey. *Anim Behav* 27:1127–41.

Hendrie, C. A., S. M. Weiss, and D. Eilam. 1998. Behavioural response of wild rodents to the calls of an owl: A comparative study. *J Zool* 245:439–46.

Hennessy, D. F. 1986. On the deadly risk of predator harassment. *Ethology* 72:72–74.

Hennessy, D. F., and D. H. Owings. 1978. Snake species in discrimination and the role of olfactory cues in the snake-directed behavior of the California ground squirrel. *Behaviour* 65:115–24.

———. 1988. Rattlesnakes create a context for localizing their search for potential prey. *Ethology* 77:317–29.

Hennessy, D. F., D. H. Owings, M. P. Rowe, R. G. Coss, and D. W. Leger. 1981. The information afforded by a variable signal: Constraints on snake-elicited tail flagging by California ground squirrels. *Behaviour* 78:188–226.

Herman, C. S., and T. J. Valone. 2000. The effect of mammalian predator scent on the foraging behavior of *Diplodomys merriami*. *Oikos* 91:139–45.

Herrera, E. A., and D. W. Macdonald. 1993. Aggression, dominance and mating success in capybaras, *Hydrochaeris hydrochaeris*. *Behav Ecol* 4:114–19.

Hersek, M. J., and D. H. Owings. 1993. Tail flagging by adult California ground squirrels: A tonic signal that serves different functions for males and females. *Anim Behav* 46:129–38.

Herter, K. 1965. *Hedgehogs*. Wittenberg and Lutherstadt: J. M. Dent & Sons and A. Ziemsen Verlag.

Hertz, P. E., R. B. Huey, and E. Nevo. 1982. Fight versus flight: Body temperature influences defensive responses of lizards. *Anim Behav* 30:676–79.

Herzog, M., and S. Hopf. 1984. Behavioral responses to species-specific warning calls in infant squirrel monkeys reared in social isolation. *Am J Primatol* 7:99–106.

Heth, G., A. Beilers, and E. Nevo. 1988. Adaptive variation of pelage color within and between species of the subterranean mole rat (*Spalax ehrenbergi*) in Israel. *Oecologia* 74:617–22.

Heymann, E. W. 1990. Reactions of wild tamarins, *Saguinus mystax* and *Saguinus fuscicollis* to avian predators. *Int J Primatol* 11:327–37.

Hickerson, H. 1965. The Virginia deer and intertribal buffer zones in the upper Mississippi Valley. In *Man's culture and animals*, ed. A. Leeds and A. P. Vayda, 43–66. Publication no. 8. Washington, DC: American Association Advancement of Science.

Hieber, C. S., R. S. Wilcox, J. Boyle, and G. W. Uetz. 2002. The spider and the fly revisited: Ploy-counterploy behavior in a unique predator-prey system. *Behav Ecol Sociobiol* 53:51–60.

Hilden, O., and S. Vuolanto. 1972. Breeding biology of the red-necked phalarope in Finland. *Ornis Fenn* 49:57–85.

Hileman, K. S., and E. D. Brodie Jr. 1994. Survival strategies of the salamander *Desmognathus ochrophaeus*: Interaction of predator-avoidance and anti-predator mechanisms. *Anim Behav* 47:1–6.

Hill, D. A., 1984. Clutch predation in relation to nest density in mallard and tufted duck. *Wildfowl* 35:151–56.

Hill, G. E. 1986. The function of distress calls given by tufted titmice (*Parus bicolor*): An experimental approach. *Anim Behav* 34:590–98.

Hill, R. A., and G. Cowlishaw. 2002. Foraging female baboons exhibit similar patterns of anti-predator vigilance across two populations. In *Eat or be eaten: Predator sensitive foraging among primates* ed. L. E. Miller, 187–204. Cambridge: Cambridge University Press.

Hill, R. A., and R. I. M. Dunbar. 1998. An evaluation of the roles of predation rate and predation risk as selective pressures on primate grouping behaviour. *Behaviour* 135:411–30.

Hill, R. A., and P. C. Lee. 1998. Predation risk as an influence on group size in cercopithecoid primates: Implications for social structure. *J Zool (Lond)* 245:447–56.

Hilton, G. M., W. Cresswell, and G. D. Ruxton. 1999. Intraflock variation in the speed of escape-flight response on attack by an avian predator. *Behav Ecol* 10:391–95.

Hinde, R. A. 1954. Factors governing the changes in strength of a partially inborn response, as shown by the mobbing behaviour of the chaffinch (*Fringilla coelebs*). I. The nature of the response, and an examination of its course. *Proc R Soc Lond Ser B Biol Sci* 142:306–31.

Hindwood, K. A. 1959. Bird/wasp nesting associations. *Emu* 55:263–74.

Hingston, R. W. G. 1932. *A naturalist in the Guiana forest.* New York: Longmans, Green & Co.

Hinton, H. E. 1967. *Mongooses.* Berkeley and Los Angeles: University of California Press.

Hinton, M. A. C. 1947. Notes and exhibitions. *Proc Zool Soc Lond* 116:737–38.

Hiraiwa-Hasegawa, M., R. W. Byrne, H. Takasaki, and J. M. E. Byrne. 1986. Aggression toward large carnivores by wild chimpanzees of Mahale Mountains National Park, Tanzania. *Folia Primatol* 47:8–13.

Hirsch, B. T. 2002. Social monitoring and vigilance behavior in brown capuchin monkeys (*Cebus apella*). *Behav Ecol Sociobiol* 52:458–64.

Hirsch, S. M., and R. C. Bolles. 1980. On the ability of prey to recognize predators. *Z Tierpsychol* 54:71–84.

Hirth, D. H., and D. R. McCullough. 1977. Evolution of alarm signals in ungulates with special reference to white-tailed deer. *Am Nat* 111:31–42.

Hoare, D. J., I. D. Couzin, J-G. J. Godin, and J. Krause. 2004. Context-dependent group size choice in fish. *Anim Behav* 67:155–64.

Hobson, K. A., M. L. Bouchart, and S. G. Sealy. 1988. Responses of naïve yellow warblers to a novel nest predator. *Anim Behav* 36:1823–30.

Hodgdon, H. E., and J. S. Larson. 1973. Some sexual differences in behaviour within a colony of marked beavers (*Castor canadensis*). *Anim Behav* 21:147–52.

Hodl, W., and G. Gollman. 1986. Distress calls in neotropical frogs. *Amphibia-Reptilia* 7:11–21.

Hofer, H., and M. East. 1995. Population dynamics, population size, and the commuting system of Serengeti spotted hyenas. In *Serengeti II: Dynamics, management, and conservation of an ecosystem*, ed. A. R. E. Sinclair and P. Arcese, 332–63. Chicago: University of Chicago Press.

Hogstad, O. 1988. Advantages of social foraging of willow tits Parus montanus. *Ibis* 130:275–83.

———. 1995a. Alarm calling by willow tits, *Parus montanus*, as mate investment. *Anim Behav* 49:221–25.

———. 1995b. Do avian and mammalian nest predators select for different nest dispersion patterns of fieldfares *Turdus pilaris*? A 15-year study. *Ibis* 137:484–89.

Hogstedt, G. 1983. Adaptation unto death: Function of fear screams. *Am Nat* 121:562–70.

Hohmann, G. 1989. Vocal communication of wild bonnet macaques (*Macaca radiata*). *Primates* 30:325–45.

Hoi, H., and H. Winkler. 1994. Predation on nests: A case of apparent competition. *Oecologia* 98:436–40.

Holley, A. J. F. 1993. Do brown hares signal to foxes? *Ethology* 94:21–30.

Holmes, W. G. 1984. Predation risk and foraging behavior of the hoary marmot in Alaska. *Behav Ecol Sociobiol* 15:293–301.

———. 1991. Predator risk affects foraging behaviour of pikas: Observational and experimental evidence. *Anim Behav* 42:111–19.

Holst, D. V. 1985. The primitive eutherians I: Orders Insectivora, Macroscelidea, and Scandentia. In *Social odours in mammals*, ed. R. E. Brown and D. W. Macdonald, 105–54. Oxford: Clarendon Press.

Holt, R. D., and B. P. Kotler. 1987. Short-term apparent competition. *Am Nat* 130:412–30.

Holway, D. A. 1991. Nest-site selection and the importance of nest concealment in the black-throated blue warbler. *Condor* 93:575–81.

Hoogland, J. L. 1979. The effect of colony size on individual alertness of prairie dogs (Sciuridae: Cynomys spp.). *Anim Behav* 27:394–407.

———. 1981. The evolution of coloniality in white-tailed and black-tailed prairie dogs (Sciuridae: *Cynomys leucurus* and *C. ludovicianus*). *Ecology* 62:252–72.

———. 1983. Nepotism and alarm calling in the black-tailed prairie dog (*Cynomys ludovicianus*). *Anim Behav* 31:472–79.

———. 1995. *The black-tailed prairie dog: Social life of a burrowing mammal.* Chicago: University of Chicago Press.

———. 1996. Why do Gunnison's prairie dogs give anti-predator calls? *Anim Behav* 51:871–80.

Hoogland, J. L., and P. W. Sherman. 1976. Advantages and disadvantages of bank swallow (*Riparia riparia*) coloniality. *Ecol Monogr* 46:33–58.

Horn, H. S. 1968. The adaptive significance of colonial nesting in the Brewer's blackbird (*Euphagus cyancocephalus*). *Ecology* 49:682–94.

Hornocker, M. G. 1969. Defensive behavior in female bighorn sheep. *J Mammal* 50:128.

Horrocks, J. A., and W. Hunte. 1986. Sentinel behaviour in vervet monkeys: Who sees whom first? *Anim Behav* 34:1566–67.

Hough, D., B. Wong, G. Bennett, K. Brettschneider, M. Spinocchia, N. Manion, and R. Heinsohn. 1998. Vigilance and group size in emus. *Emu* 98:324–27.

Houston, A. I., and N. B. Davies. 1985. The evolution of cooperation and life history in the dunnock *Prunella modularis*. In *Behavioural ecology*, ed. R. M. Sibley and R. H. Smith, 471–87. Oxford: Blackwell Scientific Publications.

Houston, A. I., J. M. McNamara, and J. M. C. Hutchinson. 1993. General results concerning the trade-off between gaining energy and avoiding predation. *Philos Trans R Soc Lond B Biol Sci* 341:375–97.

Howard, D. F., M. S. Blum, and H. M. Fales. 1983. Defense in thrips: Forbidding fruitiness of a lactone. *Science* 220:335–36.

Howland, H. C. 1974. Optimal strategies for predator avoidance: The relative importance of speed and manoeuvrability. *J Theor Biol* 47:333–50.

Howlett, J. S., and B. J. Stutchbury. 1996. Nest concealment and predation in hooded warblers: Experimental removal of nest cover. *Auk* 113:1–9.

Hudson, P. J., and D. Newborn. 1990. Brood defence in a precocial species: Variations in the distraction displays of red grouse *Lagopus lagopus scoticus. Anim Behav* 40:254–61.

Hudson, W. H. 1892. *The naturalist in La Plata.* London: J. M. Dent & Sons.

Hughes, J. J., and D. Ward. 1993. Predation risk and distance to cover affect foraging behaviour in Namib Desert gerbils. *Anim Behav* 46:1243–45.

Hughes, P. M., and J. M. V. Rayner. 1991. Addition of artificial loads to long-eared bats *Pleocotus auritus*: Handicapping flight performance. *J Exp Biol* 161:285–98.

Hugie, D. M. 2003. The waiting game: A "battle of waits" between predator and prey. *Behav Ecol* 14:807–17.

Huheey, J. E. 1988. Mathematical models of mimicry. In *Mimicry and the evolutionary process*, ed. L. P. Brower, 22–41. Chicago: University of Chicago Press.

Huhta, E., T. Aho, A. Jantti, P. Suorsa, M. Knitunen, A. Nikula, and H. Hakkarainen. 2004. Forest fragmentation increases nest predation in the Eurasian tree creeper. *Conserv Biol* 18:148–55.

Humphries, D. A., and P. M. Driver. 1967. Erratic display as a device against predators. *Science* 156:1767–68.

———. 1970. Protean defence by prey animals. *Oecologia* 5:285–302.

Hunter, L. T. B., and J. D. Skinner. 1998. Vigilance behaviour in African ungulates: The role of predation pressure. *Behaviour* 135:195–211.

Huntly, N. J. 1987. Influence of refuging consumers (pikas: *Ochotona princeps*) on subalpine meadow vegetation. *Ecology* 68:1856–62.

Hurd, C. R. 1996. Interspecific attraction to the mobbing calls of black-capped chickadees (*Parus atricapillus*). *Behav Ecol Sociobiol* 38:287–92.

Illius, A. W., and C. FitzGibbon. 1994. Costs of vigilance in foraging ungulates. *Anim Behav* 47:481–84.

Imber, M. J. 1976. The origin of petrel stomach oils—A review. *Condor* 78:366–69.

Inglis, I. R., M. R. Fletcher, C. J. Feare, P. W. Greig-Smith, and S. Land. 1982. The incidence of distress calling among British birds. *Ibis* 124:351–55.

Inglis, I. R., and A. J. Isaacson. 1978. The responses of dark-bellied brent geese to models of geese in various postures. *Anim Behav* 26:953–58.

Inglis, I. R., and J. Lazarus. 1981. Vigilance and flock size in brent geese: The edge effect. *Z Tierpsychol* 57:193–200.

Inman, A. J., and Krebs, J. 1987. Predation and group living. *Trends Ecol Evol* 2:31–32.

Iriarte-Diaz, J. 2002. Differential scaling of locomotor performance in small and large terrestrial mammals. *J Exp Biol* 205:2897–2908.

Irschick, D. J., and T. Garland Jr. 2001. Integrating function and ecology in studies of adaptation: Investigations of locomotor capacity as a model system. *Annu Rev Ecol Syst* 32:367–96.

Irwin, R. E. 1994. The evolution of plumage dichromatism in the new world blackbirds: Social selection on female brightness? *Am Nat* 144:890–907.

Isbell, L. A. 1990. Sudden short-term increase in mortality of vervet monkeys (*Cercopithecus aethiops*) due to leopard predation in Amboseli National Park, Kenya. *Am. J Primatol* 21:41–52.

———. 1994. Predation on primates: Ecological patterns and evolutionary consequences. *Evol Anthropol* 3:61–71.

Isbell, L. A., D. L. Cheney, and R. M. Seyfarth. 1993. Are immigrant vervet monkeys, *Cercopithecus aethiops*, at greater risk of mortality than residents? *Anim Behav* 45:729–34.

Isbell, L. A., and T. P. Young. 1993a. Human presence reduces predation in a free-ranging vervet monkey population in Kenya. *Anim Behav* 45:1233–35.

———. 1993b. Social and ecological influences on activity budgets of vervet monkeys, and their implications for group living. *Behav Ecol Sociobiol* 32:377–85.

Ishihara, M. 1987. Effect of mobbing toward predators by the damselfish *Pomacentrus coelestis* (Pisces: Pomacentridae). *J Ethol* 5:43–52.

Ivins, B. L., and A. T. Smith. 1983. Responses of pikas (*Ochotona princeps*, Lagomorpha) to naturally occurring terrestrial predators. *Behav Ecol Sociobiol* 13:277–85.

Iwamoto, T., A. Mori, M. Kawai, and A. Bekele. 1996. Anti-predator behavior of gelada baboons. *Primates* 37:389–97.

Jablonski, P. G. 1999. A rare predator exploits prey escape behavior: The role of tail-fanning and plumage contrast in foraging of the painted redstart (*Myioborus pictus*). *Behav Ecol* 10:7–14.

Jackson, S. L., D. S. Hik, and R. F. Rockwell 1988. The influence of nesting habitat on reproductive success of the lesser snow goose. *Can J Zool* 66:1699–1703.

Jacobsen, N. K. 1979. Alarm bradycardia in white-tailed deer fawns (*Odocoileus virginianus*). *J Mammal* 60:343–49.

Jacobsen, O. W., and M. Ugelvik. 1994. Effects of presence of waders on grazing and vigilance behaviour in breeding wigeon, *Anas penelope*. *Anim Behav* 47:488–90.

Jakobsen, P. J., and G. H. Johnsen. 1988. Size-specific protection against predation by fish in swarming waterfleas, *Bosmina longispina*. *Anim Behav* 36:986–90.

Jakobsson, S., O. Brick, and C. Kullberg. 1995. Escalated fighting behaviour incurs increased predation risk. *Anim Behav* 49:235–39.

James, R., P. G. Bennett, and J. Krause. 2004. Geometry for mutualistic and selfish herds: The limited domain of danger. *J Theor Biol* 228:107–13.

Jansen, D. H. 1980. When is it coevolution? *Evolution* 34:611–12.

Janson, C. H. 1992. Evolutionary ecology of primate social structure. In *Evolutionary ecology and human behavior*, ed. E. A. Smith and B. Winterhalder, 95–130. New York: Aldine de Gruyter.

———. 1998. Testing the predation hypothesis for vertebrate sociality: Prospects and pitfalls. *Behaviour* 135:389–410.

———. 2000. Primate socio-ecology: The end of a golden age. *Evol Anthropol* 9:73–86.

Janson, C. H., and M. L. Goldsmith. 1995. Predicting group size in primates: Foraging costs and predation risks. *Behav Ecol* 6:326–36.

Jansson, L., and M. Enquist. 2003. Receiver bias for colourful signals. *Anim Behav* 66:965–71.

Jarman, P. J. 1974. The social organization of antelope in relation to their ecology. *Behaviour* 48:215–67.

———. 1987. Group size and activity in eastern grey kangaroos. *Anim Behav* 35:1044–50.

Jarman, P. J., and S. M. Wright. 1993. Macropod studies at Wallaby Creek. IX. Exposure and responses of eastern grey kangaroos to dingoes. *Austr Wildl Res* 20:833–43.

Jarvi, T., B. Sillen-Tullberg, and C. Wiklund. 1981. The cost of being aposematic. An experimental study of predation on larvae of *Papilio machaon* by the great tit *Parus major*. *Oikos* 36:267–72.

Jayne, B. C., and A. F. Bennett. 1990. Scaling of speed and endurance in garter snakes: A comparison of cross-sectional and longitudinal allometries. *J Zool* 220:257–77.

Jedrzejewski, W., and B. Jedrzejewska. 1990. Effect of a predator's visit on the spatial distribution of bank voles: Experiments with weasels. *Can J Zool* 68:660–66.

Jedrzejewski, W., L. Rychlik, and B. Jedrzejewska. 1993. Responses of bank voles to odours of seven species of predators: Experimental data and their relevance to natural predator-vole relationships. *Oikos* 68:251–57.

Jennings, T., and S. M. Evans. 1980. Influence of position in the flock and flock size on vigilance in the starling *Sturnus vulgaris. Anim Behav* 28:634–35.

Jennions, M. D., and M. Petrie. 1997. Variation in mate choice and mating preferences: A review of causes and consequences. *Biol Rev* 72:283–327.

Jetz, W., C. Rowe, and T. Guilford. 2001. Non-warning odors trigger innate color aversions—as long as they are novel. *Behav Ecol* 12:134–39.

Johnsgard, P. A. 2002. *North American owls: Biology and natural history.* Washington, DC: Smithsonian Institution.

Johnsingh, A. J. T. 1983. Large mammalian prey-predators in Bandipur. *J Bombay Nat Hist Soc* 80:3–49.

Johnson, C. E. 1925. Kingfisher and Copper's hawk. *Auk* 42:585–86.

Johnsson, K. 1994. Colonial breeding and nest predation in the jackdaw *Corvus monedula* using old black woodpecker *Dryocopus martius* holes. *Ibis* 136:313–17.

Johnston, C. E. 1921. The "hand-stand" habit of the spotted skunk. *J Mammal* 2:87–89.

Johnstone, R. A. 1996. Multiple displays in animal communication: "Backup signals" and "multiple messages." *Philos Trans R Soc Lond B Biol Sci* 351:329–38.

Jones, G., and J. Rydell. 1994. Foraging strategy and predation risk as factors influencing emergence time in echolocating bats. *Philos Trans R Soc Lond B Biol Sci* 346:445–55.

Jones, K. J., and W. L. Hill. 2001. Auditory perception of hawks and owls for passerine alarm calls. *Ethology* 107:717–26.

Jones, M. E. 1998. The function of vigilance in sympatric marsupial carnivores: The eastern quoll and the Tasmanian devil. *Anim Behav* 56:1279–84.

Jones, M. E., G. C. Smith, and S. M. Jones. 2004. Is anti-predator behaviour in Tasmanian eastern quolls (*Dasyurus viverrinus*) effective against introduced predators? *Anim Conserv* 7:155–60.

Jones, R. B. 1986. Conspecific vocalizations, tonic immobility and fearfulness in the domestic fowl. *Behav Proc* 13:217–25.

Jonsson, P., E. Koskela, and T. Mappes. 2000. Does risk of predation by mammalian predators affect the spacing behaviour of rodents? Two large-scale experiments. *Oecologia* 122:487–92.

Jori, F., M. Lopez-Bejar, and P. Houben. 1998. The biology and use of the African brush-tailed porcupine (*Atherurus africanus*, Gray, 1842) as a food animal. A review. *Biodivers Conserv* 7:1417–26.

Jourdain, F. C. R. 1936. The so-called injury feigning in birds. *Oologists' Rec* 16:25–37.

———. 1937. The so-called injury-feigning in birds. Part III. *Oologists' Rec* 17:14–16.

Joyce, F. J. 1993. Nesting success of rufous-naped wrens (*Campylorhynchus rufinucha*) is greater near wasp nests. *Behav Ecol Sociobiol* 32:71–77.

Jurisevic, M. A., and K. J. Sanderson. 1988. Acoustic discrimination of passerine anti-predator signals by Australian raptors. *Austr J Zool* 46:369–79.

———. 1994. Alarm vocalizations in Australian birds: Convergent characteristics and phylogenetic differences. *Emu* 94:69–77.

———. 1998. A comparative analysis of distress call structure in Australian passerine and non-passerine species: Influence of size and phylogeny. *J Avian Biol* 29:61–71.

Kaby, U., and J. Lind. 2003. What limits predator detection in blue tits (*Parus caeruleus*): Posture, task or orientation? *Behav Ecol Sociobiol* 54:534–38.

Kaitala, V., K. Lindstrom, and E. Ranta. 1989. Foraging, vigilance and risk of predation in birds—A dynamic game study of ESS. *J Theor Biol* 138:329–45.

Kapan, D. D. 2001. Three-butterfly system provides a field test of mullerian mimicry. *Nature* 409:338–40.

Kats, L. B., and L. M. Dill. 1998. The scent of death: Chemosensory assessment of predation risk by prey animals. *Ecoscience* 5:361–94.

Kaufman, D. W. 1974. Adaptive coloration in *Peromyscus polionotus*: Experimental selection by owls. *J Mammal* 55:271–83.

Kaufmann, J. H. 1974. The ecology and evolution of social organization in the kangaroo family (Macropodidae). *Am Zool* 14:51–62.

Kauppinen, J., and J. Mappes. 2003. Why are wasps so intimidating: Field experiments on hunting dragonflies (Odonata: *Aeshna grandis*). *Anim Behav* 66:505–11.

Kear, J. 1970. The adaptive radiation of parental care in waterfowl. In *Social behaviour in birds and mammals: Essays on the social ethology of animals and man*, ed. J. H. Crook, 357–91. London: Academic Press.

Kelly, J. P. 1993. The effect of nest predation on habitat selection by dusky flycatchers in limber pine-juniper woodland. *Condor* 95:83–93.

Kemp, A. 1995. *The hornbills*. Oxford: Oxford University Press.

Kenward, R. E. 1978. Hawks and doves: Factors affecting success and selection in goshawk attacks on woodpigeons. *J Anim Ecol* 47:449–60.

Kerfoot, W. C., and A. Sih, eds. 1987. *Predation: Direct and indirect impacts on aquatic communities.* Hanover, NH: University of New England Press.

Kerlinger, P., and P. H. Lehrer. 1982. Owl recognition and anti-predator behaviour of sharp-shinned hawks. *Z Tierpsychol* 58:163–73.

Keverne, E. B., R. A. Leonard, D. M. Scruton, and S. K. Young. 1978. Visual monitoring in social groups of talapoin monkeys (*Miopithecus talapoin*). *Anim Behav* 26:933–44.

Keys, G. C., and L. A. Dugatkin. 1990. Flock size and position effects on vigilance, aggression, and prey capture in the European starling. *Condor* 92:151–59.

Kildaw, S. D. 1995. The effect of group size manipulations on the foraging behavior of black-tailed prairie dogs. *Behav Ecol* 6:353–58.

Kilham, L. 1978. Alarm call of crested guan when attacked by ornate hawk-eagle. *Condor* 80:347–48.

Kilmon, J. A. Sr. 1976. High tolerance to snake venom by the Virginia possum, Didelphis virginiana. *Toxicon* 14:337–40.

Kilpi, M. 1987. Do herring gulls (*Larus argentatus*) invest more in offspring defence as the breeding season advances? *Ornis Fenn* 64:16–20.

Kiltie, R. A. 1985. Evolution and function of horns and hornlike organs in female ungulates. *Biol J Linn Soc* 24:299–320.

———. Countershading: Universally deceptive or deceptively universal? *Trends Ecol Evol* 3:21–23.

———. 1989a. Wildfire and the evolution of dorsal melanism in fox squirrels, *Sciurus niger*. *J Mammal* 70:726–39.

———. 1989b. Testing Thayer's countershading hypothesis: An image processing approach. *Anim Behav* 38:542–44.

———. 1992a. Camouflage comparisons among fox squirrels from Mississippi River delta. *J Mammal* 73:906–13.

———. 1992b. Tests of hypotheses on predation as a factor maintaining polymorphic melanism in coastal-plain fox squirrels (*Sciurus niger* L.). *Biol J Linn Soc* 45:17–37.

———. 2000. Scaling of visual acuity with body size in mammals and birds. *Funct Ecol* 14:226–34.

Kiltie, R. A., and A. F. Laine. 1992. Visual textures, machine vision and animal camouflage. *Trends Ecol Evol* 7:163–66.

King, J. A. 1955. Social behavior, social organization, and population dynamics in a black-tailed prairie dog town in the Black Hills of South Dakota. *Contrib Lab Vert Biol Univ Michigan* 67:1–123.

Kitchen, D. W. 1974. Social behavior and ecology of the pronghorn. *Wildl Monogr* 38:3–96.

Kleindorfer, S., B. Fessl, and H. Hoi. 2005. Avian nest defence behaviour: Assessment in relation to predator distance and type, and nest height. *Anim Behav* 69:307–13.

Kleindorfer, S., H. Hoi, and B. Fessl. 1996. Alarm calls and chick reactions in the moustached warbler, *Acrocephalus melanopogon*. *Anim Behav* 51:1199–1206.

Klemola, T., E. Korpimaki, and K. Norrdahl. 1998. Does avian predation risk depress reproduction of voles? *Oecologia* 115:149–53.

Klump, G. M., and E. Curio. 1983. Reactions of blue tits *Parus caerulus* to hawk models of different sizes. *Bird Behav* 4:78–81.

Klump, G. M., E. Kretzschmar, and E. Curio. 1986. The hearing of an avian predator and its avian prey. *Behav Ecol Sociobiol* 18:317–23.

Klump, G. M., and M. D. Shalter. 1984. Acoustic behaviour of birds and mammals in the predator context. I. Factors affecting the structure of alarm calls. II. The functional significance and evolution of alarm signals. *Z Tierpsychol* 66:189–226.

Knight, R. L., and S. A. Temple. 1986a. Methodological problems in studies of avian nest defence. *Anim Behav* 34:561–66.

———. 1986b. Nest defence in the American goldfinch. *Anim Behav* 34:887–97.

———. 1986c. Why does intensity of avian nest defense increase during the nesting cycle? *Auk* 103:318–27.

———. 1988. Nest-defense in the red-winged blackbird. *Condor* 90:193–200.

Knight, S. K., and R. L. Knight. 1986. Vigilance patterns of bald eagles feeding in groups. *Auk* 103:263–72.

Kobayashi, T., and M. Watanabe. 1986. An analysis of snake-scent application behaviour in Siberian chipmunks (*Eutamias sibiricus asiaticus*). *Ethology* 72:40–52.

Koenig, W. D., M. T. Stanback, P. N. Hooge, and R. L. Mumme. 1991. Distress calls in the acorn woodpecker. *Condor* 93:637–43.

Koeppl, J. W., R. S. Hoffmann, and C. F. Nadler. 1978. Pattern analysis of acoustical behavior in four species of ground squirrels. *J Mamm* 59:677–96.

Kohlmann, S. G., D. M. Muller, and P. U. Alkon. 1996. Antipredator constraints on lactating Nubian ibexes. *J Mammal* 77:1122–31.

Koivunen, V., E. Korpimaki, and H. Hakkarainen. 1998. Refuge sites of voles under owl predation risk: Priority of dominant individuals? *Behav Ecol* 9:261–66.

Komdeur, J., and R. H. K. Kats. 1999. Predation risk affects trade-off between nest guarding and foraging in Seychelles warblers. *Behav Ecol* 10:648–58.

Konishi, M. 1973. Locatable and nonlocatable acoustic signals for barn owls. *Am Nat* 107:775–85.

Kono, H., P. J. Reid, and A. C. Kamil. 1998. The effect of background cuing on prey detection. *Anim Behav* 963–72.

Korpimaki, E., V. Koivunen, and H. Hakkarainen. 1996. Microhabitat use and behavior of voles under weasel and raptor predation risk: Predator facilitation? *Behav Ecol* 7:30–34.

Korpimaki, E., K. Norrdahl, and J. Valkama. 1994. Reproductive investment under fluctuating predation risk: Microtine rodents and small mustelids. *Evol Ecol* 8:357–68.

Koskela, E., T. J. Horne, T. Mappes, and H. Ylonen. 1996. Does risk of small mustelid predation affect the oestrus cycle in the bank vole, *Clethrionomys glareolus*? *Anim Behav* 51:1159–63.

Koskela, E., and H. Ylonen. 1994. Suppressed breeding in the field vole (*Microtus agrestis*): An adaptation to cyclically fluctuating predation risk. *Behav Ecol* 6:311–15.

Kotler, B. P. 1984a. Risk of predation and the structure of desert rodent communities. *Ecology* 65:689–701.

———. 1984b. Harvesting rates and predatory risk in desert rodents: A comparison of two communities on different continents. *J Mammal* 65:91–96.

———. 1985. Owl predation on desert rodents which differ in morphology and behavior. *J Mammal* 66:824–28.

———. 1997. Patch use by gerbils in a risky environment: Manipulating food and safety to test four models. *Oikos* 78:274–82.

Kotler, B. P., Y. Ayal, and A. Subach. 1994. Effects of predatory risk and resource renewal on the timing of foraging activity in a gerbil community. *Oecologia* 100:391–96.

Kotler, B. P., L. Blaustein, and J. S. Brown. 1992. Predator facilitation: The combined effect of snakes and owls on the foraging behavior of gerbils. *Ann Zool Fenn* 29:199–206.

Kotler, B. P., and J. S. Brown. 1988. Environmental heterogeneity and the coexistence of desert rodents. *Annu Rev Ecol Syst* 19:281–307.

Kotler, B. P., J. S. Brown, S. R. X. Dall, S. Gresser, D. Ganey, and A. Bouskila. 2002. Foraging games between gerbils and their predators: Temporal dynamics of resource depletion and apprehension in gerbils. *Evol Ecol Res* 4:495–518.

Kotler, B. P, J. S. Brown, and O. Hasson. 1991. Factors affecting gerbil foraging behavior and rates of owl predation. *Ecology* 72:2249–60.

Kotler, B. P., J. S. Brown, and W. A. Mitchell. 1994. The role of predation in shaping the behaviour, morphology and community organisation of desert rodents. *Aust J Zool* 42:449–66.

Kotler, B. P., J. S. Brown, R. H. Slotow, W. L. Goodfriend, and M. Strauss. 1993. The influence of snakes on the foraging behavior of gerbils. *Oikos* 67:309–16.

Kotler, B. P., J. S. Brown, R. J. Smith, and W.O. Wirtz II. 1988. The effects of morphology and body size on rates of owl predation on desert rodents. *Oikos* 53:145–52.

Kotler, B. P., and R. D. Holt. 1989. Predation and competition: The interaction of two types of species interactions. *Oikos* 54:256–60.

Krakauer, D. C. 1995. Groups confuse predators by exploiting perceptual bottlenecks: A connectionist model of the confusion effect. *Behav Ecol Sociobiol* 36:421–29.

Krama, T., and I. Krams. 2005. Cost of mobbing call to breeding pied flycatcher, *Fidecula hypoleuca*. *Behav Ecol* 16:37–40.

Kramer, D. L., and M. Bonenfant. 1997. Direction of predator approach and the decision to flee to a refuge. *Anim Behav* 54:289–95.

Kramer, G., and U. von St Paul. 1951. Uber angeborenes und erworbenes Feinderkennen beim Gimpel (*Pyrrhula pyrrhula* L.). *Behaviour* 3:243–55.

Krams, I. 2000. Length of feeding day and body weight of great tits in a single- and two-predator environment. *Behav Ecol Sociobiol* 48:147–53.

———. Perch selection by singing chaffinches: A better view of surroundings and the risk of predation. *Behav Ecol* 12:295–300.

———. 2002. Mass-dependent take-off ability in wintering great tits (*Parus major*): Comparison of top-ranked adult males and subordinate juvenile females. *Behav Ecol Sociobiol* 51:345–49.

Kratzig, H. 1940. Untersuchungen zur Lebensweise des Moorschneehuhns (*Lagopus l. lagopus* L.) wahrend der Jugendentwicklung. *J Ornithol* 88:139–65.

Krause, J. 1993a. The relationship between foraging and shoal position in a mixed shoal of roach (*Rutilus rutilus*) and chub (*Leuciscus cephalus*): A field study. *Oecologia* 93:356–59.

———. 1993b. The effect of "Schreckstoff" on the shoaling behaviour of the minnow: A test of Hamilton's selfish herd theory. *Anim Behav* 45:1019–24.

————. 1993c. Positioning behaviour in fish shoals: A cost-benefit analysis. *J Fish Biol* 43 (Suppl. A):309–14.

————. 1994. Differential fitness returns in relation to spatial position in groups. *Biol Rev* 69:187–206.

Krause, J., D. Bumann, and D. Todt. 1992. Relationship between the position preference and nutritional state of individuals in schools of juvenile roach (*Rutilus rutilus*). *Behav Ecol Sociobiol* 30:177–80.

Krause, J., and J-G.J. Godin. 1994. Shoal choice in the banded killifish (*Fundulus diaphanus*, Teleostei, Cyprinodontidae): Effects of predation risk, fish size, species composition and size of shoals. *Ethology* 98:128–36.

————. 1995. Predator preferences for attacking particular prey group sizes: Consequences for predator hunting success and prey predation risk. *Anim Behav* 50:465–73.

————. 1996. Influence of prey foraging posture on flight behavior and predation risk: Predators take advantage of unwary prey. *Behav Ecol* 7:264–71.

Krause, J., and G. D. Ruxton. 2002. *Living in groups.* Oxford: Oxford University Press.

Krause, J., G. D. Ruxton, and D. Rubenstein. 1998. Is there always an influence of shoal size on predator hunting success? *J Fish Biol* 52:494–501.

Krebs, J. R. 1971. Territory and breeding density in the great tit (*Parus major* L.). *Ecology* 52:2–22.

————. 1973. Social learning and the significance of mixed-species flocks of chickadees (*Parus* spp.). *Can J Zool* 51:1275–88.

————. 1979. Bird colours. *Nature* 282:14–16.

————. 1980. Optimal foraging, predation risk, and territory defense. *Ardea* 68:83–90.

Krieber, M., and C. Barrette. 1984. Aggregation behaviour of harbour seals at Forillon National Park, Canada. *J Anim Ecol* 53:913–28.

Krivan, V. 1998. Effects of optimal antipredator behavior of prey on predator-prey dynamics: The role of the refuges. *Theoret Popul Biol* 53:131–42.

Kruuk, H. 1964. Predators and anti-predator behaviour of the black-headed gull (*Larus ridibundus* L.). *Behaviour* Suppl. 11:1–129.

————. 1972. *The spotted hyena: A study of predation and social behavior.* Chicago: University of Chicago Press.

————. 1976. The biological function of gull's attraction towards predators. *Anim Behav* 24:146–53.

Kullberg, C. 1998. Does diurnal variation in body mass affect take-off ability in wintering willow tits? *Anim Behav* 56:227–33.

Kullberg, C., T. Fransson, and S. Jakobsson. 1996. Impaired predator evasion in fat blackcaps (*Sylvia atricapilla*). *Proc R Soc Lond Ser B Biol Sci* 263:1671–75.

Kullberg, C., D. C. Houston, and N. B. Metcalfe. 2002. Impaired flight ability—A cost of reproduction in female blue tits. *Behav Ecol* 13:575–79.

Kullberg, C., S. Jakobsson, and T. Fransson. 1998. Predator-induced take-off strategy in great tits (*Parus major*). *Proc R Soc Lond Ser B Biol Sci* 265:1659–64.

————. 2000. High migratory fuel loads impair predator evasion in sedge warblers. *Auk* 117:1034–38.

Kullberg, C., and J. Lind. 2002. An experimental study of predator recognition in great tit fledglings. *Ethology* 108:429–41.

Kullberg, C., N. B. Metcalfe, and D. C. Houston. 2002. Impaired flight ability during incubation in the pied flycatcher. *J Avian Biol* 33:179–83.

Kulling, D., and M. Milinski. 1992. Size-dependent predation risk and partner quality in predator inspection of sticklebacks. *Anim Behav* 44:949–55.

Kunkel, K. E., T. K. Ruth, D. H. Pletcher, and M. D. Hornocker. 1999. Winter prey selection by wolves and cougars in and near Glacier National Park, Montana. *J Wildl Manag* 63:901–10.

Kus, B. E. 1986. Attack patterns of merlins hunting flocking sandpipers. In *Behavioural ecology and population biology*, ed. L. E. Drickamer, 133–38. Toulouse: Privat, I.E.C.

Lack, D. 1958. The significance of the colour of turdive eggs. *Ibis* 100:145–66.

———. 1968. *Ecological adaptations for breeding in birds*. London: Metheun.

LaGory, K. E. 1981. The possible communicative role of tail-flicking in white-tailed deer. *Anim Behav* 29:966.

———. 1986. Habitat, group size, and the behaviour of white-tailed deer. *Behaviour* 98:168–79.

———. 1987. The influence of habitat and group characteristics on the alarm and flight response of white-tailed deer. *Anim Behav* 35:20–25.

Lagos, V. O., L. C. Contreras, P. L. Meserve, J. R. Gutierrez, and F. M. Jaksic. 1995. Effects of predation risk on space use by small mammals: A field experiment with a neotropical rodent. *Oikos* 74:259–64.

Lahti, D. C. 2001. The "edge effect on nest predation" hypothesis after twenty years. *Biol Conserv* 99:365–74.

Lambrechts, M. M., B. Prieur, A. Caizergues, O. Dehorter, M-J. Galan, and P. Perret. 2000. Risk-taking restraints in a bird with reduced egg-hatching success. *Proc R Soc Lond Ser B Biol Sci* 267:333–38.

Lamprecht, J. 1978. On diet, foraging behaviour and interspecific food competition of jackals in the Serengeti National Park, East Africa. *Z Saeugetierk* 43:210–33.

Landeau, L., and J. Terborgh. 1986. Oddity and the "confusion effect" in predation. *Anim Behav* 34:1372–80.

Langley, C. M. 1996. Search images: Selective attention to specific visual features of prey. *J Exp Psychol Anim Behav Process* 22:152–63.

Lankford, T. E., J. M. Billerbeck, and D. O. Conover. 2001. Evolution of intrinsic growth and energy acquisition rates. II. Trade-offs with vulnerability to predation in *Menidia menidia*. *Evolution* 55:1873–81.

Lariviere, S. 1999. Reasons why predators cannot be inferred from nest remains. *Condor* 1 01:718–21.

Lariviere, S., and F. Messier. 1996. Aposematic behaviour in the striped skunk, *Mephitis mephitis*. *Ethology* 102:986–92.

———. 1998. Effect of density and nearest neighbours on simulated waterfowl nests: Can predators recognize high-density nesting patches? *Oikos* 83:12–20.

Larsen, T. 1991. Anti-predator behaviour and mating systems in waders: Aggressive nest defence selects for monogamy. *Anim Behav* 41:1057–62.

———. 2000. Influence of rodent density on nesting associations involving the bar-tailed godwit *Limosa lapponica*. *Ibis* 142:476–81.

Larsen, T., and S. Grundetjern. 1997. Optimal choice of neighbor: Predator protection among tundra birds. *J Avian Biol* 28:303–8.

Larsen, T., T. A. Sordahl, and I. Byrkjedal. 1996. Factors related to aggressive nest protection behaviour: A comparative study of Holarctic waders. *Biol J Linn Soc* 58:409–39.

Latimer, W. 1977. A comparative study of the songs and alarm calls of some *Parus* species. *Z Tierpsychol* 45:414–33.

Laundre, J. W., L. Hernandez, and K. B. Altendorf. 2001. Wolves, elk, and bison: Reestablishing the "landscape of fear" in Yellowstone National Park, U.S.A. *Can J Zool* 79:1401–9.

Laurenson, M. K. 1993. Early maternal behavior of cheetahs in the wild: Implications for captive husbandry. *Zoo Biol* 12:31–45.

————. 1994. High juvenile mortality in cheetahs (*Acinonyx jubatus*) and its consequences for maternal care. *J Zool (Lond)* 234:387–408.

Law, R. 1979. Optimal life histories under age-specific predation. *Am Nat* 114:399–417.

Lawrence, E. S. 1985a. Evidence for search image in blackbirds *Turdus merula* L.: Long-term learning. *Anim Behav* 33:1301–9.

————. 1985b. Vigilance during "easy" and "difficult" foraging tasks. *Anim Behav* 33:1373–75.

————. 1988. Why blackbirds overlook cryptic prey: Search rate or search image? *Anim Behav* 37:157–60.

Lawrence, E. S., and J. A. Allen. 1983. On the term "search image." *Oikos* 40:313–14.

Lazarus, J. 1978. Vigilance, flock size and domain of danger size in the white-fronted goose. *Wildfowl* 29:135–45.

————. 1979. The early warning function of flocking in birds: An experimental study with captive quelea. *Anim Behav* 27:855–65.

Lazarus, J., and I. R. Inglis. 1978. The breeding behaviour of the pink-footed goose: Parental care and vigilant behaviour during the fledgling period. *Behaviour* 65:62–88.

————. 1986. Shared and unshared parental investment, parent offspring conflict and brood size. *Anim Behav* 34:1791–1804.

Lazarus, J., and M. Symonds. 1992. Contrasting effects of protective and obstructive cover on avian vigilance. *Anim Behav* 43:519–21.

Leader, N., and Y. Yom-Tov. 1998. The possible function of stone ramparts at the nest entrance of the blackstart. *Anim Behav* 56:207–17.

Leal, M. 1999. Honest signaling during prey-predator interactions in the lizard *Anolis cristatellus*. *Anim Behav* 58:521–26.

Leal, M., and J. A. Rodriguez-Robles. 1995. Antipredator responses of *Anolis cristatellus* (Sauria: Polychrotidae). *Copeia* 1995:155–61.

————. 1997a. Antipredator responses of the Puerto Rican Giant anole, *Anolis cuvieri* (Squamata: Polychrotidae). *Biotropica* 29:372–75.

————. 1997b. Signalling displays during predator-prey interactions in a Puerto Rican anole, *Anolis cristatellus*. *Anim Behav* 54:1147–54.

Lee, S. J., M. S. Witter, I. C. Cuthill, and A. R. Goldsmith. 1996. Reduction in escape performance as a cost of reproduction in gravid starlings, *Sturnus vulgaris*. *Proc R Soc Lond Ser B Biol Sci* 263:619–24.

Leech, S. M., and M. L. Leonard. 1997. Begging and the risk of predation in nestling birds. *Behav Ecol* 8:644–46.

Leger, D. W., S. D. Berney-Key, and P. W. Sherman. 1984. Vocalizations of Belding's ground squirrels (*Spermophilus beldingi*). *Anim Behav* 32:753–64.

Leger, D. W., and D. H. Owings. 1978. Responses to alarm calls by California ground squirrels: Effects of call structure and maternal status. *Behav Ecol Sociobiol* 3:177–86.

Leger, D. W., D. H. Owings, and L. M. Boal. 1979. Contextual information and differential responses to alarm whistles in California ground squirrels. *Z Tierpsychol* 49:142–55.

Leger, D. W., D. H. Owings, and R. G. Coss. 1983. Behavioral ecology of time allocation in California ground squirrels (*Spermophilus beecheyi*): Microhabitat effects. *J Comp Psychol* 97:283–91.

Leger, D. W., D. H. Owings, and D. L. Gelfand. 1980. Single-note vocalizations of California ground squirrels: Graded signals and situation-specificity of predator and socially evoked calls. *Z Tierpsychol* 52:227–46.

Leimar, O., M. Enquist, and B. Sillen-Tullberg. 1986. Evolutionary stability of aposematic coloration and prey unprofitability: A theoretical analysis. *Am Nat* 128:469–90.

Lemmetyinen, R. 1971. Nest defence behaviour of common and arctic terns and its effects on the success achieved by predators. *Ornis Fenn* 48:13–24.

Lendrem, D. W. 1984a. Flocking, feeding and predation risk: Absolute and instantaneous feeding rates. *Anim Behav* 32:298–99.

———. 1984b. Sleeping and vigilance in birds: II. An experimental study of the barbary dove (*Streptopelia risoria*). *Anim Behav* 243–48.

Lent, P. C. 1974. Mother-infant relationships in ungulates. In *The behaviour of ungulates and its relation to management*, ed. V. Geist and F. Walther, 14–55. Morges, Switzerland: International Union for the Conservation of Nature and Natural Resources.

Lenti Boero, D. 1992. Alarm calling in Alpine marmot (*Marmota marmota* L.): evidence for semantic communication. *Ethol Ecol Evol* 4:125–38.

Le Roux, A., T. P. Jackson, and M. Cherry. 2001. Does Brant's whistling rat (*Parotomys brantsii*) use an urgency-based alarm system in reaction to aerial and terrestrial predators? *Behaviour* 138:757–73.

Lessels, C. M. 1987. Parental investment, brood size and time budgets: Behaviour of lesser snow goose families. *Ardea* 75:189–203.

———. 1991. The evolution of life-histories. In *Behavioural ecology: An evolutionary approach*, ed. J. R. Krebs and N. B. Davies, 32–68. Oxford: Blackwell Scientific Publications.

Lessels, C. M., K. R. Oddie, and A. C. Mateman. 1998. Parental behaviour is unrelated to experimentally manipulated great tit brood sex ratio. *Anim Behav* 56:385–93.

Leuthold, W. 1977. *African ungulates: A comparative review of their ethology and behavioural ecology.* New York: Springer-Verlag.

Lewis, D. C., E. S. Metallinos-Katsaras, and L. E. Grivetti. 1987. Coturnism: Human poisoning by European migratory quail. *J Cult Geog* 7:51–65.

Lewis, M. A., and J. D. Murray. 1993. Modelling territoriality and wolf-deer interactions. *Nature* 366:738–40.

Li, G., U. Roze, and D.C. Locke. 1997. Warning odor of the North American porcupine (*Erethizon dorsatum*). *J Chem Ecol* 23:2737–54.

Li, P., and T. E. Martin. 1991. Nest-site selection and nesting success of cavity-nesting birds in high elevation forest drainages. *Auk* 108:405–18.

Lilliendahl, K. 1997. The effect of predator presence on body mass in captive greenfinches. *Anim Behav* 53:75–81.

———. 1998. Yellowhammers get fatter in the presence of a predator. *Anim Behav* 55:1335–40.

———. 2000. Daily accumulation of body reserves under increased predation risk in captive greenfinches *Carduelis chloris*. *Ibis* 142:587–95.

Lima, S. L. 1985. Maximizing feeding efficiency and minimizing time exposed to predators: A trade-off in the black-capped chickadee. *Oecologia* 66:60–67.

———. 1986. Predation risk and unpredictable feeding conditions: Determinants of body mass in birds. *Ecology* 67:377–85.

———. 1987a. Vigilance while feeding and its relation to risk of predation. *J Theor Biol* 124:303–16.

———. 1987b. Distance to cover, visual obstructions, and vigilance in house sparrows. *Behaviour* 102:231–38.

———. 1987c. Clutch size in birds: A predation perspective. *Ecology* 68:1062–70.

———. 1988a. Initiation and termination of daily feeding in dark-eyed juncos: Influences of predation risk and reserves. *Oikos* 53:3–11.

———. 1988b. Vigilance and diet selection: The classical diet model reconsidered. *J Theor Biol* 132:127–43.

———. 1988c. Vigilance during the initiation of daily feeding in dark-eyed juncos. *Oikos* 53:12–16.

———. 1989. Iterated Prisoner's Dilemma: An approach to evolutionary stable co-operation. *Am Nat* 134:828–34.

———. 1990a. The influence of models on the interpretation of vigilance. In *Interpretation and explanation in the study of animal behaviour*, ed. M. Bekoff and D. Jamieson, 246–67. Boulder, CO: Westview Press.

———. 1990b. Protective cover and the use of space: Different strategies in finches. *Oikos* 58:151–58.

———. 1992a. Life in a multi-predator environment: Some considerations for anti-predatory vigilance. *Ann Zool Fen* 29:217–26.

———. 1992b. Strong preferences for apparently dangerous habitats? A consequence of differential escape from predators. *Oikos* 64:597–600.

———. 1993. Ecological and evolutionary perspectives on escape from predatory attack: A survey of North American birds. *Wilson Bull* 105:1–47.

———. 1994a. Collective detection of predatory attack by birds in the absence of alarm signals. *J Avian Biol* 25:319–26.

———. 1994b. On the personal benefits of anti-predatory vigilance. *Anim Behav* 48:734–36.

———. 1995a. Back to basics of anti-predatory vigilance: The group-size effect. *Anim Behav* 49:11–20.

———. 1995b. Collective detection of predatory attack by social foragers: Fraught with ambiguity? *Anim Behav* 50:1097–1108.

———. 1998. Stress and decision making under the risk of predation: Recent developments from behavioral, reproductive, and ecological perspectives. *Adv Study Behav* 27:215–90.

———. 2002. Putting predators back into behavioral predator-prey interactions. *Trends Ecol Evol* 17:70–75.

Lima, S. L., and P. A. Bednekoff. 1999a. Temporal variation in danger drives antipredator behavior: The predation risk allocation hypothesis. *Am Nat* 153:649–59.

———. 1999b. Back to basics of antipredatory vigilance: Can nonvigilant animals detect attack? *Anim Behav* 58:537–43.

Lima, S. L., and L. M. Dill. 1990. Behavioral decisions made under the risk of predation: A review and prospectus. *Can J Zool* 68:619–40.

Lima, S. L., and T. J. Valone. 1986. Influence of predation risk on diet selection: A simple example in the grey squirrel. *Anim Behav* 34:536–44.

Lima, S. L., T. J. Valone, and T. Caraco. 1985. Foraging efficiency-predation risk tradeoff in the gray squirrel. *Anim Behav* 33:1555–65.

Lima, S. L., K. L. Wiebe, and L. M. Dill. 1987. Protective cover and the use of space by finches: Is closer better? *Oikos* 50:225–30.

Lima, S. L., and P. A. Zollner. 1996. Anti-predatory vigilance and the limits to collective detection: Visual and spatial separation between foragers. *Behav Ecol Sociobiol* 38:355–63.

Lima, S. L., P. A. Zollner, and P. A. Bednekoff. 1999. Predation, scramble competition, and the vigilance group size effect in dark-eyed juncos (*Junco hyemalis*). *Behav Ecol Sociobiol* 46:110–16.

Lind, J., T. Fransson, S. Jakobsson, and C. Kullberg. 1999. Reduced take-off ability in robins (*Erithacus rubecula*) due to migratory fuel load. *Behav Ecol Sociobiol* 46:65–70.

Lind, J., L. Hollen, E. Smedberg, U. Svensson, A. Vallin, and S. Jakobsson. 2003. Detection distance influences escape behaviour in two parids, *Parus major* and *P. caerulus*. *J Avian Biol* 34:233–36.

Lind, J., U. Kaby, and S. Jakobsson, 2002. Split-second escape decisions in blue tits (*Parus caerulus*). *Naturwissenschaften* 89:420–23.

Lindell, C. 1996. Benefits and costs to plain-fronted thornbirds (*Phacellodomus rufifrons*) of interactions with avian nest associates. *Auk* 113:565–77.

Lindstedt, S. L., J. F. Hokanson, D. J. Wells, S. D. Swain, H. Hoppeler, and V. Navarro. 1991. Running energetics in the pronghorn antelope. *Nature* 353:748–50.

Lindstrom, A. 1989. Finch flock size and risk of hawk predation at a migratory stopover site. *Auk* 106:225–32.

Lindstrom, L., R. V. Alatalo, A. Lyytinen, and J. Mappes. 2001. Predator experience on cryptic prey affects the survival of conspicuous aposematic prey. *Proc R Soc Lond Ser B Biol Sci* 268:357–61.

Lindstrom, L., R. V. Alatalo, J. Mappes, M. Rippi, and L. Vertainen. 1999. Can aposematic signals evolve by gradual change? *Nature* 397:249–51.

Lindstrom, L., R. V. Alatalo, and J. Mappes. 1999. Reactions of hand-reared and wild caught predators toward warningly colored, gregarious, and conspicuous prey. *Behav Ecol* 10:317–22.

Lingle, S. 1992. Escape gaits of white-tailed deer, mule deer and their hybrids: Gaits observed and patterns of limb coordination. *Behaviour* 122:153–81.

———. 1993. Escape gaits of white-tailed deer, mule deer and their hybrids: Body configuration, biomechanics and function. *Can J Zool* 71:708–24.

———. 2001. Anti-predator strategies and grouping patterns in white-tailed deer and mule deer. *Ethology* 107:295–314.

———. 2002. Coyote predation and habitat segregation of white-tailed deer and mule deer. *Ecology* 83:2037–48.

Lingle, S., and S. M. Pellis. 2002. Fight or flight? Antipredator behavior and the escalation of coyote encounters with deer. *Oecologia* 131:154–64.

Lingle, S., and W. F. Wilson. 2001. Detection and avoidance of predators in white-tailed deer (*Odocoileus virginianus*) and mule deer (*O. hemionus*). *Ethology* 107:125–47.

Lipetz, V. E., and M. Bekoff. 1980. Possible functions of predator harassment in pronghorn antelopes. *J Mammal* 61:741–43.

———. 1982. Group size and vigilance in pronghorns. *Z Tierpsychol* 58:203–16.

Lishak, R. S. 1984. Alarm vocalizations of adult gray squirrels. *J Mammal* 65:681–84.

Listoen, C., R. F. Karlsen, and T. Slagsvold. 2000. Risk taking during parental care: A test of the harm-to-offspring hypothesis. *Behav Ecol* 11:40–43.

Lloyd, M., and H. S. Dybas. 1966. The periodical cicada problem. *Evolution* 30:133–44.

Lloyd, P., E. Plaganyi, D. Lepage, R. M. Little, and T. M. Crowe. 2000. Nest-site selection, egg pigmentation and clutch predation in the ground-nesting namaqua sandgrouse. *Pterocles namaqua*. *Ibis* 142:123–31.

Lockard, R. B., and D. H. Owings. 1974. Moon-related surface activity of bannertail (*Dipodomys spectabilis*) and Fresno (*D. nitratoides*) kangaroo rats. *Anim Behav* 22:262–73.

Loman, J., and G. Goransson. 1978. Egg shell dumps and crow (*Corvus cornix*) predation on simulated bird's nests. *Oikos* 30:461–66.

Longland, W. S. 1991. Risk of predation and food consumption by black-tailed jackrabbits. *J Range Manag* 44:447–50.

Longland, W. S., and M. V. Price. 1991. Direct observations of owls and heteromyid rodents: Can predation risk explain microhabitat use? *Ecology* 72:2261–73.

Lord, A., J. R. Waas, J. Innes, and M. J. Whittingham. 2001. Effects of human approaches to nests of northern New Zealand dotterels. *Biol Conserv* 98:233–40.

Lorenz, K. 1939. Vergleichende Verhaltensforschung. *Zool Anz* Suppl. 12:69–102.

Losos, J. B. 1990. The evolution of form and function: Morphology and locomotor performance in West Indian *Anolis* lizards. *Evolution* 44:1189–1203.

Losos, J. B., P. L. F. N. Mouton, R. Bickel, I Cornelius, and L. Ruddock. 2002. The effect of body armature on escape behaviour in cordylid lizards. *Anim Behav* 64:313–21.

Lotem, A., R. H. Wagner, and S. Balshine-Earn. 1999. The overlooked signaling component of non-signaling behavior. *Behav Ecol* 10:209–12.

Loughry, W. J. 1987a. The dynamics of snake harassment by black-tailed prairie dogs. *Behaviour* 103:27–48.

———. 1987b. Differences in experimental and natural encounters of black-tailed prairie dogs with snakes. *Anim Behav* 35:1568–70.

———. 1988. Population differences in how black-tailed prairie dogs deal with snakes. *Behav Ecol Sociobiol* 22:61–67.

———. 1993. Mechanisms of change in the ontogeny of black-tailed prairie dog time budgets. *Ethology* 95:54–64.

Loughry, W. J., and C. M. McDonough. 1988. Calling and vigilance in California ground squirrels: A test of the tonic communication hypothesis. *Anim Behav* 36:1533–40.

———. 1989. Calling and vigilance in California ground squirrels: Age, sex and seasonal differences in responses to calls. *Am Midl Nat* 121:312–21.

Lovegrove, B. G. 2001. The evolution of body armor in mammals: Plantigrade constraints of large body size. *Evolution* 55:1464–73.

Lyon, B. E., and R. D. Montgomerie. 1985. Conspicuous plumage of birds: Sexual selection or unprofitable prey? *Anim Behav* 33:1038–40.

Lythgoe, J. N. 1979. *The ecology of vision.* Oxford: Oxford University Press.

Macdonald, D. W. 1977. On food preference in red fox. *Mammal Review* 7:7–23.

———, ed. 1984. *The encyclopedia of mammals.* London: Allen & Unwin.

———. 1985. The carnivores: Order Carnivora. In *Social odours in mammals*, ed. R. E. Brown and D. W. Macdonald, 619–722. Oxford: Clarendon Press.

MacDougall, A., and M. S. Dawkins. 1998. Predator discrimination error and the benefits of Mullerian mimicry. *Anim Behav* 55:1281–88.

Macedonia, J. M. 1990. What is communicated in the antipredator calls of lemurs: Evidence from playback experiments with ringtailed and ruffed lemurs. *Ethology* 86:177–90.

Macedonia, J. M., and C. S. Evans. 1993. Variation among mammalian alarm call systems and the problem of meaning in alarm signals. *Ethology* 93:177–97.

MacHutchon, A. G., and A. S. Harestad. 1990. Vigilance behaviour and use of rocks by Columbian ground squirrels. *Can J Zool* 68:1428–32.

MacLaren, P. I. R. 1950. Bird-ant nesting associations. *Ibis* 92:564–66.

MacWhirter, R. B. 1991. Effects of reproduction on activity and foraging behaviour of adult female Columbian ground squirrels. *Can J Zool* 69:2209–16.

———. 1992. Vocal and escape responses of Columbian ground squirrels to simulated terrestrial and aerial predator attacks. *Ethology* 91:311–25.

Maestripieri, D. 1992. Functional aspects of maternal aggression in mammals. *Can J Zool* 70:1069–77.

———. 1993. Vigilance costs of allogrooming in macaque mothers. *Am Nat* 141:744–53.

Magnhagen, C. 1991. Predation risk as a cost of reproduction. *Trends Ecol Evol* 6:183–86.

Magurran, A. E. 1989. Acquired recognition of predator odour in the European minnow (*Phoxinus phoxinus*). *Ethology* 82:216–23.

———. 1990a. The inheritance and development of minnow anti-predator behaviour. *Anim Behav* 39:834–42.

———. 1990b. The adaptive significance of schooling as an anti-predator defence in fish. *Ann Zool Fenn* 27:51–66.

————. 1999. The causes and consequences of geographic variation in antipredator behavior: Perspectives from fish populations. In *Geographic variation in behavior: Perspectives on evolutionary mechanisms*, ed. S. A. Foster and J. A. Endler, 139–63. New York: Oxford University Press.

Magurran, A. E., and S. L. Girling. 1986. Predator model recognition and response habituation in shoaling minnows. *Anim Behav* 34:510–18.

Magurran, A. E., and T. J. Pitcher. 1987. Provenance, shoal size and the sociobiology of predator evasion behaviour in minnow shoals. *Proc R Soc Lond Ser B Biol Sci* 229:439–65.

Magurran, A. E., and B. H. Seghers. 1994. Predator inspection behaviour covaries with schooling tendency amongst wild guppy, *Poecilia reticulata*, populations in Trinidad. *Behaviour* 128:121–34.

Magurran, A. E., B. H. Seghers, P. W. Shaw, and G. R. Carvalho. 1995. The behavioural diversity and evolution of guppy, *Poecilia reticulata* populations in Trinidad. *Adv Study Behav* 24:155–202.

Maier, V. 1982. Acoustic communication in the Guinea fowl (*Numida meleagris*): Structure and use of vocalizations, and the principle of message coding. *Z Tierpsychol* 59:29–83.

Maier, V., O. A. E. Rasa, and H. Scheich. 1983. Call-system similarity in a ground-living social bird and a mammal in the bush habitat. *Behav Ecol Sociobiol* 12:5–9.

Majerus, M. E. N. 1996. *Melanism: Evolution in action*. Oxford: Oxford University Press.

Major, P. F. 1978. Predator-prey interactions in two schooling fishes, *Caranx ignobilis* and *Stolephorous purpureus*. *Anim Behav* 26:760–77.

Major, R. E., and C. E. Kendal. 1996. The contribution of artificial nest experiments to understanding avian reproductive success: A review of methods and conclusions. *Ibis* 138:298–307.

Maklakov, A. A. 2002. Snake-directed mobbing in a cooperative breeder: Anti-predator behaviour or self-advertisement for the formation of dispersal coalitions? *Behav Ecol Sociobiol* 52:372–78.

Mallet, J., and M. Juron. 1999. Evolution of diversity in warning color and mimicry: Polymorphisms, shifting balance, and speciation. *Annu Rev Ecol Syst* 30:201–33.

Mallet, J., and M. C. Singer. 1987. Individual selection, kin selection, and the shifting balance in the evolution of warning colours: The evidence from butterflies. *Biol J Linn Soc* 32:337–50.

Maloney, R. F., and I. G. McLean. 1995. Historical and experimental learned predator recognition in free-living New Zealand robins. *Anim Behav* 50:1193–1201.

Manser, M. B. 1999. Response of foraging group members to sentinel calls in suricates, *Suricata suricatta*. *Proc R Soc Lond Ser B Biol Sci* 266:1013–19.

Mappes, J., and R. V. Alatalo. 1997. Effects of novelty and gregariousness in survival of aposematic prey. *Behav Ecol* 8:174–77.

Mappes, T., E. Koskela, and H. Ylonen. 1998. Breeding suppression in voles under predation risk of small mustelids: Laboratory or methodological artifact? *Oikos* 82:365–69.

Marchetti, K., T. Price, and A. Richman. 1995. Correlates of wing morphology with foraging behaviour and migration distance in the genus *Phylloscopus*. *J Avian Biol* 26:177–81.

Marden, J. H. 1987. Maximum lift production during takeoff in flying animals. *J Exp Biol* 130:235–58.

Marini, M. A., S. K. Robinson, and E. J. Heske. 1995. Edge effects on nest predation in the Shawnee National Forest, Southern Illinois. *Biol Conserv* 64:203–14.

Marion, K. R., and O. J. Sexton. 1979. Protective behavior by male pronghorn, *Antilocapra americana* (Artiodactyla). *Southwest Nat* 24:709–10.

Marler, P. 1955. Characteristics of some animal calls. *Nature* 176:6–8.

————. 1956. The voice of the chaffinch and its function as a language. *Ibis* 98:231–61.

————. 1957. Specific distinctiveness in communication signals of birds. *Behaviour* 11:13–39.

————. 1959. Developments in the study of animal communication. In *Darwin's biological work*, ed. P. R. Bell, 150–206. Cambridge: Cambridge University Press.

Marler, P., and W. J. Hamilton III. 1966. *Mechanisms of animal behavior.* New York: John Wiley.

Marples, N. M., and D. J. Kelly. 1999. Neophobia and dietary conservatism: Two distinct processes? *Evol Ecol* 13:641–53.

Marples, N. M., and T. J. Roper. 1996. Effects of novel colour and smell on the response of naïve chicks towards food and water. *Anim Behav* 51:1417–24.

Marples, N. M., T. J. Roper, and D. G. C. Harper. 1998. Responses of wild birds to novel prey: Evidence of dietary conservatism. *Oikos* 83:161–65.

Marsh, R. L. 1988. Ontogenesis of contractile properties of skeletal muscle and sprint performance in the lizard *Dipsosaurus dorsalis. J Exp Biol* 137:119–39.

Marshall, C. D., and J. F. Eisenberg. 1996. *Hemicentetes semispinosus. Mamm Species* 541:1–4.

Marten, K. and P. Marler. 1977. Sound transmission and its significance for animal vocalization. I. Temperate habitats. *Behav Ecol Sociobiol* 2:271–90.

Marten, K., D. Quine, and P. Marler. 1977. Sound transmission and its significance for animal vocalization. II. Tropical forest habitats. *Behav Ecol Sociobiol* 2:291–302.

Martin, D. J. 1973. A spectrographic analysis of burrowing owl vocalizations. *Auk* 90:564–78.

Martin, J., and P. Lopez. 1999. When to come out from a refuge: Risk-sensitive and state-dependent decisions in an alpine lizard. *Behav Ecol* 10:487–92.

Martin, K. 1984. Reproductive defence priorities of male willow ptarmigan (*Lagopus lagopus*): Enhancing mate survival or extending paternity options? *Behav Ecol Sociobiol* 16:57–63.

Martin, K., and A. G. Horn. 1993. Clutch defense by male and female willow ptarmigan *Lagopus lagopus. Ornis Scand* 24:261–66.

Martin, P. S., and T. E. Martin. 2001. Ecological and fitness consequences of species coexistence: A removal experiment with wood warblers. *Ecology* 82:189–206.

Martin, P. S., and C. R. Szuter. 1999. War zones and game sinks in Lewis and Clark's west. *Cons Biol* 13:36–45.

Martin, T. E. 1987. Artificial nest experiments: Effects of nest appearance and type of predator. *Condor* 89:925–28.

———. 1988a. Processes organizing open-nesting bird assemblages: Competition or nest predation? *Evol Ecol* 2:37–50.

———. 1988b. On the advantages of being different: Nest predation and the coexistence of bird species. *Proc Natl Acad Sci USA* 85:2196–99.

———. 1992a. Breeding productivity considerations: What are the appropriate habitat features for management? In *Ecology and conservation of neotropical migrant landbirds*, ed. J. M. Hagan and D. W. Johnston, 455–73. Washington, DC: Smithsonian Institution Press.

———. 1992b. Interaction of nest predation and food limitation in reproductive strategies. *Curr Ornithol* 9:163–97.

———. 1993a. Nest predation among vegetation layers and habitat types: Revising the dogmas. *Am Nat* 141:897–913.

———. 1993b. Nest predation and nest sites: New perspectives on old patterns. *BioScience* 43:523–32.

———. 1995. Avian life history evolution in relation to nest sites, nest predation and food. *Ecol Monogr* 65:101–27.

———. 1996. Fitness costs of resource overlap among coexisting bird species. *Nature* 380:338–40.

———. 1998. Are microhabitat preferences of coexisting species under selection and adaptive? *Ecology* 79:656–70.

Martin, T. E., and A. V. Badyaev. 1996. Sexual dichromatism in birds: Importance of nest predation and nest location for females versus males. *Evolution* 50:2454–60.

Martin, T. E., and C. K. Ghalambor. 1999. Males feeding females during incubation. I. Required by microclimate or constrained by nest predation? *Am Nat* 153:131–39.

Martin, T. E., and P. Li. 1992. Life history traits of open- vs cavity-nesting birds. *Ecology* 73:579–92.

Martin, T. E., P. R. Martin, C. R. Olson, B. J. Heidinger, and J. J. Fontaine. 2000. Parental care and clutch sizes in North and South American birds. *Science* 287:1482–85.

Martin, T. E., and J. J. Roper. 1988. Nest predation and nest-site selection of a western population of the hermit thrush. *Condor* 90:51–57.

Martin, T. E., J. Scott, and C. Menge. 2000. Nest predation increases with parental activity: Separating nest site and parental activity effects. *Proc R Soc Lond Ser B Biol Sci* 267:2287–93.

Marzluff, J. M. 1988. Do pinyon jays alter nest placement based on prior experience? *Anim Behav* 36:1–10.

Mason, P., and S. I. Rothstein. 1987. Crypsis versus mimicry and the color of shiny cowbird eggs. *Am Nat* 130:161–67.

Mateo, J. M. 1996a. The development of alarm-call response behaviour in free-living juvenile Belding's ground squirrels. *Anim Behav* 52:489–505.

———. 1996b. Early auditory experience and the ontogeny of alarm-call discrimination in Belding's ground squirrel (*Spermophilus beldingi*). *J Comp Psychol* 110:115–24.

Mateo, J. M., and W. G. Holmes. 1997. Development of alarm-call responses in Belding's ground squirrels: The role of dams. *Anim Behav* 54:509–24.

———. 1999a. Plasticity of alarm-call response development in Belding's ground squirrels (*Spermophilus beldingi*, Sciuridae). *Ethology* 105:193–206.

———. 1999b. How rearing history affects alarm-call responses of Belding's ground squirrels (*Spermophilus beldingi*, Sciuridae). *Ethology* 105:207–22.

Matessi, G., and G. Bogliani. 1999. Effects of nest features and surrounding landscape on predation rates of artificial nests. *Bird Study* 46:184–94.

Mathis, A., and D. P. Chivers. 2003. Overriding the oddity effect in mixed-species aggregations: Group choice by armored and nonarmored prey. *Behav Ecol* 14:334–39.

Matocha, K. G. 1977. The vocal repertoire of *Spermophilus tridecemlineatus*. *Am Midl Nat* 98:482–87.

Matsuda, H., P. A. Abrams, and M. Hori. 1993. The effect of adaptive anti-predator behavior on exploitative competition and mutualism between predators. *Oikos* 68:549–59.

Matsuoka, S. 1980. Pseudo warning call in titmice. *Tori* 29:87–90.

Maynard Smith, J. 1965. The evolution of alarm calls. *Am Nat* 99:59–63.

———. 1977. Parental investment: A prospective analysis. *Anim Behav* 25:1–9.

Maynard Smith, J., and D. Harper. 2003. *Animal signals.* Oxford: Oxford University Press.

McAdam, A. G., and D. L. Kramer. 1998. Vigilance as a benefit of intermittent locomotion in small mammals. *Anim Behav* 55:109–17.

McDonough, C. M., and W. J. Loughry. 1995. Influences on vigilance in nine-banded armadillos. *Ethology* 100:50–60.

McGowan, K. J., and G. E. Woolfenden. 1989. A sentinel system in the Florida scrub jay. *Anim Behav* 37:1000–1006.

McLean, E. B., and J-G. J. Godin. 1989. Distance to cover and fleeing from predators in fish with different amounts of defensive armour. *Oikos* 55:281–90.

McLean, I. G. 1983. Paternal behaviour and infanticide in Arctic ground squirrels. *Anim Behav* 31:32–44.

———. 1987. Response to a dangerous enemy: Should a brood parasite be mobbed? *Ethology* 75:235–45.

McLean, I. G., C. Holzer, and B. J. S. Srudholme. 1999. Teaching predator-recognition to a naïve bird: Implications for management. *Biol Conserv* 87:123–30.

McLean, I. G., and G. Rhodes. 1991. Enemy recognition and response in birds. *Curr Ornithol* 8:173–211.

McLean, I. G., J. N. M. Smith, and K. G. Stewart. 1986. Mobbing behaviour, nest exposure, and breeding success in the American robin. *Behaviour* 96:171–86.

McLeery, R. H., and C. M. Perrins. 1991. Effects of predation on the numbers of great tits Parus major. In *Bird population studies: Relevance to conservation and management*, ed. C. M. Perrins, J.-D. Lebreton, and G. J. M. Hirons, 129–47. Oxford: Oxford University Press.

McManus, J. J. 1970. Behavior of captive opossums, *Didelphis marsupialis virginiana*. *Am Midl Nat* 84:144–69.

McNamara, J. M. 1990. The starvation-predation trade-off and some behavioural and ecological consequences. In *Behavioural mechanisms of food selection*, ed. R. N. Hughes, 39–59. Berlin: Springer-Verlag.

McNamara, J. M., and A. I. Houston. 1986. The common currency for behavioral decisions. *Am Nat* 127:358–78.

———. 1990. The value of fat reserves and the tradeoff between starvation and predation. *Acta Biotheor* 38:37–61.

———. 1992. Evolutionarily stable levels of vigilance as a function of group size. *Anim Behav* 43:641–58.

McNamara, J. M., A. I. Houston, and S. L. Lima. 1994. Foraging routines of small birds in winter: A theoretical investigation. *J Avian Biol* 25:287–302.

McNeil, R., P. Drapeau, and R. Pierotti. 1993. Nocturnality in colonial waterbirds: Occurrence, special adaptations, and suspected benefits. *Curr Ornithol* 10:187–246.

McNicholl, M. K. 1973. Habituation of aggressive responses to avian predators by terns. *Auk* 90:902–4.

Mech, L. D. 1970. *The wolf: The ecology and behavior of an endangered species*. Minneapolis: University of Minnesota Press.

———. 1977. Wolf-pack buffer zones as prey reservoirs. *Science* 198:320–21.

Mech, L. D., and R. O. Peterson. 2003. Wolf-prey relations. In *Wolves*, ed. L. D. Mech and L. Boitani, 131–60. Chicago: University of Chicago Press.

Meinertzhagen, R. 1959. *Pirates and predators*. Edinburgh: Oliver & Boyd.

Melchior, H. R. 1971. Characteristics of arctic ground squirrel alarm calls. *Oecologia* 7:184–90.

Merilaita, S. 1998. Crypsis through disruptive coloration in an isopod. *Proc R Soc Lond Ser B Biol Sci* 256:1–6.

Merilaita, S., J. Tuomi, and V. Jormalainen. 1999. Optimization of cryptic coloration in heterogeneous habitats. *Biol J Linn Soc* 67:151–61.

Merriam, C. H. 1890. Results of a biological survey of the San Francisco Mountain region and desert of the Little Colorado, Arizona III: Annotated list of mammals, with descriptions of new species. *US Dep Agric Div Ornith and Mamm, N Amer Fauna* 3:43–86.

Messier, F., and C. Barrette. 1985. The efficiency of yarding behaviour by white-tailed deer as an antipredator strategy. *Can J Zool* 63:785–89.

Metcalfe, N. B. 1984a. The effects of habitat on the vigilance of shorebirds: Is visibility important? *Anim Behav* 32:981–85.

———. 1984b. The effects of mixed-species flocking on the vigilance of shorebirds: Who do they trust? *Anim Behav* 32:986–93.

Metcalfe, N. B., and S. E. Ure. 1995. Diurnal variation in flight performance and hence potential predation risk in small birds. *Proc R Soc Lond Ser B Biol Sci* 261:395–400.

Metz, K. J., and C. D. Ankney. 1991. Are brightly coloured male ducks selectively shot by duck hunters? *Can J Zool* 69:279–82.

Metzgar, L. H. 1967. An experimental comparison of screech owl predation on resident and transient white-footed mice (*Peromyscus leucopus*). *J Mammal* 48:387–91.

Michener, G. R. 1976. Tail autotomy as an escape mechanism in *Rattus rattus*. *J Mammal* 57:600–603.

Michl, G., J. Torok, L. Z. Garamszegi, and L. Toth. 2000. Sex-dependent risk taking in the collared flycatcher, *Ficedula albicollis*, when exposed to a predator at the nestling stage. *Anim Behav* 59:623–28.

Mikolajewski, D. J., and F. Johansson. 2004. Morphological and behavioral defenses in dragonfly larvae: Trait compensation and cospecialization. *Behav Ecol* 15:614–20.

Milinski, M. 1977a. Experiments on the selection by predators against spatial oddity of their prey. *Z Tierpsychol* 43:311–25.

———. 1977b. Do all members of a swarm suffer the same predation? *Z Tierpsychol* 45:373–88.

———. 1984. A predator's costs of overcoming the confusion-effect of swarming prey. *Anim Behav* 32:1157–62.

———. 1987. TIT FOR TAT in sticklebacks and the evolution of cooperation. *Nature* 325:433–37.

Milinski, M., and P. Boltshauser. 1995. Boldness and predator deterrence: A critique of Godin & Davis. *Proc R Soc Lond Ser B Biol Sci* 262:103–5.

Milinski, M., and R. Heller. 1978. Influence of a predator on the optimal foraging behaviour of sticklebacks (*Gasterosteus aculeatus* L.). *Nature* 275:642–44.

Milinski, M., J. H. Luthi, R. Eggler, and G. A. Parker. 1997. Cooperation under predation risk: Experiments on costs and benefits. *Proc R Soc Lond Ser B Biol Sci* 264:831–37.

Milinski, M., and G. A. Parker. 1991. Competition for resources. In *Behavioural ecology: An evolutionary approach*, ed. J. R. Krebs and N. B. Davies, 3rd ed., 137–68. Oxford: Blackwell Scientific Publications.

Miller, D. B., G. Hicinbothom, and C. F. Blaich. 1990. Alarm call responsivity of mallard ducklings: Multiple pathways in behavioural development. *Anim Behav* 39:1207–12.

Miller, F. L., and A. Gunn. 1984. Muskox defense formations in response to helicopters in the Canadian high arctic. *Biol Pap Univ Alsk Spec Rep* 4:123–26.

Miller, R. C. 1922. The significance of the gregarious habit. *Ecology* 3:122–26.

Mineka, S., and M. Cook. 1988. Social learning and the acquisition of snake fear in monkeys. In *Social learning: Psychological and biological perspectives*, ed. T. R. Zentall and B. G. Galef Jr., 51–73. Hillsdale, NJ: Lawrence Erlbaum Associates.

Moen, A. N., M. A. DellaFera, A. L. Heller, and B. A. Buxton. 1978. Heart rates of white-tailed deer fawns in response to recorded wolf howls. *Can J Zool* 56:1207–10.

Moholt, R. K., and C. Trost. 1989. Self-advertisement: Relations to dominance in black-billed magpies. *Anim Behav* 38:1079–88.

Moller, A. P. 1984. Parental defence of offspring in the barn swallow. *Bird Behaviour* 5:110–17.

———. 1987a. Egg predation as a selective factor for nest design: An experiment. *Oikos* 50:91–94.

———. 1987b. Advantages and disadvantages of coloniality in the swallow, *Hirundo rustica*. *Anim Behav* 35:819–32.

———. 1988a. Nest predation and nest site choice in passerine birds in habitat patches of different size: A study of magpies and blackbirds. *Oikos* 53:215–21.

———. 1988b. False alarm calls as a means of resource usurpation in the great tit *Parus major*. *Ethology* 79:25–30.

———. 1989a. Nest site selection across field-woodland ecotones: The effect of nest predation. *Oikos* 56:240–46.

———. 1989b. Deceptive use of alarm calls by male swallows, *Hirundo rustica*: A new paternity guard. *Behav Ecol* 1:1–16.

———. 1990. Nest predation selects for small nest size in the blackbird. *Oikos* 57:237–40.

Molvar, E. M., and R. T. Bowyer. 1994. Costs and benefits of group living in a recently social ungulate: The Alaskan moose. *J Mammal* 75:621–30.

Monaghan, P., and N. B. Metcalfe. 1985. Group foraging in wild brown hares: Effects of resource distribution and social status. *Anim Behav* 33:993–99.

Montevecchi, W. A. 1976. Field experiments on the adaptive significance of avian eggshell pigmentation. *Behaviour* 58:26–39.

Montgomerie, R., B. Lyon, and K. Holder. 2001. Dirt plumage: Behavioral modification of conspicuous male plumage. *Behav Ecol* 12:429–38.

Montgomerie, R. D., and P. J. Weatherhead. 1988. Risks and rewards of nest defence by parent birds. *Q Rev Biol* 63:167–87.

Mooring, M. S., and B. J. Hart. 1995. Costs of allogrooming in impala: Distraction from vigilance. *Anim Behav* 49:1414–16.

Moran, G. 1984. Vigilance behaviour and alarm calls in a captive group of meerkats, *Suricata suricatta*. *Z Tierpsychol* 65:228–40.

Morgan, M. J., and J-G. J. Godin. 1985. Antipredator benefits of schooling behaviour in a cyprinodontid fish, the banded killifish (*Fundulus diaphanus*). *Z Tierpsychol* 70:236–46.

Morrison, D. W. 1978. Lunar phobia in a neotropical fruit bat, *Artibeus jamaicensis* (Chiroptera: Phyllostomidae). *Anim Behav* 26:852–55.

Morse, D. H. 1977. Feeding behavior and predator avoidance in heterospecific groups. *BioScience* 27:332–39.

Morton, E. S. 1975. Ecological sources of selection of avian sounds. *Am Nat* 109:17–34.

Morton, E. S., and M. D. Shalter. 1977. Vocal responses to predators in pair-bonded Carolina wrens. *Condor* 79:222–27.

Morton, T. L., J. W. Haefner, V. Nugala, R. D. Decino, and L. Mendes. 1994. The selfish herd revisited: Do simple movement rules reduce relative predation risk? *J Theor Biol* 167:73–79.

Mottram, J. C. 1915. Some observations on pattern-blending with reference to obliterative shading and concealment of outline. *Proc Zool Soc Lond* 1915:679–92.

Mougeot, F., and V. Bretagnolle. 2000. Predation as a cost of sexual communication in nocturnal seabirds: An experimental approach using acoustic signals. *Anim Behav* 60:647–56.

Moynihan, M. 1962. The organization and probable evolution of some mixed species flocks of neotropical birds. *Smithson Misc Collect* 143:1–140.

Mueller, H. C. 1971. Oddity and specific searching image more important than conspicuousness in prey selection. *Science* 233:345–46.

———. 1974. Factors influencing prey selection in the American kestrel. *Auk* 91:705–21.

———. 1975. Hawks select odd prey. *Science* 188:953–54.

Mueller, H. C., and P. G. Parker. 1980. Naïve ducklings show different cardiac response to hawk than to goose models. *Behaviour* 74:101–13.

Muller-Schwarze, D. 1972. Responses of young black-tailed deer to predator odors. *J Mammal* 53:393–94.

Munn, C. A. 1986a. Birds that "cry wolf." *Nature* 319:143–45.

———. 1986b. The deceptive use of alarm calls by sentinel species in mixed-species flocks of neotropical birds. In *Deception: Perspectives in human and nonhuman deceit*, ed. R. W. Mitchell and N. S. Thompson, 169–75. Albany: State University of New York.

Munn, C. A., and J. W. Terborgh. 1980. Multi-species territoriality in neotropical foraging flocks. *Condor* 81:338–47.

Munro, J., and J. Bedard. 1977. Gull predation and creching behaviour in the common eider. *J Anim Ecol* 46:799–810.

Murcia, C. 1995. Edge effects in fragmented forests: Implications for conservation. *Trends Ecol Evol* 10:58–62.

Murie, A. 1944. *The wolves of Mount McKinley.* Fauna of the National Parks of the U.S. Series no. 5.

Murphy, M. T., C. T. Cummings, and M. S. Palmer. 1997. Comparative analysis of habitat selection, nest-site and nest success by cedar waxwings (*Bombycilla cedrorum*) and eastern kingbirds (*Tyrannus tyrannus*). *Am Midl Nat* 138:344–56.

Murray, D. L., S. Boutin, M. O'Donoghue, and V. O. Nams. 1995. Hunting behaviour of a sympatric felid and canid in relation to vegetative cover. *Anim Behav* 50:1203–10.

Myers, J. G. Jr. 1935. Nesting associations with social insects. *Trans Entomol Soc Lond* 83:11–22.

Naguib, M., R. Mundry, R. Ostreiher, H. Hultsch, L. Schrader, and D. Todt. 1999. Cooperatively breeding Arabian babblers call differently when mobbing in different predator-induced situations. *Behav Ecol* 10:636–40.

Neal, E., and C. Cheeseman. 1996. *Badgers.* London: T & AD Poyser.

Neill, S. R. St. J., and J. M. Cullen. 1974. Experiments on whether schooling by their prey affects the hunting behaviour of cephalopods and fish predators. *J Zool (Lond)* 172:549–69.

Nelson, M. E., and L. D. Mech. 1985. Observation of a wolf killed by a deer. *J Mammal* 66:187–88.

Neuchterlein, G. L. 1981. "Information parasitism" in mixed colonies of western grebes and Forster's terns. *Anim Behav* 29:985–89.

Neudorf, D. L., and S. G. Sealy. 2002. Distress calls of birds in a neotropical cloud forest. *Biotropica* 34:118–26.

Newman, C., C. D. Buesching, and J. D. Wolff. 2005. The function of facial masks in "midguild" carnivores. *Oikos* 108:623–33.

Newman, J. A., and T. Caraco. 1987. Foraging, predation hazard and patch use in grey squirrels. *Anim Behav* 35:1804–13.

Newman, J. A., G. M. Recer, S. M. Zwicker, and T. Caraco. 1988. Effects of predation hazard on foraging "constraints": Patch-use strategies in grey squirrels. *Oikos* 53:93–97.

Newton, I. 1979. *Population ecology of raptors.* Vermillion, SD: Buteo Books.

Newton, P. N. 1989. Associations between langur monkeys (*Presbytis entellus*) and chital deer (*Axis axis*): Chance encounters or a mutualism? *Ethology* 83:89–120.

Nice, M. M. 1943. Studies in the life history of the song sparrow. II. The behavior of the song sparrow and other passerines. *Trans Linn Soc NY* 6:1–328.

———. 1957. Nest success in altricial birds. *Auk* 74:305–21.

Nicholson, M. C., R. T. Bowyer, and J. G. Kie. 1997. Habitat selection and survival of mule deer: Tradeoffs associated with migration. *J Mammal* 78:483–504.

Nilsson, S. G. 1984. The evolution of nest-site selection among hole-nesting birds: The importance of nest predation and competition. *Ornis Scand* 15:167–75.

Nilsson, S. G., K. Jonhsson, and M. Tjernberg. 1991. Is avoidance by black woodpeckers of old nest holes due to predators? *Anim Behav* 41:439–41.

Nisbet, I. C. T. 1975. Selective effects of predation in a tern colony. *Condor* 77:221–26.

Noe, R., and R. Bshary. 1997. The formation of red colobus-diana monkey associations under predation pressure from chimpanzees. *Proc R Soc Lond Ser B Biol Sci* 264:253–59.

Nolan, V. Jr. 1963. Reproductive success of birds in a deciduous scrub habitat. *Ecology* 44:305–13.

———. 1978. The ecology and behavior of the prairie warbler *Dendroica discolor. Ornithol Monogr* 26. American Ornithologists Union.

Nolte, D. L., J. R. Mason, G. Epple, E. Aronov, and D. L. Campbell. 1994. Why are predator urines aversive to prey? *J Chem Ecol* 20:1505–16.

Nonacs, P., and L. M. Dill. 1990. Mortality risk vs. food quality trade-offs in a common currency: Ant patch preferences. *Ecology* 71:1886–92.

Nordstrom, C. A. 2002. Haul-out selection by Pacific harbor seals (*Phoca vitulina richardii*): Isolation and perceived predation risk. *Mar Mamm Sci* 18:194–205.

Norrdahl, K., J. Suhonen, O. Hemminki, and E. Korpimaki. 1995. Predator presence may benefit: Kestrels protect curlew nests against nest predators. *Oecologia* 101:105–9.

Nowak, R. M. 1999. *Walker's mammals of the world*. 6th ed. Baltimore: Johns Hopkins University Press.

Noyes, D. H., and W. G. Holmes. 1979. Behavioral responses of free-living hoary marmots to a model golden eagle. *J Mamm* 60:408–11.

Nuechterlein, G. L. 1981. "Information parasitism" in mixed colonies of western grebes and Forster's terns. *Anim Behav* 29:985–89.

Nunez, R., B. Miller, and F. Lindzey. 2000. Food habits of jaguars and pumas in Jalisco, Mexico. *J Zool* 252:373–79.

Nur, N., and O. Hasson. 1984. Phenotypic plasticity and the handicap principle. *J Theor Biol* 110:275–97.

Oda, R. 1998. The responses of Verreaux's sifakas to anti-predator alarm calls given by sympatric ring-tailed lemurs. *Folia Primatol* 69:357–60.

Oda, R., and N. Masataka. 1996. Interspecific responses of ringtailed lemurs to playback of antipredator alarm calls given by Verreaux's sifakas. *Ethology* 102:441–53.

O'Donald, P. 1983. *The arctic skua: A study of the ecology and evolution of a seabird*. Cambridge: Cambridge University Press.

OED. 1973. *The Shorter Oxford English Dictionary on Historical Principles*. 3rd ed. Prepared by W. Little, H. W. Fowler, and J. Coulson; revised and edited by C. T. Onions. Oxford: Clarendon Press.

O'Farrell, M. J. 1974. Seasonal activity patterns of rodents in a sagebrush community. *J Mammal* 55:809–23.

Ohguchi, O. 1978. Experiments on the selection against colour oddity of water fleas by three-spined sticklebacks. *Z Tierpsychol* 47:254–67.

———. 1981. *Prey density and selection against oddity by three-spined sticklebacks*. Berlin: Verlag Paul Parey.

Olupot, W., and P. M. Waser. 2001. Activity patterns, habitat use and mortality risks of mangabey males living outside social groups. *Anim Behav* 61:1227–35.

Oniki, S. 1979. Is nesting success of birds low in the tropics? *Biotropica* 11:60–69.

Oniki, Y. 1985. Why robin eggs are blue and birds build nests: Statistical tests for Amazonian birds. In *Neotropical ornithology. Ornithological monographs 36*, ed. P. A. Buckley, M. S. Foster, E. S. Morton, R. S. Ridgely, and F. G. Buckley, 536–45. Washington, DC: American Ornithologists' Union.

Onnebrink, H., and E. Curio. 1991. Brood defense and age of young: A test of the vulnerability hypothesis. *Behav Ecol Sociobiol* 29:61–68.

O'Reilly, P., and S. J. Hannon. 1989. Predation of simulated willow ptarmigan nests: The influence of density and cover on spatial and temporal patterns of predation. *Can J Zool* 67:1263–67.

Orrock, J. L., B. J. Danielson, and R. J. Brinkerhoff. 2004. Rodent foraging is affected by indirect, but not by direct, cues of predation risk. *Behav Ecol* 15:433–37.

Orsdol, K. G. van. 1984. Foraging behaviour and hunting success of lions in Queen Elizabeth National Park, Uganda. *Afr J Ecol* 22:79–99.

Ortolani, A. 1999. Spots, stripes, tail tips and dark eyes: Predicting the function of carnivore colour patterns using the comparative method. *Biol J Linn Soc* 67:433–76.

Ortolani, A., and T. M. Caro. 1996. The adaptive significance of coat patterns in carnivores: Phylogenetic tests of classic hypotheses. In *Carnivore behavior, ecology and evolution*, ed. J. Gittleman, 132–88. Ithaca, NY: Cornell University Press.

Osgood, W. H. 1909. A revision of the mice of the American genus *Peromyscus*. *US Dep Agric Bur Biol Surv N Amer Fauna* 28.

Osterholm, H. 1964. The significance of distance receptors in the feeding behaviour of the fox, *Vulpes vulpes* L. *Acta Zool Fenn* 106:3–31.

Owen, D. 1980. *Camouflage and mimicry*. Chicago: University of Chicago Press.

Owen, M. 1972. Some factors affecting food intake and selection in white-fronted geese. *J Anim Ecol* 41:79–92.

Owen-Smith, R. N. 1988. *Megaherbivores: The influence of very large body size on ecology*. Cambridge: Cambridge University Press.

Owens, N. 1977. Responses of wintering brent geese to human disturbance. *Wildfowl* 28:5–14.

Owens, N. W., and J. D. Goss-Custard. 1976. The adaptive significance of alarm calls given by shorebirds on their winter feeding grounds. *Evolution* 30:397–98.

Owings, D. H., and R. G. Coss. 1977. Snake mobbing by California ground squirrels: Adaptive variation and ontogeny. *Behaviour* 62:50–69.

Owings, D. H., R. G. Coss, D. McKernon, M. P. Rowe, and P. C. Arrowood. 2001. Snake-directed antipredator behavior of rock squirrels (*Spermophilus variegates*): Population differences and snake-species discrimination. *Behaviour* 138:575–95.

Owings, D. H., and D. F. Hennessy. 1984. The importance of variation in scuirid visual and vocal communication. In *The biology of ground-dwelling squirrels*, ed. J. O. Murie and G. R. Michener, 169–200. Lincoln: University of Nebraska Press.

Owings, D. H., D. F. Hennessy, D. W. Leger, and A. B. Gladney. 1986. Different functions of "alarm" calling for different time scales: A preliminary report on ground squirrels. *Behaviour* 99:101–16.

Owings, D. H., and D. W. Leger. 1980. Chatter vocalizations of California ground squirrels: Predator- and social-role specificity. *Z Tierpsychol* 54:163–84.

Owings, D. H., and W. J. Loughry. 1985. Variation in snake-elicited jump-yipping by black-tailed prairie dogs: Ontogeny and snake-specificity. *Z Tierpsychol* 70:177–200.

Owings, D. H., and S. C. Owings. 1979. Snake-directed behavior by black-tailed prairie dogs (*Cynomys ludovicianus*). *Z Tierpsychol* 49:35–54.

Owings, D. H., and R. A. Virginia. 1978. Alarm calls of California ground squirrels (*Spermophilus beecheyi*). *Z Tierpsychol* 46:58–70.

Packer, C. 1979. Male dominance and reproductive activity in *Papio anubis*. *Anim Behav* 27:37–45.

———. 1983. Sexual dimorphism: The horns of African antelopes. *Science* 221:1191–93.

Packer, C., and P. Abrams. 1990. Should co-operative groups be more vigilant than selfish groups? *J Theor Biol* 142:341–57.

Packer, C., D. Scheel, and A. E. Pusey. 1990. Why lions form groups: Food is not enough. *Am Nat* 136:1–19.

Page, G. W., L. E. Stenzel, D. W. Winkler, and C. W. Swarth. 1983. Spacing out at Mono lake: Breeding success, nest density and predation in the snowy plover. *Auk* 100:13–24.

Page, G., and D. F. Whitacre. 1975. Raptor predation on wintering shorebirds. *Condor* 77:73–83.

Palmer, W. 1909. Instinctive stillness in birds. *Auk* 26:23–36.

Palomares, F., and T. M. Caro. 1999. Interspecific killing among mammalian carnivores. *Am Nat* 153:492–508.

Parker, G. A., and P. Hammerstein. 1985. Game theory and animal behavior. In *Essays in honour of John Maynard Smith*, ed. P. J. Greenwood, P. H. Harvey, and M. Slatkin, 73–94. New York: Cambridge University Press.

Parker, G. R. 1977. Morphology, reproduction, diet, and behavior of the Arctic hare (*Lepus arcticus monstrabilis*) on Axel Heiberg Island, Northwest Territories. *Can Field-Nat* 91:8–18.

Parrish, J. K. 1989. Re-examining the selfish herd: Are central fish safer? *Anim Behav* 38:1048–53.

———. 1993. Comparison of the hunting behavior of four piscine predators attacking schooling prey. *Ethology* 95:233–46.

Parrish, K., S. W. Strand, and J. L. Lott. 1989. Predation on a school of flat-iron herring, *Harengula thrissina*. *Copeia* 1989:1089–91.

Parsons, G. J., and S. Bondrup-Nielsen. 1996. Experimental analysis of behaviour of meadow voles (*Microtus pennsylvanicus*) to odours of the short-tailed weasel (*Mustela erminea*). *Ecoscience* 3:63–69.

Partridge, L. 1978. Habitat selection. In *Behavioral ecology: An evolutionary approach*, ed. J. R. Krebs and N. B. Davies, 351–76. Oxford: Blackwell Scientific Publications.

Passamani, M. 1995. Field observation of a group of Geoffroy's marmosets mobbing a margay cat. *Folia Primatol* 64:163–66.

Paton, P. W. C. 1994. The effect of edge on avian nest success: How strong is the evidence? *Cons Biol* 8:17–26.

Patterson, I. J. 1965. Timing and spacing of broods in the black-headed gull. *Ibis* 107:433–59.

Patterson, T. L., L. Petrinovich, and D. K. James. 1980. Reproductive value and appropriateness of response to predators by white-crowned sparrows. *Behav Ecol Sociobiol* 7:227–31.

Pavel, V., and S. Bures. 2001. Offspring age and nest defence: Test of the feedback hypothesis in the meadow pipit. *Anim Behav* 61:297–303.

Pavel, V., S. Bures, K. Weidinger, and P. Kovarik. 2000. Distraction displays in meadow pipit (*Anthus pratensis*) females in central and northern Europe. *Ethology* 106:1007–19.

Pavey, C. R., and A. K. Smyth. 1998. Effects of avian mobbing on roost use and diet of powerful owls, *Ninox strenua*. *Anim Behav* 55:313–18.

Pearl, R. 1911. Data on the relative conspicuousness of barred and self-colored fowls. *Am Nat* 45:107–17.

Pearson, D. L. 1989. What is the adaptive significance of multicomponent defensive repertoires? *Oikos* 54:251–53.

Pedersen, H. C., and J. B. Steen. 1985. Parental care and chick production in a fluctuating population of willow ptarmigan. *Ornis Scand* 16:270–76.

Pedley, T. J. 1977. *Scale effects in animal locomotion.* London: Academic Press.

Pennycuick, C. J. 1975. Mechanics of flight. In *Avian biology*, ed. D. S. Farner and J. R. King, 1–75. New York: Academic Press.

———. 1997. Actual and "optimum" flight speeds: Field data reassessed. *J Exp Biol* 200:2355–61.

Percival, A. B. 1924. *A game ranger's note book.* London: Nisbet & Co. Ltd.

Pereira, M. E., and J. M. Macedonia. 1991. Ringtailed lemur anti-predator calls denote predator class, not response urgency. *Anim Behav* 41:543–44.

Peres, C. A. 1993. Anti-predation benefits in a mixed-species group of Amazonian tamarins. *Folia Primatol* 61:61–76.

Perez, J. C., W. C. Haws, V. E. Garcia, and B. M. Jennings III. 1978. Resistance of warm-blooded animals to snake venoms. *Toxicon* 16:375–83.

Perez, J. C., S. Pichyangkul, and V. E. Garcia. 1979. The resistance of three species of warm-blooded animals to western diamondback rattlesnake (*Crotalus atrox*) venom. *Toxicon* 17:601–7.

Perla, B. S., and C. N. Slobodchikoff. 2002. Habitat structure and alarm call dialects in Gunnison's prairie dog (*Cynomys gunnisoni*). *Behav Ecol* 13:844–50.

Perrins, C. M. 1965. Population fluctuations and clutch size in the great tit, *Parus major* L. *J Anim Ecol* 34:601–47.

———. 1968. The purpose of the high-intensity alarm call in small passerines. *Ibis* 110:200–201.

Perrone, M. Jr. 1980. Factors affecting the incidence of distress calls in passerines. *Wilson Bull* 92:404–8.

Perrone, M. Jr., and D. R. Paulson. 1979. Incidence of distress calls in mist-netted birds. *Condor* 81:423–24.

Perry, R. 1938. *At the turn of the tide.* London: Lindsay Drummond.

Peters, R. H. 1983. *The ecological implications of body size.* Cambridge: Cambridge University Press.

Petit, D. R., and K. L. Bildstein.1987. Effect of group size and location within the group on the foraging behaviour of white ibises. *Condor* 89:602–9.

Petit, K. E., L. J. Petit, and D. R. Petit. 1989. Fecal sac removal: Do the pattern and distance of dispersal affect the chance of nest predation? *Condor* 91:479–82.

Pettifor, R. A. 1990. The effects of avian mobbing on a potential predator, the European kestrel, *Falco tinnunculus. Anim Behav* 821–27.

Picman, J. 1988. Experimental study of predation on eggs of ground-nesting birds: Effects of habitat and nest distribution. *Condor* 90:124–31.

Picman, J., M. Leonard, and A. Horn. 1988. Antipredation role of clumped nesting by marsh-nesting red-winged blackbirds. *Behav Ecol Sociobiol* 22:9–15.

Pielowski, Z. 1959. Studies of the relationship: Predator (goshawk)-prey (pigeon). *Bull Acad Pol Sci Ser Sci Biol* Cl II, 7:401–3.

Pienkowski, M., and P. R. Evans. 1982. Breeding behaviour, productivity and survival of colonial and non-colonial shelducks *Tadorna tadorna. Ornis Scand* 13:101–16.

Pierce, B. M., V. C. Bleich, and R. T. Bowyer. 2000. Selection of mule deer by mountain lions and coyotes, effects of hunting style, body size and reproductive status. *J Mammal* 81:461–72.

Pierce, B. M., W. S. Longland, and S. H. Jenkins. 1992. Rattlesnake predation on desert rodents: Microhabitat and species-specific effects on risk. *J Mammal* 73:859–65.

Pietrewicz, A. T., and A. C. Kamil. 1979. Search image formation in the blue jay (*Cyanocitta cristata*). *Science* 204:1332–33.

Pinheiro, C. E. G. 1996. Palatability and escaping ability in neotropical butterflies: Tests with wild kingbirds (*Tyrannus melancholicus*, Tyrannidae). *Biol J Linn Soc* 59:351–65.

Pitcher, T. J. 1980. Some ecological consequences of fish school volumes. *Freshw Biol* 10:539–44.

Pitcher, T. J., D. A. Green, and A. E. Magurran. 1986. Dicing with death: Predator inspection behaviour in minnow shoals. *J Fish Biol* 28:439–48.

Pitcher, T. J., and J. K. Parrish. 1993. Functions of shoaling behaviour in teleosts. In *Behaviour of teleost fishes*, 2nd ed., ed. T. J. Pitcher, 363–439. London: Chapman & Hall.

Pitcher, T. J., and J. R. Turner. 1986. Danger at dawn: Experimental support for the twilight hypothesis in shoaling minnows. *J Fish Biol* 29 Suppl. A:59–70.

Pius, S. M., and P. L. Leberg. 1998. The protector species hypothesis: Do black skimmers find refuge from predators in gull-billed tern colonies? *Ethology* 104:273–84.

Plaistead, K. C., and N. J. Mackintosh. 1995. Visual search for cryptic stimuli in pigeons: Implications for the search image and search rate hypotheses. *Anim Behav* 50:1219–32.

Poduschka, V. W. 1974. Augendrusensekretionen bei den Tenreciden *Setifer setosus* (Froriep 1806), *Echinops telfairi* (Martin 1838), *Microgale dobsoni* (Thomas 1918) und *Microgale talazaci* (Thomas 1918). *Z Tierpsychol* 35:303–19.

Pongracz, P. and V. Altbacker. 2000. Ontogeny of the responses of European rabbits (*Oryctolagus cuniculus*) to aerial and ground predators. *Can J Zool* 78:655–65.

Poole, J. H. 1989. Announcing intent: The aggressive state of musth in African elephants. *Anim Behav* 37:140–52.

Popp, J. W. 1988. Scanning behavior of finches in mixed-species groups. *Condor* 90:510–12.

Poran, N. S., and R. G. Coss. 1990. Development of antisnake defenses in California ground squirrels (Spermophilus beecheyi): I. Behavioral and immunological relationships. *Behaviour* 112:222–45.

Poran, N. S., R. G. Coss, and E. Benjamini. 1987. Resistance of California ground squirrels (*Spermophilus beecheyi*) to the venom of the northern pacific rattlesnake (*Crotalus viridis oreganos*): A study of adaptive radiation. *Toxicon* 25:767–77.

Post, W., and C. A. Seals. 1993. Nesting associations of least bitterns and boat-tailed grackles. *Condor* 95:139–44.

Pough, F.H. 1988. Mimicry of vertebrates: Are the rules different? *Am Nat* 131:S67–S102.

Poulton, E. B. 1890. *The colours of animals: Their meaning and use especially in the case of insects.* London: Kegan Paul, Trench, Trubner & Co.

———. 1898. Natural selection: The cause of mimetic resemblance and common warning colours. *J Linn Soc Lond Zool* 26:558–612.

———. 1902. The meaning of the white undersides of animals. *Nature* 65:596.

Powell, F., and P. B. Banks. 2004. Do house mice modify their foraging behaviour in response to predator odours and habitat? *Anim Behav* 67:753–59.

Powell, G. V. N. 1974. Experimental analysis of the social value of flocking by starlings (*Sturnus vulgaris*) in relation to predation and foraging. *Anim Behav* 22:501–5.

Powell, L. A., and L. L. Frasch. 2000. Can nest predation and predator type explain variation in dispersal of adult birds during the breeding season? *Behav Ecol* 11:437–43.

Powell, R. A. 1982. Evolution of black-tipped tails in weasels: Predator confusion. *Am Nat* 1982:126–31.

Powell, R. A., and R. B. Brander. 1977. Adaptations of fishers and porcupines to their predator prey system. In *Proceedings of the 1975 Predatory Symposium*, ed. R. L. Phillips and C. Jonkel, 45–53. Missoula: University of Montana.

Poysa, H. 1985. Changes in predator surveillance in a foraging great tit *Parus major* in response to presence and group size of yellow-hammers *Emberiza citrinella*. *Ornis Fenn* 62:140–42.

———. 1987a. Feeding-vigilance trade-off in the teal (*Anas crecca*): Effects of feeding method and predation risk. *Behaviour* 103:108–22.

———. 1987b. Costs and benefits of group foraging in the teal (*Anas crecca*). *Behaviour* 103:123–40.

———. 1994. Group foraging, distance to cover and vigilance in the teal, *Anas crecca*. *Anim Behav* 48:921–28.

Pravosudov, V. V., and T. C. Grubb Jr. 1998a. Body mass, ambient temperature, time of day, and vigilance in tufted titmice. *Auk* 115:221–23.

———. 1998b. Management of fat reserves in tufted titmice *Baelophus bicolor* in relation to risk of predation. *Anim Behav* 56:49–54.

———. 1999. Effects of dominance on vigilance in avian social groups. *Auk* 116:241–46.

Price, M. V., Waser, N. M., and Bass, T. A. 1984. Effects of moonlight on microhabitat use by desert rodents. *J Mammal* 65:353–56.

Prins, H. H. T., and G. R. Iason. 1989. Dangerous lions and nonchalant buffalo. *Behaviour* 108:262–96.

Proctor, C. J., M. Broom, and G. D. Ruxton. 2001. Modelling antipredator vigilance and flight response in group foragers when warning signals are ambiguous. *J Theor Biol* 211:409–17.

Promislow, D. E. L., and P. H. Harvey. 1990. Living fast and dying young: A comparative analysis of life-history variation among mammals. *J Zool (Lond)* 220:417–37.

Promislow, D. E. L., R. Montgomerie, and T. E. Martin. 1992. Mortality costs of sexual dimorphism in birds. *Proc R Soc Lond Ser B Biol Sci* 250:143–50.

Pugesek, B. H. 1983. The relationship between parental age and reproductive effort in the California gull (*Larus californicus*). *Behav Ecol Sociobiol* 13:161–71.

Pulliam, H. R. 1973. On the advantages of flocking. *J Theor Biol* 38:419–22.

Pulliam, H. R., G. H. Pyke, and T. Caraco. 1982. The scanning behavior of juncos: A game-theoretical approach. *J Theor Biol* 95:89–103.

Pyare, S., and J. Berger. 2003. Beyond demography and delisting: Ecological recovery for Yellowstone's grizzly bears and wolves. *Biol Conserv* 113:63–73.

Quenette, P-Y. 1990. Functions of vigilance behaviour in mammals: A review. *Acta Oecologica* 11:801–18.

Quick, H. F. 1953. Occurrence of porcupine quills in carnivorous mammals. *J Mammal* 34:256–59.

Quinn, J. L., and W. Cresswell. 2004. Predator hunting behaviour and prey vulnerability. *J Anim Ecol* 73:143–54.

Quinn, J. L., and Y. Kokorev. 2002. Trading-off risks from predators and from aggressive hosts. *Behav Ecol Sociobiol* 51:455–60.

Quinn, J. L., J. Prop, Y. Kokorev, and J.M. Black. 2003. Predator protection or similar habitat selection in red-breasted goose nesting associations: Extremes along a continuum. *Anim Behav* 65:297–307.

Rachlow, J. L., and R. T. Bowyer. 1998. Habitat selection by Dall's sheep (*Ovis dalli*): Maternal trade-offs. *J Zool* 245:457–65.

Radford, A. N., and J. K. Blakey. 2000. Intensity of nest defence is related to offspring sex ratio in the great tit *Parus major*. *Proc R Soc Lond Ser B Biol Sci* 267:535–38.

Ralls, K., K. Kranz, and B. Lundrigan.1986. Mother-young relationships in captive ungulates: Variability and clustering. *Anim Behav* 34:134–45.

Ralls, K., B. Lundrigan, and K. Kranz. 1987. Mother-young relationships in captive ungulates: Behavioral changes over time. *Ethology* 75:1–14.

Ramakrishnan, U., and R. G. Coss. 2000a. Recognition of heterospecific alarm vocalizations by bonnet macaques (*Macaca radiata*). *J Comp Psychol* 114:3–12.

———. 2000b. Age differences in the responses to adult and juvenile alarm calls by bonnet macaques (*Macaca radiata*). *Ethology* 106:131–44.

———. 2002. Strategies used by bonnet macaques (*Macaca radiata*) to reduce predation risk while sleeping. *Primates* 42:193–206.

Randall, J. A., and D. K. Boltas King. 2001. Assessment and defence of solitary kangaroo rats under risk of predation by snakes. *Anim Behav* 61:579–87.

Randall, J. A., S. M. Hatch, and E. R. Hekkala. 1995. Inter-specific variation in anti-predator behavior in sympatric species of kangaroo rat. *Behav Ecol Sociobiol* 36:243–50.

Randall, J. A., and M. D. Matocq. 1997. Why do kangaroo rats (*Dipodomys spectabilis*) footdrum at snakes? *Behav Ecol* 8:404–13.

Randall, J. A., K. A. Rogovin, and D. M. Shier. 2000. Antipredator behavior of a social desert rodent: Footdrumming and alarm calling in the great gerbil, *Rhombomys opiums*. *Behav Ecol Sociobiol* 48:110–18.

Randall, J. A., and C. M. Stevens. 1987. Footdrumming and other anti-predator responses in the bannertail kangaroo rat (*Dipodomys spectabilis*). *Behav Ecol Sociobiol* 20:187–94.

Rands, S. A., and I. C. Cuthill. 2001. Separating the effects of predation risk and interrupted foraging upon mass changes in the blue tit *Parus caeruleus. Proc R Soc Lond Ser B Biol Sci* 268:1783–90.

Rangen, S. A., R. G. Clark, and K. A. Hobson. 1999. Influence of nest-site vegetation and predator community on the success of artificial songbird nests. *Can J Zool* 77:1676–81.

———. 2000. Visual and olfactory attributes of artificial nests. *Auk* 117:136–46.

Ranta, E., K. Lindstrom, and N. Peuhkuri. 1992. Size matters when three-spined sticklebacks go to school. *Anim Behav* 43:160–62.

Ranta, E., N. Peuhkuri, H. Hirvonen, and C. J. Barnard. 1998. Producers, scroungers and the price of a free meal. *Anim Behav* 55:737–44.

Rasa, O. A. E. 1983. Dwarf mongoose and hornbill mutualism in the Taru desert, Kenya. *Behav Ecol Sociobiol* 12:181–90.

———. 1986. Coordinated vigilance in dwarf mongoose family groups: The "Watchman's Song" hypothesis and the costs of guarding. *Ethology* 71:340–44.

———. 1987. The dwarf mongoose: A study of behaviour and social structure in relation to ecology in a small carnivore. *Adv Study Behav* 17:121–60.

———. 1989a. The costs and effectiveness of vigilance behaviour in the dwarf mongoose: Implications for fitness and optimal group size. *Ethol Ecol Evol* 1:265–82.

———. 1989b. Behavioural parameters of vigilance in the dwarf mongoose: Social acquisition of sex-biased role. *Behaviour* 110:125–45.

Ratner, S. C. 1967. Comparative aspects of hypnosis. In *Handbook of clinical and experimental hypnosis*, ed. J. E. Gordon, 550–87. New York: Macmillan.

———. 1975. Animal's defenses: Fighting in predator-prey relations. In *Nonverbal communication of aggression*, ed. P. Pliner, L. Krames, and T. Alloway, 175–90. New York: Plenum Press.

Ratner, S. C., and R. W. Thompson. 1960. Immobility reactions (fear) of domestic fowl as a function of age and prior experience. *Anim Behav* 8:186–91.

Rattenborg, N. C., S. L. Lima, and C. J. Amlaner. 1999a. Half-awake to the risk of predation. *Nature* 397:397–98.

———. 1999b. Facultative control of avian unihemispheric sleep under risk of predation. *Behav Brain Res* 105:163–72.

Ratti, J. T., and K. P. Reese. 1988. Preliminary test of the ecological trap hypothesis. *J Wildl Manag* 52:484–91.

Rayner, J. M. V. 1993. On aerodynamics and the energetics of vertebrate flapping flight. In *Fluid dynamics in biology*, ed. A. Y. Cheer and C. P. van Dam, 351–400. Providence, RI: American Mathematical Society.

Rayor, L. S., and G. W. Uetz. 1990. Trade-offs in foraging success and predation risk with spatial position in colonial spiders. *Behav Ecol Sociobiol* 27:77–85.

———. 1993. Ontogenetic shifts within the selfish herd: Predation risk and foraging trade-offs change with age in colonial web-building spiders. *Oecologia* 95:1–8.

Reby, D., B. Cargnelutti, and A. J. M. Hewison. 1999. Contexts and possible functions of barking in roe deer. *Anim Behav* 57:1121–28.

Redondo, T. 1989. Avian nest defence: Theoretical models and evidence. *Behaviour* 110:161–95.

Redondo, T., and L. Arias de Reyna. 1988. Locatability of begging calls in nestling altricial birds. *Anim Behav* 36:653–61.

Redondo, T., and J. Carranza. 1989. Offspring reproductive value and nest defense in the magpie (*Pica pica*). *Behav Ecol Sociobiol* 25:369–78.

Reeve, H. K., and P. W. Sherman. 1993. Adaptation and the goals of evolutionary research. *Q Rev Biol* 68:1–32.

Reeve, N. 1994. *Hedgehogs*. London: T & AD Poyser.

Regelmann, K., and E. Curio. 1983. Determinants of brood defence in the great tit *Parus major* L. *Behav Ecol Sociobiol* 13:131–45.

———. 1986a. Why do great tit (*Parus major*) males defend their brood more than females do? *Anim Behav* 34:1206–1214.

———. 1986b. How do great tit (*Parus major*) pair mates cooperate in brood defence? *Behaviour* 97:10–36.

Reichman, O. J., and S. Aitchison. 1981. Mammal trails on mountain slopes: Optimal paths in relation to slope angle and body weight. *Am Nat* 117:416–20.

Reid, J. B. 1984. Bird coloration: Predation, conspicuousness and the unprofitable prey model. *Anim Behav* 32:294–95.

Reid, M. L., and R. D. Montgomerie. 1985. Seasonal patterns of nest defence by Baird's sandpipers. *Can J Zool* 63:2207–11.

Reimchen, T. E. 1980. Spine deficiency and polymorphism in a population of *Gasterosteus aculeatus*: An adaptation to predators? *Can J Zool* 58:1232–44.

Renouf, D., and J. W. Lawson. 1986. Harbour seal vigilance: Watching for predators or mates? *Biol Behav* 11:44–49.

Repasky, R. R. 1996. Using vigilance behavior to test whether predation promotes habitat partitioning. *Ecology* 77:1880–87.

Reznick, D. A., H. Bryga, and J. A. Endler. 1990. Experimentally induced life history evolution in a natural population. *Nature* 346:357–59.

Rhine, R. J., and B. J. Westlund. 1981. Adult male positioning in baboon progressions: Order and chaos revisited. *Folia Primatol* 35:77–116.

Rhisiart, A. ap. 1989. Communication and anti-predator behaviour. Ph.D. thesis, University of Oxford.

Richards, D. G., and R. H. Wiley. 1980. Reverberations and amplitude fluctuations in the propagation of sound in a forest: Implications for animal communication. *Am Nat* 115:381–99.

Richardson, D. S., and G. M. Bolen. 1999. A nesting association between semi-colonial Bullock's orioles and yellow-billed magpies: Evidence for the predator protection hypothesis. *Behav Ecol Sociobiol* 46:373–80.

Richardson, L. W., H. A. Jacobson, R. J. Muncy, and C. J. Perkins. 1983. Acoustics of white-tailed deer (*Odocoileus virginianus*). *J Mammal* 64:245–52.

Ricklefs, R. E. 1969a. Preliminary models for growth rates in altricial birds. *Ecology* 50:1031–39.

———. 1969b. An analysis of nesting mortality in birds. *Smithson Contrib Zool* 9:1–48. Washington, DC: Smithsonian Institution Press.

———. 1984. The optimization of growth rate in altricial birds. *Ecology* 65:1602–16.

Ridley, M., and A. Grafen. 1981. Are green beard genes outlaws? *Anim Behav* 29:954–55.

Rippi, M., R. V. Alatalo, L. Lindstrom, and J. Mappes. 2001. Multiple benefits of gregariousness cover detectability costs in aposematic aggregations. *Nature* 413:512–14.

Risenhoover, K. L., and J. A. Bailey. 1985. Relationships between group size, feeding time, and agonistic behavior of mountain goats. *Can J Zool* 63:2501–6.

Robel, R. J. 1969. Nesting activities and brood movements of black grouse (*Lyrurus tetrix*) in Scotland. *Ibis* 111:395–99.

Roberts, G. 1995. A real-time response of vigilance behaviour to changes in group size. *Anim Behav* 50:1371–74.

———. 1996. Why individual vigilance declines as group size increases. *Anim Behav* 51:1077–86.

————. 1997. How many birds does it take to put a flock to flight? *Anim Behav* 54:1517–22.

Roberts, G., and P. R. Evans. 1993. Responses of foraging sanderlings to human approaches. *Behaviour* 126:29–43.

Roberts, S. C. 1988. Social influences on vigilance in rabbits. *Anim Behav* 36:905–13.

Roberts, S. C., L. M. Gosling, E. A. Thorton, and J. McClung. 2001. Scent marking by male mice under risk of predation. *Behav Ecol* 12:698–705.

Robertson, G. J. 1995. Factors affecting nest site selection and nesting success in the common eider *Somateria mollissima. Ibis* 137:109–15.

Robertson, R. J. 1973. Optimal niche space of the redwinged blackbird: Spatial and temporal patterns of nesting activity and success. *Ecology* 54:1085–93.

Robertson, R. J., and G. C. Biermann. 1979. Parental investment strategies determined by expected benefits. *Z Tierpsychol* 50:124–28.

Robinette, R. L., and J. C. Ha. 2001. Social and ecological factors influencing vigilance by northwestern crows, *Corvus caurinus. Anim Behav* 62:447–52.

Robins, C. R., R. M. Bailey, C. E. Bond, J. R. Brooker, E. A. Lancher, R. N. Lea, and W. B. Scott. 1991. *Common and scientific names of fishes from the United States and Canada.* 5th ed. Bethseda, MD: US Fish and Wildlife Service.

Robinson, I. 1990. The effect of mink odour on rabbits and small mammals. In *Chemical signals in vertebrates 5,* ed. D. W. Macdonald, D. Muller-Schwarze, and S. E. Natynczuk, 566–72. Oxford: Oxford University Press.

Robinson, J. G. 1981. Spatial structure in foraging groups of wedge-capped capuchin monkeys *Cebus nigrivittatus. Anim Behav* 29:1036–56.

Robinson, J. G., and E. L. Bennett, eds. 2000. *Hunting for sustainability in tropical forests.* New York: Columbia University Press.

Robinson, M. H. 1969. Defenses against visually hunting predators. *Evol Biol* 3:225–59.

Robinson, S.K. 1985. Coloniality in the yellow-rumped cacique as a defense against nest predators. *Auk* 102:506–19.

Robinson, S. K., F. R. Thompson III, T. M. Donovan, D. R. Whitehead, and J. Faaborg. 1995. Regional forest fragmentation and the nesting success of migratory birds. *Science* 267:1987–90.

Robinson, S. R. 1980. Antipredator behaviour and predator recognition in Belding's ground squirrels. *Anim Behav* 28:840–52.

————. 1981. Alarm communication in Belding's ground squirrels. *Z Tierpsychol* 56:150–68.

Rodriguez-Girones, M. A., and R. A. Vasquez. 2002. Evolutionary stability of vigilance coordination among social foragers. *Proc R Soc Lond Ser B Biol Sci* 269:1803–10.

Roell, A., and I. Bossema. 1982. A comparison of nest defence by jackdaws, rooks, magpies and crows. *Behav Ecol Sociobiol* 11:1–6.

Rohwer, S., S. D. Fretwell, and R. C. Tuckfield. 1976. Distress screams as a measure of kinship in birds. *Am Midl Nat* 96:418–30.

Rolando, A., R. Caldoni, A. De Sanctis, and P. Laiolo. 2001. Vigilance and neighbour distance in foraging flocks of red-billed choughs, *Pyrrhocorax pyrrhocorax. J Zool* 253:225–32.

Rolland, C., E. Danchin, and M. de Fraipont. 1998. The evolution of coloniality in birds in relation to food, habitat, predation, and life-history traits: A comparative analysis. *Am Nat* 151:514–29.

Romey, W. L. 1995. Position preferences within groups: Do whirligigs select positions which balance feeding opportunities with predator avoidance? *Behav Ecol Sociobiol* 37:195–200.

Ronkainen, H., and H. Ylonen. 1994. Behaviour of cyclic bank voles under risk of mustelid predation: Do females avoid copulations? *Oecologia* 97:377–81.

Rood, J. P. 1983. Banded mongoose rescues pack member from eagle. *Anim Behav* 31:1261–62.

Roper, J. J., and R. R. Goldstein. 1997. A test of the Skutch hypothesis: Does activity at nests increase nest predation risk? *J Avian Biol* 28:111–16.

Roper, T. J. 1990. Responses of domestic chicks to artificially coloured insect prey: Effects of previous experience and background colour. *Anim Behav* 39:466–73.

———. 1993. Effects of novelty on taste-avoidance learning in chicks. *Behaviour* 125:265–81.

———. 1994. Conspicuousness of prey retards reversal of learned avoidance. *Oikos* 69:115–18.

Roper, T. J., and S. E. Cook. 1989. Responses of chicks to brightly coloured insect prey. *Behaviour* 110:276–93.

Roper, T. J., and S. Redston. 1987. Conspicuousness of distasteful prey affects the strength and durability of one-trial avoidance learning. *Anim Behav* 35:739–47.

Roper, T. J., and R. Wistow. 1986. Aposematic colouration and avoidance learning in chicks. *Q J Exp Psychol* 38B:141–49.

Rose, L. M., and L. M. Fedigan. 1995. Vigilance in white-faced capuchins, *Cebus capucinus*, in Costa Rica. *Anim Behav* 49:63–70.

Rosell, F. 2001. Effectiveness of predator odors as gray squirrel repellents. *Can J Zool* 79:1719–23.

Rosenzweig, M. L. 1966. Community structure in sympatric Carnivora. *J Mammal* 47:602–12.

———. 1973. Habitat selection experiments with a pair of coexisting heteromyid rodent species. *Ecology* 54:111–17.

Ross, C. 1993. Predator mobbing by an all-male band of hanuman langurs (*Presbytis entellus*). *Primates* 34:105–7.

Ross, C., and K. E. Jones. 1999. Socioecology and the evolution of primate reproductive rates. In *Comparative primate socioecology*, ed. P. C. Lee, 73–110. Cambridge: Cambridge University Press.

Rovee-Collier, C., L. W. Kaufman, and P. Farina. 1980. The critical cues for diurnal death feigning in young chicks: A functional analysis. *Am J Psychol* 93:259–68.

Rowe, C. 1999. Receiver psychology and the evolution of multicomponent signals. *Anim Behav* 58:921–31.

Rowe, C., and T. Guilford. 1996. Hidden colour aversions in domestic chicks triggered by pyrazine odours of insect warning displays. *Nature* 383:520–22.

———. 1999. Novelty effects in a multimodal warning signal. *Anim Behav* 57:341–46.

Rowe, C., L. Lindstrom, and A. Lyytinen. 2004. The importance of pattern similarity between Mullerian mimics in predator avoidance learning. *Proc Roy Soc Lond Ser B Biol Sci* 271:407–13.

Rowe, M. P., R. G. Coss, and D. H. Owings. 1986. Rattlesnake rattles and burrowing owl hisses: A case of acoustic Batesian mimicry. *Ethology* 72:53–71.

Rowe, M. P., and D. H. Owings. 1978. The meaning of the sound of rattling by rattlesnakes to California ground squirrels. *Behaviour* 252–67.

Rowe-Rowe, D. T. 1974. Flight behaviour and flight distances of blesbok. *Z Tierpsychol* 34:208–11.

Roze, U. 2002. A facilitated release mechanism for quills of the North American poprcupine (*Erethizon dorsatum*). *J Mammal* 83:381–85.

Rudnicky, T. C., and M. L. Hunter. 1993. Avian nest predation in clearcuts, forests, and edges in a forest-dominated landscape. *J Wildl Manag* 57:358–64.

Rudolph, D. C., H. Kyle, and R. N. Conner. 1990. Red-cockaded woodpeckers vs rat snakes: The effectiveness of the resin barrier. *Wilson Bull* 102:14–22.

Ruiter, J. R. de. 1986. The influence of group size on predator scanning and foraging behaviour of wedge capped capuchin monkeys (*Cebus olivaceous*). *Behaviour* 98:240–78.

Ruiter, L. de. 1952. Some experiments on the camouflage of stick caterpillars. *Behaviour* 4:222–32.

———. 1956. Countershading in caterpillars: An analysis of its adaptive significance. *Arch Neerlandaises de Zoologie* 11:285–341.

Rundle, S. D., and C. Bronmark. 2001. Inter- and intraspecific trait compensation of defence mechanisms in freshwater snails. *Proc R Soc Lond Ser B Biol* 268:1463–68.

Russ, J. M., G. Jones, I. J. Mackie, and P. A. Racey. 2004. Interspecific responses to distress calls in bats (Chiroptera: Vespertilionidae): A function for convergence in call design? *Anim Behav* 67:1005–14.

Russ, J. M., P. A. Racey, and G. Jones. 1998. Intraspecific responses to distress calls of the pipistrelle bat, *Pipistrellus pipistrellus. Anim Behav* 55:705–13.

Russell, E. S. 1938. *The behaviour of animals.* London: E.Arnold & Co.

Rutberg, A. T. 1987. Adaptive hypotheses of birth synchrony in ruminants; an interspecific test. *Am Nat* 130:692–710.

Ruxton, G. D. 1996. Group size and anti-predator vigilance: A simple model requiring limited monitoring of other group members. *Anim Behav* 51:478–81.

Ruxton, G. D., and S. L. Lima. 1997. Predator-induced breeding suppression and its consequences for predator-prey population dynamics. *Proc R Soc Lond Ser B Biol Sci* 264:409–15.

Ruxton, G. D., and G. Roberts. 1999. Are vigilance sequences a consequence of intrinsic chaos or internal changes. *Anim Behav* 57:493–95.

Ruxton, G. D., T. N. Sherratt, and M. P. Speed. 2004. *Avoiding attack: The evolutionary ecology of crypsis, warning signals and mimicry.* Oxford: Oxford University Press.

Ruxton, G. D., M. Speed, and D. J. Kelly. 2004. What, if anything, is the adaptive significance of countershading? *Anim Behav* 68:445–51.

Ryan, J. M., D. B. Clark, and J. A. Lackey. 1985. Response of Artibeus lituratus (Chiroptera: Phyllostomidae) to distress calls of conspecifics. *J Mammal* 66:179–81.

Ryan, M. J., and E. A. Brenowitz. 1985. The role of body size, phylogeny, and ambient noise in the evolution of bird song. *Am Nat* 126:87–100.

Rytkonen, S. 2002. Nest defence in great tits *Parus major*: Support for parental investment theory. *Behav Ecol Sociobiol* 52:379–84.

Rytkonen, S., K. Koivula, and M. Orell. 1990. Temporal increase in nest defence intensity of the willow tit (*Parus montanus*): Parental investment or methodological artifact? *Behav Ecol Sociobiol* 27:283–86.

Rytkonen, S., P. Kuokkanen, M. Hukkanen, and K. Huhtala. 1998. Prey selection by sparrowhawks *Accipter nisus* and characteristics of vulnerable prey. *Ornis Fenn* 75:77–87.

Rytkonen, S., M. Orell, and K. Koivula. 1993. Sex-role reversal in willow tit nest defence. *Behav Ecol Sociobiol* 33: 275–82.

———. 1995. Pseudo concorde fallacy in the willow tit? *Anim Behav* 49:1017–28.

Rytkonen, S., M. Orell, K. Koivula, and M. Soppela. 1995. Correlation between two components of parental investment: Nest defence intensity and nestling provisioning effort of willow tits. *Oecologia* 104:386–93.

Sadedin, S. R., and M. A. Elgar. 1998. The influence of flock size and geometry on the scanning behaviour of spotted turtle doves, *Streptopelia chinensis. Austr J Ecol* 23:177–80.

Sage, B. L. 1962. Albinism and melanism in birds. *Br Birds* 55:201–25.

Sargeant, A. B., and L. E. Eberhardt. 1975. Death feigning by ducks in response to predation by red foxes (*Vulpes fulva*). *Am Midl Nat* 94:108–19.

Sauther, M. L. 1989. Antipredator behavior in troops of free-ranging *Lemur catta* at Beza Mahajaly special reserve, Madagascar. *Int J Primatol* 10:595–606.

Savalli, U. M. 1995. The evolution of bird coloration and plumage elaboration: A review of hypotheses. *Curr Ornithol* 12:141–90.

Saville, D. B. O. 1957. Adaptive evolution in the avian wing. *Evolution* 11:212–24.

Scaife, M. 1976a. The response to eye-like shapes by birds I. The effect of context: A predator and a strange bird. *Anim Behav* 24:195–99.

———. 1976b. The response to eye-like shapes by birds II. The importance of staring, pairedness and shape. *Anim Behav* 24:200–206.

Scannell, J., G. Roberts, and J. Lazarus 2001. Prey scan at random to evade observant predators. *Proc R Soc Lond Ser B Biol Sci* 268:541–47.

Schaik, C. P. van. 1983. Why are diurnal primates living in groups? *Behaviour* 87:120–44.

Schaik, C. P. van, and J. A. R. A. M. van Hooff. 1983. On the ultimate causes of primate social systems. *Behaviour* 87:120–43.

Schaik, C. P. van, and M. Hostermann. 1994. Predation risk and the number of adult males in a primate group: A comparative test. *Behav Ecol Sociobiol* 35:261–72.

Schaik, C. P. van, and M. A. van Noordwijk. 1985. Evolutionary effect of the absence of felids on the social organization of the macaques on the island of Simeulue (*Macaca fascicularis fusca*, Miller 1903). *Folia Primatol* 44:138–47.

———. 1989. The special role of male *Cebus* monkeys in predation avoidance and its effect on group composition. *Behav Ecol Sociobiol* 24:265–76.

Schaik, C. P. van, M. A. van Noordwijk, B. Warsano, and E. Sutriono. 1983. Party size and early detection of predators in Sumatran forest primates. *Primates* 24:211–21.

Schall, A., and P. Ropartz. 1985. Le comportement de surveillance chez le daim (*Dama dama*). *Comptes Rendus a l'Academie des Sciences, Paris* 301:731–36.

Schall, J. J., and E. R. Pianka. 1980. Evolution of escape behavior diversity. *Am Nat* 115:551–66.

Schaller, G. B. 1964. Breeding behavior of the white pelican at Yellowstone Lake, Wyoming. *Condor* 66:3–23.

———. 1967. *The deer and the tiger: A study of wildlife in India*. Chicago: University of Chicago Press.

———. 1972. *The Serengeti lion: A study of predator-prey relations*. Chicago: University of Chicago Press.

Schaller, G. B., and J. T. Emlen Jr. 1961. The development of visual discrimination patterns in the crouching reactions of nestling grackles. *Auk* 78:125–37.

Schantz, T. von, G. Goransson, G. Andersson, I. Froberg, M. Grahn, A. Helgee, and H. Wittzell. 1989. Female choice selects for a viability-based male trait in pheasants. *Nature* 337:166–69.

Scheel, D. 1993a. Watching for lions in the grass: The usefulness of scanning and its effects during hunts. *Anim Behav* 46:695–704.

———. 1993b. Profitability, encounter rates, and prey choice of African lions. *Behav Ecol* 4:90–97.

Schiek, J. O., and S. J. Hannon. 1993. Clutch predation, cover, and the overdispersion of nests of the willow ptarmigan. *Ecology* 74:743–50.

Schindler, M., and J. Lamprecht. 1987. Increase of parental effort with brood size in a nudifugous bird. *Auk* 104:688–93.

Schleidt, W. M. 1961. Reaktionen von Truthyhnern auf fliegende Raubvogel und Versuche zur Analyse ihrer AAM's. *Z Tierpsychol* 18:534–60.

———. 1973. Tonic communication: Continual effects of discrete signs in animal communication systems. *J Theor Biol* 42:359–86.

Schlenoff, D. H. 1985. The startle responses of blue jays to *Catocala* (Lepidoptera: Noctuidae) prey models. *Anim Behav* 33:1057–67.

Schmidt, K. A. 2003. Nest predation and population declines in Illinois songbirds: A case for mesopredator effects. *Cons Biol* 17:1141–50.

Schmidt, K. A., and J. S. Brown. 1996. Patch assessment in fox squirrels: The role of resource density, patch size, and patch boundaries. *Am Nat* 147:360–80.

Schmidt, K. A., and C. J. Whelan. 1998. Predator-mediated interactions between and within guilds of nesting songbirds: Experimental and observational evidence. *Am Nat* 152:393–402.

———. 1999. Nest predation on woodland songbirds: When is nest predation density dependent? *Oikos* 87:65–74.

Schmidt, K. P., and R. F. Inger. 1957. *Living reptiles of the world.* London: Hamish Hamilton.

Schmidt-Nielsen, K. 1984. *Scaling: Why is animal size so important?* Cambridge: Cambridge University Press.

Schoener, T. W. 1968. The *Anolis* lizards of Bimini: Resource partitioning in a complex fauna. *Ecology* 49:704–26.

———. 1974. Resource partitioning in ecological communities. *Science* 185:27–39.

Schooley, R. L., P. B. Sharpe, and B. van Horne. 1996. Can shrub cover increase predation risk for a desert rodent? *Can J Zool* 74:157–63.

Schradin, C. 2000. Confusion effect in a reptilian and primate predator. *Ethology* 106:691–700.

Schuetz, J. G. 2005. Common waxbills use carnivore scat to reduce the risk of nest predation. *Behav Ecol* 16:133–37.

Schuler, W., and E. Hesse. 1985. On the function of warning coloration: A black and yellow pattern inhibits pre-attack by naïve domestic chicks. *Behav Ecol Sociobiol* 16:249–55.

Schuler, W., and T. J. Roper. 1992. Responses to warning coloration in avian predators. *Adv Study Behav* 21:111–46.

Schulte-Hostedde, A. I., and J. S. Millar. 2002. Effects of body size and mass on running speed of male yellow-pine chipmunks (*Tamias amoenus*). *Can J Zool* 80:1584–87.

Schwagmeyer, P. L. 1980. Alarm calling behavior of the thirteen-lined ground squirrel, *Spermophilus tridecemlineatus. Behav Ecol Sociobiol* 7:195–200.

Schwagmeyer, P., and C. H. Brown. 1981. Conspecific reaction to playback of thirteen-lined ground squirrel vocalizations. *Z Tierpsychol* 56:25–32.

Sedinger, J. S., and D. G. Raveling. 1990. Parental behavior of cackling Canada geese during brood rearing: Division of labor within pairs. *Condor* 92:174–81.

Seeley, T. D., R. H. Seeley, and P. Akratanakul. 1982. Colony defense strategies of the honeybees in Thailand. *Ecological Monogr* 52:43–64.

Seitz, L. C., and D. A. Zegers. 1993. An experimental study of nest predation in adjacent deciduous, coniferous and successional habitats. *Condor* 95:297–304.

Selous, F. C. 1908. *African nature notes and reminiscences.* London: MacMillan.

Sentell, S. W., and W. C. Compton. 1987. Evidence of sensory awareness in death-feigning opossums (*Didelphis marsupialis*). *VA J Sci* 38:200–203.

Servedio, M. R. 1999. The effects of predator learning, forgetting, and recognition errors on the evolution of warning coloration. *Evolution* 54:751–63.

Seyfarth, R. M., and D. L. Cheney. 1980. The ontogeny of vervet monkey alarm calling behavior: A preliminary report. *Z Tierpsychol* 54:37–56.

———. 1986. Vocal development in vervet monkeys. *Anim Behav* 34:1640–58.

———. 1990. The assessment by vervet monkeys of their own and another species' alarm calls. *Anim Behav* 40:754–64.

Seyfarth, R. M., D. L. Cheney, and P. Marler. 1980a. Monkey responses to three different alarm calls: Evidence of predator classification and semantic communication. *Science* 210:801–3.

———. 1980b. Vervet monkey alarm calls: Semantic communication in a free-ranging primate. *Anim Behav* 28:1070–94.

Shadle, A. R. 1947. Porcupine spine penetration. *J Mammal* 28:180–81.

Shadle, A. R., and D. Po-Chedley. 1949. Rate of penetration of a porcupine spine. *J Mammal* 30:172–73.

Shalter, M. D. 1978. Localization of passerine seeet and mobbing calls by goshawks and pigmy owls. *Z Tierpsychol* 46:260–67.

Shalter, M. D., and W. M. Schleidt. 1977. The ability of barn owls *Tyto alba* to discriminate and localize avian alarm calls. *Ibis* 119:22–27.

Shank, C. C. 1977. Cooperative defense by bighorn sheep. *J Mammal* 58:243–44.

Shargal, E., L. Rath-Wolfson, N. Kronfeld, and T. Dayan. 1999. Ecological and histological aspects of tail loss in spiny mice (Rodentia: Muridae, *Acomys*) with a review of its occurrence in rodents. *J Zool (Lond)* 249:187–93.

Sharpe, S. T., and J. S. Millar. 1990. Relocation of nest sites by female deer mice, *Peromyscus maniculatus borealis*. *Can J Zool* 68:2364–67.

Shelley, E. L., and D. T. Blumstein. 2005. The evolution of vocal alarm communication in rodents. *Behav Ecol* 16:169–77.

Sherman, P.W. 1977. Nepotism and the evolution of alarm calls. *Science* 197:1246–53.

———. 1980a. The limits of ground squirrel nepotism. In *Sociobiology: Beyond nature/nurture?* ed. G. W. Barlow and J. Silverberg, 505–44. AAAS Selected Symposium 35. Boulder, CO: Westview Press Inc.

———. 1980b. The meaning of nepotism. *Am Nat* 116:604–6.

———. 1981. Kinship, demography, and Belding's ground squirrel nepotism. *Behav Ecol Sociobiol* 8:251–59.

———. 1985. Alarm calls of Belding's ground squirrels to aerial predators: Nepotism or self-preservation? *Behav Ecol Sociobiol* 17:313–23.

Sherratt, T. N. 2002. The coevolution of warning signals. *Proc R Soc Lond Ser B Biol Sci* 269:741–46.

———. 2003. State-dependent risk-taking by predators in systems with defended prey. *Oikos* 103:93–100.

Sherratt, T. N., and C. D. Beatty. 2003. The evolution of warning signals as reliable indicators of prey defense. *Am Nat* 162:377–89.

Sherratt, T. N., and A. D. MacDougall. 1995. Some population consequences of variation in preference among individual predators. *Biol J Linn Soc* 55:93–107.

Shields, W. M. 1980. Ground squirrel alarm calls: Nepotism or parental care? *Am Nat* 116:599–603.

———. 1984. Barn swallows mobbing: Self-defence, collateral kin defence, group defence, or parental care? *Anim Behav* 32:132–48.

Shriner, W. M. 1998. Yellow-bellied marmot and golden-mantled ground squirrel responses to heterospecific alarm calls. *Anim Behav* 55:529–36.

Shultz, S., C. Faurie, and R. Noe. 2003. Behavioural responses of Diana monkeys to male long-distance calls: Changes in ranging, association patterns and activity. *Behav Ecol Sociobiol* 53:238–45.

Sibly, R. M., D. Collett, D. E. L. Promislow, D. J. Peacock, and P. H. Harvey. 1997. Mortality rates of mammals. *J Zool* 243:1–12.

Siegel-Causey, D., and G. L. Hunt Jr. 1981. Colonial defense behavior in double-crested and pelagic cormorants. *Auk* 98:522–31.

Siegfried, W. R. 1980. Vigilance and group size in sprongbok. *Madoqua* 12:151–54.

Siegfried, W. R., and L. G. Underhill. 1975. Flocking as an anti-predator strategy in doves. *Anim Behav* 23:504–08.

Sih, A. 1980. Optimal behavior: Can foragers balance two conflicting demands? *Science* 210:1041–43.

———. 1982. Foraging strategies and the avoidance of predation by an aquatic insect, *Notonecta hoffmanni*. *Ecology* 63:786–96.

———. 1987. Predator and prey lifestyles: An evolutionary and ecological overview. In *Predation:*

Direct and indirect impacts on aquatic communities, ed. W. C. Kerfoot and A. Sih, 203–24. Hanover, NH: University Press of New England.

———. 1995. Predation risk and the evolutionary ecology of reproductive behaviour. *J Fish Biol* Suppl. 45:111–30.

Sih, A., and B. Christensen. 2001. Optimal diet theory: When does it work, and when and why does it fail? *Anim Behav* 61:379–90.

Sih, A., P. Crowley, M. McPeek, J. Petranka, and K. Strohmeier. 1985. Predation, competition, and prey communities: A review of field experiments. *Annu Rev Ecol Syst* 16:269–311.

Sih, A., G. Englund, and D. Wooster. 1998. Emergent impacts of multiple predators on prey. *Trends Ecol Evol* 13:350–55.

Sih, A., L. B. Kats, and E. F. Maurer. 2003. Behavioural correlations across situations and the evolution of antipredator behaviour in a sunfish-salamander system. *Anim Behav* 65:29–44.

Sih, A., and T. M. McCarthy. 2002. Prey responses to pulses of risk and safety: Testing the risk allocation hypothesis. *Anim Behav* 63:437–43.

Silberglied, R. E., A. Aiello, and D. M. Windsor. 1980. Disruptive coloration in butterflies: Lack of support in *Anartia fatina*. *Science* 209:617–19.

Sillen-Tullberg, B. 1985. Higher survival of an aposematic than of a cryptic form of a distasteful bug. *Oecologia* 67:411–15.

———. 1988. Evolution of gregariousness in aposematic butterfly larvae: A phylogenetic analysis. *Evolution* 42:293–305.

———. 1990. Do predators avoid groups of aposematic prey? An experimental test. *Anim Behav* 40:856–60.

———. 1993. The effect of biased inclusion of taxa on the correlation between discrete characters in phylogenetic trees. *Evolution* 47:1182–91.

Sillen-Tullberg, B., and E. H. Bryant. 1983. The evolution of aposematic coloration in distasteful prey: An individual selection model. *Evolution* 37:993–1000.

Sillen-Tullberg, B., and O. Leimar. 1988. The evolution of gregariousness in distasteful insects as a defense against predators. *Am Nat* 132:723–34.

Silva, J. da, and J. M. Terhune. 1988. Harbour seal grouping as an anti-predator strategy. *Anim Behav* 36:1309–16.

Simonetti, J. A. 1989. Microhabitat use by small mammals in central Chile. *Oikos* 56:309–18.

Simmons, K. E. L. 1952. The nature of the predator-reactions of breeding birds. *Behaviour* 4:161–71.

———. 1955. The nature of the predator-reactions of waders towards humans; with special reference to the role of the aggressive escape, and brooding drives. *Behaviour* 8:130–73.

Simpson, K., J. N. M. Smith, and J. P. Kelsall. 1987. Correlates and consequences of coloniality in great blue herons. *Can J Zool* 65:572–77.

Sinclair, A. R. E., S. Mduma, and J. S. Brashares. 2003. Patterns of predation in a diverse predator-prey system. *Nature* 425:288–90.

Skead, C. J. 1959. A study of the Cape Penduline Tit *Anthoscopus minutus minutus*. *Proc 1ˢᵗ Pan Afr Ornithol Congr Ostrich* (Suppl.) 3:274–88.

Skinner, M. P. 1928. Kingfisher and sharp-shinned hawk. *Auk* 45:100–101.

Skutch, A. F. 1949. Do tropical birds rear as many young as they can nourish? *Ibis* 91:430–55.

———. 1966. A breeding bird census and nesting success in Central America. *Ibis* 108:1–16.

Slagsvold, T. 1980. Habitat selection in birds: On the presence of other bird species with special regard to Turdus pilaris. *J Anim Ecol* 49:523–36.

———. 1982. Clutch size variation in passerine birds: The nest predation hypothesis. *Oecologia* 54:159–69.

————. 1984a. Clutch size variation of birds in relation to nest predation: On the cost of reproduction. *J Anim Ecol* 53:945–53.

————. 1984b. The mobbing behaviour of the hooded crow *Corvus corone corvix*: Anti-predator
defence or self-advertisement? *Fauna Norv Ser C Cinclus* 7:127–31.

————. 1985. Mobbing behaviour of the hooded crow *Corvus corone corvix* in relation to age, sex,
size, season, temperature and kind of enemy. *Fauna Norv Ser C Cinclus* 8:9–17.

Slagsvold, T., and S. Dale. 1996. Disappearance of female pied flycatchers in relation to breeding
stage and experimentally induced molt. *Ecology* 77:461–71.

Slagsvold, T., S. Dale, and A. Kruszewicz. 1995. Predation favours cryptic coloration in breeding
male pied flycatchers. *Anim Behav* 50:1109–21.

Slobodchikoff, C. N., and R. Coast. 1980. Dialects in the alarm calls of prairie dogs. *Behav Ecol Sociobiol* 7:49–53.

Slobodchikoff, C. N., J. Kiriazis, C. Fischer, and E. Creff. 1991. Semantic information distinguishing individual predators in the alarm calls of Gunnison's prairie dogs. *Anim Behav*
42:713–19.

Slotow, R., and N. Coumi. 2000. Vigilance in bronze manikin groups: The contributions of predation risk and intra-group competition. *Behaviour* 137:565–78.

Slotow, R., and S. I. Rothstein. 1995. Influence of social status, distance from cover, and group size
on feeding and vigilance in white-crowned sparrows. *Auk* 112:1024–31.

Small, M. F., and M. L. Hunter. 1988. Forest fragmentation and avian nest predation in forested
landscapes. *Oecologia* 76:62–64.

Smallwood, J. A. 1989. Prey preferences of free-ranging American kestrels, *Falco sparverius*. *Anim
Behav* 38:712–28.

Smith, A. C., S. Kelez, and H. M. Buchanan-Smith. 2004. Factors affecting vigilance within wild
mixed-species troops of saddleback (*Saguinus fuscicollis*) and moustached tamarins (*S. mystax*). *Behav Ecol Sociobiol* 56:18–25.

Smith, C. C. 1978. Structure and function of the vocalizations of tree squirrels (*Tamiasciurus*). *J
Mamm* 59:793–808.

Smith, D. G. 1997. Ecological factors influencing the antipredator behaviors of the ground skink,
Scincella lateralis. *Behav Ecol* 8:622–29.

Smith, E. N., and R. A. Woodruff. 1980. Fear bradycardia in free-ranging woodchucks, *Marmota
monax*. *J Mammal* 61:747–50.

Smith, J. N. M. 1974. The food searching behaviour of two European thrushes. I. Description and
analysis of search paths. *Behaviour* 48:276–302.

Smith, M. J., and H. B. Graves. 1978. Some factors influencing mobbing behavior in barn swallows (*Hirundo ristica*). *Behav Biol* 23:355–72.

Smith, R. J. F. 1986. Evolution of alarm signals: Role of benefits of retaining group members or
territorial neighbors. *Am Nat* 128:604–10.

————. 1997. Avoiding and deterring predators. In *Behavioural ecology of teleost fishes*, ed. J-G.J.
Godin, 163–90. Oxford: Oxford University Press.

Smith, S. F. 1978. Alarm calls, their origin and use in *Eutamias sonomae*. *J Mamm* 59:888–93.

Smith, S. M.. 1975. Innate recognition of coral snake pattern by a possible avian predator. *Science*
187:759–60.

————. 1977. Coral-snake pattern recognition and stimulus generalization by naïve great
kiskadees (Aves: Tyrannidae). *Nature* 265:535–36.

Smith, W. J., S. L. Smith, J. G. Devilla, and E. C. Oppenheimer. 1976. The jump-yip display of the
black-tailed prairie dog *Cynomys ludovicianus*. *Anim Behav* 24:609–21.

Smith, W. J., S. L. Smith, E. C. Oppenheimer, and J. G. Devilla. 1977. Vocalizations of the black-tailed prairie dog *Cynomys ludovicianus*. *Anim Behav* 25:152–64.

Smith, W. P. 1987. Maternal defense in Columbian white-tailed deer: When is it worth it? *Am Nat* 130:310–16.

Smith, W. P. 1991. Ontogeny and adaptiveness of tail-flagging behavior in white-tailed deer. *Am Nat* 138:190–200.

Snow, D. W. 1981. The nest as a factor determining clutch-size in tropical birds. *J Ornithol* 119:227–30.

Snyder, R. L. 1975. Some prey preference factors for a red-tailed hawk. *Auk* 92:547–52.

Snyder, R. L., W. Jenson, and C. D. Cheney. 1976. Environmental familiarity and activity: Aspects of prey selection for a ferruginous hawk. *Condor* 78:138–39.

Soderstrom, B., T. Part, and J. Ryden. 1998. Different nest predator faunas and nest predation risk on ground and shrub nests at forest ecotones: An experiment and a review. *Oecologia* 117:108–18.

Solis, J. C., and F. de Lope. 1995. Nest and egg crypsis in the ground-nesting stone curlew *Burhinus oedicnemus*. *J Avian Biol* 26:135–38.

Solonen, T. 1997. Effect of sparrowhawk predation on forest birds in southern Finland. *Ornis Fenn* 74:1–14.

Sonerud, G. A. 1985a. Brood movements in grouse and waders as defence against win-stay search in their predators. *Oikos* 44:287–300.

———. 1985b. Nest hole shift in Tengmalm's owl *Aegolius funereus* as defence against nest predation involving long-term memory in the predator. *J Anim Ecol* 54:179–92.

———. 1988. To distract display or not: Grouse hens and foxes. *Oikos* 51:233–37.

———. 1989. Reduced predation by pine martens on nests of Tengmalm's owl in relocated boxes. *Anim Behav* 37:332–43.

Sordahl, T. A. 1990. The risks of avian mobbing and distraction behavior: An anecdotal review. *Wilson Bull* 102:349–52.

Southern, H. N. 1954. Tawny owls and their prey. *Ibis* 96:384–408.

Speakman, J. R. 1991. Why do insectivorous bats in Britain not fly in daylight more frequently? *Funct Ecol* 5:518–24.

Spear, L., and D. G. Ainley. 1993. Kleptoparasitism by Kermadec petrels, jaegers, and skuas in the Eastern tropical Pacific: Evidence of mimicry by two species of *Pterodroma*. *Auk* 110:222–33.

Speed, M. P. 1993. Muellerian mimicry and the psychology of predation. *Anim Behav* 45:571–80.

———. 2000. Warning signals, receiver psychology and predator memory. *Anim Behav* 60:269–78.

———. 2001. Can receiver psychology explain the evolution of aposematism? *Anim Behav* 61:205–16.

Speed, M. P., D. J. Kelly, A. M. Davidson, and G. D. Ruxton. 2005. Countershading enhances crypsis with some bird species but not others. *Behav Ecol* 16:327–34.

Sproat, T. M., and G. Ritchison. 1993. The nest defense behavior of eastern screech-owls: Effects of nest stage, sex, nest type and predator location. *Condor* 95:288–96.

Stacey, P. B. 1986. Group size and foraging efficiency in yellow baboons. *Behav Ecol Sociobiol* 18:175–87.

Stamp, N. E. 1980. Egg deposition patterns in butterflies: Why do some species cluster their eggs rather than deposit them singly? *Am Nat* 115:367–80.

Stanford, C. B. 1995. The influence of chimpanzee predation on group size and anti-predator behaviour in red colobus monkeys. *Anim Behav* 49:577–87.

——. 1998. *Chimpanzee and red colobus: The ecology of predator and prey*. Cambridge, MA: Harvard University Press.

Stankowich, T. 2003. Marginal predation methodologies and the importance of predator preferences. *Anim Behav* 66:589–99.

Starkey, E. E., and J. F. Starkey. 1973. Description of an aerial-predator alarm call for mallard (*Anas platyrhynchos*) ducklings. *Condor* 75:364–66.

Starrett, A. 1993. Adaptive resemblance: A unifying concept for mimicry and crypsis. *Biol J Linn Soc* 48:299–317.

Steen, J. B., K. E. Erikstad, and K. Hoidal. 1992. Cryptic behaviour in moulting hen willow ptarmigan *Lagopus l. lagopus* during snow melt. *Ornis Scand* 23:101–4.

Steenbeek, R., R. C. Piek, M. van Buul, and J. A. R. A. M. van Hooff. 1999. Vigilance in wild Thomas's langurs (*Presbytis thomasi*): The importance of infanticide risk. *Behav Ecol Sociobiol* 45:137–50.

Stefanski, R. A., and J. B. Falls. 1972a. A study of distress calls of song, swamp, and white-throated sparrows (Aves: Fringillidae). I. Intraspecific responses and functions. *Can J Zool* 50:1501–12.

——. 1972b. A study of distress calls of song, swamp, and white-throated sparrows (Aves: Fringillidae). II. Interspecific responses and properties used in recognition. *Can J Zool* 50:1513–25.

Stein, R. A., and J. J. Magnuson. 1976. Behavioral response of crayfish to a fish predator. *Ecology* 57:751–61.

Stephens, S. E., D. N. Koons, J. Y. Rotella, and D. W. Willey. 2003. Effects of habitat fragmentation on avian nesting success: A review of the evidence at multiple spatial scales. *Biol Conserv* 115:101–10.

Stephenson, T. R., and V. V. Ballenberghe. 1995. Defense of one twin calf against wolves, *Canis lupus*, by a female moose, *Alces alces*. *Can Field Nat* 109:251–53.

Sterck, E. H. M. 2002. Predator sensitive foraging in Thomas langurs. In *Eat or be eaten: Predator sensitive foraging among primates*, ed. L. E. Miller, 74–91. Cambridge: Cambridge University Press.

Stirling, I. 1977. Adaptations of Weddell and ringed seals to exploit the polar fast ice habitat in the absence or presence of surface predators. In *Adaptations within Antarctic ecosystems. Proceedings of the Third SCAR Symposium on Antarctic Biology*, ed. G. A. Liano, 741–48. Washington, DC: Smithsonian Institution.

Stoddart, D.M. 1976. Effect of the odour of weasels (*Mustela nivalis* L.) on trapped samples of their prey. *Oecologia* 22:439–41.

——. 1982. Does trap odour influence estimation of population size of the short-tailed vole, *Microtus agrestis*? *J Anim Ecol* 51:375–86.

Stokes, A. W. 1961. Voice and social behavior of the chukar partridge. *Condor* 63:111–27.

Stone, E., and C. H. Trost. 1991. Predators, risks and context for mobbing and alarm calls in black-billed magpies. *Anim Behav* 41:633–38.

Stoner, C. J., O. R. P. Bininda-Emonds, and T. Caro. 2003. The adaptive significance of coloration in lagomorphs. *Biol J Linn Soc* 79:309–28.

Stoner, C. J., T. M. Caro, and C. M. Graham. 2003. Ecological and behavioral correlates of coloration in artiodactyls: Systematic attempts to verify conventional hypotheses. *Behav Ecol* 14:823–40.

Strauss, R. E. 2001. Cluster analysis and the identification of aggregations. *Anim Behav* 61:481–88.

Struhsaker, T. T. 1981. Polyspecific associations among tropical rain-forest primates. *Z Tierpsychol* 57:268–304.

Struhsaker, T. T., and M. Leakey. 1990. Prey selectivity by crowned hawk-eagles on monkeys in the Kibale Forest, Uganda. *Behav Ecol Sociobiol* 26:435–43.

Studd, M., R. D. Montgomerie, and R. J. Robertson. 1983. Group size and predator surveillance in foraging house sparrows (*Passer domesticus*). *Can J Zool* 61:226–31.

Sugden, L. G., and G. W. Beyersbergen. 1996. Effect of density and concealment on American crow predation of simulated duck nests. *J Wildl Manag* 50:9–14.

Suhonen, J. 1993. Predation risk influences the use of foraging sites by tits. *Ecology* 74:1197–1203.

Sullivan, K. A. 1984. Information exploitation by downy woodpeckers in mixed-species flocks. *Behaviour* 91:294–311.

———. 1985a. Vigilance patterns in downy woodpeckers. *Anim Behav* 33:328–30.

———. 1985b. Selective alarm calling by downy woodpeckers in mixed-species flocks. *Auk* 102:184–86.

———. 1988. Ontogeny of time budgets in yellow-eyed juncos: Adaptation to ecology constraints. *Ecology* 69:118–24.

Sullivan, M. S., and N. Hillgarth. 1993. Mating system correlates of tarsal spurs in the Phasianidae. *J Zool* 231:203–14.

Sullivan, T. P. 1985. Use of predator odours as repellants to reduce feeding damage by herbivores: Black-tailed deer (*Odocoileus hemionus columbianus*). *J Chem Ecol* 11:921–35.

Sullivan, T. P., and D. R. Crump. 1984. Influence of mustelid scent-gland compounds on suppression of feeding by snowshoe hares (*Lepus americanus*). *J Chem Ecol* 10:1809–21.

———. 1986. Feeding responses of snowshoe hares (*Lepus americanus*) to volatile constituents of red fox (*Vulpes vulpes*) urine. *J Chem Ecol* 12:729–39.

Summers, R. W., L. G. Underhill, E. E. Syroechkovski Jr., H. G. Happo, R. P. Prys-Jones, and V. Karpov. 1994. The breeding biology of dark-bellied brent geese *Branta b. bernicla* and king eiders *Somateria spectabilis* on the northeastern Taymyr peninsula, especially in relation to snowy owl *Nyctea scandiaca* nests. *Wildfowl* 45:110–18.

Sumner, F. B. 1921. Desert and lava-dwelling mice, and the problem of protective coloration in mammals. *J Mammal* 2:75–86.

Sumner, F. B., and H. S. Swarth. 1924. The supposed effects of the color tone of the background upon the coat color of mammals. *J Mammal* 5:81–113.

Sundell, J., and H. Ylonen. 2004. Behaviour and choice of refuge by voles under predation risk. *Behav Ecol Sociobiol* 56:263–69.

Svensson, P. A., I. Barber, and E. Forsgren. 2000. Shoaling behaviour of the two-spotted goby. *J Fish Biol* 56:1477–87.

Swaddle, J. P., and R. Lockwood. 1998. Morphological adaptations to predation risk in passerines. *J Avian Biol.* 29:172–76.

———. 2003. Wingtip shape and flight performance in the European starling *Sturnus vulgaris*. *Ibis* 145:457–64.

Swaddle, J. P., E. V. Williams, and J. M. V. Rayner. 1999. The effect of simulated flight feather moult on escape take-off performance in starlings. *J Avian Biol* 30:351–58.

Swaisgood, R. R., D. H. Owings, and M. P. Rowe. 1999. Conflict and assessment in a predator-prey system: Ground squirrels versus rattlesnakes. *Anim Behav* 57:1033–44.

Swaisgood, R. R., M. P. Rowe, and D. H. Owings. 1999. Assessment of rattlesnake dangerousness by California ground squirrels: Exploitation of cues from rattling sounds. *Anim Behav* 57:1301–10.

———. 2003. Antipredator responses of California ground squirrels to rattlesnakes and rattling sounds: The roles of sex, reproductive parity, and offspring age in assessment and decision-making rules. *Behav Ecol Sociobiol* 55:22–31.

Sweitzer, R. A., and J. Berger. 1992. Size-related effects of predation on habitat use and behavior of porcupines (*Erethizon dorsatum*). *Ecology* 73:867–75.

Swennen, C. 1968. Nest protection of eiderducks and shovelers by means of faeces. *Ardea* 56:248–58.

———. 1974. Observations on the effect of ejection of stomach oil by the fulmar *Fulmarus glacialis* on other birds. *Ardea* 62:111–17.

Swihart, R. K. 1991. Modifying scent marking behavior to reduce woodchuck damage to fruit trees. *Ecol Appl* 1:98–103.

Sword, G. A. 1999. Density-dependent warning coloration. *Nature* 397:217.

———. 2002. A role for phenotypic plasticity in the evolution of aposematism. *Proc R Soc Lond Ser B Biol Sci* 269:1639–44.

Sword, G. A., P. D. Lorch, and D. T. Gwynne. 2005. Migratory bands give crickets protection. *Nature* 433:703.

Sword, G. A., S. J. Simpson, O. T. M. El Hadi, and H. Wilps. 2000. Density-dependent aposematism in the desert locust. *Proc R Soc Lond Ser B Biol Sci* 267:63–68.

Swynnerton, C. F. M. 1916. On the coloration of mouths and eggs of birds.—The coloration of eggs. *Ibis* 10:529–606.

Symula, R., R. Schulte, and K. Summers. 2001. Molecular phylogenetic evidence for a mimetic radiation in Peruvian poison frogs supports a Mullerian mimicry hypothesis. *Proc R Soc Lond Ser B Biol Sci* 268:2415–21.

Szep, T., and Z. Barta. 1992. The threat to bank swallows from the hobby at a large colony. *Condor* 94:1022–25.

Tamachi, N. 1987. The evolution of alarm calls: An altruism with nonlinear effect. *J Theor Biol* 127:141–53.

Tamura, N. 1989. Snake-directed mobbing by the Formosan squirrel *Callosciurus erythraeus thaiwanensis*. *Behav Ecol Sociobiol* 24:175–80.

Tamura, N., and H-S. Yong. 1993. Vocalizations in response to predators in three species of Malaysian *Callosciurus* (Sciuridae). *J Mamm* 74:703–14.

Taraborelli, P., V. Corbalarv, and S. Giannoni. 2003. Locomotion and escape modes in rodents of the montane desert (Argentina). *Ethology* 109:475–85.

Taulman, J. F. 1977. Vocalizations of the hoary marmot, *Marmota caligata*. *J Mamm* 58:681–83.

Taylor, C. R. 1977. The energetics of terrestrial locomotion and body size in vertebrates. In *Scale effects in animal locomotion*, ed. T. J. Pedley, 127–41. London: Academic Press.

Taylor, C. R., A. Shkolnik, R. Dmpel, D. Baharav, and A. Borut. 1974. Running in cheetahs, gazelles, and goats: Energy cost and limb configuration. *Am J. Physiol* 227:848–50.

Taylor, R. H. 1962. The Adelie penguin *Pygoscelis adeliae* at Cape Royds. *Ibis* 104:176–204.

Taylor, R. J. 1976. Value of clumping to prey and the evolutionary response of ambush predators. *Am Nat* 110:13–29.

———. 1979. The value of clumping to prey when detectability increases with group size. *Am Nat* 113:229–301.

———. 1984. *Predation*. New York: Chapman & Hall.

Taylor, R. J., D. F. Balph, and M. H. Balph. 1990. The evolution of alarm calling: A cost-benefit analysis. *Anim Behav* 39:860–68.

Taylor, W. P. 1935. *Ecology and life history of the porcupine (Erethizon epixanthum) as related to the forests of Arizona and the south-western United States*. Tucson: University of Arizona.

Tchabovsky, A. V., S. V. Popov, and B. R. Krasnov. 2001. Intra- and interspecific variation in vigilance and foraging of two gerbillid rodents, *Rhombomys opimus* and *Psammomys obesus*: The effect of social environment. *Anim Behav* 62:965–72.

Temple, S. A. 1987. Do predators always capture substandard individuals disproportionately from prey populations? *Ecology* 68:669–74.

Teneza, R. 1971. Behavior and nesting success relative to nest location in Adelie penguins (*Pygoscelis adeliae*). *Condor* 73:81–92.

Teneza, R. R, and R. L. Tilson. 1977. Evolution of long-distance alarm calls in Kloss's gibbon. *Nature* 268:233–35.

Terborgh, J. 1983. *Five New World primates*. Princeton, NJ: Princeton University Press.

Terborgh, J., and C. H. Janson. 1986. The socioecology of primate groups. *Annu Rev Ecol Syst* 17:111–35.

Terhune, J. M., and S. W. Brilliant. 1996. Harbour seal vigilance decreases over time since haul out. *Anim Behav* 51:757–63.

Tewksbury, J. T., S. J. Hejl, and T. E. Martin. 1998. Breeding productivity does not decline with increasing fragmentation in a western landscape. *Ecology* 79:2890–2903.

Thayer, A. H. 1896. The law which underlies protective coloration. *Auk* 13:124–29.

———. 1902. The law which underlies protective coloration. *Nature* 65:597.

Thayer, G. H. 1909. *Concealing-coloration in the animal kingdom*. New York: MacMillan.

Theodorakis, C. W. 1989. Size segregation and the effects of oddity on predation risk in minnow schools. *Anim Behav* 38:496–502.

Thiollay, J-M. 1991. Foraging, home range size and social behaviour of a group-living rainforest raptor, the red-throated caracara *Daptrius americanus*. *Ibis* 133:382–93.

Thiollay, J-M., and M. Jullien. 1998. Flocking behaviour of foraging birds in a neotropical rain forest and the antipredator defence hypothesis. *Ibis* 140:382–94.

Thomas, R. J., N. M. Marples, I. C. Cuthill, M. Takahashi, and E. A. Gibson. 2003. Dietary conservatism may facilitate the initial evolution of aposematism. *Oikos* 101:458–66.

Thompson, D. B. A., and C. J. Barnard. 1983. Anti-predator responses in mixed-species associations lapwings, golden plovers and black-headed gulls. *Anim Behav* 31:585–93.

Thompson, F. R. III, and D. E. Burhaus. 2004. Differences in predators of artificial and real songbird nests: Evidence of bias in artificial nest studies. *Conserv Biol* 18:373–80.

Thompson, J. N. 1994. *The coevolutionary process*. Chicago: University of Chicago Press.

Thompson, R. D., C. V. Grant, E. W. Pearson, and G. W. Corner. 1968. Differential heart rate response of starlings to sound stimuli of biological origin. *J Wildl Manag* 32:888–93.

Thompson, R. K. R., R. W. Foltin, R. J. Boylan, A. Sweet, C. A. Graves, and C. E. Lowitz. 1981. Tonic immobility in Japanese quail can reduce the probability of sustained attack by cats. *Anim Learn Behav* 9:145–49.

Thompson, S. D., R. E. MacMillen, E. M. Burke, and C. R. Taylor. 1980. The energetic cost of bipedal hopping in small mammals. *Nature* 287:223–24.

Thorson, J. M., R. A. Morgan, J. S. Brown, and J. E. Norman. 1998. Direct and indirect cues of predatory risk and patch use by fox squirrels and thirteen-lined ground squirrels. *Behav Ecol* 9:151–57.

Tilson, R. 1980. Klipspringer (*Oreotragus oreotragus*) social structure and predator avoidance in a desert canyon. *Modoqua* 11:303–14.

Tilson, R., and P. M. Norton. 1981. Alarm duetting and pursuit deterrence in an African antelope. *Am Nat* 118:455–62.

Tinbergen, L. 1960. The natural control of insects in pine woods I. Factors influencing the intensity of predation by songbirds. *Arch Neerl Zool* 13:265–343.

Tinbergen, N. 1948. Social releasers and the experimental method required for their study. *Wilson Bull* 60:6–51.

———. 1951. *The study of instinct*. London: Oxford University Press (Clarendon).

———. 1962. Bullfinch escaping from cat by "playing dead." *Br Birds* 55:420–21.

———. 1963. On aims and methods in ethology. *Z Tierpsychol* 20:410–33.

———. 1965. Behavior and natural selection. In *Ideas in modern biology. Proc Int Zool Congr Washington*, ed. J. A. Morre, 6:521–42.

Tinbergen, N., G. J. Broekhuysen, E. Feekes, J. C. W. Houghton, H. Kruuk, and E. Szulc. 1962. Egg shell removal by the black-headed gull, *Larus ridibundus* L.; a behaviour component of camouflage. *Behaviour* 19:74–117.

Tinbergen, N., M. Impekoven, and D. Franck. 1967. An experiment on spacing-out as a defence against predation. *Behaviour* 28:307–21.

Tobalske, B. W., and K. P. Dial. 2000. Effects of body sizes on take-off flight performance in the Phasianidae (Aves). *J Exp Biol* 203:3319–32.

Toigo, C. 1999. Vigilance behavior in lactating female Alpine ibex. *Can J Zool* 77:1060–63.

Tolonen, P., and E. Korpimaki. 1995. Parental effort of kestrels (*Falco tinnunculus*) in nest defense: Effects of laying time, brood size, and varying survival prospects of offspring. *Behav Ecol* 6:435–41.

Towers, S. R., and R. G. Coss. 1990. Confronting snakes in the burrow: Snake-species discrimination and antisnake tactics of two California ground squirrel populations. *Ethology* 84:177–92.

Trail, P. W. 1987. Predation and antipredator behavior at Guianan cock-of-the-rock leks. *Auk* 104:496–507.

Travers, S. E., D. W. Kaufman, and G. A. Kaufman. 1988. Differential use of experimental habitat patches by foraging *Peromyscus maniculatus* on dark and bright nights. *J Mammal* 69:869–72.

Treherne, J. E., and W. A. Foster. 1981. Group transmission of predator avoidance behaviour in a marine insect: The Trafalgar effect. *Anim Behav* 29:911–17.

Treisman, M. 1969. Measurement of sensory discrimination. In *Encyclopedia of linguistics, information and control*, ed. A. R. Meetham and R. A. Hudson, 512–25. Oxford: Pergamon.

———. 1975. Predation and the evolution of gregariousness. I. Models for concealment and evasion. *Anim Behav* 23:779–800.

Treves, A. 1997. Vigilance and use of micro-habitat in solitary rainforest mammals. *Mammalia* 61:511–25.

———. 1998. The influence of group size and neighbors on vigilance in two species of arboreal monkeys. *Behaviour* 135:453–81.

———. 1999a. Within-group vigilance in red colobus and redtail monkeys. *Am J Primatol* 48:113–26.

———. 1999b. Has predation shaped the social systems of arboreal primates? *Int J Primatol* 20:35–67.

———. 2000. Theory and method in studies of vigilance and aggregation. *Anim Behav* 60:711–22.

———. 2002. Predicting predation risk for foraging arboreal monkeys. In *Eat or be eaten: Predator sensitive foraging among primates*, ed. L. E. Miller, 222–41. Cambridge: Cambridge University Press.

Treves, A., and C. A. Chapman. 1996. Conspecific threat, predation avoidance, and resource defense: Implications for grouping in langurs. *Behav Ecol Sociobiol* 39:43–53.

Treves, A., A. Drescher, and N. Ingrisano. 2001. Vigilance and aggregation in black howler monkeys (*Alouatta pigra*). *Behav Ecol Sociobiol* 50:90–95.

Trivers, R. L. 1971. The evolution of reciprocal altruism. *Q Rev Biol* 46:35–57.

———. 1972. Parental investment and sexual selection. In *Sexual selection and the descent of man*, ed. B. Campbell, 136–79. Chicago: Aldine.

———. 1974. Parent-offspring conflict. *Am Zool* 14:249–65.

Trombulak, S. C. 1989. Running speed and body mass in Belding's ground squirrels. *J Mammal* 70:194–97.

Trouilloud, W., A. Delisle, and D. L. Kramer. 2004. Head raising during foraging and pausing during intermittent locomotion as components of antipredator vigilance in chipmunks. *Anim Behav* 67:789–97.

Tsingalia, H. M., and T. E. Rowell. 1984. The behaviour of adult male blue monkeys. *Z Tierpsychol* 64:253–68.

Tucker, G. M., and J. A. Allen. 1993. The behavioural basis of apostatic selection by humans searching for computer-generated cryptic "prey." *Anim Behav* 46:713–19.

Tullberg, B. S., and A. F. Hunter. 1996. Evolution of larval gregariousness in relation to repellant defences and warning coloration in tree-feeding Macrolepidoptera: A phylogenetic analysis on independent contrasts. *Biol J Linn Soc* 57:253–76.

Tullberg, B. S., O. Leimar, and G. Gamberale-Stille. 2000. Did aggregation favour the initial evolution of warning coloration? A novel world revisited. *Anim Behav* 59:281–87.

Tulley, J. J., and F. A. Huntingford. 1987. Paternal care and the development of adaptive variation in anti-predator responses in sticklebacks. *Anim Behav* 35:1570–72.

Tullrot, A. 1994. The evolution of unpalatability and warning coloration in soft-bodied marine invertebrates. *Evolution* 48:925–28.

Tullrot, A., and P. Sundberg. 1991. The conspicuous nudibranch *Polycera quadrilineata*: Aposematic coloration and individual selection. *Anim Behav* 41:175–76.

Turner, E. R. A. 1961. Survival value of different methods of camouflage as shown in a model population. *Proc Zool Soc Lond* 136:273–84.

Turner, G. F., and T. J. Pitcher. 1986. Attack abatement: A model for group protection by combined avoidance and dilution. *Am Nat* 128:228–40.

Turner, J. R. G. 1971. Studies of Mullerian mimicry and its evolution in burnet moths and heliconid butterflies. In *Ecological genetics and evolution*, ed. R. Creed, 224–60. Oxford: Blackwell Scientific Publications.

———. 1977. Butterfly mimicry: The genetical evolution of an adaptation. *Evol Biol* 10:163–206.

Turner, J. R. G., E. P. Kearney, and L. S. Exton. 1984. Mimicry and the Monte Carlo predator: The palatability spectrum and the origins of mimicry. *Biol J Linn Soc* 23:247–68.

Turner, L. W. 1973. Vocal and escape responses of *Spermophilus beldingi* to predators. *J Mamm* 54:990–93.

Ueta, M. 1994. Azure-winged magpies, *Cyanopica cyana*, "parasitize" nest defence provided by Japanese lesser sparrowhawks, *Accipter gularis*. *Anim Behav* 48:871–74.

———. 2001. Azure-winged magpies avoid nest predation by breeding synchronously with Japanese lesser sparrowhawks. *Anim Behav* 61:1007–12.

Ulfstrand, S. 1975. Bird flocks in relation to vegetation diversification in a south Swedish coniferous plantation during winter. *Oikos* 26:65–73.

Underwood, R. 1982. Vigilance behaviour in grazing African antelopes. *Behaviour* 79:81–107.

Uster, D., and K. Zuberbuhler. 2001. The functional significance of Diana monkey "clear" calls. *Behaviour* 138:741–56.

Valone, T. J., and J. S. Brown. 1989. Measuring patch assessment abilities of desert granivores. *Ecology* 70:1800–10.

Valone, T. J., and S. L. Lima. 1987. Carrying food items to cover for consumption: The behavior of ten bird species feeding under the risk of predation. *Oecologia* 71:286–94.

Vander Haegen, W. M., and R. M. Degraaf. 1996. Predation on artificial nests in forested riparian buffer strips. *J Wildl Manag* 60:542–50.

Vanhooydonck, B., R. Van Damme, and P. Aerts. 2001. Speed and stamina trade-off in lacertid lizards. *Evolution* 55:1040–48.

Van Vuren, D., and K. B. Armitage. 1994. Survival of dispersing and philopatric yellow-bellied marmots: What is the cost of dispersal? *Oikos* 69:179–81.

Vasquez, R. A. 1994. Assessment of predation risk via illumination level: Facultative central place foraging in the cricetid rodent *Phyllotis darwini*. *Behav Ecol Sociobiol* 34:375–81.

Vasquez, R. A., L. A. Ebensperger, and F. Bozinovic. 2002. The influence of habitat on travel speed, intermittent locomotion, and vigilance in a diurnal rodent. *Behav Ecol* 2:182–87.

Veasey, J. S., N. B. Metcalfe, and D. C. Houston. 1998. A reassessment of the effect of body mass upon flight speed and predation risk in birds. *Anim Behav* 56:883–89.

Veen, I. T. van der. 1999. Effects of predation risk on diurnal mass dynamics and foraging routines of yellowhammers (*Emberiza citrinella*). *Behav Ecol* 10:545–51.

Veen, I. T. van der, and K. M. Lindstrom. 2000. Escape flights of yellowhammers and greenfinches: More than just physics. *Anim Behav* 59:593–601.

Veen, I. T. van der, and L. E. Sivars. 2000. Causes and consequences of mass loss upon predator encounter: Feeding interruption, stress or fit-for-flight? *Funct Ecol* 14:638–44.

Veen, J. 1977. Functional and causal aspects of nest distribution in colonies of the sandwich tern (*Sterna s. sandvicensis*, Lath). *Behaviour* Suppl. 20:1–193.

Veen, T., D. S. Richardson, K. Blaakmeer, and J. Komdeur. 2000. Experimental evidence for innate predator recognition in the Seychelles warbler. *Proc R Soc Lond Ser B Biol Sci* 267:2253–58.

Vega-Redondo, F., and O. Hasson. 1993. A game-theoretic model of predator-prey signaling. *J Theor Biol* 162:309–19.

Veiga, J. P. 1996. Permanent exposure versus facultative concealment of sexual traits: An experimental study in the house sparrow. *Behav Ecol Sociobiol* 39:345–52.

Vencl, F. 1977. A case of convergence in vocal signals between marmosets and birds. *Am Nat* 111:777–816.

Verbeek, N. A. M. 1972. The exploitation system of the yellow-billed magpie. *Univ Calif Publ Zool* 76:1–58.

———. 1985. Behavioural interactions between avian predators and their avian prey: Play behaviour or mobbing? *Z Tierpsychol* 67:204–14.

Verhulst, S., and O. Hogstad. 1996. Social dominance and energy reserves in flocks of willow tits. *J Avian. Biol* 27:203–8.

Vermeij, G. J. 1982. Unsuccessful predation and evolution. *Am Nat* 120:701–20.

———. 1994. The evolutionary interaction among species: Selection, escalation, and coevolution. *Annu Rev Ecol Syst* 25:219–36.

Verner, J., and M. F. Willson. 1969. Mating systems, sexual dimorphism, and the role of North American passerine birds in the nesting cycle. *Ornithol Monogr* (AOU) 9:1–76.

Vessem, J. van, and D. Draulans. 1986. The adaptive significance of colonial breeding in the grey heron Ardea cinerea: Inter- and intra-colony variability in breeding success. *Ornis Scand* 17:356–62.

Vezina, A. F. 1985. Empirical relationships between predator and prey size among terrestrial vertebrate predators. *Oecologia* 67:555–65.

Vickery, P. D., M. L. Hunter Jr., and J. V. Wells. 1992. Evidence of incidental nest predation and its effect on nests of threatened grassland birds. *Oikos* 63:281–88.

Vickery, W. L., and J. R. Bider. 1978. The effect of weather on *Sorex cinerius* activity. *Can J Zool* 56:291–99.

Vieth, W., E. Curio, and U. Ernst. 1980. The adaptive significance of avian mobbing. III. Cultural transmission of enemy recognition in blackbirds: Cross-species tutoring and properties of learning. *Anim Behav* 28:1217–29.

Vincent, J. F. V., and P. Owers. 1986. Mechanical design of hedgehog spines and porcupine quills. *J Zool* 210:55–75.

Vine, I. 1971. Risk of visual detection and pursuit by a predator and the selective advantage of flocking behaviour. *J Theor Biol* 30:405–22.

———. 1973. Detection of prey flocks by predators. *J Theor Biol* 40:207–10.

Vinuela, J., J. A. Amat, and M. Ferrer. 1995. Nest defence of nesting chinstrap penguins (*Pygoscelis antarctica*) against intruders. *Ethology* 99:323–31.

Viscido, S. V., M. Miller, and D. S. Wethey. 2001. The response of a selfish herd to an attack from outside the group perimeter. *J Theor Biol* 208:315–28.

Vitt, L. J., and J. D. Congdon. 1978. Body shape, reproductive effort, and relative clutch mass in lizards: Resolution of a paradox. *Am Nat* 112:595–608.

Vitt, L. J., and W. E. Cooper Jr. 1986. Tail loss, tail color, and predator escape in *Eumeces* (Lacertilla: Scincidae): Age-specific differences in costs and benefits. *Can J Zool* 64:583–92.

Voelke, G. 2001. Morphological correlates of migratory distance and flight display in the genus *Anthus*. *Biol J Linn Soc* 73:425–35.

Vogel, H. H. Jr. 1950. Observations on social behavior in turkey vultures. *Auk* 67:210–16.

Waage, J. K. 1981. How the zebra got its stripes: Biting flies as selective agents in the evolution of zebra coloration. *J Entomol Soc South Afr* 44:351–58.

Wada, T. 1994. Effects of height of neighboring nests on nest predation in the rufous turtle-dove (*Streptopelia orientalis*). *Condor* 96:812–16.

Wahorme, J. M., T. E. Rowell, and H. M. Tsingalia. 1993. The natural history of de Brazza's monkey in Kenya. *Int J Primatol* 14:445–66.

Wahungu, G. M., C. P. Catterall, and M. F. Olsen. 2001. Predator avoidance, feeding and habitat use in the red-necked pademelon, *Thylogale thetis*, at rainforest edges. *Austr J Zool* 49:45–48.

Wainwright, P. C. 1987. Biomechanical limits to ecological performance mollusc-crushing by the Caribbean hogfish *Lachriolaimus-maximus* Labridae. *J Zool* 213:283–98.

Wallace, A. R. 1889. *Darwinism*. London: Macmillan.

Wallin, K. 1987. Defence as parental care in tawny owls (*Strix aluco*). *Behaviour* 102:213–30.

Walsberg, G. E. 1988. Consequences of skin color and fur properties for solar heat gain and UV irradiance in two mammals. *J Comp Physiol B Biochem Syst Environ Physiol* 158:213–22.

Walters, J. R. 1990. Anti-predatory behavior of lapwings: Field evidence of discriminative abilities. *Wilson Bull* 102:49–70.

Walther, F. 1964. Einige Verhaltens beobachtungen an Thomsongazellen (*Gazella thomsoni* Gunther 1884) in Ngorongoro-Krater. *Z Tierpsychol* 22:167–208.

———. 1969. Flight behaviour and avoidance of predators in Thomson's gazelle (*Gazella thomsoni* Guenther 1884). *Behaviour* 34:184–221.

———. 1984. *Communication and expression in hooved mammals*. Bloomington: Indiana University Press.

Walton, L. R., and S. Lariviere. 1994. A striped skunk, *Mephitis mephitis*, repels two coyotes, *Canis latrans*, without scenting. *Can Field Nat* 108:492–93.

Ward, J. F., D. W. Macdonald, and C. P. Doncaster. 1997. Responses of foraging hedgehogs to badger odour. *Anim Behav* 53:709–20.

Ward, P. I. 1985. Why birds in flocks do not co-ordinate their vigilance periods. *J Theor Biol* 114:383–85.

Warham, J. 1996. *The behaviour, population biology and physiology of the petrels*. London: Academic Press.

Waring, G. H. 1970. Sound communications of black-tailed, white-tailed, and Gunnison's prairie dogs. *Am Midl Nat* 83:167–85.

Warkentin, K. J., A. T. H. Keeley, and J. F. Hare. 2001. Repetitive calls of juvenile Richardson's ground squirrels (*Spermophilus richardsonii*) communicate response urgency. *Can J Zool* 79:569–73.

Warrick, G. D., and P. R. Krausman. 1987. Foraging behavior of female mountain sheep in western Arizona. *J Wildl Manag* 51:99–104.

Waser, P. M. 1984. Chance and mixed-species associations. *Behav Ecol Sociobiol* 15:197–202.

Waser, P. M., and C. H. Brown. 1986. Habitat acoustics and primate communication. *Am J Primatol* 10:135–54.

Waser, P. M., and M. S. Waser. 1977. Experimental studies of primate vocalization: Specializations for long-distance propagation. *Z Tierpsychol* 43:239–63.

Watanuki, Y. 1986. Moonlight avoidance behavior in Leach's storm-petrels as a defense against slaty-backed gulls. *Auk* 103:14–22.

Watts, B. D. 1987. Old nest accumulation as a possible protection mechanism against search-strategy predators. *Anim Behav* 35:1566–68.

———. 1990. Cover use and predator-related mortality in song and savannah sparrows. *Auk* 107:775–78.

Weary, D. M., and D. L. Kramer. 1995. Response of eastern chipmunks to conspecific alarm calls. *Anim Behav* 49:81–93.

Weatherhead, P. J. 1979. Do savannah sparrows commit the concorde fallacy? *Behav Ecol Sociobiol* 5:373–81.

———. 1982. Risk-taking by red-winged blackbirds and the concorde fallacy. *Z Tierpsychol* 60:199–208.

———. 1989. Nest defence by song sparrows: Methodological and life history considerations. *Behav Ecol Sociobiol* 25:129–36.

———. 1990. Nest defence as shareable paternal care in red-winged blackbirds. *Anim Behav* 39:1173–78.

Weatherhead, P. J., and S. J. Sommerer. 2001. Breeding synchrony and nest predation in red-winged blackbirds. *Ecology* 82:1632–41.

Weidinger, K. 2001. Does egg colour affect predation rate on open passerine nests? *Behav Ecol Sociobiol* 49:456–64.

———. 2002. Interactive effects of concealment, parental behaviour and predators on the survival of open passerine nests. *J Anim Ecol* 71:424–37.

———. 2004. Defensive anointing: Extended chemical phenotype and unorthodox ecology. *Chemoecology* 14:1–4.

Weldon, P. J., and J. H. Rappole. 1997. A survey of birds odorous or unpalatable to humans: Possible indications of chemical defense. *J Chem Ecol* 23:2609–33.

Welham, C. V. J. 1994. Flight speeds of migrating birds: A test of maximum range speed predictions from three aerodynamic equations. *Behav Ecol* 5:1–8.

Weller, M. W. 1979. Density and habitat relationships of blue-winged teal nesting in northwestern Iowa. *J Wildl Manag* 43:367–74.

Welty, J. C. 1934. Experiments in group behavior of fishes. *Physiol Zool* 7:85–128.

Wemmer, C., and D. E. Wilson. 1983. Structure and function of hair crests and capes in African carnivora. *Spec Publ Am Soc Mamm* 7:239–64.

Wenger, C. R. 1981. Coyote-mule deer interaction observations in central Wyoming. *J Wildl Manag* 45:770–72.

Werner, E. E. 1986. Amphibian metamorphosis: Growth rate, predation risk, and the optimal size at transformation. *Am Nat* 128:319–41.

Werner, E. E., and J. F. Gilliam. 1994. The ontogenetic niche and species interactions in size-structured populations. *Annu Rev Ecol Syst* 15:393–425.

Werner, E. E., J. F. Gilliam, D. J. Hall, and G. G. Mittelbach. 1983. An experimental test of the effects of predation risk on habitat use in fish. *Ecology* 64:1540–48.

Werner, R. M., and J. A. Vick. 1977. Resistance of the opossum (*Didelphis virginiana*) to envenomation by snakes of the family Crotalidae. *Toxicon* 15:29–33.

West, P. M., and C. Packer. 2002. Sexual selection, temperature, and the lion's mane. *Science* 297:1339–43.

Westmoreland, D. 1989. Offspring age and nest defence in mourning doves: A test of two hypotheses. *Anim Behav* 38:1062–66.

Westmoreland, D., and L. B. Best. 1986. Incubation continuity and the advantage of cryptic egg coloration to mourning doves. *Wilson Bull* 98:297–300.

Westmoreland, D., and R. A. Kiltie. 1996. Egg crypsis and clutch survival in three species of blackbirds (Icteridae). *Biol J Linn Soc* 58:159–72.

Westneat, D. F. 1989. Intensity of nest defense in indigo buntings increases with stage and not number of visits. *Auk* 106:747–49.

———. 1992. Nesting synchrony by female red-winged blackbirds: Effects on predation and breeding success. *Ecology* 73:2284–94.

Westneat, D. F., and I. R. K. Stewart. 2003. Extra-pair paternity in birds: Causes, correlates, and conflict. *Annu Rev Ecol Syst* 34:365–96.

Wheelwright, N. T., J. J. Lawler, and J. H. Weinstein. 1997. Nest-site selection in savannah sparrows: Using gulls as scarecrows? *Anim Behav* 53:197–208.

White, K. S., and J. Berger. 2001. Antipredator strategies of Alaskan moose: Are maternal tradeoffs influenced by offspring activity? *Can J Zool* 79:2055–62.

Whiten, A., J. Goodall, W. C. McGrew, T. Nishida, V. Reynolds, Y. Sugiyama, C. E. G. Tutin, R. W. Wrangham, amd C. Boesch. 1999. Cultures in chimpanzees. *Nature* 399:682–85.

Whitesides, G. H. 1989. Interspecific associations of Diana monkeys, *Cercopithecus diana*, in Sierra Leone, West Africa: Biological significance or chance? *Anim Behav* 37:760–76.

Whitfield, D. P., W. Cresswell, N. P. Ashmole, N. A. Clark, and A. D. Evans. 1999. No evidence for sparrowhawks selecting redshanks according to size or condition. *J Avian. Biol* 30:31–39.

Whittingham, L. A., P. O. Dunn, and R. J. Robertson. 1993. Confidence of paternity and male parental care: An experimental study in tree swallows. *Anim Behav* 46:139–47.

Wickler, W. 1968. *Mimicry in plants and animals*. Toronto: McGraw Hill.

———. 1985. Coordination of vigilance in bird groups. The "Watchman's Song" hypothesis. *Z Tierpsychol* 69:250–53.

Wiklund, C. G. 1979. Increased breeding success for merlins *Falco columbarius* nesting among colonies of fieldfares *Turdus pilaris*. *Ibis* 121:109–11.

———. 1982. Fieldfare (*Turdus pilaris*) breeding success in relation to colony size, nest position and association with merlins (*Falco columbarius*). *Behav Ecol Sociobiol* 11:165–72.

———. 1990a. Offspring protection by merlin *Falco columbarius* females; the importance of brood size and expected offspring survival for defense of young. *Behav Ecol Sociobiol* 26:217–23.

———. 1990b. The adaptive significance of nest defence by merlin, *Falco columbarius*, males. *Anim Behav* 40:244–53.

————. 1995. Nest predation and lifespan: Components of variance in LRS among merlin females. *Ecology* 76:1994–96.

————. 1996. Breeding lifespan and nest predation determine lifetime production of fledglings by male merlins *Falco columbarius*. *Proc R Soc Lond Ser B Biol Sci* 263:723–28.

Wiklund, C., and M. Andersson. 1980. Nest predation selects for colonial breeding among fieldfares *Turdus pilaris*. *Ibis* 122:363–66.

Wiklund, C. G., and M. Andersson. 1994. Natural selection of colony size in a passerine bird. *J Anim Ecol* 63:765–74.

Wiklund, C., and T. Jarvi. 1982. Survival of distasteful insects after being attacked by naïve birds: A reappraisal of the theory of aposematic coloration evolving through individual selection. *Evolution* 36:998–1002.

Wiklund, C. G., and J. Stigh. 1983. Nest defence and evolution of reversed sexual size dimorphism in snowy owl *Nyctea scandiaca*. *Ornis Scand* 14:58–62.

Wilcove, D. S. 1985. Nest predation in forest tracts and the decline of migratory songbirds. *Ecology* 66:1211–14.

Wilcove, D. S., C. H. McClellan, and A. P. Dobson. 1986. Habitat fragmentation in the temperate zone. In *Conservation biology: The science of scarcity and diversity*, ed. M. E. Soule, 237–56. Sunderland, MA: Sinauer Assn.

Wiley, R. H. 1973. Territoriality and non-random mating in sage grouse, *Centrovercus urophasianus*. *Anim Behav Monogr* 6:87–169.

————. 1991. Associations of song properties with habitats for territorial oscine birds of eastern North America. *Am Nat* 138:973–93.

Wiley, R. H., and D. O. Richards. 1978. Physical constraints on acoustic communication in the atmostphere: Implications for the evolution of animal vocalizations. *Behav Evol Sociobiol* 3:69–94.

————. 1982. Adaptations for acoustic communication in birds: Sound transmission and signal detection. In *Acoustic communication in birds*, ed. D. E. Kroodsma, E. H. Miller, and H. Ouellet, 1:131–81. New York: Academic Press.

Wilkinson, G. S., and G. M. English-Loeb. 1982. Predation and coloniality in cliff swallows (*Petrochelidon pyrrhonota*). *Auk* 99:459–67.

Williams, G. C. 1964. Measurements of consociation among fishes and comments on the evolution of schooling. *Publ Mus Mich State Univ Biol Ser* 2:351–83.

————. 1966. *Adaptation and natural selection*. Princeton, NJ: Princeton University Press.

Williams, G. E., and P. B. Wood. 2002. Are traditional methods of determining nest predators and nest fates reliable? An experiment with wood thrushes (*Hylocichla mustelina*) using miniature video cameras. *Auk* 119:1126–32.

Williams, H. W. 1969. Vocal behavior of the adult California quail. *Auk* 86:631–59.

Williamson, K. 1950. The distraction display of certain waders. Part 1: Interpretation of "rodent-run" display. *Ibis* 92:28–33.

Wilson, D. E., and F. R. Cole. 2000. *Common names of mammals of the world*. Washington, DC: Smithsonian University Press.

Wilson, D. R., and J. F. Hare. 2004. Ground squirrel uses ultrasonic alarms. *Nature* 430:523.

Wilson, G. R., M. C. Brittingham, and L. J. Goodrich. 1998. How well do artificial nests estimate success of real nests? *Condor* 100:357–64.

Wilson, J., and A. G. Weir. 1989. Hunting behaviour and attack success of a female sparrowhawk between October 1987 and April 1988. *Scott Birds* 15:126–30.

Wilson, R. P., P. G. Ryan, A. James, and M-P. T. Wilson. 1987. Conspicuous coloration may enhance prey capture in some piscivores. *Anim Behav* 35:1558–60.

Windecker, W. 1939. *Euchelia (Hypocrita) jacobaeae* L. und das Schutztrachtenproblem. *Z Morphol Oekol Tiere* 35:84–138.

Windsor, D., and S. T. Emlen. 1975. Predator-prey interactions of adult and pre-fledgling bank swallows and American kestrels. *Condor* 77:359–61.

Windt, W., and E. Curio. 1986. Clutch defence in great tit (*Parus major*) pairs and the concorde fallacy. *Ethology* 72:236–42.

Winkler, D. W. 1987. A general model for parental care. *Am Nat* 130:526–43.

———. 1991. Parental investment decision rules in tree swallows: Parental defense, abandonment, and the so-called Concorde Fallacy. *Behav Ecol* 2:133–42.

———. 1994. Anti-predator defence by neighbours as a responsive amplifier of parental defence in tree swallows. *Anim Behav* 47:595–605.

Wirtz, P., and M. Wawra. 1986. Vigilance and group size in *Homo sapiens*. *Ethology* 71:283–86.

Wise, K. K., M. R. Conover, and F. K. Knowlton. 1999. Response of coyotes to avian distress calls: Testing the startle-predator and predator-attraction hypotheses. *Behaviour* 136:935–49.

Wit, C. A. de. 1982. Resistance of the prairie vole (*Microtus ochrogaster*) and the woodrat (*Neotoma floridana*), in Kansas, to venom of the osage copperhead (*Agkistrodon contortrix phaeogaster*). *Toxicon* 20:709–14.

Wit, C. A. de, and B. R. Westrom. 1987. Venom resistance in the hedgehog, *Erinaceus europaeus*: Purification and identification of macroglobulin inhibitors as plasma antihemorrhagic factors. *Toxicon* 25:315–23.

Witkin, S. R., and M. S. Ficken. 1979. Chickadee alarm calls: Does mate investment pay dividends? *Anim Behav* 27:1275–76.

Wittenberger, J.F., and G.L. Hunt, Jr. 1985. The adaptive significance of coloniality in birds. *Avian Biol* 8:1–78.

Witter, M. S., and I. C. Cuthill. 1993. The ecological costs of avian fat storage. *Philos Trans R Soc Lond B Biol Sci* 340:73–92.

Witter, M. S., I. C. Cuthill, and R. H. Bonser. 1994. Experimental investigations of mass-dependent predation risk in the European starling, *Sturnus vulgaris*. *Anim Behav* 48:201–22.

Wolf, C. M., B. Griffith, C. Reed, and S. A. Temple. 1996. Avian and mammalian translocations: Update and reananalysis of 1987 survey data. *Conserv Biol* 10:1142–54.

Wolf, N. G. 1985. Odd fish abandon mixed-species groups when threatened. *Behav Ecol Sociobiol* 17:47–52.

Wolfe, J. L., and C. T. Summerlin. 1989. The influence of lunar light on nocturnal activity of the old-field mouse. *Anim Behav* 37:410–14.

Wolff, J. O. 2004. Scent marking by voles in response to predation risk: A field-laboratory validation. *Behav Ecol* 15:286–89.

Wolff, J. O., and R. Davis-Born. 1997. Response of gray-tailed voles to odours of a mustelid predator: A field test. *Oikos* 79:543–48.

Wood, W. F., B. G. Sollers, G. A. Dragoo, and J. W. Dragoo. 2002. Volatile components in defensive spray of the hooded skunk, *Mephitis macoura*. *J Chem Ecol* 28:1865–70.

Woodland, D. J., Z. Jaafar, and M-L. Knight. 1980. The "pursuit deterrent" function of alarm signals. *Am Nat* 115:748–53.

Woolfenden, G. E., and J. W. FitzPatrick. 1984. *The Florida scrub jay: Demography of a cooperative-breeding bird*. Princeton, NJ: Princeton University Press.

Wourms, M. K., and F. E. Wasserman. 1985. Butterfly wing markings are more advantageous during handling than during the initial strike of an avian predator. *Evolution* 39:845–51.

Wrangham, R. W. 1980. An ecological model of female-bonded primate groups. *Behaviour* 75:262–300.

Wright, J., E. Berg, S. R. de Kort, V. Khazin, and A. A. Maklakov. 2001. Safe selfish sentinels in a cooperative bird. *J Anim Ecol* 70:1070–79.

Wright, J., A. A. Maklakov, and V. Khazin. 2001. State dependent sentinels: An experimental study in the Arabian babbler. *Proc R Soc Lond Ser B Biol Sci* 268:821–26.

Wrona, F. J., and R. W. J. Dixon. 1991. Group size and predation risk: A field analysis of encounter and dilution effects. *Am Nat* 137:186–201.

Wunderle, J. M., and K. H. Pollock. 1985. The bananaquit-wasp nesting association and a random choice model. *Ornithol Monogr* 36:595–603.

Xu, Z. J., D. M. Stoddart, H. B. Ding, and J. Jhang. 1995. Self-anointing behavior in the rice-field rat, *Rattus rattordes*. *J Mammal* 76:1238–41.

Yaber, M. C., and E. A. Herrera. 1994. Vigilance, group size and social status in capybaras. *Anim Behav* 48:1301–7.

Yachi, S., and M. Higashi. 1998. The evolution of warning signals. *Nature* 394:882–84.

Yahner, R. H. 1980. Barking in a primitive ungulate, *Muntiacus reevesi*: Function and adaptiveness. *Am Nat* 116:157–77.

Yahner, R. H., and C. G. Mahan. 1996. Effects of egg type on depredation of artificial ground nests. *Wilson Bull* 108:129–36.

Yahner, R. H., and D. P. Scott. 1988. Effects of forest fragmentation on depredation of artificial nests. *J Wildl Manag* 52:158–61.

Yasukawa, K., L. K. Whittenberger, and T. A. Nielsen. 1992. Anti-predator vigilance in the red-winged blackbird, *Agelaius phoeniceus*: Do males act as sentinels? *Anim Behav* 43:961–69.

Ydenberg, R. C., and L. M. Dill. 1986. The economics of fleeing from predators. *Adv Study Behav* 16:229–49.

Ylonen, H. 1994. Vole cycles and antipredatory behaviour. *Trends Ecol Evol* 11:426–30.

Ylonen, H., B. Jedrzejewska, W. Jedrzejewski, and J. Heikkila. 1992. Antipredatory behaviour of *Clethrionomys* voles—"David and Goliath" arms race. *Ann Zool Fenn* 39:207–16.

Ylonen, H., R. Pech, and S. Davis. 2003. Heterogeneous landscapes and the role of refuge on the population dynamics of a specialist predator and its prey. *Evol Ecol* 17:349–69.

Ylonen, H., and H. Ronkainen. 1994. Breeding suppression in the bank vole as antipredatory adaptation in a predictable environment. *Evol Ecol* 8:658–66.

Yorio, P., and F. Quintana. 1997. Predation by kelp gulls *Larus dominicanus* at a mixed-species colony of royal terns *Sterna maxima* and Cayenne terns *Sterna eurygnatha* in Patagonia. *Ibis* 139:536–41.

Young, A. D., and R. D. Titman. 1986. Costs and benefits to red-breasted mergansers nesting in tern and gull colonies. *Can J Zool* 64:2339–43.

Young, A. M. 1971. Wing coloration and reflectance in *Morpho* butterflies as related to reproductive behavior and escape from avian predators. *Oecologia* 7:209–22.

Young, B. E., M. Kaspari, and T. E. Martin. 1990. Species-specific nest site selection by birds in ant-acacia trees. *Biotropica* 22:310–15.

Zahavi, A. 1975. Mate selection—A selection for a handicap. *J Theor Biol* 53:205–14.

———. 1977. Reliability in communication systems and the evolution of altruism. In *Evolutionary ecology*, ed. B. Stonehouse and C. M. Perrins, 253–59. London: Macmillan.

Zanette, L. 2002. What do artificial nests tell us about nest predation? *Biol Conserv* 103:323–29.

Ziegler, A. P. 1971. The strange case of look-alike birds. *Animals* (London) 13:736–37.

Zimmerman, J. L. 1984. Nest predation and its relationship to habitat and density in dickcissels. *Condor* 86:68–72.

Zuberbuhler, K. 2000a. Causal knowledge of predators' behaviour in wild Diana monkeys. *Anim Behav* 59:209–20.

———. 2000b. Interspecies semantic communication in two forest primates. *Proc R Soc Lond Ser B Biol Sci* 267:713–18.

———. 2000c. Referential labeling in Diana monkeys. *Anim Behav* 59:917–27.

———. 2001. Predator-specific alarm calls in Campbell's monkeys, *Cercopithecus campbelli*. *Behav Ecol Sociobiol* 50:414–22.

———. 2002. A syntactic rule in forest monkey communication. *Anim Behav* 63:293–99.

Zuberbuhler, K., D. Jenny, and R. Bshary. 1999. The predator deterrence function of primate alarm calls. *Ethology* 105:477–90.

Zuberbuhler, K., R. Noe, and R. M. Seyfarth. 1997. Diana monkey long-distance calls: Messages for conspecifics and predators. *Anim Behav* 53:589–604.

Zuk, M., and G. R. Kolluru. 1998. Exploitation of sexual signals by predators and parasitoids. *Q Rev Biol* 73:415–38.

Figure I.0 (*Facing page*) South American sea lions bunch close together on the beach as killer whales launch attacks on them in the shallow surf. Individuals on the periphery of the sea lion group nearest the water are in most danger of being killed (reproduced by kind permission of Sheila Girling).